Signal Transduction and Smooth Muscle

Methods in Signal Transduction

Series Editors
Joseph Eichberg, Jr. and Michael Xi Zhu

Published Titles

Lipid Second Messengers, *Suzanne G. Laychock and Ronald P. Rubin*

G Proteins: Techniques of Analysis, *David R. Manning*

Signaling Through Cell Adhesion Molecules, *Jun-Lin Guan*

G Protein-Coupled Receptors, *Tatsuya Haga and Gabriel Berstein*

G Protein-Coupled Receptors: Structure, Function, and Ligand Screening, *Tatsuya Haga and Shigeki Takeda*

Calcium Signaling, Second Edition, *James W. Putney, Jr.*

Analysis of Growth Factor Signaling in Embryos, *Malcolm Whitman and Amy K. Sater*

Signal Transduction in the Retina, *Steven J. Fliesler and Oleg G. Kisselev*

Signaling by Toll-Like Receptors, *Gregory W. Konat*

Lipid-Mediated Signaling, *Eric J. Murphy and Thad A. Rosenberger*

TRP Channels, *Michael Xi Zhu*

Cyclic Nucleotide Signaling, *Xiaodong Cheng*

Gap Junction Channels and Hemichannels, *Donglin Bai and Juan C. Sáez*

Signaling Mechanisms Regulating T Cell Diversity and Function, *Jonathan Soboloff and Dietmar J. Kappes*

Lipid-Mediated Signaling Transduction, Second Edition, *Eric Murphy, Thad Rosenberger, and Mikhail Golovko*

Calcium Entry Channels in Non-Excitable Cells, *Juliusz Ashot Kozak and James W. Putney, Jr.*

Autophagy and Signaling, *Esther Wong*

Signal Transduction and Smooth Muscle, *Mohamed Trebak and Scott Earley*

For more information about this series, please visit: https://www.crcpress.com/Methods-in-Signal-Transduction-Series/book-series/CRCMETSIGTRA?page=&order=pubdate&size=12&view=list&status=published, forthcoming

Signal Transduction and Smooth Muscle

Edited by
Mohamed Trebak
Scott Earley

CRC Press
Taylor & Francis Group
Boca Raton London New York

CRC Press is an imprint of the
Taylor & Francis Group, an **informa** business

CRC Press
Taylor & Francis Group
6000 Broken Sound Parkway NW, Suite 300
Boca Raton, FL 33487-2742

Printed on acid-free paper

International Standard Book Number-13: 978-1-4987-7422-2 (Hardback)

Visit the Taylor & Francis Web site at
http://www.taylorandfrancis.com

and the CRC Press Web site at
http://www.crcpress.com

Contents

Series Preface

The concept of signal transduction is now long established as a central tenet of biological sciences. Since the inception of the field nearly 50 years ago, the number and variety of signal transduction pathways, cascades and networks have steadily increased and now constitute what is often regarded as a bewildering array of mechanisms by which cells sense and respond to extracellular and intracellular environmental stimuli. It can be confidently stated that virtually every cell function is dependent on the detection, amplification and integration of these signals. Moreover, there is increasing appreciation that in many disease states, aspects of signal transduction are critically perturbed.

Our knowledge of how information is conveyed and processed through these cellular molecular circuits and biochemical switches has increased enormously in scope and complexity since this series was initiated over 15 years ago. Such advances would not have been possible without the supplementation of older technologies, drawn chiefly from cell and molecular biology, biochemistry, physiology, pharmacology, with newer methods that make use of sophisticated genetic approaches as well as structural biology, imaging, bioinformatics and systems biology analysis.

The overall theme of this series continues to be the presentation of the wealth of up to date research methods applied to the many facets of signal transduction. Each volume is assembled by one or more editors who are pre-eminent in their specialty. In turn, the guiding principle for editors is to recruit chapter authors who will describe procedures and protocols with which they are intimately familiar in a reader-friendly format. The intent is to assure that each volume will be of maximum practical value to a broad audience, including students and researchers just entering an area, as well as seasoned investigators.

Smooth muscles occupy a large portion of the body and perform many vital functions that affect systems such as the vasculature, lung, urinary and gastrointestinal tracks, and brain. The complex and diverse signaling processes involved in regulating smooth muscle activities in different organ systems are areas of intense investigation. These often involve various ion channels, enzymes and intracellular signaling messengers, such as calcium ions, nitric oxide and reactive oxygen species. This rapidly growing area of signal transduction is constantly evolving new concepts and methodologies, many of which are covered in the current volume. It is hoped that the information contained in this compendium, as well as in other books of this series, will constitute a useful resource to the life sciences research community well into the future.

Joseph Eichberg and Michael Xi Zhu

Preface

Smooth muscle cells are the major structural component of arteries, veins, lymphatic vessels, and all hollow organs. The primary function of fully differentiated smooth muscle cells is to contract or relax to regulate motility, vascular tone, and appropriate sphincter tone. Beyond this classical role, smooth muscle cells are active participants in intercellular communications and dedifferentiated smooth muscle cells have many unexpected functions, such as acting as immune effector cells that produce and respond to a plethora of chemical and mechanical cues, contributing to physiology and pathophysiology. The goal of this book is to provide graduate students, medical students, and postdoctoral fellows who are engaged in research with a useful resource on emerging signal transduction mechanisms that regulate different smooth muscle types during health and disease. We have provided experimental approaches and strategies for studying smooth muscle, such as methods for isolating native cells and for optical and biophysical measurements of ion channel function under physiological conditions. When appropriate, pharmacological approaches and the emerging relevance of ion channels, membrane receptors and other molecules to physiological function and disease are addressed.

The chapters cover a wide variety of topics. For example, the functional roles and properties of different ion channels and receptors in a number of smooth muscle types from the systemic and cerebral vasculature, lymphatic circulation (Chapter 16), airways (Chapters 11 and 12), gastrointestinal tract, and bladder (Chapter 14) are addressed. Chapter 15 discusses the role of the Interstitial Cells of Cajal (ICC) that provide pacemaker activity and modulate smooth muscle contractility and motility of the gastrointestinal tract. Chapter 7 addresses the processes of myosin light chain phosphatase-mediated calcium sensitization, whereas Chapter 9 reviews the emerging role of mitochondria in shaping local calcium signals in smooth muscle cells. Chapter 17 addresses NADPH oxidase and reactive oxygen species signaling in smooth muscle. This topic is particularly timely, as it is increasingly recognized that function of various ion channels and other signal transduction cascades are regulated through oxidant signaling.

We are extremely grateful to all of the colleagues who have contributed chapters to this project. We realize that it takes considerable effort to produce and deliver a book chapter, which diverts energy from doing research in the laboratory. We hope that this book will provide a resource that will support and inspire students and colleagues who are performing advanced studies on different types of smooth muscle.

Mohamed Trebak
Hershey, Pennsylvania

Scott Earley
Reno, Nevada

Editors

Mohamed Trebak, PhD studied Plant Biology (BS) at the University of Rabat, Morocco. He completed his Masters and PhD in Biochemistry from the University of Liège, Belgium. He did a first postdoc in Protein Chemistry at the Wistar Institute of Philadelphia, Pennsylvania. He completed a second postdoc at the National Institute of Environmental Health Sciences, National Institutes of Health (NIEHS/NIH) in Research Triangle Park, North Carolina, where he studied Calcium Signaling and Ion Channel Pharmacology and Physiology, focusing on the transient receptor potential (TRP) superfamily of cation channels in non-excitable cells. He was appointed Assistant Professor (2007–2010) and Associate Professor (2010–2015) at Albany Medical College and subsequently at the State University of New York (SUNY). In 2015, he was promoted to professor (with tenure) and moved his laboratory to Penn State University College of Medicine in Hershey, Pennsylvania. His research interests are in the signaling and regulatory mechanisms of store-operated ORAI calcium entry channels and their role in proliferative and migratory diseases of the vasculature and airways. He has published over 100 peer-reviewed research papers and reviews and is a member of the editorial board of several leading scientific journals, including *Pflügers Archives, The Journal of Biological Chemistry, Cell Calcium* and *PLoS One.*

Scott Earley, PhD studied Electrical Engineering (BS) and Microbiology (MS) at the University of Maine, Orono, Maine. He completed his PhD in Biomedical Sciences at the University of New Mexico, Albuquerque, New Mexico. He was an NIH-funded postdoctoral fellow at the University of Vermont, Burlington, Vermont where he studied the roles of ion channels and Calcium signaling in the regulation of vascular control. He was appointed Assistant Professor (2006–2011) and Associate Professor with tenure (2011–2013) at Colorado State University in Fort Collins, Colorado. He joined the faculty of the University of Nevada, Reno School of Medicine in 2014 and was promoted to Professor in 2016. His research is focused on membrane biophysics, ion channels, and subcellular Ca^{2+} signaling cascades in the vasculature, with an emphasis on the TRP superfamily of cation channels and the cerebral circulation. He has published more than 100 peer-reviewed research papers, reviews, editorial comments, and conference proceedings and is a member of the editorial board of several leading scientific journals, including *Physiological Reviews, Microcirculation,* and *The American Journal of Physiology—Heart and Circulatory Physiology.*

Contributors

Rodney Britt Jr
Department of Anesthesiology and Perioperative Medicine
Mayo Clinic
Rochester, Minnesota

Lyubov I. Brueggemann
Department of Molecular Pharmacology & Therapeutics
Loyola University Chicago
Maywood, Illinois

Kenneth L. Byron
Department of Molecular Pharmacology & Therapeutics
Loyola University Chicago
Maywood, Illinois

Jorge A. Castorena-Gonzalez
Department of Medical Pharmacology and Physiology
University of Missouri
Columbia, Missouri

Susan Chalmers
Strathclyde Institute of Pharmacy and Biomedical Sciences
University of Strathclyde
Glasgow, United Kingdom

Eugenia Cifuentes-Pagano
Department of Pharmacology & Chemical Biology
Heart, Lung, Blood and Vascular Medicine Institute
University of Pittsburgh
Pittsburgh, Pennsylvania

William C. Cole
Department of Physiology & Pharmacology
Cumming School of Medicine
University of Calgary
Calgary, Alberta, Canada

Katelyn Cummings
Department of Anesthesiology and Perioperative Medicine
Mayo Clinic
Rochester, Minnesota

Michael J. Davis
Department of Medical Pharmacology and Physiology
University of Missouri
Columbia, Missouri

Daniel S. de Jesus
Department of Pharmacology & Chemical Biology
Heart, Lung, Blood and Vascular Medicine Institute
University of Pittsburgh
Pittsburgh, Pennsylvania

Timothy Domeier
Department of Medical Pharmacology and Physiology
University of Missouri
Columbia, Missouri

Bernard T. Drumm
Department of Physiology & Cell Biology
University of Nevada, Reno School of Medicine
Reno, Nevada

Scott Earley
Department of Pharmacology
Center for Cardiovascular Research
University of Nevada, Reno School of Medicine
Reno, Nevada

Scott Emrich
Department of Cellular and Molecular Physiology
The Pennsylvania State University College of Medicine
Hershey, Pennsylvania

John M. Girkin
Department of Physics
Centre for Advanced Instrumentation
Durham University
Durham, United Kingdom

Iain A. Greenwood
Vascular Biology Research Centre
Molecular and Clinical Sciences
St George's University of London
London, United Kingdom

and

Ion Channel Group
Heart and Circulatory Research Section
Department of Biomedical Sciences
Faculty of Health Sciences
University of Copenhagen
Copenhagen, Denmark

Maxime Guéguinou
Department of Cellular and Molecular Physiology
The Pennsylvania State University College of Medicine
Hershey, Pennsylvania

Peichun Gui
Department of Medical Pharmacology and Physiology
University of Missouri
Columbia, Missouri

Osama F. Harraz
Department of Pharmacology
Larner College of Medicine
University of Vermont
Burlington, Vermont

Ahmed M. Hashad
Department of Physiology & Pharmacology
University of Calgary
Calgary, Alberta, Canada

and

Department of Physiology & Pharmacology
University of Western Ontario
London, Ontario, Canada

William F. Jackson
Department of Pharmacology and Toxicology
Michigan State University
East Lansing, Michigan

Jonathan H. Jaggar
Department of Physiology
University of Tennessee Health Science Center
Memphis, Tennessee

M. Dennis Leo
Department of Physiology
University of Tennessee Health Science Center
Memphis, Tennessee

Justin A. MacDonald
The Smooth Muscle Research Group
Department of Biochemistry & Molecular Biology
Cumming School of Medicine
University of Calgary
Calgary, Alberta, Canada

John G. McCarron
Strathclyde Institute of Pharmacy and Biomedical Sciences
University of Strathclyde
Glasgow, United Kingdom

Manuel F. Navedo
Department of Pharmacology
University of California
Davis, California

Madeline Nieves-Cintrón
Department of Pharmacology
University of California
Davis, California

Christina M. Pabelick
Department of Anesthesiology and Perioperative Medicine
and
Department of Physiology and Biomedical Engineering
Mayo Clinic
Rochester, Minnesota

Patrick J. Pagano
Department of Pharmacology & Chemical Biology
Heart, Lung, Blood and Vascular Medicine Institute
University of Pittsburgh
Pittsburgh, Pennsylvania

Henry R. Askew Page
Vascular Biology Research Centre
Molecular and Clinical Sciences
St George's University of London
London, United Kingdom

and

Ion Channel Group
Heart and Circulatory Research Section
Department of Biomedical Sciences
Faculty of Health Sciences
University of Copenhagen
Copenhagen, Denmark

Trayambak Pathak
Department of Cellular and Molecular Physiology
The Pennsylvania State University College of Medicine
Hershey, Pennsylvania

Raymond B. Penn
Division of Pulmonary and Critical Care Medicine
Department of Medicine
Center for Translational Medicine
Jane and Leonard Korman Respiratory Institute
Thomas Jefferson University
Philadelphia, Pennsylvania

Tonio Pera
Division of Pulmonary and Critical Care Medicine
Department of Medicine
Center for Translational Medicine
Jane and Leonard Korman Respiratory Institute
Thomas Jefferson University
Philadelphia, Pennsylvania

Brian A. Perrino
Department of Physiology & Cell Biology
University of Nevada, Reno School of Medicine
Reno, Nevada

Georgi V. Petkov
Department of Pharmaceutical Sciences
College of Pharmacy
University of Tennessee Health Science Center
Memphis, Tennessee

Paulo W. Pires
Department of Pharmacology
Center for Cardiovascular Research
University of Nevada, Reno School of Medicine
Reno, Nevada

Y. S. Prakash
Department of Anesthesiology and Perioperative Medicine
and
Department of Physiology and Biomedical Engineering
Mayo Clinic
Rochester, Minnesota

Harry A. T. Pritchard
Department of Pharmacology
Center for Cardiovascular Research
University of Nevada, Reno School of Medicine
Reno, Nevada

Anne Roesler
Department of Anesthesiology and Perioperative Medicine
Mayo Clinic
Rochester, Minnesota

Sanghamitra Sahoo
Department of Pharmacology & Chemical Biology
Heart, Lung, Blood and Vascular Medicine Institute
University of Pittsburgh
Pittsburgh, Pennsylvania

Kenton M. Sanders
Department of Physiology & Cell Biology
University of Nevada, Reno School of Medicine
Reno, Nevada

L. Fernando Santana
Department of Physiology & Membrane Biology
University of California
Davis, California

Christopher Saunter
Department of Physics
Centre for Advanced Instrumentation
Durham University
Durham, United Kingdom

Sendoa Tajada
Department of Physiology & Membrane Biology
University of California
Davis, California

Michael Thompson
Department of Anesthesiology and Perioperative Medicine
Mayo Clinic
Rochester, Minnesota

Kim H. T. To
Department of Medical Pharmacology and Physiology
University of Missouri
Columbia, Missouri

Mohamed Trebak
Department of Cellular and Molecular Physiology
The Pennsylvania State University College of Medicine
Hershey, Pennsylvania

Michael P. Walsh
The Smooth Muscle Research Group
Department of Biochemistry & Molecular Biology
Cumming School of Medicine
University of Calgary
Calgary, Alberta, Canada

Donald G. Welsh
Department of Physiology & Pharmacology
University of Calgary
Calgary, Alberta, Canada

and

Department of Physiology & Pharmacology
University of Western Ontario
London, Ontario, Canada

Calum Wilson
Strathclyde Institute of Pharmacy and Biomedical Sciences
University of Strathclyde
Glasgow, United Kingdom

Yeming Xie
Department of Physiology & Cell Biology
University of Nevada, Reno School of Medicine
Reno, Nevada

Ryan Yoast
Department of Cellular and Molecular Physiology
The Pennsylvania State University College of Medicine
Hershey, Pennsylvania

Scott D. Zawieja
Department of Medical Pharmacology and Physiology
University of Missouri
Columbia, Missouri

Xuexin Zhang
Department of Cellular and Molecular Physiology
The Pennsylvania State University College of Medicine
Hershey, Pennsylvania

1 Elucidation of Vasopressin Signal Transduction Pathways in Vascular Smooth Muscle

Kenneth L. Byron and Lyubov I. Brueggemann

CONTENTS

1.1 INTRODUCTION

Arginine[8]-vasopressin (AVP), also known as antidiuretic hormone, is a peptide hormone that is synthesized by magnocellular neurons in the hypothalamus of the mammalian brain. AVP is stored in secretory vesicles within nerve terminals located in the posterior pituitary gland, from whence it can be released into the systemic circulation (Treschan and Peters 2006). One of vasopressin's primary functions, as its name implies, is to constrict the vasculature to elevate blood pressure (BP). This important function comes into play when AVP is released from the pituitary nerve terminals following an increase in plasma osmolality or a fall in BP. Its primary endocrine signaling actions are two-fold: AVP exerts antidiuretic actions by stimulating the kidneys to increase water reabsorption, and it exerts vasoconstrictor actions by stimulating contraction of vascular smooth muscle cells (VSMCs). Through this combination of actions, AVP plays an important part in restoring BP and water balance.

Plasma AVP exerts a potent antidiuretic effect on the kidney, with a maximal antidiuresis occurring, on average, at plasma [AVP] of approximately 4 pM (Baylis 1987). With a normal mean arterial pressure (MAP) of 70–105 mmHg and normal water balance (plasma osmolality approximately 280–285 mOsmol/kg), the concentrations of AVP in the systemic circulation are undetectable by conventional radioimmunoassay techniques (Baylis 1989). However, measurements conducted in human volunteers subjected to either increasing plasma osmolality or lowering of MAP revealed that either stimulus can increase circulating [AVP] in proportion to the magnitude of the stimulus (Baylis 1983). Increasing plasma osmolality from 280 to 310 mOsmol/kg increased [AVP] from undetectable levels to approximately 10 pM; by comparison, lowering BP was a more effective stimulus, elevating plasma [AVP] to as much as 500 pM when BP was lowered by 60% (Baylis 1983, Baylis and Ball 2000).

At the level of the vasculature, the BP restoring actions of circulating AVP have been attributed to the vasoconstrictor effects elicited by binding of AVP to V_{1a}-vasopressin receptors on VSMCs, particularly those located on the surface of arterial smooth muscle cells of the splanchnic circulation. The splanchnic circulation supplies oxygen and nutrients to the gastrointestinal tract via a vast arterial network of resistance vessels originating from the celiac trunk, the inferior mesenteric artery and the superior mesenteric artery. Mesenteric arterioles have been found to be highly sensitive to AVP, constricting in response to concentrations of AVP as low as 10 pM (Altura 1975). Constriction of the splanchnic arterial vasculature can significantly increase total peripheral vascular resistance and thereby provides a very effective mechanism for restoration of BP.

The signal transduction pathways whereby binding of AVP to V_{1a} receptors results in contraction of VSMCs had been extensively studied by the early to mid-1990s. Because elevation of cytosolic Ca^{2+} concentration ($[Ca^{2+}]_{cyt}$) was recognized as the primary effector of smooth muscle contraction, many studies measured changes in $[Ca^{2+}]_{cyt}$ in response to AVP, in combination with various biochemical and pharmacological approaches to elucidate the signaling pathways that might elicit this response. The V_{1a} receptor is in the family of G protein-coupled heptahelical receptors (Morel et al. 1992). It had been determined to be coupled to $G_{q/11}$ proteins (Wange et al. 1991, Thibonnier et al. 1993), which link agonist binding to activation of phospholipase C (PLC). PLC cleaves phosphatidylinositol 4,5-bisphosphate (PIP_2), a minor membrane phospholipid, into two components: diacylglycerol (DAG) and inositol 1,4,5-trisphosphate (IP_3) (Berridge 1984). The latter compound was well established as an effector of the release of intracellular Ca^{2+} stores (Berridge 1984). In VSMCs, Ca^{2+} is stored within the sarcoplasmic reticulum (SR); activation of Ca^{2+} channels on the membrane of the SR by IP_3 results in release of the stored Ca^{2+} into the cytosol where it can activate the contractile proteins. Thus, this pathway provides a reasonable mechanism to explain how AVP increases $[Ca^{2+}]_{cyt}$ to elicit smooth muscle contraction. Evidence supporting this signaling pathway included the demonstration of robust increases in $[Ca^{2+}]_{cyt}$ in VSMCs in response to AVP even in the absence of extracellular Ca^{2+} (Vallotton et al. 1986, Byron and Taylor 1995), corresponding to measurements of AVP-induced IP_3

formation (Doyle and Ruegg 1985, Aiyar et al. 1986). In addition to the release of intracellular Ca^{2+} stores, AVP was also shown to increase influx of Ca^{2+} across the plasma membrane, an effect that was largely attributed to activation of store-operated Ca^{2+} (SOC) channels (Byron and Taylor 1995) and/or non-selective cation channels, including TRPC6 (Soboloff et al. 2005, Maruyama et al. 2006), which can be activated by DAG. A hypothetical signaling pathway for AVP-induced elevation of $[Ca^{2+}]_{cyt}$ and VSMC contraction was developed to account for all of these observations (Figure 1.1). Although it was generally well accepted, there was at least one crucial discrepancy that raised questions about its validity.

What had often been overlooked or ignored in considering the vascular AVP signal transduction model is the concentration-dependence of the components. Although AVP robustly elevates $[Ca^{2+}]_{cyt}$ in VSMCs by releasing intracellular Ca^{2+} stores, this effect is half-maximal at approximately 5 nM AVP, in good agreement with measurements of AVP-stimulated IP_3 formation, which similarly requires nanomolar concentrations for half-maximal activation (Doyle and Ruegg 1985, Aiyar et al. 1986, Ito et al. 1993). Activation of SOC entry and TRPC6 currents were also reported based on exposure of VSMCs to 50–100 nM AVP (Byron and Taylor 1995, Soboloff et al. 2005, Brueggemann et al. 2006). It appears that the hypothetical Ca^{2+} signaling model (Figure 1.1) requires nanomolar concentrations of AVP even though such concentrations are at least an order of magnitude higher than the highest concentrations measured in the systemic circulation. Can this model really explain the vasoconstrictor actions of AVP observed at physiological concentrations of AVP in the 10–100 pM range? We have explored this question for more than two decades and will describe some of our approaches and their remarkable outcomes in the remainder of this chapter.

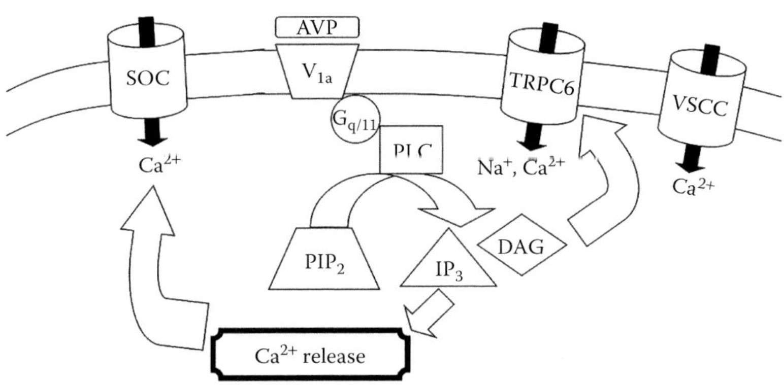

FIGURE 1.1 Hypothetical AVP Ca^{2+} signaling in VSMCs. Binding of Arg^8-vasopressin (AVP) to the V_{1a} vasopressin receptor is coupled to activation of the $G_{q/11}$ G protein α-subunit, which in turn activates phospholipase C (PLC). PLC cleaves PIP_2 to produce DAG and inositol 1,4,5-trisphosphate (IP_3). The latter activates a Ca^{2+} release channel in the SR, and the resulting depletion of intracellular Ca^{2+} stores triggers activation of SOC channels on the plasma membrane. DAG can activate TRPC6 non-selective cation channels in the plasma membrane. Influx of cations can depolarize the membrane to activate VSCCs to further stimulate Ca^{2+} influx.

1.2 CONCENTRATION-DEPENDENCE OF AVP-INDUCED Ca^{2+} RESPONSES IN VASCULAR SMOOTH MUSCLE CELLS MEASURED WITH FURA-2

In our initial attempt to investigate the concentration-dependence of AVP-stimulated Ca^{2+} responses in VSMCs, the A7r5 embryonic rat aorta smooth muscle cell line was employed. A7r5 cells had already been reported to respond robustly to AVP and to stably express many differentiated smooth muscle characteristics (Kimes and Brandt 1976); in contrast, smooth muscle-specific markers, V_{1a} receptor expression, and responses to AVP are often lost in primary cell cultures of VSMC (Thibonnier 1992, Owens 1995). We used the fluorescent Ca^{2+} indicator fura-2 to monitor $[Ca^{2+}]_{cyt}$ in single A7r5 cells or in confluent monolayers of A7r5 cells exposed to increasing concentrations of AVP. In the absence of extracellular Ca^{2+}, AVP stimulated a release of intracellular Ca^{2+} stores at supraphysiological concentrations, with an EC_{50} of approximately 5–10 nM AVP. In single cells, these responses were characterized by an abrupt rise in $[Ca^{2+}]_{cyt}$ after a variable latency (decreasing latency with increasing [AVP]), reaching a maximal peak of 862 ± 43 nM Ca^{2+} at 1 μM AVP (Figure 1.2). Neither the magnitude of the

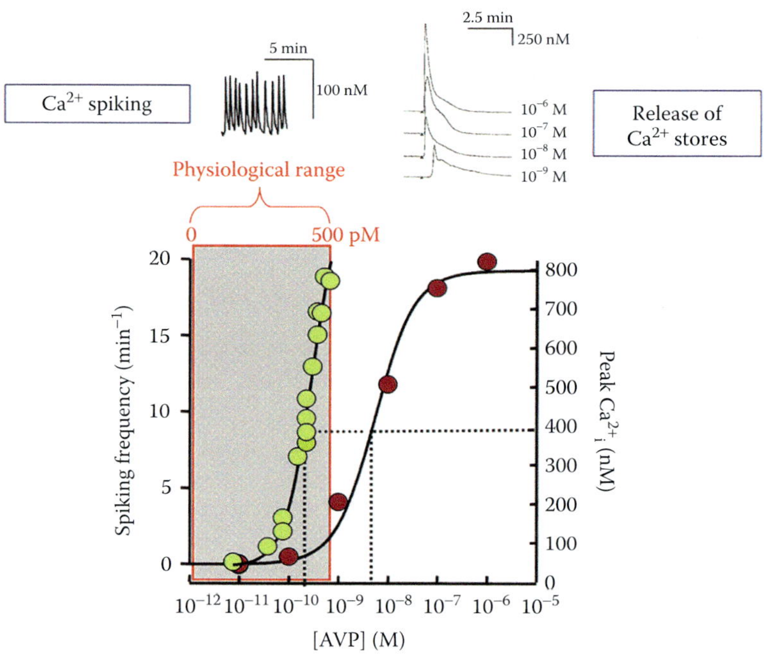

FIGURE 1.2 Ca^{2+} responses to varying [AVP] in A7r5 cells. Red symbols represent amplitude of Ca^{2+} release responses (scale bar on right). Green symbols represent frequency of Ca^{2+} spiking (scale bar on left). Representative examples of Ca^{2+} spiking and Ca^{2+} release time courses are shown above the graph. (Adapted from Byron, K. L., *Circ. Res.*, 78, 813–820, 1996.)

AVP-stimulated Ca^{2+} release signal nor its concentration-dependence was appreciably altered by inclusion of nimodipine, a blocker of L-type voltage-sensitive Ca^{2+} channels (VSCCs) (Byron 1996).

When the same A7r5 cells were exposed to much lower concentrations of AVP, starting at 10 pM (10^{-11} M) and increasing in 10 pM increments, it was immediately apparent that a different Ca^{2+} signal was elicited. Spontaneous Ca^{2+} transients (rapid elevation of $[Ca^{2+}]_{cyt}$ from a baseline of around 50 nM to a peak around 150–200 nM Ca^{2+}, referred to as *Ca^{2+} spikes*) occurred at very low frequency (~0.02/min) in the absence of AVP. When [AVP] was increased to a threshold between 10 and 30 pM, however, there was an abrupt increase in spike frequency and the frequency continued to increase in a steeply concentration-dependent manner up to a maximum of ~17/min at 500 pM AVP (EC_{50} ~ 150 pM) (Byron 1996).

Despite their strikingly divergent sensitivities to AVP, both the release of intracellular Ca^{2+} stores and the Ca^{2+} spiking responses were blocked by a V_{1a}-vasopressin receptor antagonist, suggesting that both effects are downstream of AVP binding to a single receptor subtype. However, several differences were noted in the Ca^{2+} spiking response compared with the Ca^{2+} release response: (1) Ca^{2+} spiking required extracellular Ca^{2+} and was abolished by nimodipine, whereas AVP-stimulated Ca^{2+} release occurred in the absence of extracellular Ca^{2+} and was unaffected by nimodipine; (2) the Ca^{2+} spiking response to AVP was a frequency-modulated (FM) response (spike frequency changed dramatically while the amplitude of the Ca^{2+} spiking was fairly constant over a wide range of AVP concentrations); in contrast, the release of intracellular Ca^{2+} by AVP was an amplitude-modulated (AM) response (increasing amplitude with increasing [AVP]); and (3) Ca^{2+} spiking occurred nearly synchronously among all of the cells of a confluent monolayer, whereas Ca^{2+} release responses occurred asynchronously, even among neighboring cells (Byron 1996). These exciting new observations suggested that distinct Ca^{2+} signaling mechanisms might underlie the moment-to-moment regulation of vascular tone by picomolar concentrations of AVP. Although the signaling pathway for AVP-stimulated release of Ca^{2+} stores was already described, it remained to be determined what mechanisms produce the Ca^{2+} spiking responses in the physiological range of AVP concentrations.

1.3 PROXIMAL SIGNALING EVENTS UNDERLIE DISTINCT Ca^{2+} RESPONSES TO VARYING [AVP]

The picomolar sensitivity to AVP suggested that the Ca^{2+} spiking response was elicited by binding of AVP to a small fraction of the V_{1a} receptors (i.e., a spare receptor response). This seemed unlikely to result from PLC activation, which occurs in proportion to the fraction of occupied receptors (based on EC_{50} for IP_3 formation (2–3 nM) being approximately equal to the K_D for AVP binding to V_{1a} receptors (Thibonnier et al. 1993). A previous study had reported that picomolar concentrations of AVP are sufficient to stimulate a different phospholipase in rat aortic smooth muscle cells, phospholipase A_2 (PLA_2, EC_{50} ~ 50 pM) (Ito et al. 1993). We therefore tested the hypothesis that Ca^{2+} spiking arises downstream of PLA_2 activation. In support of this hypothesis, we found that a putative PLA_2 inhibitor, ONO-RS-082, prevented AVP-stimulated Ca^{2+} spiking without inhibiting the release of Ca^{2+} stores by high [AVP] and without inhibiting Ca^{2+} responses to high extracellular [K^+], which directly stimulates Ca^{2+} influx via L-type voltage sensitive Ca^{2+} channels (Byron 1996). Furthermore, we supplemented the fura-2 measurements with measures of PLA_2 activity, measuring release of ^{3}H-arachidonic acid in response to AVP $\pm$ ONO-RS-082. These studies supported our hypothesis that AVP stimulates arachidonic acid release, which can be a direct product of PLA_2 activity, and that this effect is required for the stimulation of Ca^{2+} spiking (Byron 1996). Based on these findings, we proposed that AVP-stimulated Ca^{2+} signaling is concentration-dependent, with low concentrations evoking Ca^{2+} spiking responses via PLA_2 and arachidonic acid formation, and high concentrations eliciting release of Ca^{2+} stores via IP_3 formation (Figure 1.3).

Although our conclusions were reasonable, they were based on a limited set of experimental approaches. Further investigation with a broader set of biochemical and pharmacological tools suggested that the details of the initial hypothesis were wrong. Using thin layer chromatography (TLC) to more definitively identify the lipid second messengers generated following AVP treatment, we found that the putative PLA_2 inhibitor ONO-RS-082 did not inhibit arachidonic acid formation, but instead inhibited production of phosphatidic acid (PA), the primary product of phospholipase D (PLD) activity (Li et al. 2001). AVP had been shown previously to activate PLD in A7r5 cells (Thibonnier et al. 1991, Meier et al. 1998), though its role in Ca^{2+} signaling had not been determined. We confirmed that AVP robustly stimulates PLD activation at concentrations as low as 100 pM and found that multiple inhibitors of PLD (confirmed using TLC) prevented AVP-stimulated Ca^{2+} spiking (inactive structural analogs used as negative controls were without effect on either PLD activation or Ca^{2+} spiking). Addition of purified bacterial PLD, which had been shown previously to stimulate PA formation in A7r5 cells (Jones et al. 1994), mimicked AVP in stimulating Ca^{2+} spiking, whereas neither purified PLA_2 nor the products of PLA_2

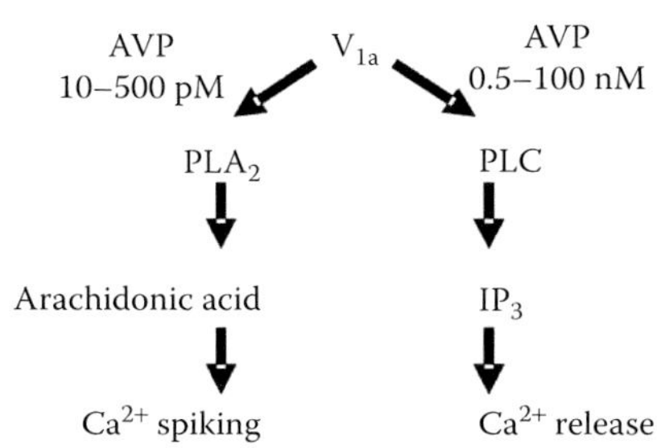

FIGURE 1.3 Revised Ca^{2+} signaling pathways. Hypothetical concentration-dependent Ca^{2+} signaling based on studies in A7r5 cells.

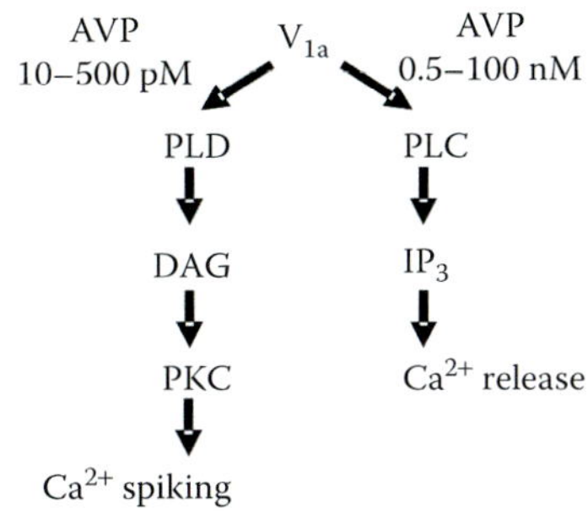

FIGURE 1.4 Further revised Ca^{2+} signaling pathways. Hypothetical concentration-dependent Ca^{2+} signaling based on studies in A7r5 cells.

(arachidonic acid and lysophospholipids) were effective over a broad range of concentrations (Li et al. 2001). We therefore replaced PLA_2 with PLD in our hypothetical signaling scheme (Figure 1.4).

1.4 ELUCIDATING DOWNSTREAM EFFECTORS OF AVP-STIMULATED Ca^{2+} SPIKING

Protein kinase C (PKC) is a serine/threonine protein kinase that had been implicated as a common downstream target in PLD signaling in many cell types. We therefore examined the possibility that PKC might be a crucial mediator of AVP-stimulated Ca^{2+} spiking. Phosphorylation of proteins on serine or threonine residues was suggested by our finding that a potent serine/threonine phosphatase inhibitor, calyculin A, sensitized A7r5 cells to AVP, such that concentrations of AVP as low as 1 pM stimulated robust Ca^{2+} spiking (calyculin A in the absence of AVP had no effect) (Fan and Byron 2000). A specific role for PKC was implicated by our demonstration that a phorbol ester, 4β-Phorbol 12-myristate 13-acetate (PMA), which is a direct activator of PKC, stimulated Ca^{2+} spiking in a concentration-dependent manner with a robust effect at concentrations as low as 100 pM; an inactive phorbol ester analog had no effect (Fan and Byron 2000). Moreover, both PMA- and AVP-stimulated Ca^{2+} spiking were completely suppressed by two structurally distinct PKC inhibitors, calphostin C and Ro-31-8220, and by chronic (24-hr) treatment with 1 μM PMA, which resulted in down-regulation of PKC-α and PKC-δ, but not PKC-ε expression (Fan and Byron 2000). PLD activation was not inhibited by Ro-31-8220 (Li et al. 2001), suggesting that PKC activation likely occurs downstream of PLD (Figure 1.4), perhaps via DAG produced by the action of PA phosphohydrolase (Lee and Severson 1994). These results were complemented with protein biochemical approaches to demonstrate expression of multiple PKC isoforms (α, β, γ, δ, ε, ζ, and λ) and measurements of the time courses of AVP-induced translocation of PKC-α, -δ, and -ε from cytosol to membrane compartments (Figure 1.5), which is a hallmark of their activation (Fan and Byron 2000).

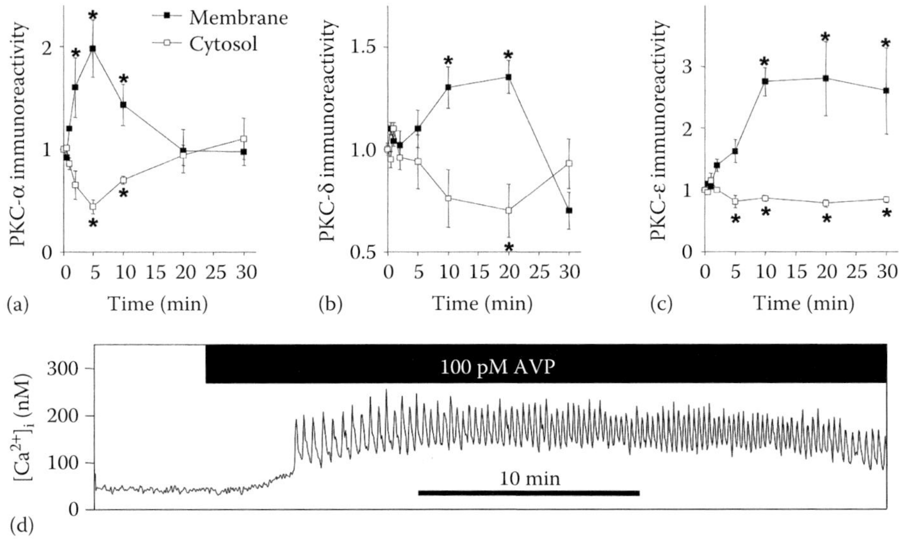

FIGURE 1.5 Time-dependent translocation of PKC-α, -δ, -ε from cytosol to membranes in response to 100 pM AVP. A7r5 cell monolayers were treated for varying times with 100 pM AVP, lysed, and separated into cytosolic and membrane fractions. The amount of PKC-α (a), PKC-δ (b), and PKC-ε (c) isoforms present in either fraction at each time point was determined by Western blot analysis and densitometry. Results from three experiments for each isoform are summarized (means ± s.e.m.; asterisks indicate significant difference from 0 min time point, $P < 0.05$, repeated measures analysis of variance/Dunnett's test). (d) A representative $[Ca^{2+}]_{cyt}$ response to 100 pM AVP (filled box) illustrating a latency of several minutes between addition of AVP and initiation of repetitive Ca^{2+} spiking. (Reproduced from Fan, J., and Byron, K. L., *J. Physiol.*, 524, 821–831, 2000. With permission.)

1.5 ION CHANNEL TARGETS IN AVP SIGNAL TRANSDUCTION

The characteristics of the Ca^{2+} spiking response (dependence on L-type VSCCs, synchronization in electrically-coupled cells) suggested an electrical signaling mechanism. Spontaneous action potential firing in A7r5 cells had been described previously (Kimes and Brandt 1976, Van Renterghem et al. 1988, Marks et al. 1990, Knot et al. 1991), leading us to hypothesize that Ca^{2+} spikes correspond to Ca^{2+}-dependent action potentials and that physiological [AVP] increases the frequency of action potential firing. Although we had solid evidence to support several upstream steps in the signaling cascade, the electrical signaling events underlying activation of L-type channels and the stimulation of action potential firing by physiological [AVP] were not known.

The stimulus for activation of L-type Ca^{2+} channels is membrane depolarization, which most commonly arises via increased influx of cations (Na^+ or Ca^{2+}), increased efflux of Cl^-, or decreased efflux of K^+. The gold standard approach for investigating contributions of ionic fluxes to cellular function is through the use of patch clamp techniques. We have previously described in detail the technical considerations for application of patch clamp methods to determine drug effects on vascular smooth muscle ion channels (Brueggemann and Byron 2012). Here, we will more briefly describe how these methods were originally applied to investigate AVP signal transduction.

Patch clamp techniques can be used to clamp membrane ionic currents (current clamp mode) and measure changes in voltage, such as action potential firing. Alternatively, voltage clamp can be employed to measure the ionic currents that pass across the cell membrane as a result of the opening of various kinds of ion channels at specific voltages. At the time of our initial studies, A7r5 cells were a popular model for VSMC electrophysiology. Spontaneous action potential firing in confluent monolayers of A7r5 cells was described as early as 1976 (Kimes and Brandt 1976), and numerous subsequent studies had reported L-type Ca^{2+} currents, Ca^{2+}-activated K^+ currents, non-selective cation currents, chloride currents, and so on, under various recording conditions (Fish et al. 1988, McCarthy and Fry 1988, Van Renterghem et al. 1988, Jiang et al. 2003). We initially sought to characterize the electrophysiological effects of physiological [AVP] under conditions similar to those in which we recorded AVP-stimulated Ca^{2+} spiking. In confluent monolayers of A7r5 cells, ruptured patch whole cell recording of membrane voltage in current clamp mode and using an external solution that was equivalent to the bath solution used for Ca^{2+} spiking studies, revealed that physiological [AVP] stimulated action potential firing with the same exquisite sensitivity as it did Ca^{2+} spiking (Figure 1.6). Under similar ionic conditions, using a perforated patch whole cell configuration to record from isolated A7r5 cells, repetitive action potential firing (Figure 1.7a) was observed following treatment with 100 pM AVP. This repetitive firing usually started abruptly following AVP-induced membrane depolarization (on average, from -55.9 ± 1.6 to -45.2 ± 1.4 mV, Figure 1.7b). These effects were associated with a significant increase in membrane electrical resistance (from 1.44 ± 0.27 to 2.26 ± 0.41 GΩ, Figure 1.7c) (Brueggemann et al. 2007). In voltage clamp mode, a mix of currents was recorded over a range of voltages. Application of 1 s voltage steps from a—74 mV holding potential to test potentials ranging from −94 to +46 mV revealed inward L-type Ca^{2+} currents that were activated transiently at voltages positive to −40 mV (peak at $+1 \pm 1.3$ mV), and outward K^+ currents that developed more slowly, but were sustained through the end of the voltage steps (Brueggemann et al. 2007). The effect of 100 pM AVP was a significant decrease in the amplitude of the outward K^+ currents; there was no effect on L-type currents under voltage clamp conditions (Brueggemann et al. 2007).

Among the most revealing of these findings was that AVP-stimulated membrane depolarization was associated with an increase in membrane resistance. This supported inhibition of K^+ channels regulating membrane voltage as a primary effect of AVP, rather than activation of cation or Cl^- channels, which would tend to decrease membrane resistance. The voltage clamp recordings also indicated suppression of K^+ currents by AVP, though with the recording conditions used, the exact nature of the AVP-sensitive K^+ currents could not be determined. Conditions had to be developed under which the AVP-sensitive K^+ currents could be recorded in relative isolation from other currents. With some trial and error, this was accomplished, revealing voltage-dependent outward K^+ currents with some surprising features.

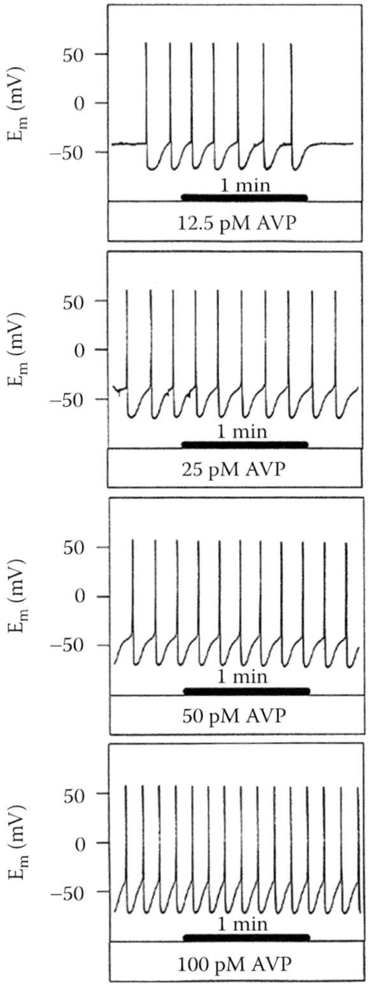

FIGURE 1.6 Whole cell current clamp measurements of action potentials in A7r5 cells. Increasing frequency of action potentials in increasing concentrations of AVP. Membrane potential was measured in a single A7r5 cell in a monolayer. Traces shown were recorded after exposure for at least 10 min in each of the indicated concentrations of AVP (no action potentials occurred spontaneously during 10 min prior to addition of AVP (not shown).

Well-isolated, outwardly rectifying K_V currents were recorded in A7r5 cells with a holding potential of −74 mV and a 5-s voltage step protocol applied to test potentials from −94 to +36 mV (Figure 1.8a). Inclusion of gadolinium chloride ($GdCl_3$, 100 μM) in the external solution eliminated contributions from inward cation currents and also, by blocking voltage-dependent Ca^{2+} channels, prevented activation of iberiotoxin-sensitive Ca^{2+}-activated K^+ channels (BK_{Ca}). Several characteristics of the K_V currents recorded under these conditions were different from what had been described in earlier studies of VSMC K_V currents (Nelson and

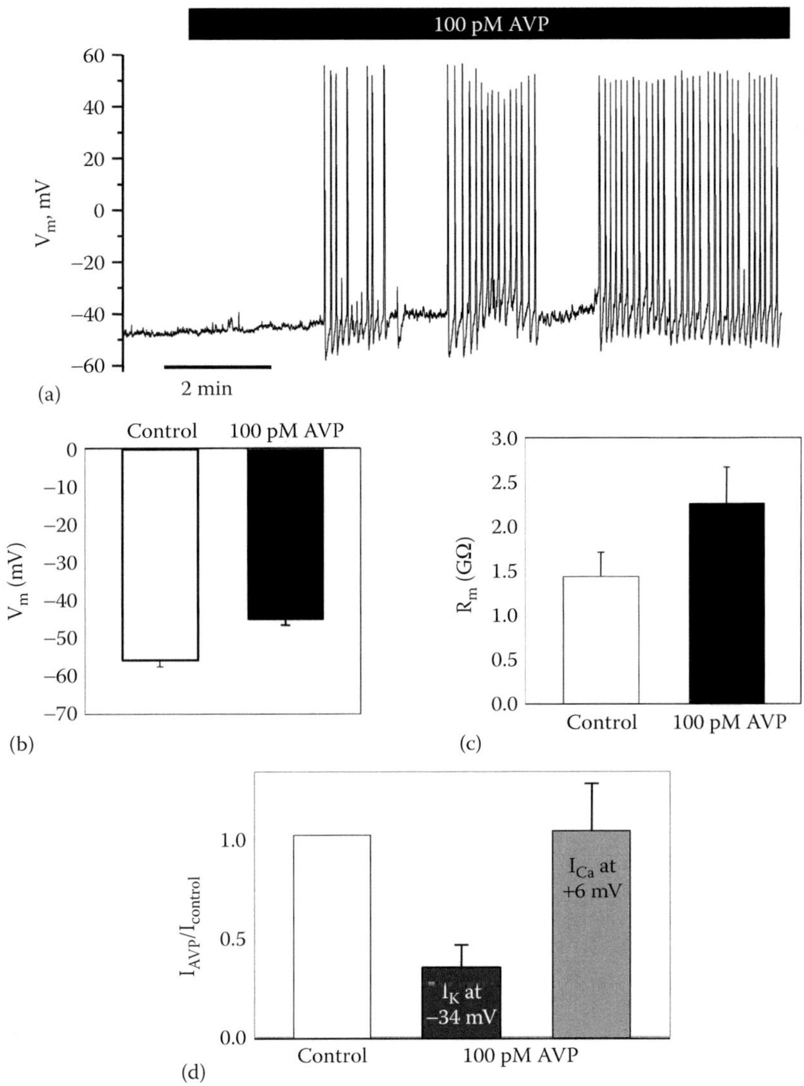

FIGURE 1.7 Patch clamp data from A7r5 cells under physiological ionic conditions. (a) Representative trace shows time course of action potential firing in a single A7r5 cell treated with 100 pM AVP. Note membrane depolarization prior to AP generation. Similar results were obtained in 6 experiments; mean time to initiation of AP firing was 5.7 ± 0.8 min ($n = 6$). V_m, membrane potential. (b) Application of 100 pM AVP significantly depolarized the membrane from −55.9 ± 1.6 to −45.2 ± 1.4 mV ($n = 8$, $P < 0.001$). (c) In the same time frame, application of 100 pM AVP significantly increased R_m from 1.44 ± 0.27 to 2.26 ± 0.41 GΩ ($n = 7$, $P = 0.005$, paired *t*-test) as calculated from the currents measured in response to 20-mV voltage steps from −64 to −44 mV (±10 mV from the resting membrane potential, V_m). (Reproduced from Brueggemann, L. I. et al., *Am. J. Physiol. Heart Circ. Physiol.*, 292, H1352–H1363, 2007. With permission.) (d) Application of 100 pM AVP suppressed (K_V) currents (I_K) measured at −34 mV by 74% without affecting L-type Ca^{2+} currents (I_{Ca}) measured at +6 mV ($n = 5$).

Quayle 1995): (1) the currents activated very slowly, requiring approximately 1 s to reach steady-state activation at positive voltages (Figure 1.8a); (2) the threshold for voltage-dependent activation of the currents was approximately −60 mV (Figure 1.8b and c), nearly 40 mV more negative than reported in earlier studies (Tanaka et al. 2006); and (3) there was no inactivation of the currents at any membrane voltages up to +46 mV, even with 5-s voltage steps (Figure 1.8a). Pharmacological characterization revealed that the currents were insensitive to blockers of several classes of vascular K^+ channels, including iberiotoxin (100 nM), which inhibits BK_{Ca} channels, glibenclamide (10 µM, K_{ATP} channel blocker), and 4-aminopyridine (4-AP, 1 mM, a K_V channel blocker) (Brueggemann et al. 2007). In contrast,

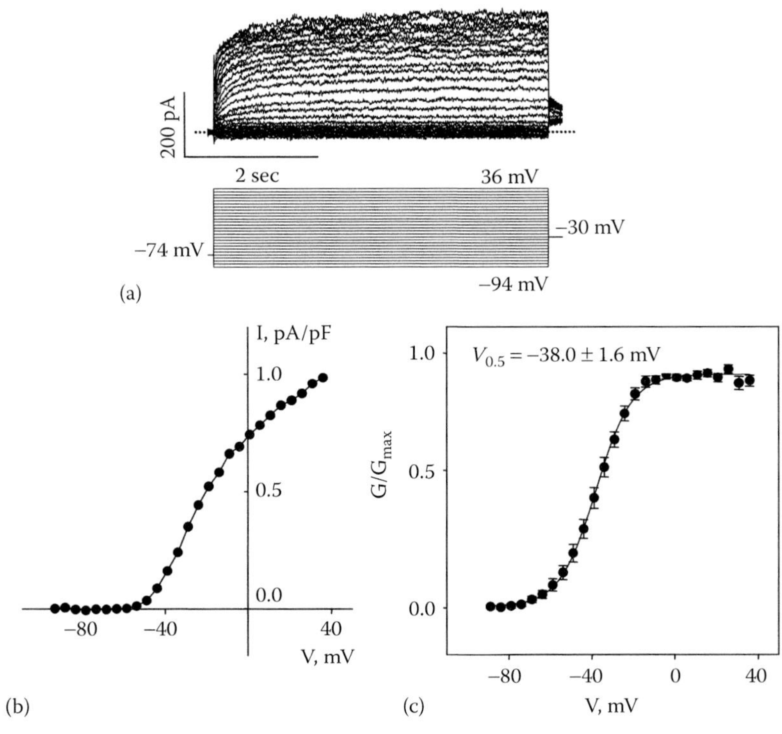

FIGURE 1.8 Isolation of AVP-sensitive K_V currents in single A7r5 cells. (a) Representative traces of current recorded in the presence of 100 µMGdCl_3 in external solution to isolate K_V component (top). With a holding potential of −74 mV, a voltage step protocol was applied to test potentials from −94 to +36 mV, followed by steps to −30 mV for tail-current recording. Dotted line indicates 0 current level. (b) Mean current recorded at the end of the pulse (1,000 points recorded over the last 500 ms were averaged) from the traces presented in (a) after leak subtraction and normalization to the cell capacitance (C = 284 pF). (c) Voltage dependence of steady-state activation fitted by the single Boltzmann function. Conductance was calculated from the tail currents measured at −30 mV based on a K^+ reversal potential of −84 mV. Voltage of half-maximal activation ($V_{0.5}$) = −38.0 ± 1.6 mV; slope factor (s) = 8.3 ± 0.4 mV (n = 21); G/G_{max}, fractional maximal conductance. *(Continued)*

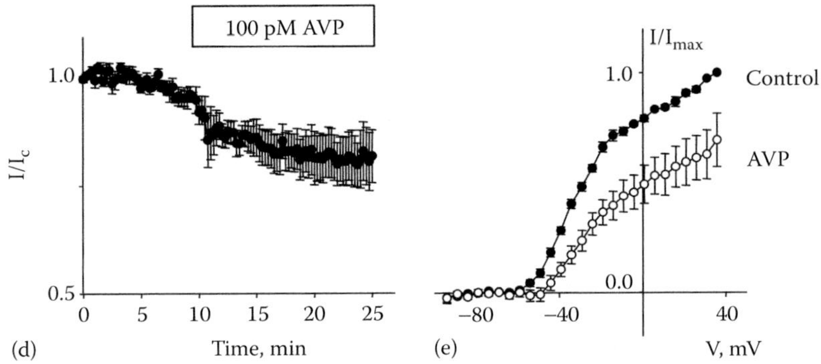

FIGURE 1.8 (Continued) Isolation of AVP-sensitive K_V currents in single A7r5 cells. (d) Time course of K_V current inhibition by 100 pM AVP recorded by applying a 5-s step to a membrane potential of 0 mV from a holding potential of –74 mV at 15-s intervals. Currents were normalized to mean control current (I/I_c) recorded for 10 min before adding AVP ($n = 7$). (e) I-V curves of mean outward current measured in the presence of 100 μMGd$_{3+}$ before (control) and after exposure to 100 pM AVP for 15 min. After leak subtraction, currents were normalized to the maximal control current (I/I_{max}) measured at +36 mV. AVP significantly reduced K_V current at all membrane potentials from −54 to +36 mV ($n = 7$, $P < 0.05$, paired Student's t-test). (Reproduced from Brueggemann, L. I. et al., *Am. J. Physiol. Heart Circ. Physiol.*, 292, H1352–H1363, 2007. With permission.)

100 μM $BaCl_2$ did effectively inhibit the currents and also mimicked AVP in the stimulation of Ca^{2+} spiking (Brueggemann et al. 2007). Despite their lack of sensitivity to 4-AP, the K_V currents were sensitive to treatment with 100 pM AVP, which induced a significant reduction in current amplitude within a few minutes (Figure 1.8d). *I-V* curves revealed that 100 pM AVP treatment for 15 min significantly reduced K_V current amplitude at all test potentials from −54 to +36 mV (Figure 1.8e) (Brueggemann et al. 2007).

Although the AVP-sensitive K_V currents did not resemble K_V currents previously recorded from VSMCs, they did resemble currents denoted as *M-currents* recorded in neurons. M-currents are slowly activating, non-inactivating delayed rectifier K_V currents, which are suppressed in response to activation of $G_{q/11}$-coupled receptors (Adams et al. 1982). The neuronal M-currents had been attributed to the activity of K_V7 family channels encoded by five *KCNQ* genes (*KCNQ1-5*). In particular, neurons primarily express channels formed as heteromeric tetramers composed of $K_V7.2$ and $K_V7.3$ subunits, or $K_V7.3$ and $K_V7.5$ subunits (Delmas and Brown 2005). When we began this work, none of the neuronal *KCNQ* genes was known to be expressed in vascular smooth muscle. However, our evidence for AVP-sensitive M-currents in A7r5 cells led us to examine a potential role for K_V7 channels in AVP signal transduction. Fortunately, drugs had been developed to target neuronal K_V7 channels with high selectivity over other classes of neuronal K^+ channels. We found that K_V7 channel blockers (linopirdine and XE991) very effectively inhibited the AVP-sensitive K_V current in A7r5 cells.

Furthermore, an activator of neuronal K_V7 channels, flupirtine, more than doubled the amplitude of the A7r5 cell K_V current, an effect that was reversed following washout of flupirtine. Polymerase chain reaction (PCR), using primers for all five *KCNQ* gene products, revealed that A7r5 cells express only *KCNQ5*, whereas freshly isolated adult rat aortic smooth muscle cells detectably expressed both *KCNQ1* and *KCNQ5*. All five *KCNQ* mRNAs were detected in adult rat brain as a positive control. Knockdown of *KCNQ5* expression by siRNA or shRNA produced corresponding decreases in the AVP-sensitive K_V current in A7r5 cells and stimulated spontaneous action potential firing (Figure 1.9) (Brueggemann et al. 2007, Mani et al. 2009). Overexpression of human *KCNQ5* in A7r5 cells produced currents with similar biophysical characteristics (kinetics of activation, voltage-dependence of activation, lack of time-dependent inactivation, etc.) and

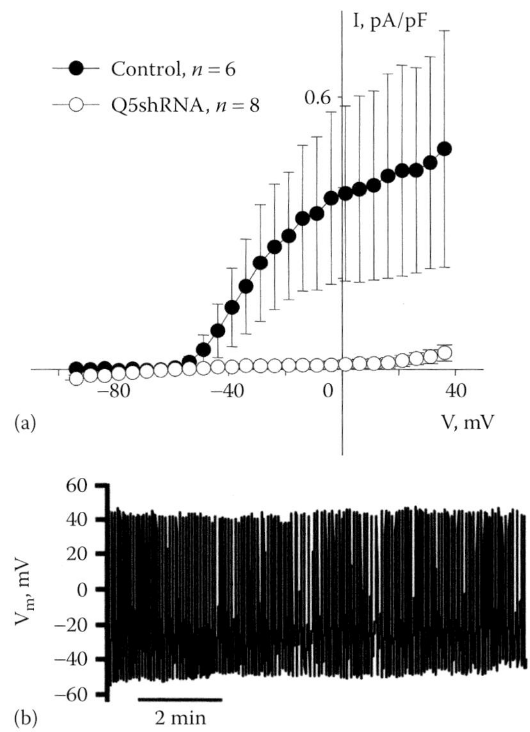

FIGURE 1.9 Expression of KCNQ5shRNA inhibits K_V7 currents and induces action potential firing in A7r5 cells. (a) IV graph shows M-currents measured in control A7r5 cells (black circles, $n = 7$) and in A7r5 cells expressing KCNQ5shRNA (open circles, $n = 8$). I, current; V, voltage. Currents in A7r5 cells expressing KCNQ5shRNA were significantly smaller than control currents at all voltages positive to −49 mV ($p < 0.05$, Rank Sum test). (b) Representative trace of membrane potential (V_m) recorded in current-clamp mode shows spontaneous electrical activity (high frequency action potential firing) in a cell expressing KCNQ5shRNA. (Reproduced from Mani, B. K. et al., *Cell Calcium*, 45, 400–411, 2009. With permission.)

sensitivity to AVP. We concluded that the native A7r5 AVP-sensitive K_V channels are homotetramers of $K_V7.5$ subunits. The importance of $K_V7.5$ channels in A7r5 cells' Ca^{2+} signaling was reinforced by the observations that either knockdown of *KCNQ5* expression or pharmacological inhibition of K_V7 channel activity induced membrane depolarization, and stimulated action potential firing and Ca^{2+} spiking, similar to the effects of physiological [AVP] (Mani et al. 2009).

Although suppression of K_V7 currents might account for AVP-induced membrane depolarization and stimulation of Ca^{2+} influx in A7r5 cells, other mechanisms, particularly activation of TRPC6 nonselective cation channels, had been suggested by other laboratories (Soboloff et al. 2005, Li et al. 2008). We recorded non-selective cation currents in A7r5 cells that were activated by 100 nM AVP and were abolished by adenoviral overexpression of TRPC6 shRNA (Mani et al. 2009). However, there was no effect of TRPC6 knockdown on 100 pM AVP-induced membrane depolarization, suggesting little contribution of TRPC6 channel activation to membrane depolarization at physiological concentrations of AVP (Mani et al. 2009).

1.6 A MORE PHYSIOLOGICAL MODEL SYSTEM

The cultured A7r5 cell model had been extraordinarily useful in identifying a novel signaling pathway activated at physiological concentrations of AVP. However, its relevance to vascular physiology remained to be determined. Our hypothetical signaling scheme predicted that physiological concentrations of AVP (10–100 pM) would exert vasoconstrictor effects via PKC activation, suppression of $K_V7.5$ channel activity, membrane depolarization, and stimulation of Ca^{2+} influx via L-type VSCCs in arterial myocytes. In contrast, [AVP] > 1 nM was predicted to elicit a vasoconstrictor response independently of these signaling events, primarily mobilizing intracellular Ca^{2+} stores via the IP_3 pathway. We tested these predictions using rat mesenteric arteries, as both a convenient tissue for experimental use (relative ease of access, abundance of arteries that can be easily dissected free of adjacent tissue structures, well established as a model system for research studies) and as the primary physiological target for endocrine AVP signaling.

Vasoconstrictor actions of varying concentrations of AVP (10^{-14}–10^{-6} M) were measured in vitro using segments of 4th-order rat mesenteric arteries (~250–400 microns in outer diameter) and pressure myography (Henderson and Byron 2007). Under approximately physiological conditions (bathed in and internally filled with physiological saline solution, maintained at 37°C, 80 mmHg internal pressure), the mesenteric artery segments constricted in response to AVP in a dose-dependent manner, with half-maximal constriction at approximately 30 pM and maximal constriction (complete occlusion of the lumen) at concentrations ≥10 nM (Henderson and Byron 2007). As predicted based on our A7r5 cell studies, vasoconstriction in response to 30 pM AVP was dependent on PKC and L-type Ca^{2+} channels (abolished in the presence of a PKC inhibitor or a blocker of L-type channels). In contrast, the vasoconstrictor responses to 10 nM AVP were unaffected by either treatment, establishing the same concentration-dependent dichotomy of signaling that was apparent in the A7r5 cell line (Henderson and Byron 2007).

We hypothesized that K_V7 channels serve as an intermediate between PKC activation and activation of L-type Ca^{2+} channels in the physiological AVP signaling pathway of mesenteric artery smooth muscle. We developed conditions for recording M-currents in freshly isolated mesenteric arterial smooth muscle cells (MASMCs) and validated these recordings based on their M-current-like electrophysiological characteristics (e.g., voltage-dependence of activation, lack of inactivation with sustained voltage steps, outward rectification), appropriate responses to a pharmacological K_V7 channel activator (flupirtine) and K_V7 channel inhibitors (XE991 and linopirdine), and insensitivity to a variety of pharmacological blockers of other classes of vascular K^+ channels (Mackie et al. 2008). In support of our hypothesis, a physiological concentration of AVP (100 pM) was sufficient to suppress the MASMC M-currents, and this effect was PKC-dependent. Both AVP (30–100 pM) and linopirdine (10 μM) caused membrane depolarization in MASMCs and constricted pressurized rat mesenteric artery segments, but these effects of AVP and linopirdine were not additive (Figure 1.10), suggesting a common mechanism

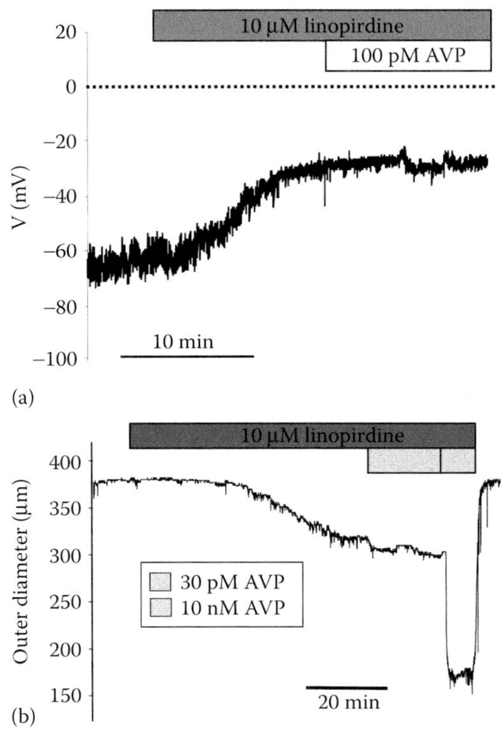

FIGURE 1.10 Membrane depolarization in MASMCs and constriction of mesenteric arteries induced by K_V7 channel blocker linopirdine. (a) Representative time course of membrane depolarization in a mesenteric artery myocyte in response to 10 μM linopirdine followed by additional application of 100 pM AVP. (b) Representative recording of mesenteric artery constriction in response to K_V7 channel blocker linopirdine (10 μM) followed by addition of 30 pM AVP and then 10 nM AVP. (Reproduced from Mackie, A. R. et al., *J. Pharmacol. Exp. Ther.*, 325, 475–483, 2008. With permission.)

(Mackie et al. 2008). The vasoconstrictor effects of linopirdine were prevented by pretreatment with verapamil, a blocker of L-type Ca^{2+} channels, further supporting our hypothesis that PKC-dependent suppression of K_V7 currents underlies the L-type Ca^{2+} channel-mediated vasoconstrictor actions of physiological [AVP].

1.7 DRILLING DOWN TO THE MOLECULAR DETAILS OF K_V7 CHANNEL MODULATION

Questions remained about the molecular identities of the channel subunits that conduct the K_V7 currents in rat mesenteric artery myocytes and which subunits are directly affected as a consequence of V_{1a}-vasopressin receptor activation. While A7r5 cells express only one neuronal *KCNQ* gene, *KCNQ5*, we found that rat mesenteric artery myocytes express both *KCNQ4* and *KCNQ5*, similar to rat arterial myocytes from many other vascular beds (Yeung et al. 2007, Mackie et al. 2008, Joshi et al. 2009). The presumed tetrameric channel structure and the potential for homomeric or heteromeric subunit assembly raised some interesting questions. Do the functional channels represent a mix of homotetrameric $K_V7.4$ and homotetrameric $K_V7.5$ channels or do $K_V7.4$ and $K_V7.5$ subunits join up to form heterotetrameric channels, and which of these tetrameric channel assemblies are the primary functional units? Which channel configuration(s) is/are regulated by AVP?

To begin to answer these questions, we attempted to silence the expression of *KCNQ4* and *KCNQ5* in the arterial myocytes but were unable to achieve more than a partial knockdown using siRNA or shRNA approaches (unpublished results). As an alternative strategy, finding a pharmacological agent that would distinguish between $K_V7.4$ and $K_V7.5$ channels was desirable. Fortunately, we found that diclofenac, a nonsteroidal anti-inflammatory drug, served as an activator of homomeric $K_V7.4$ and a blocker of homomeric $K_V7.5$ channels (Brueggemann et al. 2011). We decided to use A7r5 cells as an expression system for homomeric $K_V7.4$, homomeric $K_V7.5$, and heteromeric $K_V7.4/7.5$, to evaluate biophysical properties of the different channel subunit combinations, and to use diclofenac as a diagnostic tool to evaluate the natively expressed channels in mesenteric artery myocytes. The exogenous expression studies revealed distinct electrophysiological characteristics of homomeric $K_V7.4$, homomeric $K_V7.5$, and heteromeric $K_V7.4/7.5$ channels in A7r5 cells (Brueggemann et al. 2011). Notably, when both subunits were expressed, the resulting currents reflected primarily heteromeric channels rather than a mix of homomeric channels. Mesenteric artery myocytes, which natively express both *KCNQ4* and *KCNQ5* genes, displayed whole-cell K_V7 currents with electrophysiological properties and diclofenac responses characteristic of heteromeric $K_V7.4/7.5$ channels (Brueggemann et al. 2011). We confirmed that $K_V7.4/7.5$ heteromers are endogenously expressed in mesenteric artery myocytes using a proximity ligation assay (Brueggemann et al. 2014). In further support of the hypothesis that heteromeric $K_V7.4/7.5$ channels are the primary functional channels in mesenteric artery myocytes, we found that introduction of dominant-negative $K_V7.4$ and $K_V7.5$ subunits reduced endogenous K_V7 currents by 84% and 76%, respectively (Figure 1.11) (Brueggemann et al. 2014).

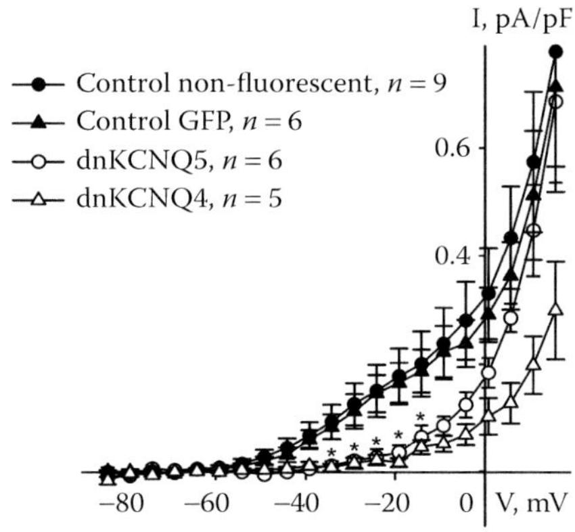

FIGURE 1.11 Dominant-negative hKCNQ4(G285S) and hKCNQ5(G278S) expressed in MASMCs suppress endogenous K_V7 currents. Mean K_V7 current amplitudes were measured at varying voltages in MASMCs expressing either dominant-negative KCNQ4 (dnKCNQ4) or dominant-negative KCNQ5 (dnKCNQ5) and compared with those measured in MASMCs expressing only the GFP vector or non-fluorescent MASMCs with no exogenous expression. * indicates a significant difference for dnKCNQ4 and dnKCNQ5 from control GFP, $p < 0.05$, two-way ANOVA. (Reproduced from Brueggemann et al. 2014. With permission.)

A crucial missing piece in our pathway between $V1_a$ receptor activation and suppression of K_V7 currents was clear evidence of PKC-dependent phosphorylation of channel subunits. Using an inducible protein kinase C alpha (PKCα) translocation system, we found that PKCα activation is sufficient to suppress endogenous K_V7 currents in A7r5 cells and rat mesenteric artery smooth muscle cells (Figure 1.12) (Brueggemann et al. 2014). We also examined AVP signal transduction in A7r5 cells overexpressing human $K_V7.5$ homomers ($hK_V7.5$), $K_V7.4$ homomers ($hK_V7.4$), or $K_V7.4/7.5$ heteromers ($hK_V7.4/7.5$). Notably, when $hK_V7.5$ was overexpressed, the currents were generally about 100-fold larger than the native $K_V7.5$ currents, but their sensitivity to 100 pM AVP was reduced by about 10-fold when examined 24–48 h after introduction of the expression vector. As time after overexpression increased, the whole cell current amplitudes remained about the same, but their sensitivity to AVP increased, such that, by 10–14 days, 100 pM AVP robustly inhibited the currents, similar to its effect on the much smaller native currents. We have observed this time-dependent sensitization repeatedly and speculated that it reflects the gradual adaptation of the cells to the overly abundant K_V7 channels. We routinely utilized the cells 10–14 days after introduction of exogenous channels to account for this effect and to more closely mimic the cells' normal AVP signaling sensitivity. Under these conditions, we found that AVP (100 and 500 pM) and the PKC activator PMA (1 nM) each inhibited $hK_V7.5$ and $hK_V7.4/7.5$, but not $hK_V7.4$ channels overexpressed in A7r5 cells. In addition, using immunoprecipitation and western blot approaches, we determined that the AVP- and PMA-induced inhibition of $hK_V7.5$ and $hK_V7.4/7.5$ current was associated with an increase in PKC-dependent phosphorylation of both

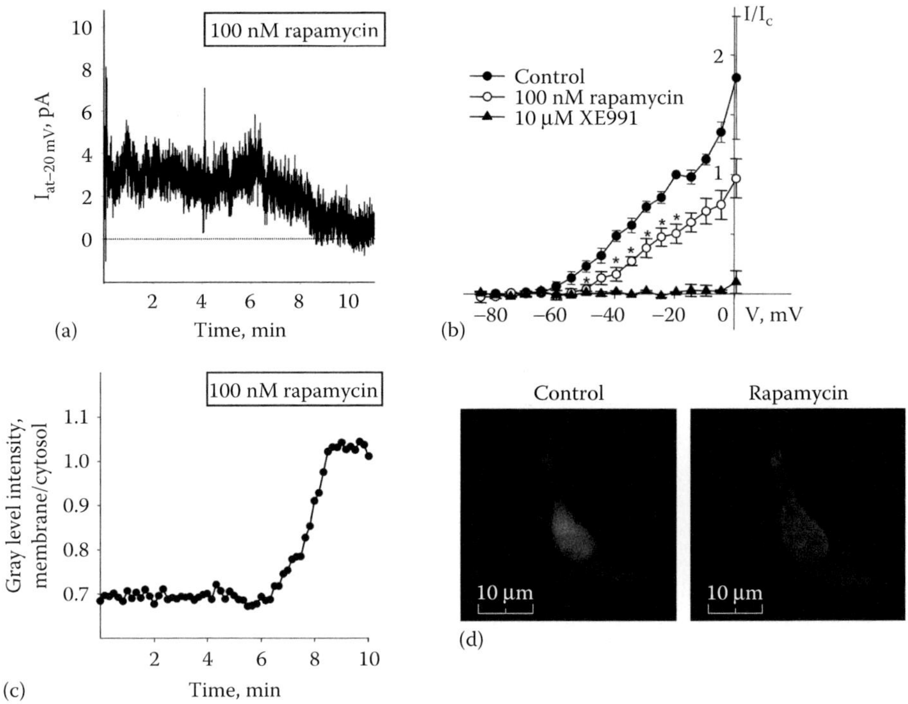

FIGURE 1.12 Suppression of endogenous K_V7 currents in MASMCs upon activation of PKCα using a rapamycin-induced PKCα translocation system. (a) Representative time course of K_V7 current inhibition in a mesenteric artery myocyte upon application of rapamycin (100 nM). (b) Mean I-V curves of steady-state K_V7 currents recorded in myocytes with a voltage step protocol before (control, filled circles), after treatment with 100 nM rapamycin for 12 min (open circles) and after treatment with 10 μM XE991 (filled triangles). Currents were normalized by the current recorded at –20 mV before application of rapamycin. * indicates significant difference from control, $p < 0.05$, Student's paired t-test, $n = 4$). (c) Representative time course of PKCα translocation recorded in an isolated arterial myocyte upon application of rapamycin (100 nM). (d) Representative images of the myocyte used for recordings for panel C before application of rapamycin (100 nM, left panel) and at the end of application of rapamycin for 10 min (right panel). (Reproduced from Brueggemann et al. 2014. With permission.)

channel subunits (Brueggemann et al. 2014). Interestingly, $hK_V7.4$ subunits were phosphorylated only when they were associated with $K_V7.5$, not when $hK_V7.4$ was overexpressed by itself (Brueggemann et al. 2014). We speculate that the latter observation may reflect a selective interaction of PKC with $K_V7.5$ subunits, and that $K_V7.5$ subunits within a $K_V7.4/K_V7.5$ heteromeric channel can recruit PKC to within sufficient proximity of $K_V7.4$ subunits (within the same tetramer), so that these subunits may then also serve as a substrate for the kinase.

1.8 CLINICAL PHARMACOLOGY OF K_V7 CHANNELS AND ITS CARDIOVASCULAR CONSEQUENCES

Many of the drugs that we had used in our *in vitro* testing of vascular K_V7 currents were being used, or had previously been used, as clinical therapeutic agents to treat human conditions, with the intention of specifically targeting neuronal M-currents to affect neuronal activity. Our findings that these drugs have robust effects on M-currents in arterial myocytes suggested a potential for cardiovascular side effects when administered for treatment of neurological disorders. Furthermore, AVP levels are altered in a number of cardiovascular diseases and AVP is also used as a therapeutic agent in certain conditions, raising possibilities of favorable or adverse interactions with K_V7 channel modulating drugs. Notably, the K_V7 channel blockers, such as linopirdine, inhibit channels formed from all five *KCNQ* gene products, whereas clinically used K_V7 channel activators (e.g., retigabine, flupirtine, celecoxib), have little if any effect on the $K_V7.1$ homotetramers found in cardiac myocytes, but these drugs do enhance the activity of channels formed from any of the other four types of K_V7 channel subunits.

Flupirtine has been used clinically for treatment of acute and chronic pain for more than 30 years. It had been reported to modestly lower systolic BPs in patients taking flupirtine chronically (Herrmann et al. 1987), though the mechanism underlying this effect had not been elucidated. Our pressure myography experiments revealed that flupirtine elicits a concentration-dependent relaxation of pressurized mesenteric arteries that had been pre-constricted with 30 pM AVP (Mackie et al. 2008). This vasodilatory effect might be expected to oppose the vasoconstrictor actions of AVP and hence to lower BP *in vivo*. We established an *in vivo* model to test this possibility, implanting a femoral artery BP catheter and an ultrasonic blood flow probe at the superior mesenteric artery in an anesthetized Sprague-Dawley rat, along with a femoral vein catheter for intravenous (i.v.) drug delivery. Using this preparation, we determined that i.v. flupirtine induced concentration-dependent decreases in both MAP and mesenteric vascular resistance (MVR), consistent with its *in vitro* vasodilatory effects. Conversely, the K_V7 channel blocker, linopirdine, had the opposite effects: it elevated MAP and increased MVR (Mackie et al. 2008). Linopirdine had also been used clinically, in this case in the hope that it would improve cognition in patients suffering from Alzheimer's disease, an effect that was ultimately not upheld in clinical trials (Rockwood et al. 1997). Our findings suggested that clinically used K_V7 channel activators and blockers have the potential to elicit cardiovascular effects in proportion to their capacity to activate or inhibit vascular K_V7 channels, effectively opposing or mimicking the actions of AVP, respectively.

1.9 DEVELOPING NEW THERAPIES FOR DISEASES OF ALTERED AVP SECRETION

AVP has been implicated in the etiology of a variety of medical conditions. Considering our finding that drugs already in clinical use have the potential to oppose or mimic the vasoconstrictor actions of AVP *in vivo*, we speculated that these drugs might be able to be re-purposed for the treatment of AVP-related cardiovascular diseases. One such disease is cerebral vasospasm, a devastating form of stroke that arises following subarachnoid hemorrhage (SAH), in which the rupture of a cerebral aneurysm results in leakage of blood into the subarachnoid space. The clotted blood induces a sustained constriction (spasm) of nearby blood vessels (most commonly the basilar artery), decreasing brain perfusion and ultimately causing ischemic brain damage. AVP has been implicated in this condition—experimental SAH in rats was found to result in increased central AVP release, which exceeded peripheral elevation of [AVP] and persisted for a longer period (Kleindienst et al. 2004). Furthermore, a V_{1a} vasopressin receptor antagonist was found to be effective in reducing cerebral vasospasm in a rat model of SAH (Trandafir et al. 2004). Circulating AVP levels were significantly elevated in patients with ischemic stroke and correlated with the severity score of the patients' neurologic deficit and their mean lesion size (Barreca et al. 2001).

SAH is associated with high morbidity and mortality, with few treatment options to prevent or reduce cerebral vasospasm. We hypothesized that K_V7 channel activating drugs might be effective anti-vasospastic agents based on our *in vitro* and *in vivo* studies of mesenteric artery contraction/relaxation. Rat basilar artery is a more difficult preparation for *in vitro* experimentation because there is only one short segment of this artery per animal and because it often has numerous small branches that must be tied off for pressure myography. Nonetheless, we were able to isolate basilar artery myocytes using a protocol that was modified from that described by Berra-Romani et al. (2005), and we determined that all five *KCNQ* genes are expressed in the smooth muscle cells. We could also record M-currents under conditions similar to those we had used for mesenteric artery myocytes, but with the addition of spermine (100 μM) to the external solution to suppress inward rectifier K^+ currents (these are often prominent in basilar artery myocytes but usually undetectable in mesenteric artery myocytes). The basilar artery myocyte M-currents had electrophysiological and pharmacological characteristics similar to those from mesenteric artery myocytes. Furthermore, these currents were significantly inhibited by 100 pM AVP, and the effects of AVP could be overcome by addition of a K_V7 channel activator, such as flupirtine, retigabine, or celecoxib (a cyclooxygenase-2 inhibitor, which we had discovered is also a K_V7 channel activator (Brueggemann et al. 2009; Mani et al. 2011, 2013). Moreover, endothelin-1 and serotonin (5-hydroxytryptamine; 5-HT), which, like AVP, are $G_{q/11}$-coupled receptor agonists and had been implicated as contributors to cerebral vasospasm, also inhibited the M-currents and induced significant membrane depolarization in basilar artery myocytes (Mani et al. 2013). In addition, in pressurized basilar arteries, constriction with any of the three spasmogens (AVP, endothelin-1, or 5-HT) could be completely reversed in a concentration-dependent manner by addition of flupirtine, retigabine, or celecoxib (Mani et al. 2011, 2013).

The *in vitro* studies of rat basilar artery supported our hypothesis that K_V7 channels are important signaling intermediates in the actions of several basilar artery spasmogens, including AVP, which had been implicated in SAH-induced cerebral vasospasm. In addition, clinically-used K_V7 channel activators were more effective than nimodipine (the only USFDA-approved drug for relief of cerebral vasospasm) in relaxation of the constricted basilar artery (Mani et al. 2011). To determine whether these same K_V7 channel activating drugs could be effective as an *in vivo* therapy for relief of cerebral vasospasm, we adopted a commonly used rat model of SAH. In this model, arterial blood is collected from the femoral artery of an anesthetized Sprague-Dawley rat and injected into the cisterna magna of the same animal using a stereotaxic apparatus. Control animals were injected with artificial cerebrospinal fluid (aCSF) instead of arterial blood. Following recovery from anesthesia, the rats were treated twice daily by intraperitoneal (i.p.) injection with retigabine (7.5 mg/kg), celecoxib (20 mg/kg), or vehicle (dimethyl sulfoxide, 1 mL/kg) and monitored for behavioral signs of ischemic brain damage. After 48 h, the rats were sacrificed and the basilar artery diameter was measured. All measurements were conducted by investigators who were blinded to the treatments.

The basilar artery diameter was significantly smaller in rats injected with arterial blood compared with controls that received aCSF, establishing the vasospastic component of the animal model. However, twice-daily i.p. injection with retigabine or celecoxib fully prevented the basilar artery constriction, relative to the vehicle controls (Figure 1.13). No behavioral deficits were observed even in the vehicle-treated rats with constricted basilar arteries—a limitation of this model in that, despite significant constriction of the basilar artery, the collateral circulation of the rat cerebral vasculature appears to adequately perfuse the brain, and in this sense does not mimic the human disease. BP and heart rate were measured telemetrically in a separate cohort of animals not subjected to SAH but treated with retigabine (7.5 mg/kg), celecoxib (50 mg/kg), or vehicle twice daily in the same manner as the SAH rats. Neither of the K_V7 channel activators (retigabine nor celecoxib) produced sustained hypotension or a sustained change in heart rate in the conscious rats. There was a transient drop in BP at the time of injection, which persisted a bit longer with celecoxib than with retigabine. By comparison, vehicle administration tended to increase BP transiently.

The animal model we used, and the limited series of experimental strategies we employed, was certainly not optimal for establishing clinical efficacy of K_V7 channel activators as anti-vasospastic agents in SAH. More definitive results might be obtained using larger animal models and strategies of multiple blood injections at daily intervals, which have demonstrated features that more closely mimic the time course of vasospasm and stroke that develops in human SAH (Megyesi et al. 2000, Gules et al. 2002). However, considering the paucity of approved treatments, the severity of the disease, and our evidence that already approved drugs for other human therapies might be safe and effective anti-vasospastic agents, a fast-track to clinical trials of these drugs in SAH patients might be warranted.

AVP is not always a villain in pathological conditions. In patients with septic, vasodilatory or hemorrhagic shock, endogenous AVP levels may be insufficient to maintain BP at a level that supports adequate cardiac output, and hence exogenous AVP is often administered therapeutically to help restore and maintain BP (Landry and Oliver 2001). We recently investigated the possibility that the K_V7 channel blockers, linopirdine and XE991, might also be useful to restore BP in hypotensive conditions, that is, during hemorrhagic shock (Nassoiy et al. 2017). In these studies, anesthetized male Sprague-Dawley rats were instrumented with arterial and venous catheters for BP monitoring, hemorrhage, and fluid resuscitation. In normal (non-hemorrhaged) rats, i.v. linopirdine (0.1–6 mg/kg) transiently increased MAP up to 15%, whereas retigabine (0.1–12 mg/kg) decreased MAP by up to 60%; the latter could be rapidly reversed by linopirdine. When rats were hemorrhaged to a MAP of 25 mmHg for 30 min, fluid

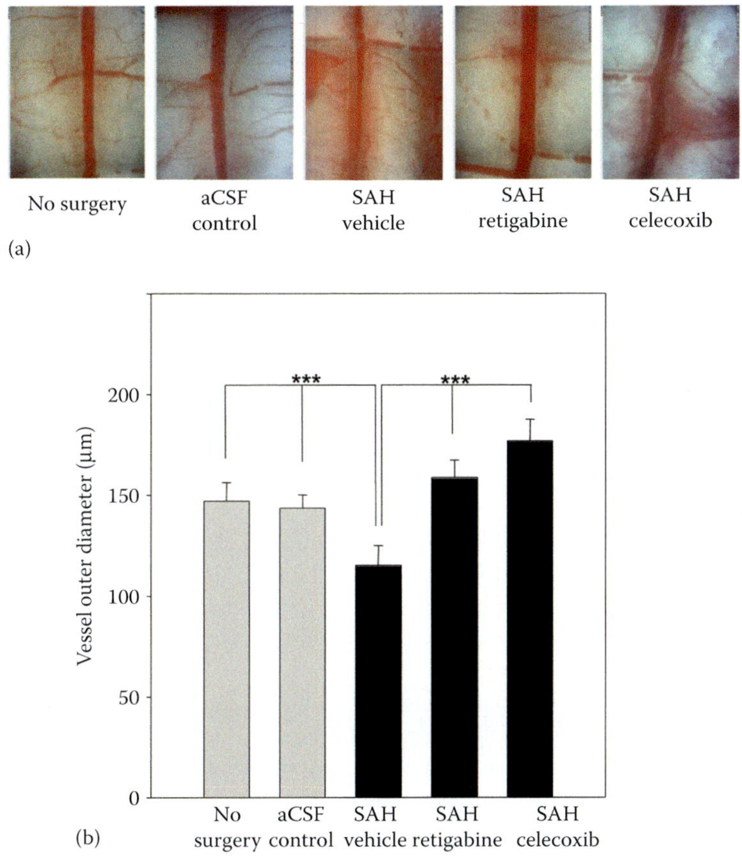

FIGURE 1.13 K_V7 channel openers attenuate basilar artery spasm in a rat model of SAH. (a) Illustrative photographs show basilar artery in the base of brain from rats in different treatment groups. Brains were photographed 48 h after injection of blood or aCSF and compared with brains of control rats (no surgery). (b) Bar graph shows the mean basilar artery outer diameter measured in each of the treatment groups—Control (no surgery), aCSF control (sham surgery), SAH rats (black bars) treated with vehicle (DMSO, 1 mL/kg, i.p., b.i.d), SAH rats treated with retigabine (7.5 mg/kg, i.p., b.i.d) and SAH rats treated with celecoxib (20 mg/kg, i.p., b.i.d). *** Significant difference between the treatment groups as indicated using one-way ANOVA followed by post hoc Holm-Sidak test, $p < 0.001$, $n = 9$. (From Mani, B. K. et al., *J. Cardiovasc. Pharmacol.*, 61, 51–62, 2013. With permission.)

resuscitation with normal saline (NS) could restore MAP to 70 mmHg. The volume of resuscitation fluid required to restore BP was dramatically and dose-dependently reduced by the K_V7 channel blockers, relative to their vehicle controls, when administered as single bolus injection (1–6 mg/kg) at the beginning of fluid resuscitation and also when NS was supplemented with 1.25–200 mg/mL linopirdine (Nassoiy et al. 2017). These findings further support the important role of K_V7 channels as regulators of vascular function and illustrate the utility of administration of K_V7 channel blockers as a novel therapeutic approach in restoration of BP under hypotensive conditions.

1.10 METHODOLOGICAL CONSIDERATIONS

Despite the recognition that the endocrine actions of AVP on the vasculature are evident at concentrations below 100 pM, earlier signal transduction studies often used concentrations several orders of magnitude higher, implicating mechanisms that are unlikely to contribute appreciably at physiological concentrations of AVP. Several methodological considerations were required to identify the vasoconstrictor signaling pathways that account for the physiological actions of circulating AVP. First, and perhaps most importantly, was the use of physiological concentrations in our assays. In our initial studies, we explored a range of [AVP], starting at femtomolar (10^{-15} M) concentrations and increasing in small increments until we established a threshold of approximately 10 pM to elicit a detectable Ca^{2+} spiking effect in A7r5 cells. Having a real-time functional read-out (fura-2 fluorescence measurements of $[Ca^{2+}]_{cyt}$), was very helpful in this regard. The initial use of a cultured cell line, though criticized by some at the time as being irrelevant to vascular physiology, was very helpful in terms of ease and throughput of experiments, and ultimately turned out to predict quite accurately the primary features of AVP signal transduction in intact arteries.

Elucidating the complexities of AVP signal transduction, from its initiation at the level of surface receptors, to activation of downstream lipid and protein signaling, and ultimately to functional responses at the cell, tissue, and whole-body levels, required a plethora of experimental approaches and a progression from the cultured cell line to freshly dissociated arterial myocytes and intact arteries. The use of very low doses of agonist (usually $\leq$100 pM AVP) presented an additional challenge because the changes measured were often quite small, necessitating meticulous attention to experimental procedures to avoid losing the small signal within the background noise.

The lipid biochemistry (TLC) and protein biochemistry (immunohistochemistry, proximity ligation assays, immunoprecipitation/western blotting) experiments were conducted in parallel with functional assays (e.g., $[Ca^{2+}]_{cyt}$ measurements, K_V7 current measurements, pressure myography, BP, and blood flow measurements) using a combination of pharmacological and molecular biological strategies to target specific steps in the hypothetical pathway. Notably, the signaling pathway we have examined is an acute signaling pathway, which turns on and off rapidly and therefore lends itself well to the real-time assays we have employed to monitor functional responses. Although the functional responses occur relatively quickly (within minutes) following our treatments, considerable patience was required to assure that we had stable recordings prior to treatment, that responses were monitored long enough to evaluate the time course over which they developed and stabilized, and also to evaluate how quickly the responses turned off following termination of the treatment. Many of our fura-2 fluorescence, patch clamp, and pressure myography experiments required recordings over one or more hours. Much of the signal transduction literature is devoted to studies of changes in gene expression that are the result of more chronic stimuli or pathologies—those studies face different challenges and often require different methods and experimental strategies conducted over longer time frames.

Proper controls have also been essential throughout our studies. We have generally attempted to use inactive analogs or vehicle controls for pharmacological studies, as well as both positive and negative regulators of the pathway elements, to reach firm conclusions. The acute time-dependent nature of the signaling pathway under investigation required a careful assessment of the time courses over which various signaling events were observed, with attendant time controls to assure that responses did not develop non-specifically over time under the recording conditions we used. We have also tried to use the lowest possible concentrations of drugs and confirmed that the drugs had the intended effect in our system (i.e., not relying on catalog descriptions or reports from the literature). For molecular biological approaches such as overexpression or knockout of expression, we also confirmed the effects and utilized empty vector or scrambled constructs as well as untreated cells as controls. Similarly, antibodies, which are essential for many of the protein biochemistry approaches, have been carefully validated to assure that they specifically recognize the intended target (e.g., based on detection of proteins at the expected molecular weight, loss of signal in knockout cells, increased signal in overexpressing cells).

1.11 SUMMARY

Our initial *in vitro* studies, using physiological concentrations of AVP, revealed a concentration-dependent dichotomy of signaling pathways. We identified a novel PLD-PKC-L-type VSCC pathway as the primary track downstream of the V_{1a}-vasopressin receptor, which elicits finely-tuned vasoconstrictor responses at the concentrations of AVP detected in the human circulation. This work ultimately led to the discovery that K_V7 potassium channels serve as essential intermediates in physiological vasoconstrictor signal transduction. Our *in vivo* studies confirmed that pharmacological activation or inhibition of these channels can robustly affect BP and blood flow. These findings may lead to more effective strategies for treatment of cardiovascular conditions in which AVP levels are altered.

REFERENCES

Adams, P. R., D. A. Brown, and A. Constanti. 1982. Pharmacological inhibition of the M-current. *J Physiol* 332 (1): 223–262.

Aiyar, N., P. Nambi, F. L. Stassen, and S. T. Crooke. 1986. Vascular vasopressin receptors mediate phosphatidylinositol turnover and calcium efflux in an established smooth muscle cell line. *Life Sci* 39 (1): 37–45.

Altura, B. M. 1975. Dose-response relationships for arginine vasopressin and synthetic analogs on three types of rat blood vessels: Possible evidence for regional differences in vasopressin receptor sites within a mammal. *J Pharmacol Exp Ther* 193 (2): 413–423.

Barreca, T., C. Gandolfo, G. Corsini, M. Del Sette, A. Cataldi, E. Rolandi, and R. Franceschini. 2001. Evaluation of the secretory pattern of plasma arginine vasopressin in stroke patients. *Cerebrovasc Dis* 11 (2): 113–118. doi:10.1159/000047622.

Baylis, P. H. 1983. Posterior pituitary function in health and disease. *Clin Endocrinol Metab* 12 (3): 747–770.

Baylis, P. H. 1987. Osmoregulation and control of vasopressin secretion in healthy humans. *Am J Physiol* 253 (5 Pt 2): R671–R678.

Baylis, P. H. 1989. Regulation of vasopressin secretion. *Baillieres Clin Endocrinol Metab* 3 (2): 313–330.

Baylis, P. H., and S. Ball. 2000. *The Neurohypophysis: Endocrinology of Vasopressin and Oxytocin.*

Berra-Romani, R., M. P. Blaustein, and D. R. Matteson. 2005. TTX-sensitive voltage-gated Na^+ channels are expressed in mesenteric artery smooth muscle cells. *Am J Physiol Heart Circ Physiol* 289 (1): H137–H145.

Berridge, M. J. 1984. Inositol trisphosphate and diacylglycerol as second messengers. *Biochem J* 220 (2): 345–360.

Brueggemann, L. I., and K. L. Byron. 2012. Use of patch clamp electrophysiology to identify off-target effects of clinically used drugs. In *Patch Clamp Technique*, F. S. Kaneez (Ed). Rijeka, Croatia: Intech.

Brueggemann, L. I., A. R. Mackie, L. L. Cribbs, J. Freda, A. Tripathi, M. Majetschak, and K. L. Byron. 2014. Differential protein kinase C-dependent modulation of $K_V7.4$ and $K_V7.5$ subunits of vascular K_V7 channels. *J Biol Chem* 289 (4): 2099–2111.

Brueggemann, L. I., A. R. Mackie, B. K. Mani, L. L. Cribbs, and K. L. Byron. 2009. Differential effects of selective cyclooxygenase-2 inhibitors on vascular smooth muscle ion channels may account for differences in cardiovascular risk profiles. *Mol Pharmacol* 76 (5): 1053–1061.

Brueggemann, L. I., A. R. Mackie, J. L. Martin, L. L. Cribbs, and K. L. Byron. 2011. Diclofenac distinguishes among homomeric and heteromeric potassium channels composed of KCNQ4 and KCNQ5 subunits. *Mol Pharmacol* 79 (1): 10–23.

Brueggemann, L. I., D. R. Markun, K. K. Henderson, L. L. Cribbs, and K. L. Byron. 2006. Pharmacological and electrophysiological characterization of store-operated currents and capacitative Ca^{2+} entry in vascular smooth muscle cells. *J Pharmacol Exp Ther* 317 (2): 488–499.

Brueggemann, L. I., C. J. Moran, J. A. Barakat, J. Z. Yeh, L. L. Cribbs, and K. L. Byron. 2007. Vasopressin stimulates action potential firing by protein kinase C-dependent inhibition of KCNQ5 in A7r5 rat aortic smooth muscle cells. *Am J Physiol Heart Circ Physiol* 292 (3): H1352–H1363.

Byron, K., and C. W. Taylor. 1995. Vasopressin stimulation of Ca^{2+} mobilization, two bivalent cation entry pathways and Ca^{2+} efflux in A7r5 rat smooth muscle cells. *J Physiol* 485 (Pt 2): 455–468.

Byron, K. L. 1996. Vasopressin stimulates Ca^{2+} spiking activity in A7r5 vascular smooth muscle cells via activation of phospholipase A_2. *Circ Res* 78 (5): 813–820.

De Groot, L. J., G. Chrousos, K. Dungan et al., Eds. *Endotext [Internet]*. South Dartmouth, MA: MDText.com, Inc.

Delmas, P., and D. A. Brown. 2005. Pathways modulating neural KCNQ/M (K_V7) potassium channels. *Nat Rev Neurosci* 6 (11): 850–862.

Doyle, V. M., and U. T. Ruegg. 1985. Vasopressin induced production of inositol trisphosphate and calcium efflux in a smooth muscle cell line. *Biochem Biophys Res Commun* 131 (1): 469–476.

Fan, J., and K. L. Byron. 2000. Ca^{2+} signalling in rat vascular smooth muscle cells: A role for protein kinase C at physiological vasoconstrictor concentrations of vasopressin. *J Physiol* 524 (Pt 3): 821–831.

Fish, R. D., G. Sperti, W. S. Colucci, and D. E. Clapham. 1988. Phorbol ester increases the dihydropyridine-sensitive calcium conductance in a vascular smooth muscle cell line. *Circ Res* 62 (5): 1049–1054.

Gules, I., M. Satoh, B. R. Clower, A. Nanda, and J. H. Zhang. 2002. Comparison of three rat models of cerebral vasospasm. *Am J Physiol Heart Circ Physiol* 283 (6): H2551–H2559.

Henderson, K. K., and K. L. Byron. 2007. Vasopressin-induced vasoconstriction: Two concentration-dependent signaling pathways. *J Appl Physiol* 102 (4): 1402–1409.

Herrmann, W. M., U. Kern, and M. Aigner. 1987. On the adverse reactions and efficacy of long-term treatment with flupirtine: Preliminary results of an ongoing twelve-month study with 200 patients suffering from chronic pain states in arthrosis or arthritis. *Postgrad Med J* 63 (Suppl 3): 87–103.

Ito, Y., O. Kozawa, H. Tokuda, J. Kotoyori, and Y. Oiso. 1993. Vasopressin induces arachidonic acid release through pertussis toxin-sensitive GTP-binding protein in aortic smooth muscle cells: Independence from phosphoinositide hydrolysis. *J Cell Biochem* 53 (2): 169–175.

Jiang, J., P. H. Backx, H. Teoh, and M. E. Ward. 2003. Role of Cl- currents in rat aortic smooth muscle activation by prostaglandin F2 alpha. *Eur J Pharmacol* 481 (2–3): 133–140.

Jones, L. G., K. M. Ella, C. D. Bradshaw, K. C. Gause, M. Dey, A. E. Wisehart-Johnson, E. C. Spivey, and K. E. Meier. 1994. Activations of mitogen-activated protein kinases and phospholipase D in A7r5 vascular smooth muscle cells. *J Biol Chem* 269 (38): 23790–23799.

Joshi, S., V. Sedivy, D. Hodyc, J. Herget, and A. M. Gurney. 2009. KCNQ modulators reveal a key role for KCNQ potassium channels in regulating the tone of rat pulmonary artery smooth muscle. *J Pharmacol Exp Ther* 329 (1): 368–376.

Kimes, B. W., and B. L. Brandt. 1976. Characterization of two putative smooth muscle cell lines from rat thoracic aorta. *Exp Cell Res* 98 (2): 349–366.

Kleindienst, A., G. Hildebrandt, S. A. Kroemer, G. Franke, M. R. Gaab, and R. Landgraf. 2004. Hypothalamic neuropeptide release after experimental subarachnoid hemorrhage: In vivo microdialysis study. *Acta Neurol Scand* 109 (5): 361–368.

Knot, H. J., M. M. de Ree, B. H. Gahwiler, and U. T. Ruegg. 1991. Modulation of electrical activity and of intracellular calcium oscillations of smooth muscle cells by calcium antagonists, agonists, and vasopressin. *J Cardiovasc Pharmacol* 18 (Suppl 10): S7–S14.

Landry, D. W., and J. A. Oliver. 2001. The pathogenesis of vasodilatory shock. *N Engl J Med* 345 (8): 588–595.

Lee, M. W., and D. L. Severson. 1994. Signal transduction in vascular smooth muscle: Diacylglycerol second messengers and PKC action. *Am J Physiol* 267 (3 Pt 1): C659–C678.

Li, M., J. Zacharia, X. Sun, and W. G. Wier. 2008. Effects of siRNA knock-down of TRPC6, InsP3R1 in vasopressin-induced Ca^{2+} oscillations of A7r5 vascular smooth muscle cells. *Pharmacol Res* 58: 308–315.

Li, Y., A. J. Shiels, G. Maszak, and K. L. Byron. 2001. Vasopressin-stimulated Ca^{2+} spiking in vascular smooth muscle cells involves phospholipase D. *Am J Physiol Heart Circ Physiol* 280 (6): H2658–H2664.

Mackie, A. R., L. I. Brueggemann, K. K. Henderson, A. J. Shiels, L. L. Cribbs, K. E. Scrogin, and K. L. Byron. 2008. Vascular KCNQ potassium channels as novel targets for the control of mesenteric artery constriction by vasopressin, based on studies in single cells, pressurized arteries, and in vivo measurements of mesenteric vascular resistance. *J Pharmacol Exp Ther* 325 (2): 475–483.

Mani, B. K., L. I. Brueggemann, L. L. Cribbs, and K. L. Byron. 2009. Opposite regulation of KCNQ5 and TRPC6 channels contributes to vasopressin-stimulated calcium spiking responses in A7r5 vascular smooth muscle cells. *Cell Calcium* 45 (4): 400–411.

Mani, B. K., L. I. Brueggemann, L. L. Cribbs, and K. L. Byron. 2011. Activation of vascular KCNQ (K_V7) potassium channels reverses spasmogen-induced constrictor responses in rat basilar artery. *Br J Pharmacol* 164: 237–249.

Mani, B. K., J. O'Dowd, L. Kumar, L. I. Brueggemann, M. Ross, and K. L. Byron. 2013. Vascular KCNQ (K_V7) potassium channels as common signaling intermediates and therapeutic targets in cerebral vasospasm. *J Cardiovasc Pharmacol* 61: 51–62.

Marks, T. N., G. R. Dubyak, and S. W. Jones. 1990. Calcium currents in the A7r5 smooth muscle-derived cell line. *Pflugers Arch* 417 (4): 433–439.

Maruyama, Y., Y. Nakanishi, E. J. Walsh, D. P. Wilson, D. G. Welsh, and W. C. Cole. 2006. Heteromultimeric TRPC6-TRPC7 channels contribute to arginine vasopressin-induced cation current of A7r5 vascular smooth muscle cells. *Circ Res* 98 (12): 1520–1527.

McCarthy, R. T., and H. K. Fry. 1988. Nitrendipine block of calcium channel currents in vascular smooth muscle and adrenal glomerulosa cells. *J Cardiovasc Pharmacol* 12 (Suppl 4): S98–S101.

Megyesi, J. F., B. Vollrath, D. A. Cook, and J. M. Findlay. 2000. In vivo animal models of cerebral vasospasm: A review. *Neurosurg* 46 (2): 448–460; discussion 460–461.

Meier, K. E., K. C. Gause, A. E. Wisehart-Johnson, A. C. Gore, E. L. Finley, L. G. Jones, C. D. Bradshaw, A. F. McNair, and K. M. Ella. 1998. Effects of propranolol on phosphatidate phosphohydrolase and mitogen-activated protein kinase activities in A7r5 vascular smooth muscle cells. *Cell Signal* 10 (6): 415–426.

Morel, A., A. M. O'Carroll, M. J. Brownstein, and S. J. Lolait. 1992. Molecular cloning and expression of a rat V1a arginine vasopressin receptor. *Nature* 356 (6369): 523–526.

Nassoiy, S. P., K. L. Byron, and M. Majetschak. 2017. K_V7 voltage-activated potassium channel inhibitors reduce fluid resuscitation requirements after hemorrhagic shock in rats. *J Biomed Sci* 24 (1): 8.

Nelson, M. T., and J. M. Quayle. 1995. Physiological roles and properties of potassium channels in arterial smooth muscle. *Am J Physiol* 268: C799–C822.

Owens, G. K. 1995. Regulation of differentiation of vascular smooth muscle cells. *Physiol Rev* 75 (3): 487–517.

Rockwood, K., B. L. Beattie, M. R. Eastwood, H. Feldman, E. Mohr, W. Pryse-Phillips, and S. Gauthier. 1997. A randomized, controlled trial of linopirdine in the treatment of Alzheimer's disease. *Can J Neurol Sci* 24 (2): 140–145.

Soboloff, J., M. Spassova, W. Xu, L. P. He, N. Cuesta, and D. L. Gill. 2005. Role of endogenous TRPC6 channels in Ca^{2+} signal generation in A7r5 smooth muscle cells. *J Biol Chem* 280 (48): 39786–39794.

Tanaka, Y, G. Tang, K. Takizawa, K. Otsuka, M. Eghbali, M. Song et al. 2006. K_V channels contribute to nitric oxide- and atrial natriuretic peptide-induced relaxation of a rat conduit artery. *J Pharmacol Exp Ther* 317 (1): 341–354.

Thibonnier, M. 1992. Signal transduction of V1-vascular vasopressin receptors. *Regul Pept* 38 (1): 1–11.

Thibonnier, M., A. L. Bayer, and Z. Leng. 1993. Cytoplasmic and nuclear signaling pathways of V_1-vascular vasopressin receptors. *Regul Pept* 45 (1–2): 79–84.

Thibonnier, M., A. L. Bayer, M. S. Simonson, and M. Kester. 1991. Multiple signaling pathways of V_1-vascular vasopressin receptors of A7r5 cells. *Endocrinol* 129 (6): 2845–2856.

Trandafir, C. C., T. Nishihashi, A. Wang, S. Murakami, X. Ji, and K. Kurahashi. 2004. Participation of vasopressin in the development of cerebral vasospasm in a rat model of subarachnoid haemorrhage. *Clin Exp Pharmacol Physiol* 31 (4): 261–266.

Treschan, T. A., and J. Peters. 2006. The vasopressin system: Physiology and clinical strategies. *Anesthesiol* 105 (3): 599–612; quiz 639–640.

Vallotton, M. B., R. P. Wuthrich, P. D. Lew, and A. M. Capponi. 1986. Effects of vasopressin and its analogs on rat aortic smooth muscle and renal medullary tubular cells: Characterization of receptor subtypes. *J Cardiovasc Pharmacol* 8 (Suppl 7): S5–S11.

Van Renterghem, C., G. Romey, and M. Lazdunski. 1988. Vasopressin modulates the spontaneous electrical activity in aortic cells (line A7r5) by acting on three different types of ionic channels. *Proc Nat Acad Sci USA* 85 (23): 9365–9369.

Wange, R. L., A. V. Smrcka, P. C. Sternweis, and J. H. Exton. 1991. Photoaffinity labeling of two rat liver plasma membrane proteins with [^{32}P]gamma-azidoanilido GTP in response to vasopressin. Immunologic identification as alpha subunits of the G_q class of G proteins. *J Biol Chem* 266 (18): 11409–11412.

Yeung, S. Y., V. Pucovsky, J. D. Moffatt, L. Saldanha, M. Schwake, S. Ohya, and I. A. Greenwood. 2007. Molecular expression and pharmacological identification of a role for K(v)7 channels in murine vascular reactivity. *Br J Pharmacol* 151 (6): 758–770.

2 Investigating Chloride Channels in Smooth Muscle

Henry R. Askew Page and Iain A. Greenwood

CONTENTS

2.1 INTRODUCTION

The membrane potential (E_m) of smooth muscle cells (SMCs) is modulated by the patency of different ion channels positioned in the cell membrane. Electrochemical gradients are established across the cell membrane by various ion transporters and pumps. Since the lipid bilayer of the cell membrane is impermeable to charged particles, the movement of ions across the cell membrane occurs through selective ion channels that open and close in response to varying and numerous stimuli. K^+ channels play a significant role in hyperpolarizing the cell membrane (E_m) by permitting the efflux of K^+ ions. Their function as resting E_m regulators has been established by a number of groups [1,2]. In contrast, the prevention of K^+ ion efflux by the closure or blockade for K channels can cause depolarization of the E_m and ultimately activate voltage-gated calcium channels (VGCCs). Furthermore, non-selective cation channels such as TRP channels and the P2X receptor can also regulate the E_m. Much of the work relating to ion channels in smooth muscle has focused on the role of cations and cation channels, a lot of which will be discussed in other chapters of this book. This chapter will focus on the role of chloride (Cl^-) channels in smooth muscle. Cl^- channels are proteinaceous pores that allow the passive distribution of negatively charged ions down their electrochemical gradients. Cl^- channels can allow the passage of various anions, some more readily than Cl^-, and could therefore conceivably be more generally referred to as anion channels. However, its nomenclature reflects the fact that Cl^- is the most abundant anion in organisms, and therefore is the most frequent permeating anion in biological systems. Cl^- channels play a role in membrane excitability as well as cell volume regulation and trans-epithelial transport. Furthermore, they contribute to the acidification of intra- and extracellular compartments, the cell cycle and apoptosis [3].

Despite wide variation in experimental conditions, it is reasonable to suggest that there is a large molecular diversity of Cl^- channels. Many mechanisms modulate the gating of Cl^- channels depending on the nature of the specific channel. However, these gating mechanisms are not always mutually exclusive to a particular type of channel. Such mechanisms include changes in transmembrane voltage and cell volume; binding of modulatory molecules, different ions, or ATP; and phosphorylation of intracellular amino acid residues by intracellular kinases. Electrophysiological experiments using the patch clamp technique have unearthed numerous anion channels with differing characteristics such as single channel conductance, anion selectivity, and activation mechanisms.

As it stands, mammalian Cl^- channels fall into five main groups based on their activation mechanism: cystic fibrosis transmembrane conductance regulator (CFTR); calcium-activated Cl^- channels (CACCs); voltage-gated Cl^- channels (ClCs); ligand-gated Cl^- channels (GABA [γ-aminobutyric acid] and glycine-activated); and volume-regulated Cl^- channels [4].

2.2 CHLORIDE ACCUMULATION

To understand the role of Cl^- in smooth muscle it is important to know how it is distributed within the intra- and extracellular spaces. The concentration of intracellular Cl^- ions ($[Cl^-]_i$) varies widely between different cell types. A comprehensive study of intracellular ion concentrations in smooth muscle was carried out in 1981 [5], which identified three main findings:

1. Intracellular Cl^- concentrations in smooth muscle are higher than expected if Cl^- ions were passively distributed.
2. Some values of $[Cl^-]_i$ were so high that, if correct, Cl^- would have to be sequestered in, or bound to, an intracellular constituent to maintain osmotic balance.
3. Suggestions of how and why Cl^- is accumulated into SMCs were sparse.

It is now thought that there are three mechanisms by which SMCs can accumulate Cl^- ions. These are Cl^-/HCO^-_3 exchange, $Na^+/K^+/Cl^-$ co-transport, and Pump III [6].

2.2.1 Chloride/Bicarbonate Exchange

Two sets of experiments performed by two separate groups provided the strongest evidence for Cl^- accumulation by Cl^-/HCO^-_3 exchange [7]. The first set of experiments, carried out on guinea-pig ureters, observed the rate of recovery of the effective concentration (activity) of Cl^- (a^i_{Cl}) by cells bathed in Cl^--free media upon restoration of extracellular Cl^-, using double-barreled microelectrodes. The recovery of a^i_{Cl} took less than ten minutes in a solution that contained Cl^- and HCO_3^-, but was three times slower in a solution that did not contain HCO_3^-. Furthermore, the recovery of a^i_{Cl} was ten times slower in the presence of 4,4′-Diisothiocyanato-2,2′-stilbenedisulfonic acid (DIDS), a drug that effectively inhibits Cl^-/HCO^-_3 exchange in other systems [9]. These findings were consolidated by Aickin and Brading [10], who used pH-sensitive microelectrodes to measure intracellular pH before and after Cl^- was removed from the external solution. In the latter environment, a^i_{Cl} fell while the pH within the same cell increased from physiological 7.06 to 7.69 [10]. This suggests that in these conditions the Cl^-/HCO^-_3 exchanger is reversed, leading to HCO_3^- being brought into the cell and thus causing the increase in pH. Conversely, when tissues had been bathed in Cl^--free media the intracellular pH was 7.8, which decreased to 7.1 when Cl^- was reintroduced to the bathing solution, suggesting extrusion of HCO_3^- from the cell. This effect was completely blocked by incubation with DIDS.

2.2.2 $Na^+/K^+/Cl^-$ Co-Transport

Kreye et al. [11] reported active Cl^- transport in rabbit aorta that could be inhibited by the loop-diuretic furosemide rather than DIDS. Furthermore, hyperpolarization of the membrane potential, and relaxation of aortic smooth muscle were caused by a furosemide-induced reduction in intracellular Cl^-. This study not only provided a clue as to how Cl^- accumulation might occur in SMCs, but it also shows that intracellular Cl^- played a significant role in the link between membrane potential and contractility [11]. A furosemide-sensitive K^+ flux, which had represented $Na^+/K^+/Cl^-$ co-transport, was also reported in a cerebral artery SMC line (BC_3H1) [12]. Nonetheless, neither of these studies conclusively identified $Na^+/K^+/Cl^-$ co-transport. More convincing evidence for $Na^+/K^+/Cl^-$ co-transport was brought to light using another SMC line, derived from embryonic rat thoracic aorta [13]. These experiments, along with others [14], show that $Na^+/K^+/Cl^-$ co-transport in smooth muscle is

1. Separate from the Na^+/K^+—ATPase pump and Cl^-/HCO^-_3 exchange.
2. Dependent on Na^+ and Cl^-.
3. Inhibited by loop diuretics.
4. Responsible for most K^+ influx in cells that is not provided by the Na^+/K^+—ATPase pump.

Application of furosemide to rabbit aorta reduced intracellular Cl^- to a level where it was in equilibrium with the resting membrane potential, suggesting that $Na^+/K^+/Cl^-$ co-transport was wholly responsible for the accumulation of Cl^- in rabbit thoracic aorta SMCs [15]. Similarly, in guinea-pig vas deferens, application of DIDS blocked Cl^- accumulation through the Cl^-/HCO_3^- exchanger but did not inhibit all Cl^- accumulation by the cell. This suggested that there was another mechanism by which SMCs of guinea-pig vas deferens accumulate Cl^-—this mechanism was shown to be $Na^+/K^+/Cl^-$ co-transport [16,17].

2.2.3 Pump III

Pump III was first hypothesized during experiments as part of a study investigating $Na^+/K^+/Cl^-$ co-transport in femoral artery SMCs of hypertensive rats [18]. The experiments showed that, even when both $Na^+/K^+/Cl^-$ co-transport and Cl^-/HCO_3^- exchange could not function, $[Cl^-]_i$ was still above equilibrium. This suggested that there is another way in which cells can accumulate Cl^- ions, separate from $Na^+/K^+/Cl^-$ co-transport and Cl^-/HCO_3^- exchange. Further experiments were carried out using the Cl^- channel blocker 5-Nitro-2-(3-phenylpropylamino) benzoic acid (NPPD) to reduce the outward Cl^- leak in addition to inhibiting the previously discovered Cl^- accumulation mechanisms [19]. The data show that there was still an increase in intracellular Cl^- and thus further demonstrates the existence of *Pump III*. This increase in intracellular Cl^- can be inhibited by 1 mMol/L acetazolamide, a carbonic anhydrase inhibitor. Since carbonic anhydrase is not expressed in vascular smooth muscle [15], it follows that this drug does not bring about its effects through this enzyme but by binding to another cellular component. Naturally, at such high concentrations it is impossible to say that the drug is specific for anything; however, it is possible that it could be inhibiting Pump III.

Finally, when all three proposed Cl^- accumulation mechanisms are rendered inoperative, the concentration of intracellular Cl^- comes into equilibrium with E_m [19]. Therefore, it is inferred that these three activities represent Cl^- accumulation *in toto* in rat vascular SMCs (VSMC) [6]. It is worth stressing that the relative impact of individual uptake mechanisms varies between arteries. Interestingly, there is evidence for increased activity of $Na^+/K^+/Cl^-$ co-transport in VSMC after application of vasoconstrictors [20].

2.3 CHLORIDE CHANNELS IN SMOOTH MUSCLE

The regulation of L-type Ca^{2+} channels by voltage- and ligand-gated K^+ channels and non-selective cation channels, and therefore of the intracellular Ca^{2+} concentration ($[Ca^{2+}]_i$), has constituted much of the research in VSMCs. In comparison, relatively little is known about the functional impact of Cl^- channels in smooth muscle. The calculated equilibrium potential for Cl^- in VSMCs is −26 mV [21]; however, this number varies between smooth muscle types and has even been reported to be as low as −6 mV [7]. These membrane potentials are high enough to increase the open probability of L-type VGCCs considerably and can ultimately contribute to contraction of SMCs. In addition, intracellular Cl^- has been implicated in the regulation of intracellular pH, cell volume, and cell growth [6].

Of the five identified types of Cl^- channels, three have been reported in vascular smooth muscle (summarized in Figure 2.1). These are the volume-regulated Cl^- channel (VRAC), CFTR, and CACCs. This chapter will explore the evidence surrounding these channels and their purported roles in smooth muscle, whilst discussing the techniques used in these investigations.

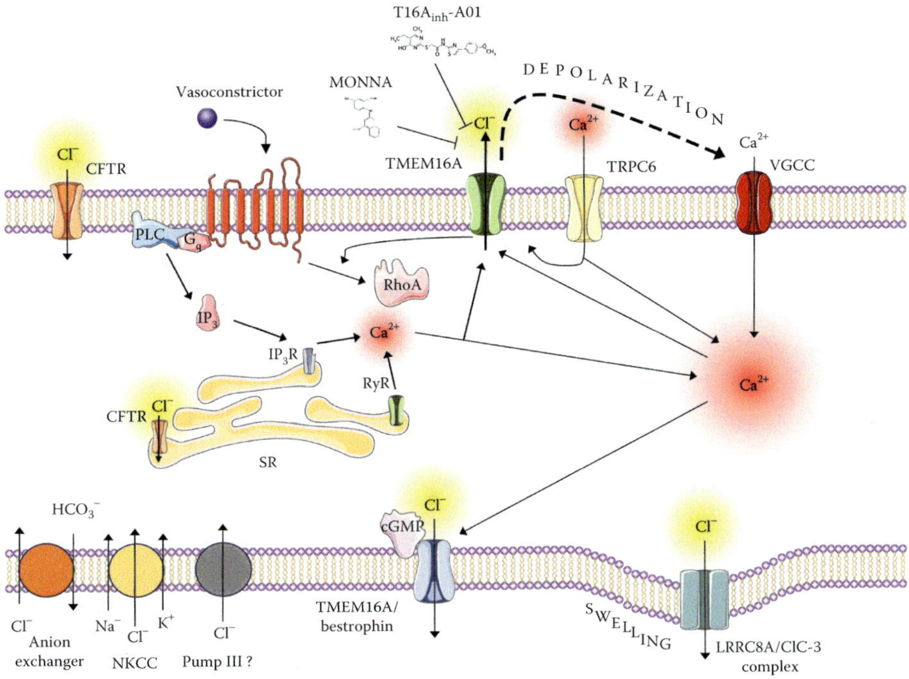

FIGURE 2.1 Chloride ions are brought into the cell by Cl^-/HCO_3^- exchange, $Na^+/K^+/Cl^-$ co-transport, and Pump III—an unidentified third accumulation mechanism. The resulting electrochemical gradient results in chloride ions leaving the cell when chloride channels open at resting membrane potentials. Ca^{2+}-activated chloride channels (CACCs) are activated by rises in intracellular Ca^{2+} originating from SR store release via IP_3 receptor, or ryanodine receptor activation, local influx via TRCP6 channels, and influx via VGCCs. The ensuing chloride efflux results in membrane potential depolarization and further Ca^{2+} influx. TMEM16A-associated CACCs are inhibited by novel, TMEM16A-specific blockers, which cause vasodilation and impede vasoconstriction. Recent evidence suggests that TMEM16A participates in smooth muscle contraction via a non-canonical pathway involving RhoA activation. cGMP-dependent CACCs are reliant on the expression of both bestrophin and TMEM16A. CFTR chloride channels are expressed in smooth muscle. Activation of this channel paradoxically results in vasorelaxation, leading to speculation that chloride ions enter the cell through CFTR channels when the membrane potential is positive in comparison to E_{Cl}. CFTR may also be important in maintaining electroneutral Ca^{2+} movement across the SR membrane. Volume-regulated chloride channels (VRACs) allow conduction of chloride ions in response to changes in cell volume. Manipulation of the expression of ClC-3 has differing effects on VRAC activity, while LRRC8A expression is essential, but not sufficient, for recording volume regulated chloride currents.

2.3.1 Cystic Fibrosis Transmembrane Conductance Regulator

CFTR, studied extensively in epithelial cells, has also been found in cardiac myocytes as well as vascular and non-vascular smooth muscle [22–25]. CFTR and cAMP pathway activators do not contract arteries and instead relax preconstricted aorta and intrapulmonary artery segments denuded of endothelium [22,23]. Under these conditions, the SMC E_m is more positive than E_{Cl} so Cl^- would move back into the cell, hyperpolarizing E_m, and oppose vasoconstriction [26,27]. However, the reason why CFTR channels only allow Cl^- ingress has not been addressed, especially as the CFTR channel has a linear current-voltage relationship. Interestingly, aortae from CFTR knockout mice contracted more robustly to application of 80 mMol/L K^+ in comparison to the wild type controls [23]. As the E_m under these conditions would be close to the theoretical E_{Cl}, there would be little Cl flux, suggesting that CFTR may influence SMC contraction independent of an effect as an ion channel. Contrasting evidence shows very little difference in contraction of aortae between CFTR knockout and wild type piglets [28]. Furthermore, no difference was seen in angiotensin II-induced pulmonary artery contraction in CFTR knockout mice, although hypoxic vasoconstriction in the perfused lung was reduced, albeit it by a Cl^- independent mechanism [29]. Interestingly, newborn piglets that express the most prevalent CFTR mutant, ΔF508, show reduced vasoconstrictor responses and decreased blood pressure [28,30]. Since this mutant CFTR does not traffic to the cell membrane, the phenotype of these mice can't be explained by a direct E_m modifying role for CFTR. Human CFTR patients also have decreased blood pressure [31,32] although this has not been attributed directly to an effect on the vasculature. CFTR serves multiple cellular functions in addition to being a conduit for Cl^- flux and these may contribute to its vasomodulatory function, especially as there is no published or anecdotal evidence for cAMP-activated Cl^- currents from VSMCs. A recent study indicates that CFTR may have a role in maintaining electroneutral Ca^{2+} reuptake into the sarcoplasmic reticulum (SR) of porcine airway smooth muscle [33], while other studies suggested that CFTR interacts with other ion channels to regulate smooth muscle function including endogenous CACCs [29,34]. Given the conflicting evidence, the fact that CFTR staining in airway smooth muscle is localized to the intracellular compartment [33], and the lack of electrophysiological data in smooth muscle, CFTR's role continues to be enigmatic.

2.3.2 Volume-Regulated Anion Channel

A crucial aspect of cell physiology is the ability of a cell to regulate volume. Cells employ several proteins to control the osmotic balance across their membrane. K^+, Cl^-, and H_2O efflux all occur as a result of cell swelling, allowing cellular volume to return to pre-stimulus levels. Volume-sensitive Cl^- currents ($I_{Cl,\ vol}$), evoked by cell swelling or stretching, have been recorded in a variety of vascular [35–38] and non-vascular SMCs [39,40]. As well as volume regulation, $I_{Cl,\ vol}$ is implicated in cell proliferation [41,42] and contraction of smooth muscle. Pressure-induced depolarization and constriction of cerebral arteries is sensitive to DIDS but not niflumic acid (NFA), suggesting this current was mediated by a channel distinct from CACCs [43].

Furthermore, Cl^- efflux has been measured in rat cerebral arteries during myogenic contraction [44]. In rat coronary arteries, nitric oxide inhibition caused contraction that was blocked by $I_{Cl, vol}$ antagonists [45], however, since many blockers of $I_{Cl, vol}$ are also blockers of Ca^{2+}-activated Cl^- currents ($I_{Cl(Ca)}$), it is difficult to say whether this Cl^- efflux is due to VRAC or CACC. As is the case with a number of Cl^- currents, the molecular correlate responsible for $I_{Cl, vol}$ in smooth muscle had evaded identification for many years. Whilst several candidates such as bestrophins [46], CLCs [47,48], and anoctamins [49] were shown to be sensitive either directly or indirectly to changes in cell volume, the unique characteristics of native $I_{Cl, vol}$ were sufficiently distinct from the currents produced by these proposed correlates that controversy arose as to whether they were the direct molecular correlate to $I_{Cl, vol}$. ClC-3 was a lead candidate as a component of VRAC in smooth muscle. ClC-3 mRNA was present in canine VSMCs in which a hypotonic solution elicited a Cl^- current [36]. Furthermore, overexpression of ClC-3 increased $I_{Cl, vol}$ in aortic VSMCs [50], while intracellular dialysis of ClC-3 antibodies [51] and knockdown of ClC-3 both inhibited $I_{Cl, vol}$ in canine pulmonary artery VSMCs and a rat aortic SMC line, respectively [52,53]. There is debate over whether ClC-3 forms a functional ion channel [48], especially as there is no difference in $I_{Cl, vol}$ in a variety of cell types, including VSMC, in ClC-3 knockout mice [54–58]. However, interestingly, an increase in ClC-1 and ClC-2 mRNA was recorded in the atrial myocytes of the ClC-3 knockout mice, perhaps indicating a compensatory upregulation of ClC-3 family members. Regardless, the bountiful evidence of ClC-3's contribution to $I_{Cl, vol}$ could possibly point, at the very least, to an ancillary role whereby it forms a complex with other VRAC components to produce $I_{Cl, vol}$ [57].

More recently, LRRC8A was identified as an essential constituent of VRAC using genome-wide loss of function studies and a yellow fluorescent protein (YFP) quenching assay [59,60]. LRRC8A is a member of the *leucine-rich repeat containing 8 family* (A–E) and is essential for VRAC (Figure 2.1). LRRC8A must form a heteromer with another member in order to produce $I_{Cl, vol}$ [61] and evidence is emerging that it is the pore-forming unit in VRAC [59,60,62,63]. However, overexpression of LRRC8 did not increase $I_{Cl, vol}$ amplitude, an observation that has led the authors to postulate the necessity for other VRAC components [61] including TMEM16A [64]. Currently, there is scant evidence of LRRC8A forming VRAC in smooth muscle, with one report of a reduction in $I_{Cl, vol}$ in cultured aortic SMCs after LRRC8A siRNA knockdown [65]. Despite the limited evidence surrounding LRRC8A in smooth muscle, it is possible that ClC-3 and LRRC8 work together, as individual components of VRAC, to mediate $I_{Cl, vol}$.

2.3.3 Calcium-Activated Chloride Channels

Cl^- currents activated by a rise in intracellular Ca^{2+} were first recorded in salamander rod cells and *Xenopus laevis* oocytes [66–68]. Later, in SMCs, it was noted that the equilibrium potential of the noradrenaline-induced depolarization of rat anococcygeus SMCs was similar to the Cl^- equilibrium potential measured in guinea-pig vas deferens [69,70]. Byrne and Large subsequently showed a noradrenaline-induced increase in rat anococcygeus SMC membrane Cl^- conductance that was preceded by a rise in the intracellular Ca^{2+} concentration. In the case of agonist-induced contractions, G_q protein-coupled receptor activation causes IP_3-induced Ca^{2+} mobilization, which subsequently activates $I_{Cl(Ca)}$ [71,72]. In SMCs, $I_{Cl(Ca)}$ can be activated by IP_3- and ryanodine receptor-mediated Ca^{2+} mobilization [73–75], spontaneously released Ca^{2+} from juxtaposing SR [76,77], Ca^{2+} entry though VGCCs [74,78,79], Ca^{2+} influx via reverse mode Na^+/Ca^{2+} exchanger [80], and as recent evidence suggests, Ca^{2+} entry through neighboring store-operated cation channels [81] or TRPC6 channels [82]. A wide variety of biophysical characteristics have been reported for these channels. SMC $I_{Cl(Ca)}$ idiosyncrasies include high Ca^{2+} sensitivity (Kd ~ 250 nM) [83], outward rectification, distinct voltage-dependence, small unitary conductance (~1–3pS), and modulation by Ca^{2+} dependent phosphorylation [84]. Some of this variation could be assigned to inconsistencies in experimental conditions; however, it is likely that they highlight a more complex picture where many of these variations are due to channels differing in their molecular architecture. Functionally, the activation of $I_{Cl(Ca)}$, by receptor-mediated Ca^{2+}-mobilization generates sufficient depolarization to increases the open probability of VGCCs, consequently allowing more Ca^{2+} into the cell, which can further activate $I_{Cl(Ca)}$, and contraction. This was supported by isomeric tension studies that showed impeded responses to vasoconstricting agonists in Cl^- free solutions or in the presence of NFA—a blocker of $I_{Cl(Ca)}$ [85–88].

2.4 MOLECULAR IDENTIFICATION OF CALCIUM-ACTIVATED Cl^- CHANNELS IN SMOOTH MUSCLE

Several different proteins were proposed as the molecular correlate to $I_{Cl(Ca)}$, including bestrophins, ClCs, and CLCAs. Despite similar activation characteristics, and a sensitivity to DIDS [89], currents generated by CLCA incorporated into lipid bilayers showed different biophysical characteristics to those of classical $I_{Cl(Ca)}$ [90]. Similarly, over-expression of ClC-3 produces $I_{Cl(Ca)}$ with poor Ca^{2+} sensitivity and a voltage-dependence too dissimilar to that of the native, classical $I_{Cl(Ca)}$ [91]. The expression product of the vitelliform macular dystrophy type 2 gene, bestrophin, was a promising candidate for SMC CACCs. Expression of bestrophins in heterologous expression systems gave rise to Cl^- channels that were sensitive to intracellular Ca^{2+} at concentrations similar to native $I_{Cl(Ca)}$ [92,93]; however, the characteristics of the currents differed from the *classical* $I_{Cl(Ca)}$ that showed outward rectification and voltage-dependent kinetics at submaximal Ca^{2+} concentrations. The currents generated by bestrophins in heterologous expression systems did, however, share many characteristics with the cGMP-dependent $I_{Cl(Ca)}$, characterized in smooth muscle contemporaneously [94,95]. Indeed, downregulation of the bestrophin family member *Best-3* in both cultured cells and VSMCs was accompanied by a decrease in the cGMP-dependent $I_{Cl(Ca)}$ [96]. The cGMP-dependent $I_{Cl(Ca)}$ mediated by bestrophin in VSMCs was later shown to be integral to the initiation of vasomotion in mesenteric arteries, the rhythmic contractions of arteries implicated in the regulation of blood pressure [97].

Nearly thirty years after the first characterization of $I_{Cl(Ca)}$ in salamander photoreceptor cells and Xenopus ooctyes [66–68], TMEM16A (also referred to as anoctamin 1 [ANO1]) expression was found to be essential for the generation of the classical $I_{Cl(Ca)}$ by three independent groups. Each group used distinct strategies: Yang et al. searched in databases for genes which may encode channel-like proteins with unknown function [98]; Caputo et al. used the knowledge that IL-4-treated airway epithelial cells exhibited greater CACC activity, in combination with gene expression and manipulation studies [99]; and Schroeder et al. used the fact that the $I_{Cl(Ca)}$ is essential for the prevention of polyspermic fertilization in Xenopus oocytes and compared them with those of axolotl, which naturally allow polyspermic fertilization [100]. Following its identification, TMEM16A was shown to be expressed by VSMCs, where it functions as a native CACC [101–103]. The biophysical characteristics of TMEM16A-associated $I_{Cl(Ca)}$ are akin to those of SMC *classical* $I_{Cl(Ca)}$ [101,103–105]. As far as bestrophin is concerned, siRNA-mediated downregulation of TMEM16A in rat mesenteric arteries causes a suppression of both the cGMP-dependent and the classical $I_{Cl(Ca)}$, whereas bestrophin downregulation only reduces the cGMP-dependent $I_{Cl(Ca)}$ [96,106]. The authors of these papers suggest that TMEM16A may be the main component of the complex that forms both cGMP-dependent and classical CACCs in VSMCs (Figure 2.1), with other proteins, such as bestrophins, acting as characteristic-altering subunits when co-expressed [107].

2.5 THE PHYSIOLOGICAL ROLE OF TMEM16A IN SMOOTH MUSCLE CELLS

Consistent with the early studies of CACCs in smooth muscle, TMEM16A is implicated in the development of contraction in a variety of smooth muscle types, vascular in particular. In gastrointestinal (GI) smooth muscle TMEM16A is expressed by the interstitial cells of Cajal (ICCs) where it is involved in the electrical pacemaker activity responsible for the initiation of peristaltic contractions [108]. While a role for TMEM16A has been established in GI smooth muscle, the presence and function of TMEM16A in uterine smooth muscle is contentious. Conflicting studies show both the presence and absence of TMEM16A immunohistochemical staining in murine myometrium [109,110]. However, application of classical CACC inhibitors reduced the frequency of spontaneous contractions [109,110]. In the murine oviduct, a group of GI-ICC-like cells, responsible for the contractions involved in oocyte transport, exhibit TMEM16A-mediated $I_{Cl(Ca)}$ and slow waves (the propagation of electrical events responsible for contractions along the uterine tube) were absent in mice with a homozygous null allele for TMEM16A [111]. Finally, in rat urinary bladder, TMEM16A is expressed at an mRNA and protein level in SMCs and a subpopulation of cells that co-express vimentin—a marker of interstitial cells. Again, pharmacological inhibition with classical CACC blockers attenuated the phasic contractions of both intact and denuded bladder strips [112]. Downstream, sheep, rat, and murine urethral SMCs express TMEM16A and contractions in the urethra of the species studied were NFA-sensitive [113].

In the vasculature, where the majority of TMEM16A-mediated $I_{Cl(Ca)}$ research has been carried out, there is a plethora of evidence suggestive of TMEM16A's involvement in smooth muscle contraction, and indeed, the regulation of blood pressure.

In recent years, several small molecular weight inhibitors for TMEM16A-associated $I_{Cl(Ca)}$ have been identified. Notably T16A$_{inh}$-A01 and, more recently, MONNA show selectivity for TMEM16A $I_{Cl(Ca)}$ in overexpression systems and have subsequently been used by researchers to try and elucidate TMEM16A's function [114,115]. These new pharmacological tools, as well as molecular knockdown techniques, have helped substantiate the evidence surrounding TMEM16A's role in contraction of VSMCs. Pharmacological inhibition with T16A$_{inh}$-A01 significantly reduced $I_{Cl(Ca)}$ and agonist-induced contraction in VSMCs from a number of vascular beds, including human abdominal visceral adipose arteries and hyperpolarized noradrenaline-stimulated mouse mesenteric arteries [116–118]. Reports of MONNA in the literature are few, however, in line with unpublished data from our group in rat coronary arteries, Boedtkjer et al. reported that MONNA relaxed preconstricted mouse mesenteric arteries, inhibited vasoconstriction, and hyperpolarized mouse mesenteric arteries [117]. A newly identified inhibitor of TMEM16A-associated $I_{Cl(Ca)}$, Ani9, shows selectivity for TMEM16A over TMEM16B, with the highest potency over other TMEM16A inhibitors to date [119] but no functional data in smooth muscle has been published yet. As for activators of TMEM16A, very little data has been published with regards to their effect on smooth muscle function. E$_{act}$ and F$_{act}$, respectively an activator and potentiator of TMEM16A, were reported in 2011 by the same group that identified T16A$_{inh}$-A01 [120]. E$_{act}$ increased Cl^- conductance at 0 Ca^{2+} in TMEM16A expressing Fisher rat thyroid (FRT) cells, while F$_{act}$ reduced the EC_{50} for Ca^{2+}-dependent activation. The currents that were activated/enhanced by these drugs were sensitive to application of T16A$_{inh}$-A01. While no contractile data is available regarding these drugs' actions in smooth muscle, E$_{act}$ activated $I_{Cl(Ca)}$ in rat cerebral artery SMCs [121] and enhanced $I_{Cl(Ca)}$ in rabbit pulmonary artery SMCs [116]. More recently, the major bioactive component of the traditional Chinese medicine, Ginseng, GRb1, has been reported to directly activate TMEM16A in a Ca^{2+}- dependent manner and enhance guinea pig ileum contractions [122].

2.5.1 TMEM16A and Vascular Contraction

In line with the model for CACCs outlined previously, there are a number of papers where TMEM16A has been implicated in vascular tone. Mice with an inducible, VSMC-specific knockout of TMEM16A were hypotensive and less susceptible to increases in blood pressure in response to chronic angiotensin II infusion [123]. On closer inspection, no significant attenuation of agonist-induced contraction of small mesenteric arteries was seen in these mice, while there was attenuation in retinal arterioles. Similarly, a reduction in the response to U46619, a thromboxane A_2 receptor agonist, was observed in isolated perfused hind limb experiments. Conversely, pulmonary arteries from rats with monocrotaline- or chronic hypoxia-induced pulmonary hypertension have increased amounts of TMEM16A protein and increased $I_{Cl(Ca)}$ density [124,125], while siRNA knockdown of TMEM16A prevented the development of hypertension in spontaneously hypertensive rats (SHRs) [118], and reduced agonist-induced contraction of rat small mesenteric arteries [106]. These studies contrast with that of Wang et al. [126], who observed hypertension-induced cerebrovascular remodeling and a reduction in TMEM16A protein and $I_{Cl(Ca)}$ in basilar artery SMCs isolated from 2-kidney, 2-clip renohypertensive rats [126]. The authors postulated that the remodeling of the basilar artery was due to the down-regulation of TMEM16A, noting TMEM16A's role as a negative regulator of cell proliferation.

Studying CACCs has always been hampered by the lack of either potent or selective modulators. All of the classical Cl^- channel blockers used in the characterization of $I_{Cl(Ca)}$, such as NFA, NPPB, and DIDS, have significant effects on other ion channels, with the pharmacology of large conductance, Ca^{2+}-activated K^+ channels overlapping in particular [127]. In addition, these agents are known to paradoxically enhance $I_{Cl(Ca)}$ under certain conditions [128,129]. The identification of TMEM16A as a component of CACCs has led to the development of new, more specific reagents. However, a recent study of new selective inhibitors concluded that $T16A_{inh}$-A01, $CaCC_{inh}$-A01, and MONNA were not selective in vascular tissues at concentrations that inhibit isolated $I_{Cl(Ca)}$ [117]. There is considerable debate over this subject and there seems to be a degree of artery-specific effects. However, it could be possible that TMEM16A does not function exclusively as a CACC in smooth muscle. Indeed, TMEM16A has been shown to participate in AngII-mediated RhoA/ROCK signaling in basilar artery constriction [130]. Likewise, a concomitant downregulation of bestrophins and VGCCs was seen upon TMEM16A knockdown [107], perhaps alluding to a role as a modulator of the expression of other ion channels. Either way, a dearth of pharmacological TMEM16A modulators befalls this area of research.

In spite of this desolate pharmacological tool box, there is a relative abundance of evidence outlining the regulation of TMEM16A in various cell types, including smooth muscle. In the case of molecular regulation, recent investigations have uncovered calmodulin [131–133], protons [134], cholesterol [135], and phosphoinositides [136] to control TMEM16A's function. Furthermore, as well as the canonical sources of Ca^{2+}, discussed earlier, TRPC6 has been shown to provide local Ca^{2+} influx capable of activating $I_{Cl(Ca)}$ in cerebral artery SMCs [82]. Aside from intracellular regulation, TMEM16A is also sensitive to thermal and mechanical provocations [137,138]. With regards to the molecular diversity of TMEM16A protein, several isoforms have been identified. Alternative splicing and alternative transcription start sites give rise to at least four different alternatively spliced segments that have differing channel properties [99,139,140]. A comprehensive review of the regulation of TMEM16A was recently published [141]; however, it is important to remember, with the information in the review in mind, that TMEM16A's function can vary drastically from tissue to tissue. Furthermore, understanding the complexity of TMEM16A isoform expression could provide a valuable insight into TMEM16A's various roles in the physiology.

2.6 CLOSING REMARKS

Whilst there is considerable evidence for CACCs' role in a simple schematic of receptor mediated contraction (see earlier, Figure 2.1 [72]), there are ample complexities and incongruities for the channel to continue to be enigmatic and for the role of Cl^- within smooth muscle to remain complex. Future studies of identified Cl^- channels should be wary of the various purported functions of the proteins.

With LRRC8A and TMEM16A recently reported to activate in parallel [64], and with TMEM16A's known role as a regulator of bestrophin and VGCC expression in vascular smooth muscle, it is important to retain an open mind when investigating these channels. The generation of new, more specific modulators of Cl^- channel function, combined with modern genetic manipulation techniques, should be at the center of future studies.

REFERENCES

1. Ko, E. A., J. Han, I. D. Jung, and W. S. Park. 2008. Physiological roles of K+ channels in vascular smooth muscle cells. *J Smooth Muscle Res* 44 (2): 65–81.
2. Stott, J. B., T. A. Jepps, and I. A. Greenwood. 2014. K(V)7 potassium channels: A new therapeutic target in smooth muscle disorders. *Drug Discov Today* 19 (4): 413–424. doi:10.1016/j.drudis.2013.12.003.
3. Nilius, B., and G. Droogmans. 2003. Amazing chloride channels: An overview. *Acta Physiol Scand* 177 (2): 119–147. doi:10.1046/j.1365-201X.2003.01060.x.
4. Verkman, A. S., and L. J. V. Galietta. 2009. Chloride channels as drug targets. *Nat Rev Drug Discov* 8 (2): 153–171. doi:10.1038/nrd2780.
5. Brading, A. F. 1981. Ionic distribution and mechanisms of transmembrane ion movements in smooth muscle. In *Smooth Muscle, an Assessment of Current Knowledge*, E. Bülbring, A. F. Brading, A. W. Jones, and T. Tomita (Eds.). London, UK: Edward Arnold, pp. 65–92.
6. Chipperfield, A. R., and A. A. Harper. 2000. Chloride in smooth muscle. *Prog Biophys Mol Biol* 74 (3–5): 175–221. doi:10.1016/S0079-6107(00)00024-9.
7. Aickin, C. C. 1990. Chloride transport across the sarcolemma of vertebrate smooth and skeletal muscle. In *Chloride Channels and Carriers in Nerve, Muscle and Glial Cells*, F. J. Alvarez-Leefmans and J. M. Russell (Eds.). New York: Plenum Press, pp. 209–249.
8. Aickin, C. C., and N. A. Vermuë. 1983. Microelectrode measurement of intracellular chloride activity in smooth muscle cells of guinea-pig ureter. *Pflügers Arch* 397 (1): 25–28.
9. Hoffmann, E. K., and L. O. Simonsen. 1989. Membrane mechanisms in volume and pH regulation in vertebrate cells. *Physiol Rev* 69 (2): 315–382.
10. Aickin, C. C., and A. F. Brading. 1984. The role of chloride-bicarbonate exchange in the regulation of intracellular chloride in guinea-pig vas deferens. *J Physiol* 349: 587–606.
11. Kreye, V. A., P. K. Bauer, and I. Villhauer. 1981. Evidence for furosemide-sensitive active chloride transport in vascular smooth nuscle. *Eur J Pharmacol* 73 (1): 91–95.
12. Aiton, J. F., A. R. Chipperfield, J. F. Lamb, P. Ogden, and N. L. Simmons. 1981. Occurrence of passive furosemide-sensitive transmembrane potassium transport in cultured cells. *Biochim Biophys Acta* 646 (3): 389–398.
13. Owen, N. E. 1984. Regulation of Na/K/Cl cotransport in vascular smooth muscle cells. *Biochem Biophys Res Commun* 125 (2): 500–508.
14. O'Donnell, M. E., and N. E. Owen. 1994. Regulation of ion pumps and carriers in vascular smooth muscle. *Physiol Rev* 74 (3): 683–721.
15. Gerstheimer, F. P., M. Mühleisen, D. Nehring, and V. A. Kreye. 1987. A chloride-bicarbonate exchanging anion carrier in vascular smooth muscle of the rabbit. *Pflügers Arch* 409 (1–2): 60–66.
16. Aickin, C. C., and A. F. Brading. 1990a. Effect of Na^+ and K^+ on Cl^- distribution in guinea-pig vas deferens smooth muscle: Evidence for Na^+, K^+, Cl^- co-transport. *J Physiol* 421: 13–32.

17. Aickin, C. C., and A. F. Brading. 1990b. The effect of loop diuretics on Cl^- transport in smooth muscle of the guinea-pig vas deferens and taenia from the caecum. *J Physiol* 421: 33–53.
18. Davis, J. P., A. R. Chipperfield, and A. A. Harper. 1993. Accumulation of intracellular chloride by (Na-K-Cl) co-transport in rat arterial smooth muscle is enhanced in deoxycorticosterone acetate (DOCA)/salt hypertension. *J Mol Cell Cardiol* 25 (3): 233–237. doi:10.1006/jmcc.1993.1029.
19. Chipperfield, A. R., J. P. Davis, and A. A. Harper. 1993. An acetazolamide-sensitive inward chloride pump in vascular smooth muscle. *Biochem Biophys Res Commun* 194 (1): 407–412. doi:10.1006/bbrc.1993.1834.
20. Akar, F., G. Jiang, R. J. Paul, and W. C. O'Neill. 2001. Contractile regulation of the Na^+-K^+-$2Cl^-$ cotransporter in vascular smooth muscle. *Am J Physiol Cell Physiol* 281 (2): C579–C584.
21. Casteels, R., K. Kitamura, H. Kuriyama, and H. Suzuki. 1977. The membrane properties of the smooth muscle cells of the rabbit main pulmonary artery. *J Physiol* 271 (1): 41–61.
22. Robert, R., V. Thoreau, C. Norez, A. Cantereau, A. Kitzis, Y. Mettey, C. Rogier, and F. Becq. 2004. Regulation of the cystic fibrosis transmembrane conductance regulator channel by beta-adrenergic agonists and vasoactive intestinal peptide in rat smooth muscle cells and its role in vasorelaxation. *J Biol Chem* 279 (20): 21160–21168. doi:10.1074/jbc. M312199200.
23. Robert, R., C. Norez, and F. Becq. 2005. Disruption of CFTR chloride channel alters mechanical properties and cAMP-dependent Cl^- transport of mouse aortic smooth muscle cells. *J Physiol* 568 (Pt 2): 483–495. doi:10.1113/jphysiol.2005.085019.
24. Robert, R., J.-P. Savineau, C. Norez, F. Becq, and C. Guibert. 2007. Expression and function of cystic fibrosis transmembrane conductance regulator in rat intrapulmonary arteries. *Eur Respir J* 30 (5): 857–864. doi:10.1183/09031936.00060007.
25. Vandebrouck, C., P. Melin, C. Norez, R. Robert, C. Guibert, Y. Mettey, and F. Becq. 2006. Evidence that CFTR is expressed in rat tracheal smooth muscle cells and contributes to bronchodilation. *Respir Res* 7: 113. doi:10.1186/1465-9921-7-113.
26. Bulley, S., and J. H. Jaggar. 2014. Cl^- channels in smooth muscle cells. *Pflügers Arch* 466 (5): 861–872. doi:10.1007/s00424-013-1357-2.
27. Hübner, C. A., B. C. Schroeder, and H. Ehmke. 2015. Regulation of vascular tone and arterial blood pressure: Role of chloride transport in vascular smooth muscle. *Pflügers Arch* 467 (3): 605–614. doi:10.1007/s00424-014-1684-y.
28. Guo, J. J., D. A. Stoltz, V. Zhu, K. A. Volk, J. L. Segar, P. B. McCray, and R. D. Roghair. 2014. Genotype-specific alterations in vascular smooth muscle cell function in cystic fibrosis piglets. *J Cyst Fibros* 13 (3): 251–259. doi:10.1016/j.jcf.2013.10.009.
29. Tabeling, C., H. Yu, L. Wang, H. Ranke, N. M. Goldenberg, D. Zabini, E. Noe et al. 2015. CFTR and sphingolipids mediate hypoxic pulmonary vasoconstriction. *Proc Natl Acad Sci USA* 112 (13): E1614–E1623. doi:10.1073/pnas.1421190112.
30. Peotta, V. A., P Bhandary, U. Ogu, K. A. Volk, R. D. Roghair, M. P. Rogan, D. A. Stoltz et al. 2014. Reduced blood pressure of CFTR-F508del carriers correlates with diminished arterial reactivity rather than circulating blood volume in mice. Edited by J Karhausen. *PLoS One* 9 (5): e96756. doi:10.1371/journal.pone.0096756.
31. Lieberman, J., and S. Rodbard. 1975. Low blood pressure in young adults with cystic fibrosis: An effect of chronic salt loss in sweat? *Ann Intern Med* 82 (6): 806–808.
32. Super, M., A. Irtiza-Ali, S. A. Roberts, M. Schwarz, M. Young, A. Smith, T. Roberts, J. Hinks, and A. Heagerty. 2004. Blood pressure and the cystic fibrosis gene: Evidence for lower pressure rises with age in female carriers. *Hypertension* 44 (6): 878–883. doi:10.1161/01.HYP.0000145901.81989.46.

33. Cook, D. P., M. V. Rector, D. C. Bouzek, A. S. Michalski, N. D. Gansemer, L. R. Reznikov, X. Li et al. 2016. Cystic fibrosis transmembrane conductance regulator in sarcoplasmic reticulum of airway smooth muscle: Implications for airway contractility. *Am J Respir Crit Care Med* 193 (4): 417–426. doi:10.1164/rccm.201508-1562OC.
34. Wei, L., A. Vankeerberghen, H. Cuppens, J. Eggermont, J. J. Cassiman, G. Droogmans, and B. Nilius. 1999. Interaction between calcium-activated chloride channels and the cystic fibrosis transmembrane conductance regulator. *Pflügers Arch* 438 (5): 635–641.
35. Greenwood, I. A., and W. A. Large. 1998. Properties of a Cl^- current activated by cell swelling in rabbit portal vein vascular smooth muscle cells. *Am J Physiol* 275 (5 Pt 2): H1524–H1532.
36. Yamazaki, J., D. Duan, R. Janiak, K. Kuenzli, B. Horowitz, and J. R. Hume. 1998. Functional and molecular expression of volume-regulated chloride channels in canine vascular smooth muscle cells. *J Physiol* 507 (Pt 3): 729–736.
37. Ellershaw, D. C., I. A. Greenwood, and W. A. Large. 2000. Dual modulation of swelling-activated chloride current by NO and NO donors in rabbit portal vein myocytes. *J Physiol* 528 (Pt 1): 15–24.
38. Ellershaw, D. C., I. A. Greenwood, and W. A. Large. 2002. Modulation of volume-sensitive chloride current by noradrenaline in rabbit portal vein myocytes. *J Physiol* 542 (Pt 2): 537–547.
39. Xu, W. X., S. J. Kim, I. So, T. M. Kang, J. C. Rhee, and K. W. Kim. 1997. Volume-sensitive chloride current activated by hyposmotic swelling in antral gastric myocytes of the guinea-pig. *Pflügers Arch* 435 (1): 9–19. doi:10.1007/s004240050478.
40. Dick, G. M., K. K. Bradley, B. Horowitz, J. R. Hume, and K. M. Sanders. 1998. Functional and molecular identification of a novel chloride conductance in canine colonic smooth muscle. *Am J Physiol Cell Physiol* 275 (4): C940–C950.
41. Qian, J.-S., R.-P. Pang, K.-S. Zhu, D.-Y. Liu, Z.-R. Li, C.-Y. Deng, and S.-M. Wang. 2009. Static pressure promotes rat aortic smooth muscle cell proliferation via upregulation of volume-regulated chloride channel. *Cell Physiol Biochem* 24 (5–6): 461–470. doi:10.1159/000257485.
42. Liang, W., L. Huang, D. Zhao, J. Z. He, P. Sharma, J. Liu, A. O. Gramolini, M. E. Ward, H. C. Cho, and P. H. Backx. 2014. Swelling-activated Cl^- currents and intracellular CLC-3 are involved in proliferation of human pulmonary artery smooth muscle cells. *J Hypertens* 32 (2): 318–330. doi:10.1097/HJH.0000000000000013.
43. Nelson, M. T., M. A. Conway, H. J. Knot, and J. E. Brayden. 1997. Chloride channel blockers inhibit myogenic tone in rat cerebral arteries. *J Physiol* 502 (Pt 2 (July): 259–264.
44. Doughty, J. M., and P. D. Langton. 2001. Measurement of chloride flux associated with the myogenic response in rat cerebral arteries. *J Physiol* 534 (Pt 3): 753–761. doi:10.1111/j.1469-7793.2001.t01-1-00753.x.
45. Graves, J. E., I. A. Greenwood, and W. A. Large. 2000. Tonic regulation of vascular tone by nitric oxide and chloride ions in rat isolated small coronary arteries. *Am J Physiol Heart Circ Physiol* 279 (6): H2604–H2611.
46. Fischmeister, R., and H. Criss Hartzell. 2005. Volume sensitivity of the bestrophin family of chloride channels. *J Physiol* 562 (2): 477–491. doi:10.1113/jphysiol.2004.075622.
47. Gründer, S., A. Thiemann, M. Pusch, and T. J. Jentsch. 1992. Regions involved in the opening of CIC-2 chloride channel by voltage and cell volume. *Nature* 360 (6406): 759–762. doi:10.1038/360759a0.
48. Jentsch, T. J., T. Friedrich, A. Schriever, and H. Yamada. 1999. The CLC chloride channel family. *Pflügers Arch* 437 (6): 783–795.
49. Juul, C. A., S. Grubb, K. A. Poulsen, T. Kyed, N. Hashem, I. H. Lambert, E. H. Larsen, and E. K. Hoffmann. 2014. Anoctamin 6 differs from VRAC and VSOAC but is involved in apoptosis and supports volume regulation in the presence of Ca^{2+}. *Pflügers Arch* 466 (10): 1899–1910. doi:10.1007/s00424-013-1428-4.

50. Matsuda, J. J., M. S. Filali, J. G. Moreland, F. J. Miller, and F. S. Lamb. 2010. Activation of swelling-activated chloride current by tumor necrosis factor-alpha requires ClC-3-dependent endosomal reactive oxygen production. *J Biol Chem* 285 (30): 22864–22873. doi:10.1074/jbc.M109.099838.
51. Wang, G.-X., W. J. Hatton, G. L. Wang, J. Zhong, I. Yamboliev, D. Duan, and J. R. Hume. 2003. Functional effects of novel anti-ClC-3 antibodies on native volume-sensitive osmolyte and anion channels in cardiac and smooth muscle cells. *Am J Physiol Heart Circ Physiol* 285 (4): H1453–H1463. doi:10.1152/ajpheart.00244.2003.
52. Wang, L., L. Chen, and T. J. C. Jacob. 2000. The role of ClC-3 in volume-activated chloride currents and volume regulation in bovine epithelial cells demonstrated by antisense inhibition. *J Physiol* 524 (1): 63–75. doi:10.1111/j.1469-7793.2000.t01-1-00063.x.
53. Zhou, J.-G., J.-L. Ren, Q.-Y. Qiu, H. He, and Y.-Y. Guan. 2005. Regulation of intracellular Cl^- concentration through volume-regulated ClC-3 chloride channels in A10 vascular smooth muscle cells. *J Biol Chem* 280 (8): 7301–7308. doi:10.1074/jbc.M412813200.
54. Stobrawa, S. M., T. Breiderhoff, S. Takamori, D. Engel, M. Schweizer, A. A. Zdebik, M. R. Bösl et al. 2001. Disruption of ClC-3, a chloride channel expressed on synaptic vesicles, leads to a loss of the hippocampus. *Neuron* 29 (1): 185–96. doi:10.1016/S0896-6273(01)00189-1.
55. Arreola, J., T. Begenisich, K. Nehrke, H.-V. Nguyen, K. Park, L. Richardson, B. Yang, B. C. Schutte, F. S. Lamb, and J. E. Melvin. 2002. Secretion and cell volume regulation by salivary acinar cells from mice lacking expression of the Clcn3 Cl^- channel gene. *J Physiol* 545 (Pt 1): 207–216.
56. Gong, W., H. Xu, T. Shimizu, S. Morishima, S. Tanabe, T. Tachibe, S. Uchida, S. Sasaki, and Y. Okada. 2004. ClC-3-independent, PKC-dependent activity of volume-sensitive Cl channel in mouse ventricular cardiomyocytes. *Cell Physiol Biochem* 14 (4–6): 213–224. doi:10.1159/000080330.
57. Yamamoto-Mizuma, S., G.-X. Wang, L. L. Liu, K. Schegg, W. J. Hatton, D. Duan, The Late B Horowitz, F S Lamb, and J R Hume. 2004. Altered properties of volume-sensitive osmolyte and anion channels (VSOACs) and membrane protein expression in cardiac and smooth muscle myocytes from Clcn3–/– Mice. *J Physiol* 557 (Pt 2): 439–456. doi:10.1113/jphysiol.2003.059261.
58. Wang, J., H. Xu, S. Morishima, S. Tanabe, K. Jishage, S. Uchida, S. Sasaki, Y. Okada, and T. Shimizu. 2005. Single-channel properties of volume-sensitive Cl^- channel in ClC-3–deficient cardiomyocytes. *Jpn J Physiol* 55 (6): 379–383. doi:10.2170/jjphysiol.S655.
59. Qiu, Z., A. E. Dubin, J. Mathur, B. Tu, K. Reddy, L. J. Miraglia, J. Reinhardt, A. P. Orth, and A. Patapoutian. 2014. SWELL1, a plasma membrane protein, is an essential component of volume-regulated anion channel. *Cell* 157 (2): 447–458. doi:10.1016/j.cell.2014.03.024.
60. Voss, F. K., F. Ullrich, J. Münch, K. Lazarow, D. Lutter, N. Mah, M. A. Andrade-Navarro, J. P. von Kries, T. Stauber, and T. J. Jentsch. 2014. Identification of LRRC8 heteromers as an essential component of the volume-regulated anion channel VRAC. *Science* 344 (6184): 634–638.
61. Jentsch, T. J., D. Lutter, R. Planells-Cases, F. Ullrich, and F. K. Voss. 2016. VRAC: Molecular identification as LRRC8 heteromers with differential functions. *Pflügers Arch* 468 (3): 385–393. doi:10.1007/s00424-015-1766-5.
62. Abascal, F., and R. Zardoya. 2012. LRRC8 proteins share a common ancestor with pannexins, and may form hexameric channels involved in cell-cell communication. *BioEssays* 34 (7): 551–560. doi:10.1002/bies.201100173.
63. Planells-Cases, R., D. Lutter, C. Guyader, N. M. Gerhards, F. Ullrich, D. A. Elger, A. Kucukosmanoglu et al. 2015. Subunit composition of VRAC channels determines substrate specificity and cellular resistance to Pt-based anti-cancer drugs. *EMBO J* 34 (24): 2993–3008. doi:10.15252/embj.201592409.

64. Benedetto, R., L. Sirianant, I. Pankonien, P. Wanitchakool, J. Ousingsawat, I. Cabrita, R. Schreiber, M. Amaral, and K. Kunzelmann. 2016. Relationship between TMEM16A/anoctamin 1 and LRRC8A. *Pflügers Arch* 468 (10): 1751–1763. doi:10.1007/s00424-016-1862-1.
65. Choi, H., N. Ettinger, J. Rohrbough, A. Dikalova, H. N. Nguyen, and F. S. Lamb. 2016. LRRC8A channels support TNFα-induced superoxide production by nox1 which is required for receptor endocytosis. *Free Radic Biol Med* 101: 413–423. doi:10.1016/j.freeradbiomed.2016.11.003.
66. Bader, C. R., D. Bertrand, and E. A. Schwartz. 1982. Voltage-activated and calcium-activated currents studied in solitary rod inner segments from the salamander retina. *J Physiol* 331: 253–284.
67. Miledi, R. 1982. A calcium-dependent transient outward current in xenopus laevis oocytes. *Proc R Soc Lond Ser B* 215 (1201): 491–497.
68. Barish, M. E. 1983. A transient calcium-dependent chloride current in the immature xenopus oocyte. *J Physiol* 342: 309–325.
69. Aickin, C. C., and A. F. Brading. 1982. Measurement of intracellular chloride in guinea-pig vas deferens by ion analysis, 36chloride efflux and micro-electrodes. *J Physiol* 326: 139–154.
70. Byrne, N. G., and W. A. Large. 1987. Action of noradrenaline on single smooth muscle cells freshly dispersed from the rat anococcygeus muscle. *J Physiol* 389: 513–525.
71. Large, W. A., and Q. Wang. 1996. Characteristics and physiological role of the $Ca(2^+)$-activated Cl^- conductance in smooth muscle. *The American Journal of Physiology* 271 (2 Pt 1): C435–C454.
72. Kitamura, K., and J. Yamazaki. 2001. Chloride channels and their functional roles in smooth muscle tone in the vasculature. *Jpn J Pharmacol* 85 (4): 351–357.
73. Ito, S., T. Ohta, and Y. Nakazato. 1993. Inward current activated by carbachol in rat intestinal smooth muscle cells. *J Physiol* 470: 395–409.
74. Lamb, F. S., K. A. Volk, and E. F. Shibata. 1994. Calcium-activated chloride current in rabbit coronary artery myocytes. *Circul Res* 75 (4): 742–750.
75. Saleh, S. N., and I. A. Greenwood. 2005. Activation of chloride currents in murine portal vein smooth muscle cells by membrane depolarization involves intracellular calcium release. *Am J Physiol Cell Physiol* 288 (1): C122–C131. doi:10.1152/ajpcell.00384.2004.
76. Wang, Q., R. C. Hogg, and W. A. Large. 1992. Properties of spontaneous inward currents recorded in smooth muscle cells isolated from the rabbit portal vein. *J Physiol* 451: 525–537.
77. Greenwood, I. A., R. M. Helliwell, and W. A. Large. 1997. Modulation of $Ca(2^+)$-activated Cl^- currents in rabbit portal vein smooth muscle by an inhibitor of mitochondrial Ca^{2+} uptake. *J Physiol* 505 (Pt 1): 53–64.
78. Pacaud, P., G. Loirand, G. Grégoire, C. Mironneau, and J. Mironneau. 1992. Calcium-dependence of the calcium-activated chloride current in smooth muscle cells of rat portal vein. *Pflügers Arch* 421 (2–3): 125–130.
79. Greenwood, I. A., and W. A. Large. 1996. Analysis of the time course of calcium-activated chloride "tail" currents in rabbit portal vein smooth muscle cells. *Pflügers Arch* 432 (6): 970–979.
80. Leblanc, N., and P. M. Leung. 1995. Indirect stimulation of $Ca(2^+)$-activated Cl^- current by Na^+/Ca^{2+} exchange in rabbit portal vein smooth muscle. *Am J Physiol* 268 (5 Pt 2): H1906–H1917.
81. Angermann, J. E., A. S. Forrest, I. A. Greenwood, and N. Leblanc. 2012. Activation of Ca^{2+}-activated Cl^- channels by store-operated Ca^{2+} entry in arterial smooth muscle cells does not require reverse-mode Na^+/Ca^{2+} exchange. *Can J Physiol Pharmacol* 90 (7): 903–921. doi:10.1139/y2012-081.

82. Wang, Q., M. Dennis Leo, D. Narayanan, K. P. Kuruvilla, and J. H. Jaggar. 2016. Local coupling of TRPC6 to ANO1/TMEM16A channels in smooth muscle cells amplifies vasoconstriction in cerebral arteries. *Am J Physiol Cell Physiol*. doi:10.1152/ajpcell.00092.2016.
83. Angermann, J. E., A. R. Sanguinetti, J. L. Kenyon, N. Leblanc, and I. A. Greenwood. 2006. Mechanism of the inhibition of Ca^{2+}-activated Cl^{-}- currents by phosphorylation in pulmonary arterial smooth muscle cells. *J Gen Physiol* 128 (1): 73–87. doi:10.1085/jgp.200609507.
84. Leblanc, N., J. Ledoux, S. Saleh, A. Sanguinetti, J. Angermann, K. O'Driscoll, F. Britton, B. A. Perrino, and I. A. Greenwood. 2005. Regulation of calcium-activated chloride channels in smooth muscle cells: A complex picture is emerging. *Can J Physiol Pharmacol* 83 (7): 541–556. doi:10.1139/y05-040.
85. Criddle, D. N., R. S. de Moura, I. A. Greenwood, and W. A. Large. 1996. Effect of niflumic acid on noradrenaline-induced contractions of the rat aorta. *Br J Pharmacol* 118 (4): 1065–1071.
86. Criddle, D. N., R. S. de Moura, I. A. Greenwood, and W A Large. 1997. Inhibitory action of niflumic acid on noradrenaline- and 5-hydroxytryptamine-induced pressor responses in the isolated mesenteric vascular bed of the rat. *Br J Pharmacol* 120 (5): 813–818. doi:10.1038/sj.bjp.0700981.
87. Yuan, X. J. 1997. Role of calcium-activated chloride current in regulating pulmonary vasomotor tone. *Am J Physiol* 272 (5 Pt 1): L959–L968.
88. Lamb, F. S., and T. J. Barna. 1998. Chloride ion currents contribute functionally to norepinephrine-induced vascular contraction. *Am J Physiol Heart Circ Physiol* 275 (1): H151–H160.
89. Cunningham, S. A., M. S. Awayda, J. K. Bubien, I. I. Ismailov, M. P. Arrate, B. K. Berdiev, D. J. Benos, and C. M. Fuller. 1995. Cloning of an epithelial chloride channel from bovine trachea. *J Biol Chem* 270 (52): 31016–31026.
90. Fuller, C. M., and D. J. Benos. 2000. Ca(2+)-activated Cl(–) channels: A newly emerging anion transport family. *News Physiol Sci* 15: 165–171.
91. Hartzell, C., I. Putzier, and J. Arreola. 2005. Calcium-activated chloride channels. *Ann Rev Physiol* 67: 719–758. doi:10.1146/annurev.physiol.67.032003.154341.
92. Sun, H., T. Tsunenari, K.-W. Yau, and J. Nathans. 2002. The vitelliform macular dystrophy protein defines a new family of chloride channels. *Proc Natl Acad Sci USA* 99 (6): 4008–4013. doi:10.1073/pnas.052692999.
93. Qu, Z., R. W. Wei, W. Mann, and H. Criss Hartzell. 2003. Two bestrophins cloned from xenopus laevis oocytes express $Ca(2^{+})$-activated Cl(–) currents. *J Biol Chem* 278 (49): 49563–49572. doi:10.1074/jbc.M308414200.
94. Matchkov, V. V., C. Aalkjaer, and H. Nilsson. 2004. A cyclic GMP-dependent calcium-activated chloride current in smooth-muscle cells from rat mesenteric resistance arteries. *J Gen Physiol* 123 (2): 121–134. doi:10.1085/jgp.200308972.
95. Piper, A. S., and W. A. Large. 2004. Single cGMP-activated Ca(+)-Dependent Cl(–) channels in rat mesenteric artery smooth muscle cells. *J Physiol* 555 (Pt 2): 397–408. doi:10.1113/jphysiol.2003.057646.
96. Matchkov, V. V., P. Larsen, E. V. Bouzinova, A. Rojek, D. M. Briggs Boedtkjer, V. Golubinskaya, F. S. Pedersen, C. Aalkjaer, and H. Nilsson. 2008. Bestrophin-3 (Vitelliform Macular Dystrophy 2-like 3 Protein) is essential for the cGMP-dependent calcium-activated chloride conductance in vascular smooth muscle cells. *Circ Res* 103 (8): 864–872. doi:10.1161/CIRCRESAHA.108.178517.
97. Broegger, T., J. C. Brings Jacobsen, V. Secher Dam, D. M. Briggs Boedtkjer, H. Kold-Petersen, F. S. Pedersen, C. Aalkjaer, and V. V. Matchkov. 2011. Bestrophin is important for the rhythmic but not the tonic contraction in rat mesenteric small arteries. *Cardiovasc Res* 91 (4): 685–693. doi:10.1093/cvr/cvr111.

98. Yang, Y. D., H. Cho, J. Y. Koo, M. H. Tak, Y. Cho, W.-S. Shim, S. P. Park et al. 2008. TMEM16A confers receptor-activated calcium-dependent chloride conductance. *Nature* 455 (7217): 1210–1215. doi:10.1038/nature07313.
99. Caputo, A., E. Caci, L. Ferrera, N. Pedemonte, C. Barsanti, E. Sondo, U. Pfeffer, R. Ravazzolo, O. Zegarra-Moran, and L. J. V. Galietta. 2008. TMEM16A, a membrane protein associated with calcium-dependent chloride channel activity. *Science (New York, N.Y.)* 322 (5901): 590–594. doi:10.1126/science.1163518.
100. Schroeder, B. C., T. Cheng, Y. N. Jan, and L. Y. Jan. 2008. Expression cloning of TMEM16A as a calcium-activated chloride channel subunit. *Cell* 134 (6): 1019–1029. doi:10.1016/j.cell.2008.09.003.
101. Davis, A. J., A. S. Forrest, T. A. Jepps, M. L. Valencik, M. Wiwchar, C. A. Singer, W. R. Sones, I. A. Greenwood, and N. Leblanc. 2010. Expression profile and protein translation of TMEM16A in murine smooth muscle. *Am J Physiol Cell Physiol* 299 (5): C948–C9459. doi:10.1152/ajpcell.00018.2010.
102. Manoury, B., A. Tamuleviciute, and P. Tammaro. 2010. TMEM16A/anoctamin 1 protein mediates calcium-activated chloride currents in pulmonary arterial smooth muscle cells. *J Physiol* 588 (Pt 13): 2305–2314. doi:10.1113/jphysiol.2010.189506.
103. Thomas-Gatewood, C., Z. P. Neeb, S. Bulley, A. Adebiyi, J. P. Bannister, M. Dennis Leo, and J. H. Jaggar. 2011. TMEM16A channels generate Ca^{2+}-activated Cl^- currents in cerebral artery smooth muscle cells. *Am J Physiol Heart Circ Physiol* 301 (5): H1819–H1827. doi:10.1152/ajpheart.00404.2011.
104. Ohshiro, J., H. Yamamura, T. Saeki, Y. Suzuki, and Y. Imaizumi. 2014a. The multiple expression of Ca^{2+}-activated Cl^- channels via homo- and hetero-dimer formation of TMEM16A splicing variants in murine portal vein. *Biochem Biophys Res Commun* 443 (2): 518–523. doi:10.1016/j.bbrc.2013.11.117.
105. Ohshiro, J., H. Yamamura, Y. Suzuki, and Y. Imaizumi. 2014b. Modulation of TMEM16A-channel activity as Ca^{2+} activated Cl^- conductance via the interaction with actin cytoskeleton in murine portal vein. *J Pharmacol Sci* 125 (1): 107–111.
106. Dam, V. S., D. M. B. Boedtkjer, J. Nyvad, C. Aalkjaer, and V. Matchkov. 2014b. TMEM16A knockdown abrogates two different Ca^{2+}-activated Cl^- currents and contractility of smooth muscle in rat mesenteric small arteries. *Pflügers Arch* 466 (7): 1391–1409. doi:10.1007/s00424-013-1382-1.
107. Dam, V. S., D. M. B. Boedtkjer, C. Aalkjaer, and V. Matchkov. 2014a. The bestrophin- and TMEM16A-associated Ca^{2+} activated Cl^- channels in vascular smooth muscles. *Channels (Austin, Tex.)* 8 (4): 361–369. doi:10.4161/chan.29531.
108. Sanders, K. M., M. H. Zhu, F. Britton, S. D. Koh, and S. M. Ward. 2012. Anoctamins and gastrointestinal smooth muscle excitability. *Exper Physiol* 97 (2): 200–206. doi:10.1113/expphysiol.2011.058248.
109. Bernstein, K., J. Y. Vink, X. Wen Fu, H. Wakita, J. Danielsson, R. Wapner, and G. Gallos. 2014. Calcium-activated chloride channels anoctamin 1 and 2 promote murine uterine smooth muscle contractility. *Am J Obstet Gynecol* 211 (6): 688.e1–688.e10. doi:10.1016/j.ajog.2014.06.018.
110. Dodds, K. N., V. Staikopoulos, and E. A. H. Beckett. 2015. Uterine contractility in the nonpregnant mouse: Changes during the estrous cycle and effects of chloride channel blockade. *Biol Reproduct* 92 (6): 141. doi:10.1095/biolreprod.115.129809.
111. Dixon, R. E., G. W. Hennig, S. A. Baker, F. C. Britton, B. D. Harfe, J. R. Rock, K. M. Sanders, and S. M. Ward. 2012. Electrical slow waves in the mouse oviduct are dependent upon a calcium activated chloride conductance encoded by *Tmem16a*. *Biol Reproduct* 86 (1): 1–7. doi:10.1095/biolreprod.111.095554.
112. Bijos, D. A., M. J. Drake, and B. Vahabi. 2014. Anoctamin-1 in the juvenile rat urinary bladder. Edited by J Reisert. *PLoS One* 9 (9): e106190. doi:10.1371/journal.pone.0106190.

113. Sancho, M., A. García-Pascual, and D. Triguero. 2012. Presence of the Ca^{2+}-activated chloride channel anoctamin 1 in the urethra and its role in excitatory neurotransmission. *Am J Physiol Renal Physiol* 302 (3): F390–F400. doi:10.1152/ajprenal.00344.2011.
114. Namkung, W., P.-W. Phuan, and A. S. Verkman. 2011a. TMEM16A inhibitors reveal TMEM16A as a minor component of calcium-activated chloride channel conductance in airway and intestinal epithelial cells. *J Biol Chem* 286 (3): 2365–2374. doi:10.1074/jbc.M110.175109.
115. Oh, S.-J., S. J. Hwang, J. Jung, K. Yu, J. Kim, J. Y. Choi, H. Criss Hartzell, E. J. Roh, and C. Justin Lee. 2013. MONNA, a potent and selective blocker for transmembrane protein with unknown function 16/anoctamin-1. *Mol Pharmacol* 84 (5): 726–735. doi:10.1124/mol.113.087502.
116. Davis, A. J., J. Shi, H. A. T. Pritchard, P. S. Chadha, N. Leblanc, G. Vasilikostas, Z. Yao, A. S. Verkman, A. P. Albert, and I. A. Greenwood. 2013. Potent vasorelaxant activity of the TMEM16A inhibitor $T16A_{inh}$ -A01. *Br J Pharmacol* 168 (3): 773–784. doi:10.1111/j.1476-5381.2012.02199.x.
117. Boedtkjer, D. M. B., S. Kim, A. B. Jensen, V. M. Matchkov, and K. E. Andersson. 2015. New selective inhibitors of calcium-activated chloride channels—$T16A_{inh}$-A01, $CaCC_{inh}$ -A01, and MONNA—what do they inhibit? *Br J Pharmacol.* doi:10.1111/bph.13201.
118. Wang, B., C. Li, R. Huai, and Z. Qu. 2015. Overexpression of ANO1/TMEM16A, an arterial Ca^{2+}-activated Cl^- channel, contributes to spontaneous hypertension. *J Mol Cell Cardiol* 82: 22–32. doi:10.1016/j.yjmcc.2015.02.020.
119. Seo, Y., H. K. Lee, J. Park, D.-K. Jeon, S. Jo, M. Jo, and W. Namkung. 2016. Ani9, a novel potent small-molecule ANO1 inhibitor with negligible effect on ANO2. *PLoS One* 11 (5): e0155771. doi:10.1371/journal.pone.0155771.
120. Namkung, W., Z. Yao, W. E. Finkbeiner, and A. S. Verkman. 2011b. Small-molecule activators of TMEM16A, a calcium-activated chloride channel, stimulate epithelial chloride secretion and intestinal contraction. *FASEB J* 25 (11): 4048–4062. doi:10.1096/fj.11-191627.
121. Burris, S. K., Q. Wang, S. Bulley, Z. P. Neeb, and J. H. Jaggar. 2015. 9-phenanthrol inhibits recombinant and arterial myocyte TMEM16A channels. *Br J Pharmacol.* doi:10.1111/bph.13077.
122. Guo, S., Y. Chen, C. Pang, X. Wang, J. Qi, L. Mo, H. Zhang, H. An, and Y. Zhan. 2017. Ginsenoside Rb1, a novel activator of the TMEM16A chloride channel, augments the contraction of guinea pig ileum. *Pflügers Arch* 469 (5–6): 681–692. doi:10.1007/s00424-017-1934-x.
123. Heinze, C., A. Seniuk, M. V. Sokolov, A. K. Huebner, A. E. Klementowicz, I. A. Szijártó, J. Schleifenbaum et al. 2014. Disruption of vascular Ca^{2+}-activated chloride currents lowers blood pressure. *J Clin Invest* 124 (2): 675–686. doi:10.1172/JCI70025.
124. Forrest, A. S., T. C. Joyce, M. L. Huebner, R. J. Ayon, M. Wiwchar, J. Joyce, N. Freitas, et al. 2012. Increased TMEM16A-encoded calcium-activated chloride channel activity is associated with pulmonary hypertension. *Am J Physiol Cell Physiol* 303 (12): C1229–C12243. doi:10.1152/ajpcell.00044.2012.
125. Sun, H., Y. Xia, O. Paudel, X.-R. Yang, and J. S. K. Sham. 2012. Chronic hypoxia-induced upregulation of Ca^{2+}-activated Cl^- channel in pulmonary arterial myocytes: A mechanism contributing to enhanced vasoreactivity. *J Physiol* 590 (Pt 15): 3507–3521. doi:10.1113/jphysiol.2012.232520.
126. Wang, M., H. Yang, L.-Y. Zheng, Z. Zhang, Y.-B. Tang, G.-L. Wang, Y.-H. Du et al. 2012. Downregulation of TMEM16A calcium-activated chloride channel contributes to cerebrovascular remodeling during hypertension by promoting basilar smooth muscle cell proliferation. *Circulation* 125 (5): 697–707. doi:10.1161/CIRCULATIONAHA.111.041806.

127. Greenwood, I. A., and N. Leblanc. 2007. Overlapping pharmacology of Ca^{2+}-activated Cl^- and K^+ channels. *Trends Pharmacol Sci* 28 (1): 1–5. doi:10.1016/j.tips.2006.11.004.
128. Piper, A. S., and I. A. Greenwood. 2003. Anomalous effect of anthracene-9-carboxylic acid on calcium-activated chloride currents in rabbit pulmonary artery smooth muscle cells. *Br J Pharmacol* 138 (1): 31–38. doi:10.1038/sj.bjp.0705000.
129. Wiwchar, M., R. Ayon, I. A. Greenwood, and N. Leblanc. 2009. Phosphorylation alters the pharmacology of Ca(2+)-activated Cl channels in rabbit pulmonary arterial smooth muscle cells. *Br J Pharmacol* 158 (5): 1356–1365. doi:10.1111/j.1476-5381.2009.00405.x.
130. Li, R.-S., Y. Wang, H.-S. Chen, F.-Y. Jiang, Q. Tu, W.-J. Li, and R.-X. Yin. 2016. TMEM16A contributes to angiotensin II-induced cerebral vasoconstriction via the RhoA/ROCK signaling pathway. *Mol Med Rep* 13 (4): 3691–3699. doi:10.3892/mmr.2016.4979.
131. Tian, Y., P. Kongsuphol, M. Hug, J. Ousingsawat, R. Witzgall, R. Schreiber, and K. Kunzelmann. 2011. Calmodulin-dependent activation of the epithelial calcium-dependent chloride channel TMEM16A. *FASEB J* 25 (3): 1058–1068. doi:10.1096/fj.10-166884.
132. Jung, J., J. H. Nam, H. W. Park, U. Oh, J.-H. Yoon, and M. G. Lee. 2013. Dynamic modulation of ANO1/TMEM16A HCO3(⁻) permeability by Ca^{2+}/calmodulin. *Proc Natl Acad Sci USA* 110 (1): 360–365. doi:10.1073/pnas.1211594110.
133. Vocke, K., K. Dauner, A. Hahn, A. Ulbrich, J. Broecker, S. Keller, S. Frings, and F. Möhrlen. 2013. Calmodulin-dependent activation and inactivation of anoctamin calcium-gated chloride channels. *J Gen Physiol* 142 (4): 381–404. doi:10.1085/jgp.201311015.
134. Chun, H., H. Cho, J. Choi, J. Lee, S. M. Kim, H. Kim, and U. Oh. 2015. Protons inhibit anoctamin 1 by competing with calcium. *Cell Calcium* 58 (5): 431–441. doi:10.1016/j.ceca.2015.06.011.
135. Sones, W. R., A. J. Davis, N. Leblanc, and I. A. Greenwood. 2010. Cholesterol depletion alters amplitude and pharmacology of vascular calcium-activated chloride channels. *Cardiovasc Res* 87 (3): 476–484. doi:10.1093/cvr/cvq057.
136. Pritchard, H. A. T., N. Leblanc, A. P. Albert, and I. A. Greenwood. 2014. Inhibitory role of phosphatidylinositol 4,5-bisphosphate on TMEM16A-encoded calcium-activated chloride channels in rat pulmonary artery. *Br J Pharmacol* 171 (18): 4311–4321. doi:10.1111/bph.12778.
137. Bulley, S., Z. P. Neeb, S. K. Burris, J. P. Bannister, C. M. Thomas-Gatewood, W. Jangsangthong, and J. H. Jaggar. 2012. TMEM16A/ANO1 channels contribute to the myogenic response in cerebral arteries. *Circ Res* 111 (8): 1027–1036. doi:10.1161/CIRCRESAHA.112.277145.
138. Cho, H., Y. D. Yang, J. Lee, B. Lee, T. Kim, Y. Jang, S. K. Back et al. 2012. The calcium-activated chloride channel anoctamin 1 acts as a heat sensor in nociceptive neurons. *Nat Neurosci* 15 (7): 1015–1021. doi:10.1038/nn.3111.
139. Ferrera, L., A. Caputo, I. Ubby, E. Bussani, O. Zegarra-Moran, R. Ravazzolo, F. Pagani, and L. J. V. Galietta. 2009. Regulation of TMEM16A chloride channel properties by alternative splicing. *J Biol Chem* 284 (48): 33360–33368. doi:10.1074/jbc. M109.046607.
140. O'Driscoll, K. E, R. A. Pipe, and F. C. Britton. 2011. Increased complexity of Tmem16a/anoctamin 1 transcript alternative splicing. *BMC Mol Biol* 12: 35. doi:10.1186/1471-2199-12-35.
141. Ma, K., H. Wang, J. Yu, M. Wei, and Q. Xiao. 2017. New insights on the regulation of Ca^{2+} -activated chloride channel TMEM16A. *J Cell Physiol* 232 (4): 707–716. doi:10.1002/jcp.25621.

3 Methods for the Isolation of and Study of Ca^{2+} Signaling in Arteriolar Smooth Muscle Cells

William F. Jackson

CONTENTS

3.1 INTRODUCTION

Arterioles in the microcirculation significantly contribute to peripheral vascular resistance, regulate the distribution of blood flow within a tissue, and participate in the regulation of capillary pressure and, hence, fluid balance.[1,2] These microvessels have internal diameters less than 75 μm and a single, circumferential layer of smooth muscle cells (SMCs). While much has been learned about the function of arteriolar SMCs from intravital microscopy (e.g.,[3–9]) and from pressure myography of isolated cannulated arterioles,[10–19] progress in understanding the vascular biology of arteriolar

SMCs has lagged behind that of vascular SMCs from larger vessels, in part because of the difficulty in obtaining single-cell model systems. For example, information concerning the presence, function, and regulation of plasma membrane and intracellular ion channels that contribute to Ca^{2+} signaling in vascular SMCs in the walls of arterioles remains largely unknown because of the lack of suitable single-cell models, despite established differences in Ca^{2+} signaling modalities between SMCs in feed arteries and SMCs in downstream arterioles.[15,16] Studies of membrane potential and gross, macroscopic currents can and have been performed on arterioles in situ[20] as well as in vitro.[21] However, interpretation of these potentials and currents is very complicated because SMCs and endothelial cells are coupled to each other by gap junctions[22,23] such that potentials and currents observed in one cell can be influenced by events in other cells of the same or different type. Furthermore, techniques such as dialysis of cells with regulatory molecules of large molecular weight (see[24] for example) cannot be performed in these multicellular preparations due to the limits imposed on cell-cell diffusion through gap junctions (see[23] for example). Thus, single-cell models are essential to complete the study of the physiology and biophysics of arteriolar SMCs.

Methods for the bulk isolation of arteriolar SMCs by enzyme perfusion of tissues such as brain,[25] kidney,[26] and heart[21] have been reported. While these methods have proven to be useful in the study of the cell biology and electrophysiology of arteriolar SMCs from these tissues, they lack the potential to study segmental differences among arteriolar SMCs from different orders of arterioles, because the vessels are isolated *en masse*. Thus, it is difficult to correlate the functional responses of single vascular SMCs isolated by these methods with the responses of distinct orders of arterioles in the intact microcirculation. Therefore, we set out to develop a method to enzymatically isolate relaxed and viable arteriolar SMCs from single, identifiable arterioles (internal diameters = 25–75 μm) removed from tissues that are routinely used for both intravital microscopy studies and *in vitro* pressure myography studies. We initially selected the cremaster muscle of rats and hamsters as the sources of arterioles because these systems have been used extensively to study the physiology[3–8,10–13] and pathophysiology[27–32] of the microcirculation in skeletal muscle. We succeeded in our efforts and have previously reported on the use of this method to study the functional expression of K^+ channels,[33–35] agonist-induced Ca^{2+} signals,[36] and expression of ion channels[16] in arteriolar SMCs. We have also extended this approach to the isolation and study of SMCs from several resistance arteries.[15,16,37]

3.2 METHODS FOR DISSECTION OF SINGLE ARTERIOLES

3.2.1 Equipment and Instrument Preparation

3.2.1.1 Dissection Table

We have modified and optimized the methods developed by Duling et al. for dissection of arterioles from hamster cheek pouches.[38] The dissection station (Figure 3.1) consists of a stable table of a suitable height (e.g., NPS-SLT2448; School Outfitters; Cincinnati, OH) such that the user can sit comfortably at the table with arms, wrists,

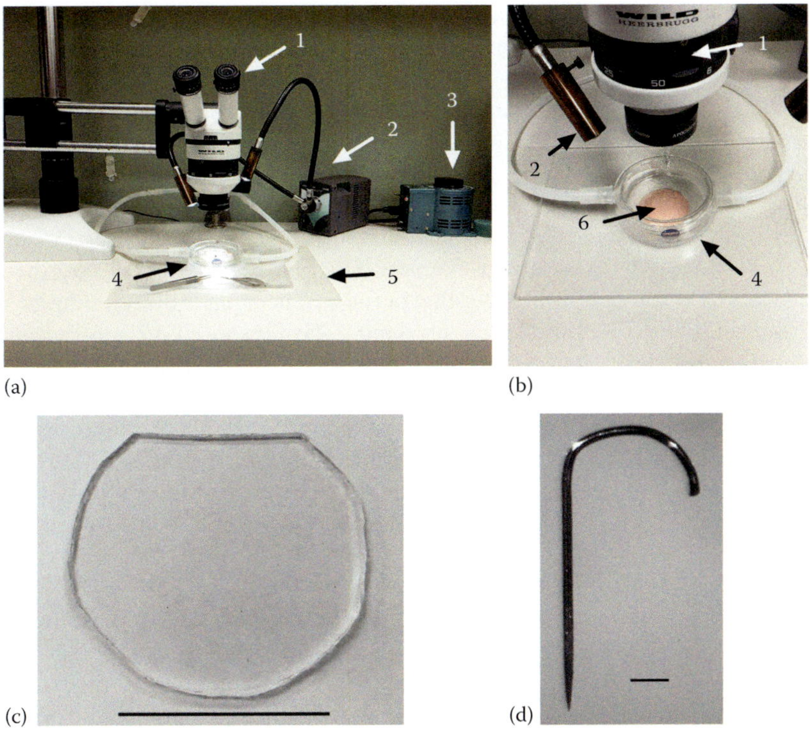

FIGURE 3.1 Dissection station components. Panel (a) shows an image of a typical dissection station used to isolate arterioles for enzymatic dissociation. The station consists of a research-grade stereo microscope mounted on a boom stand (#1). Incident illumination is provided by a fiber-optic light source with semi-rigid, goose-neck fiber optic light guides (#2). Transillumination is provided by a halogen light source mounted on the floor under the table (not shown), the intensity of which is controlled by a Variac transformer (#3). A water-jacketed dissection dish (#4) is glued to a piece of polycarbonate, as shown, and rests on another piece of polycarbonate (#5) that covers the transillumination hole in the table (#6 in Panel [b]). Panel (b) more clearly shows the hole in the table that is used for transillumination of preparations (#6). The numbers in Panel (b) refer to the same components as in Panel (a). Panel (c) shows a Sylgard pad that is inserted into the dissection chamber, and onto which tissues are pinned. Scale bar = 5 cm. Panel (c) shows an example of the pins used to immobilize tissues on the Sylgard pads shown in Panel (c). Scale bar = 1 mm. See text for more information.

and the outer surface of the hands resting on the surface of the table to eliminate tremor from the use of large muscle groups involved in supporting the arms. The table should have a water and chemical-resistant top through which a 5–8 cm diameter hole can be bored approximately 24 cm from the table's front edge at a lateral position that allows the user to rest both arms to the elbows on the table top (Figure 3.1). Using silicon adhesive, we adhere a 75 mm diameter opal diffusing glass (#46-662, Edmund optics; Barrington, NJ) to the underside of the table, covering the hole. The opal diffusing glass serves two purposes, the primary of which is to diffuse light from a transillumination light source to the preparations. The second is to seal the hole to prevent inadvertent fluid leakage from the table surface through the hole. Either a tempered glass plate or a polycarbonate sheet of appropriate dimensions (e.g., 28 × 36 cm; Figure 3.1) is placed on the table top to prevent solutions from entering the hole and to provide a surface that is transparent and can be easily cleaned.

3.2.1.2 Illumination

The hole in the table top serves as the light path for transillumination of the preparations from which arterioles are to be dissected. A simple goose neck halogen lamp (WU-41310-30; Cole-Parmer; Vernon Hills, IL) placed on the floor beneath the hole serves as the light source for this purpose. The lamp is powered by a Variac transformer (3PN1010B; ISE, Inc., Cleveland, OH; Figure 3.1). Transillumination is critical, particularly for dissecting second-order and smaller arterioles with maximal diameters of <50 μm. Transillumination is also used during instrument sharpening/polishing. Oblique illumination is provided by a goose-neck fiber-optic light source (LMI-6000; Dolan-Jenner Industries; Boxborough, MA; Figure 3.1). This can increase contrast and improve visibility at high magnifications required for dissection of small arterioles and sharpening/polishing surgical instruments.

3.2.1.3 Dissection Chamber

All dissections are performed at 4°C using a water-jacketed dissection dish (#158401; Radnoti Glass; Monrovia, Ca; Figure 3.1) coupled to a circulating cooling bath (Thermo Neslab RTE7; Thermo-Fisher) by suitable tubing and quick-connect couplers to allow the dissection dish to be removed for daily cleaning. The dissection dish is adhered to the center of a piece of 20 × 25 × 0.63 cm polycarbonate using silicone adhesive (Figure 3.1). This provides a stable base for the dish, allows the user's hands to weigh down the dish to prevent inadvertent movement, and reduces the chance of fluid spills from entering the hole cut in the table.

Tissues are pinned to a Sylgard pad placed in the bottom of the chamber. Placing a thoroughly dry pad into the dry chamber before cooling allows the pad to adhere to the glass bottom of the chamber and reduces pad movement once the chamber is filled with solution and cooled. Rather than filling the bottom of the chamber with Sylgard elastomer, we pre-form 0.5 cm thick Sylgard pads in disposable 8 cm diameter petri dishes, and trim these to size with scissors, a scalpel blade, or razor blade (Figure 3.1). This allows the pads to be thoroughly cleaned each day and placed in a 60°C drying oven overnight to dry. The drying oven step reduces microbe growth and allows the pads to be used for several weeks. We use stainless steel #000 insect pins (26001-25; Fine Science Tools [USA]; Foster City, CA) to pin tissues to the Sylgard pads. The pins are bent into a J-shape with the stem of the "J" approximately 0.5 cm long and cut to length with jeweler's wire cutters (Figure 3.1). This allows the pins to be easily handled with fine forceps and allows sufficient force to be applied to insert the pins into the Sylgard pads without slippage.

3.2.1.4 Dissection Instruments

Dissection of small arterioles from tissues such as cremaster muscles, hamster cheek pouch, or rodent mesenteries requires forceps and scissors with very fine tips. We purchase Dumont #5 forceps (11251-23; Fine Science Tools [USA]) and spring scissors

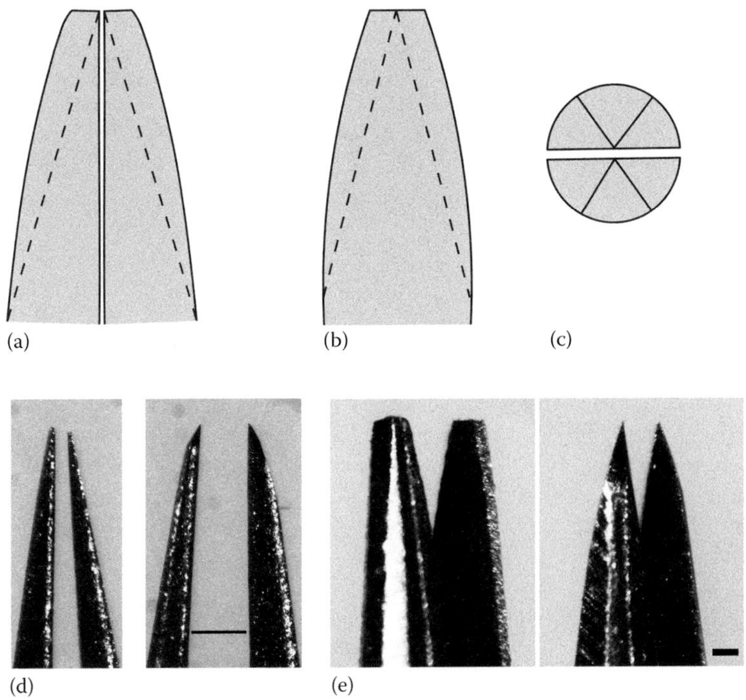

FIGURE 3.2 Dissection instruments. Panels (a–c) show schematics of the tip of forceps and how they are reformed using a sharpening stone and lapping paper. Panel (a) shows a side-view, and Panel (b) a top-view of the jaws of a pair of Dumont #5 forceps as purchased from a supplier. The dashed lines indicate the shape of the forceps after metal is ground away and the surface polished. Panel (c) shows an end-on view of the final shape of the forceps. Panel (d) shows images of the tips of forceps as purchased from a supplier (left panel) and after reshaping and polishing (right panel). Panel (e) shows images of Vannas-style scissors as purchased from a supplier (left panel) and after reshaping and polishing (right panel).

(15009-08; Fine Science Tools [USA]). Then, using a fine sharpening stone (29008-12; Fine Science Tools [USA]) and graded grit-sized lapping film (12, 3, and 0.3 μm grit size; 3-M St. Paul, MN), we shape the tips of the forceps and scissors as shown in Figure 3.2, sharpen the scissors, and polish the instruments to a mirror finish with the aid of a research-grade stereomicroscope (see section Dissecting Microscope, below). Forceps must be sharpened so that their tips meet at a fine point. This allows adventitial connective tissue to be grabbed and held without touching the underlying vascular SMCs. Similarly, the scissors must come to a fine point and must cut all the way to the tips so that adventitial connective tissue can be severed to release the vessels from the surrounding parenchymal cells and extracellular matrix. The instruments must be polished to a mirror finish (using 0.3 μm lapping film) to prevent them from sticking to the arterioles and connective tissue. The albumin included in the Dissection Solution (Table 3.1) also aids in minimizing tissue adherence to the forceps and scissors. The lapping film comes in adhesive-backed sheets that can be adhered to a glass

TABLE 3.1
Composition of Solutions Used to Enzymatically Isolate Arteriolar SMCs

	Dissection Solution	Dissociation Solution	PSS[a]
NaCl	140 mM	140 mM	135 mM
KCl	5 mM	5 mM	4 mM
$MgCl_2$	1 mM	1 mM	1 mM
$CaCl_2$	0	0.1 mM	1.8 mM
HEPES[b]	10 mM	10 mM	10 mM
Glucose	10 mM	10 mM	10 mM
Albumin	1 mg/mL	1 mg/mL	0
Sodium nitroprusside	10 μM	10 μM	0
Diltiazem	10 μM	10 μM	0
pH @ 20°C	7.4 with NaOH	7.4 with NaOH	7.4 with NaOH

[a] PSS = Physiological salt solution.
[b] HEPES = 4-(2-Hydroxyethyl)piperazine-1-ethanesulfonic acid, N-(2-Hydroxyethyl)piperazine-N′-(2-ethanesulfonic acid).

microscope slide. We typically put 3 μm film on one side and 0.3 μm film on the other side of the slide so that we can easily switch between the two during routine maintenance of the surgical instruments.

3.2.1.5 Dissecting Microscope

The most important component of the system to dissect arterioles is the stereomicroscope used to image the vessels during dissection and surgical instruments during sharpening and polishing. The microscope should have adjustable magnifications from 10X to at least 50X (higher is better), and sufficient working distance (~10 cm) to allow comfortable hand placement under the scope during dissection without contact with the microscope and allow the user to sit comfortably at the dissections station with arms resting on the table surface, and without shoulder strain. We have used both Wild M5A (Figure 3.1) and Nikon SMZ800 microscopes for these purposes; any similar microscope with research-grade optics, light transmission, working distance, and magnification range should be acceptable. We prefer to mount the dissecting scopes on boom stands (Figure 3.1) to allow free access under the microscope and to allow ease of positioning the microscope over the dissection chamber. Sufficient working distance is also required for sharpening and polishing forceps and scissors.

3.2.2 Dissection of Arterioles from Cremaster Muscles

On the day of an experiment, Dissection Solution (Table 3.1) is placed in the dissection chamber and the cooling, circulating bath turned on and allowed to cool to 4°C (approximately 30 min). Cremaster muscles from male hamsters, mice, or rats are prepared by methods similar to those described for intravital microscopy of this tissue and will not be covered here.[3,39,40] Once freed from the testis and overlying skin, the proximal portion of the muscle is severed from the animal, the muscle is placed in the Dissection Solution-filled chamber (Figure 3.1) and pinned at the edges with J-shaped pins (Figure 3.1) such that the muscle fibers are uniformly stretched. Sufficient tension is applied so that sarcomeres appear straight and perpendicular to the long axis of the skeletal muscle fibers (not wavy), as they are in vivo.

We have found that exclusion of Ca^{2+} from the Dissection Solution and inclusion of vasodilators (10 μM diltiazem and 10 μM sodium nitroprusside) are essential for maintaining the arteriolar SMCs in the cremaster muscles (or other tissues) in a relaxed state. Because the vasculature in the excised tissue is not pressurized, there is no distending force on the SMCs. Thus, if the SMCs become activated during the dissection due to damage to skeletal muscle fibers (unavoidable issue that floods the area around the damaged site with K^+ and Ca^{2+}), or due to direct mechanical stimulation of the arteriolar SMCs, the cells will contract and remain in a shortened state when enzymatically isolated from the arteriolar wall. The low Ca^{2+} concentration in the Dissection Solution (~1 μM), along with the vasodilators reduces this problem. Bovine serum albumin (BSA, 1 mg/mL; 10856; Affymetrix, Cleveland, OH) is included in the Dissection Solution to reduce tissue adherence to the surgical instruments and to reduce the potential for proteolytic degradation of membrane proteins due to release of enzymes from tissues damaged during the dissection process. It should be noted that we have tried several BSA preparations and have found that the preparation from Affymetrix (formerly US Biochemicals) is the only product that was not, on its own, vasoactive. Even with the low Ca^{2+}, vasodilators, and BSA, care must be taken to limit stimulation of the arteriolar SMCs during dissection. We have found that longitudinal stretch of the arterioles is better tolerated than radial distortion.

Using 12–25X magnification (dependent on the visual acuity of the surgeon), dissection of arterioles begins. A bundle of skeletal muscle fibers that run over the arteriole in question is selected, grasped with fine forceps at least 10 arteriole diameters away from the arteriole, and severed. If sufficient tension was applied to the preparation, the severed muscle fibers will retract beyond the arteriole and expose the upper surface of the vessel. This procedure is repeated along the length of the arteriole until a sufficient length has been exposed for use in the enzymatic isolation process (typically, 1–2 mm lengths of vessels). Care must to taken during this step not to grab and sever muscle fibers that lie in a plane below the arteriole as this will make successful excision difficult.

Once a sufficient length of arteriole has been exposed, the adventitial connective tissue on the upper surface of the arteriole is grabbed with the very tips of fine forceps (Figure 3.2) and lifted upwards. With the scissor blades open, the tips of the scissors are inserted under the vessel and the connective tissue cut, using just the tips of the scissor blades to nibble away at the connective tissue. With continued upward tension on the vessel via the connective tissue, this process is continued along the length of the vessel proximal and distal to the mid-point. The entire preparation is then rotated 180° (by freeing the Sylgard pad from the glass bottom of the chamber and rotating the pad). The connective tissue adjacent to the midpoint of the arteriole is grasped with the tips of the forceps, upward tension applied and the connective tissue along the other side of the vessel and under the vessel severed to free the entire length of the arteriole. To provide better optics for visualization of the adventitial connective tissue and to allow additional removal of excess connective tissue, the

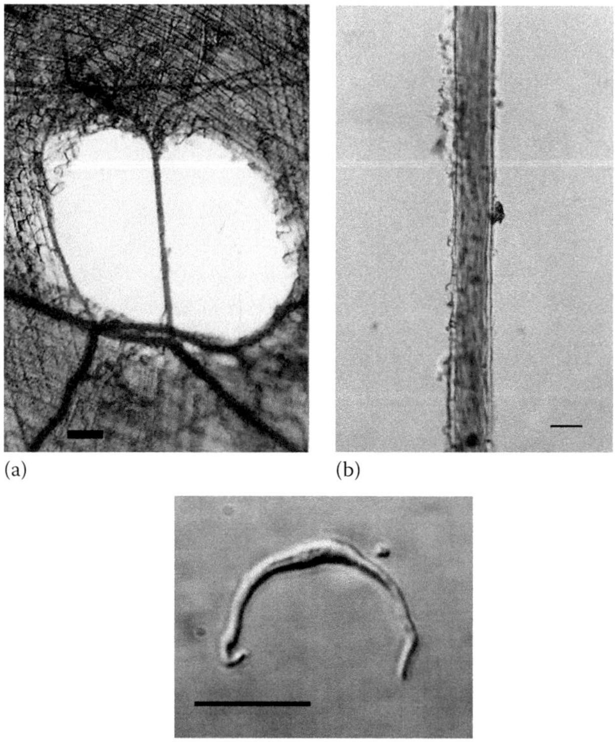

(a) (b) (c)

FIGURE 3.3 Dissection of cremaster arterioles. Panel (a) shows an image of an aparenchymal arteriolar segment spanning a hole cut in a hamster cremaster muscle after dissection of the skeletal muscle fibers away from the arteriole. Scale bar = 100 μm. Panel (b) shows an image of the arteriole shown in Panel (a) at a higher magnification. Scale bar = 25 μm. Panel (c) shows a contrast-enhanced image of a smooth muscle cell isolated from a cremaster arteriole using the methods described in the text. The smaller object is a red blood cell. Scale bar = 50 μm.

skeletal muscle fibers under the arteriole are cut to produce a hole in the cremaster muscle spanned by the vessel (Figure 3.3). An image of a well-cleaned cremaster arteriole is shown in Figure 3.3b, demonstrating that vessels down to ~25 μm (maximal diameter) can be isolated using this approach. At this point, the arteriole can be severed from the cremaster at its proximal and distal ends. We also use this dissection technique to prepare arterioles for pressure myography experiments.[15,16,41,42] For enzymatic isolation, arterioles are cut into 500–800 μm lengths, and these vessel segments are transferred to a suitable container for short-term (2–4 h) storage at 4°C. We use 2 mL autoanalyzer cups (product number 02-544-131, Fisher Scientific) for this purpose as they are transparent, fit well under a stereomicroscope, and are wide enough to allow access for retrieval of the arteriole for later use. For enzymatic dissociation, arterioles are transferred to a 12 × 75 mm borosilicate glass culture tube containing 1 mL of Dissociation Solution (Table 3.1) for enzymatic dissociation. The arteriolar segments are transferred using Drummond Wiretrol II positive displacement pipettes (50–100 μL volume; Drummond, Broomal, PA), with transfer verified under a stereomicroscope. The glass barrels of the pipettes are coated with albumin by aspirating a small volume of Dissection Solution (Table 3.1) containing 10% BSA just prior to their use.

3.3 METHODS FOR ENZYMATIC DISSOCIATION OF SINGLE ARTERIOLES

After dissection, the arteriolar segments are incubated in Dissociation Solution (Table 3.1) for 10 min at room temperature (20°C–25°C). All but 100–200 μL of this solution is then aspirated using a 1000 μL Eppendorf-style pipettor. During the aspiration process, the segments are visualized in the bottom of the tube by holding the tube up to the overhead room lights. One milliliter of Dissociation Solution containing 0.62 mg/mL papain (10–21 U/mg protein; P-4762; Sigma, St. Louis, MO) and 1 mg/mL dithioerythritol (D-8255; Sigma) at room temperature is carefully added to the tube to not excessively disturb the vessel segments. The tube is incubated for 35 min in a dry-bath heater (Curtin Matheson Scientific, Inc., Woodale, IL) at 37°C. At the end of this interval, the papain-containing solution is carefully removed with verification that the segments are still present, and this solution replaced with 1 mL of Dissociation Solution containing 1.5 mg/mL collagenase (Sigma Blend Type H, ≥1.0 FALGPA units/mg solid), 22.5–45 U/mL elastase (324682-EMD Millipore through Sigma), and 1 mg/mL soybean trypsin inhibitor (Type 1-S; T-9003, Sigma), and the arteriolar segments are further incubated for 10–25 min at 37°C. The duration of incubation in the collagenase/elastase/trypsin inhibitor-solution is determined by trial and error and tends to vary with batches of animals and lots of enzymes. Mouse cremaster arterioles require a lower concentration of elastase (22.5 U/mL) than do hamster or rat arterioles (45 U/mL): higher concentrations of elastase lead to SMC activation and loss of ability to isolate relaxed cells. We have found that elastase is an essential enzyme in the process of cell isolation. Lack of inclusion of this enzyme results in lack of release of SMCs from the internal and external elastic lamina found in these vessels.[43]

At the end of the second incubation period, approximately 1 mL of the collagenase/elastase/trypsin inhibitor-containing solution is removed, with care being taken not to disturb the arteriolar segments. With the culture tube held at a 45° angle, 4 mL of ice-cold Dissociation Solution is slowly added to dilute the remaining enzymes and slow their reaction. The segments are then allowed to settle in the tube for approximately 10 min. All but 100–200 μL of the Dissociation Solution are then removed, and 0.5–1 mL of room temperature Dissociation Solution is rapidly added using a 1000 μL pipettor to disperse cells from the digested vessel. This is a critical step in the isolation process: under trituration results in lack of release of SMCs from the extracellular matrix of the arteriolar wall, excessive trituration activates the SMCs causing them to contract such that only myoballs (excessively contracted SMCs) are isolated. This arteriolar SMC-containing solution is then triturated once or twice using a 100–1000 μL Eppendorf-style pipettor set to 1000 μL, and the solution transferred to a 1.5 mL siliconized-polypropylene microcentrifuge tube for storage at room temperature for up to 4 h.

3.4 THE STUDY OF Ca^{2+} SIGNALING IN ARTERIOLAR SMOOTH MUSCLE CELLS

We have used SMCs isolated by the methods outlined earlier for patch-clamp,[33–36,41] Ca^{2+} imaging,[36,42] expression[16,42] and single SMC contraction studies,[33–36] all of which can be directed at the study of Ca^{2+} signaling in arteriolar SMCs.

3.4.1 Patch-Clamp

For patch-clamp experiments, 100–200 μL of SMC-containing solution is pipetted onto the clean, cover-glass bottom of a recording chamber using a Wiretrol II pipettor. The cells are allowed to settle for 10–15 min and then are superfused with physiological salt solution (PSS) containing 1.8 mM Ca^{2+}. Standard whole-cell or perforated-patch recording methods are then applied to study membrane currents as described.[33–37,41]

We have used isolated cremasteric arteriolar SMCs and whole-cell patch-clamp methods to study a variety of K^+ channels that control SMC membrane potential, and hence the open-state probability of voltage-gated Ca^{2+} channels (VGCCs), including voltage-gated K^+ channels,[33] large-conductance Ca^{2+}-activated K^+ channels,[34] ATP-sensitive K^+ channels[35] and inward-rectifier K^+ channels.[41] Using patch-clamp techniques we have also studied Ba^{2+} currents through VGCCs in arteriolar SMCs, have shown that nifedipine-sensitive L-type VGCC are the dominant VGCC in cremasteric arteriolar SMCs[36] (Figure 3.4), and that, for example, hypoxia does not inhibit currents through these channels.[36] We have also shown that norepinephrine increases the magnitude of Ba^{2+} currents through VGCC in these cells (Figure 3.4), supporting a major role for VGCC in the mechanism of action of this catecholamine in cremaster arteriolar SMCs.

In the hamster cheek pouch, we have shown that arteriolar constriction induced by elevated PO_2 is mediated by cysteinyl leukotrienes and activation of nifedipine-sensitive VGCC (see [44] for review). Therefore, we have also studied nifedipine-sensitive Ba^{2+} currents through VGCC in SMCs isolated from hamster cheek pouch arterioles and have shown that vasoconstrictor cysteinyl leukotrienes, such as LTD_4, augment currents through these channels (Figure 3.4). These data are consistent with a role for VGCC in the mechanism of action of these inflammatory mediators on arterioles in the peripheral microcirculation, and further demonstrate the feasibility of studying Ca^{2+} currents in arteriolar SMCs.

3.4.2 Ca^{2+} Imaging

Isolated cremasteric arteriolar SMCs also have been used to study global Ca^{2+} signals induced by norepinephrine,[36] and the role-played by VGCCs in this process. Enzymatically isolated arteriolar SMCs do not adhere firmly to glass, so to eliminate movement artifacts during the study of agonist-induced Ca^{2+} transients, we have used Cell-Tak tissue adhesive (Corning, Bedford, MA) to attach SMCs to pieces of glass coverslips (~5 × 10 mm). The coverslip pieces are coated

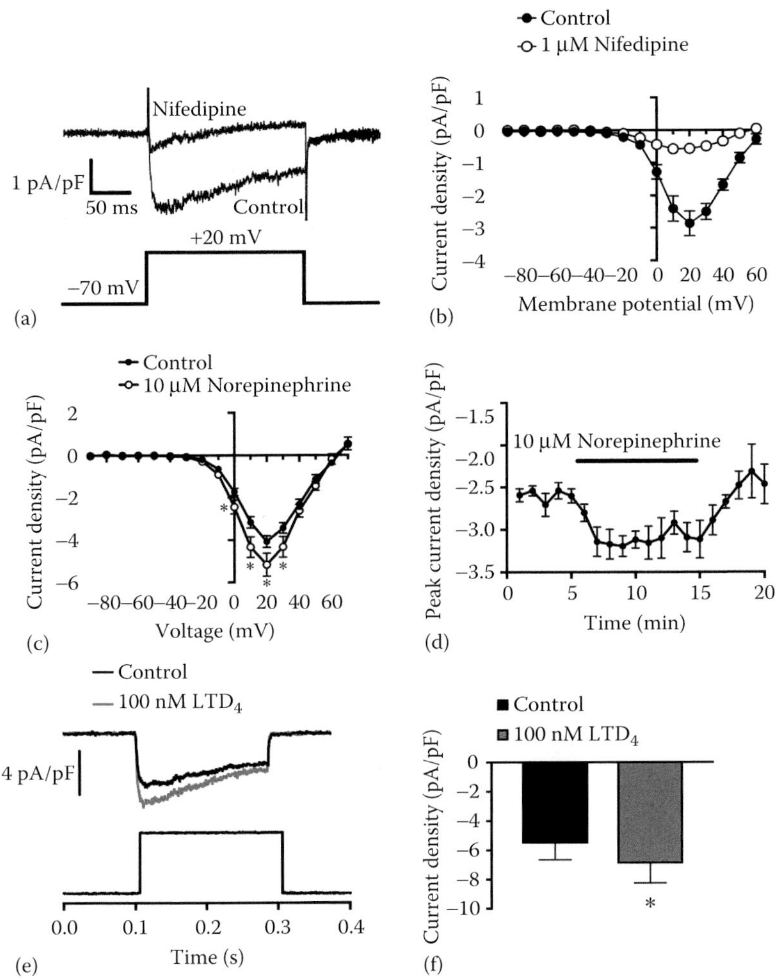

FIGURE 3.4 Ca^{2+} currents in arteriolar SMCs. Shown are examples of the use of enzymatically isolated arteriolar SMCs to study the function of VGCCs, and by inference, other Ca^{2+} permeable channels by patch-clamp methods. Panel (a) shows typical records of nifedipine-sensitive currents using 10 mM Ba^{2+} as the charge carrier. Current was elicited by stepping the holding potential from −70 mV to +20 mV for 200 ms. Nifedipine (1 μM) substantially inhibited the elicited current, demonstrating the presence and function of L-type VGCCs in cremasteric arteriolar SMCs. Panel (b) shows summary data for experiments as in (a), showing means ± SEM ($n = 8$) current densities in the absence and presence of nifedipine, as indicated. (Data in Panels [a] and [b] are from Cohen, K.D. and Jackson, W.F., *Microcirculation*, 10, 133–141, 2003. With permission.) Panel (c) shows mean ± SEM ($n = 9$) current densities for nifedipine-sensitive Ba^{2+} currents in cremasteric arteriolar muscle cells in the absence and presence of 10 μM norepinephrine, as indicated. Panel (d) shows the time course of the increase in current elicited by stepping membrane potential from −70 to +20 mV upon application of 10 μM norepinephrine. Panel (e) shows typical records of currents using 10 mM Ba^{2+} as the charge carrier as in A, in a hamster cheek pouch arteriolar SMC, before and after exposure to 100 nM leukotriene D_4 (LTD_4). Panel (f) shows summary means ± SEM ($n = 3$) for peaks current densities from experiments as in Panel (e). * = $p < 0.05$.

with Cell-Tak according to the manufacturer's protocol. They are then placed in the bottom of a recording chamber. A 100–200 μL aliquot of cell-containing solution is then pipetted onto the Cell-Tak-coated surface, the cells allowed to adhere for ~15 min. The chamber is then perfused with PSS containing 5 μm Fura 2-acetoxymethylester (AM; ThermoFisher Scientific, Waltham, MA). We include 0.1% BSA in this solution to facilitate solubility of Fura 2-AM, because we have found that arteriolar SMCs are intolerant of pluronic F-127 that is often used for this purpose. The SMCs are incubated in the Fura 2-AM solution for 30 min, then superfused with fresh Fura 2-AM solution and incubated for an additional 30 min. The chamber is then perfused with PSS for 30 min to wash away bath Fura 2-AM and allow the cells to deesterify the Fura 2. The SMCs are then illuminated, sequentially with 340 and 380 nm light using a DeltaRam X high speed monochromator (Photon Technologies International, Birmingham, NJ) coupled to a Nikon TE 300 inverted microscope (Nikon Instruments Inc., Melville, NY, USA) equipped with a X40 (N.A. 0.75) Plan Fluor long working-distance objective, emitted 510 nm light is collected by either a D-104 photomultiplier photometer system (Photon Technologies International) using Felix software (Photon Technologies International) or an intensified CCD-camera (XR-Mega10, Stanford Photonics, Palo Alto, CA) using Piper software (Stanford Photonics).

Using this approach, we have shown that norepinephrine elicits typical Ca^{2+} transients from single cremasteric arteriolar SMCs (Figure 3.5), consisting of a rapid peak that decays to a steady plateau. The VGCC blocker, diltiazem, blunts the peak, and abolishes the plateau supporting a major role for VGCCs in the Ca^{2+} transients

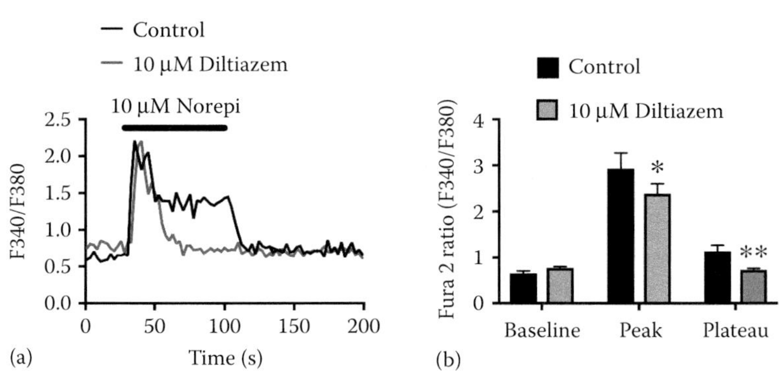

FIGURE 3.5 Norepinephrine-induced Ca^{2+} transients in arteriolar SMCs. Panel (a) shows a typical record of global Ca^{2+} transients in a single cremasteric arteriolar SMC elicited by exposure of the cell to 10 μM norepinephrine in the absence and presence of the L-type VGCC blocker, diltiazem, recorded using Fura 2 as described. (From Cohen, K.D. and Jackson, W.F., *Microcirculation*, 10, 133–141, 2003.) Diltiazem had only a small effect on the peak of the Ca^{2+} transient but abolished the plateau phase. Panel (b) shows summary means ± SE ($n = 5$) for Fura 2 ratios before (baseline), at the peak and during the plateau-phase of norepinephrine-induced Ca^{2+} transients as shown in Panel (a). * = significantly different from Control, $p < 0.05$. ** = significantly different from Control Plateau, but not significantly different from Control or diltiazem baseline ratios, $p < 0.05$.

induced by this catecholamine (Figure 3.5). We have also used this approach to show that hypoxia has no effect of norepinephrine-induced Ca^{2+} transients in cremasteric arteriolar SMCs.[36]

Enzymatically-isolated cremaster arteriolar SMCs do not display ryanodine-receptor-mediated Ca^{2+} sparks[45,46] that have been reported in SMCs isolated from, for example, cerebral arteries (~200 μm internal diameter)[46] and intact piglet cerebral resistance arteries and arterioles (200–50 μm).[47] Lack of Ca^{2+} sparks in cremasteric arteriolar SMCs does not appear to be an artefact of the isolation procedure, because SMCs in intact cremasteric arterioles (50–25 μm) studied by pressure myography also do not display ryanodine-sensitive Ca^{2+} sparks.[15,16] Lack of Ca^{2+} sparks in SMCs in intact cerebral parenchymal arterioles (~20 μm) also has been reported,[48] although in this instance, acidosis resulted in recruitment of Ca^{2+} spark activity. These examples highlight that differences in Ca^{2+} signaling modalities exist between arterial and arteriolar SMCs and likely between arteriolar SMCs from different vascular beds. Regional heterogeneity in Ca^{2+} signaling mechanisms, as we have reported in SMCs of cremaster arterioles versus SMCs in upstream feed arteries,[15,16] provide a strong rationale for the application of methods to specifically study arterioles in the microcirculation. Studies utilizing total internal reflectance fluorescence (TIRF) to image Ca^{2+} sparklets in enzymatically isolated arteriolar SMCs have yet to be reported, but such studies are encouraged.

3.4.3 Ion Channel Expression

We have used enzymatically isolated arteriolar SMCs to explore the cell-specific expression of mRNA for ion channels involved in arteriolar Ca^{2+} signaling.[16] After enzymatic digestion of arterioles or resistance arteries, the solution containing SMCs is placed into a chamber on the stage of an inverted microscope and 50 SMCs are aspirated into borosilicate glass capillary tubes (1.0 mm OD/0.58 mm ID; # 1B100-3; World Precision Instruments [WPI], Sarasota, FL) with heat-polished tips (ID, 40–50 μm) using a Nanoliter Injector (#A203XVY; WPI) and a MICRO1 controller (WPI), and then ejected into a nuclease-free microfuge tube containing 50 μL of lysis buffer from a TaqMan® PreAmp Cells-to-CT™ Kit (Applied Biosystems, Foster City, CA, USA, no. 4387299). The cells are then processed per manufacturer's instructions using TaqMan Gene Expression Master Mix (no. 4369016) and inventoried TaqMan Gene Expression assays. All samples are run in triplicate (14 cycles pre-amp + 40 cycles of amplification) on an ABI 7500 Thermocycler as instructed by the manufacturer, including no reverse transcriptase controls.[16]

Using this approach, we have shown (Figure 3.6a) that: (1) cremaster arteriolar SMCs express predominantly ryanodine receptor (RyR) 2, less RyR3 and no RyR1, (2) arteriolar muscle cell RyR2 expression is less than that in SMCs from upstream feed arteries, (3) the ratio of mRNA expression for RyR2/RyR3 is greater in arteriolar SMCs than in SMCs from upstream feed arteries, and (4) this expression profile may help, in part, to explain the lack of Ca^{2+} sparks observed in cremaster arteriolar SMCs (see Ca^{2+} imaging Section for more information).[16] Similarly, we compared the mRNA expression of IP_3 receptors in cremaster arteriolar SMCs with SMCs

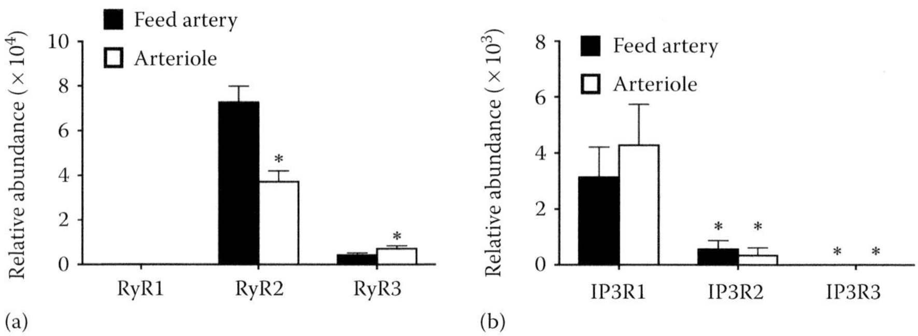

FIGURE 3.6 Heterogeneity in the expression of mRNA for ryanodine receptors but not IP_3 receptors between arteriolar and feed artery SMCs. Data are mean relative abundances ± SEM for ryanodine receptors (RyR) 1–3 (Panel [a]) or IP_3 receptors 1–3 (Panel [b]) relative to the expression of message for α-smooth muscle actin in samples of 50 SMCs. In Panel (a), * = $p < 0.05$ versus expression of same isoform in feed artery SMCs, $n = 6–8$ samples from the same number of mice. In Panel (b), * = $p < 0.05$ versus expression of IP_3R1 in the same vessel type, $n = 4–5$ samples from the same number of mice. (Data in Panels [a] and [b] are from Westcott, E.B. et al., *J Physiol.*, 590, 1849–1869, 2012. With permission.)

from an upstream feed artery and showed that the expression of this Ca^{2+}-release channel does not significantly differ between the two SMCs (Figure 3.6b).[16]

Isolated arteriolar SMCs can also be used to examine protein expression using immunolocalization approaches, as shown in Figure 3.7. Cells are isolated, as described in Section 3.3, except that the BSA is eliminated from all steps to reduce background staining and improve cell adhesion. Aliquots (~300 µL) of SMC-containing solution are then applied to coated slides (Cytoslides, ThermoFisher) using a CytoSpin 4 cytocentrifuge (ThermoFisher) using a slow spin rate (750 rpm) to limit mechanical activation of the SMCs. The cells then are fixed with 4% paraformaldehyde in phosphate-buffered saline, blocked with donkey serum (10%), permeabilized (0.3% Triton-X 100) and immunolabeled with appropriate primary and secondary antibodies using standard methods. We have used this approach to examine association of Ca_V 1.2, L-type VGCCs with large-conductance Ca^{2+}-activated K^+ (BK_{Ca}) channels in cremaster arteriolar SMCs as shown in Figure 3.7. We have consistently found a high degree of association of Ca_V 1.2 and the α-subunit of BK_{Ca} channels: analysis of cytofluorograms (Figure 3.7c) of similar images from 4 animals (7 cells analyzed) showed significant correlations (Pearson's correlation coefficient = 0.76 ± 0.03 [mean ± SE, $p < 0.05$ for each]), suggesting that Ca_V 1.2 VGCCs and BK_{Ca} channels may be colocalized. This finding supports functional and Ca^{2+} imaging data that we have reported, which suggest that Ca^{2+} influx through Ca_V 1.2 VGCCs is the primary source of Ca^{2+} for activation of BK_{Ca} channels in cremaster SMCs,[15,16] in contrast to ryanodine receptor mediated Ca^{2+} sparks that control BK_{Ca} channel activity in SMCs from, for example, cerebral arteries.[45,46] Application of higher resolution approaches to study the cellular localization of Ca^{2+} channels and accessory proteins in cremaster arteriolar SMCs, such as proximity ligation assays[49] or immunofluorescence resonance energy transfer (immuno-FRET),[50] have yet to be reported. However, for

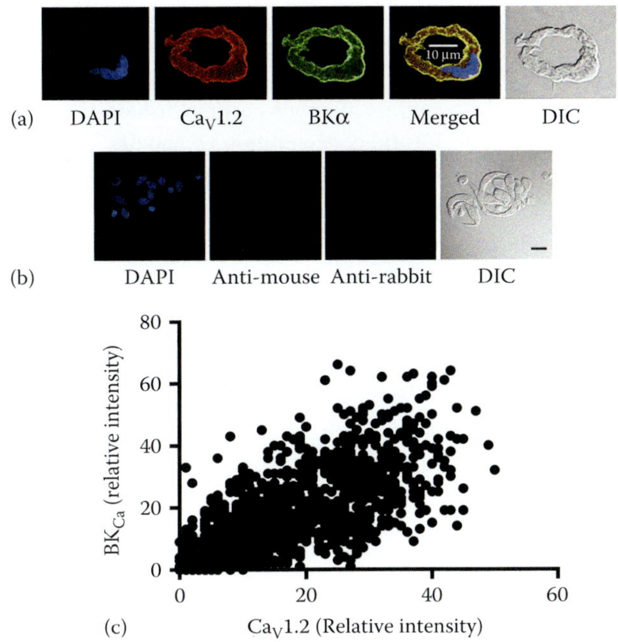

FIGURE 3.7 Immunolabeling and colocalization of $Ca_V1.2$ VGCCs and BK_{Ca} channels in cremaster arteriolar SMCs. Panel (a) shows a single SMC in which the nucleus was stained with DAPI (4′,6- diamidino-2-phenylindole) and immunostained with an anti-Ca_V 1.2 1° antibody (mouse monoclonal, NeuroMab/Antibodies Incorporated, Davis, CA) and 1° antibodies against the α-subunit of BK_{Ca} channels (rabbit polyclonal, Alamone Labs, Jerusalem, Israel), as indicated. The scale bar shown in the merged image applies to all images in Panel (a). The merged image is an overlay of the first three images, with yellow representing regions of colocalization. DIC = differential interference contrast image of the SMC. Panel (b) shows an example of a control experiment in which SMCs were exposed only to the 2° antibodies directed at either mouse or rabbit antibodies, as indicated, demonstrating lack of non-specific staining. Scale bar in the DIC image is 10 μm and applies to all images in Panel (b). Panel (c) shows the cytofluorogram for the membrane region of the immunostained SMC shown in Panel (a). The Pearson correlation-coefficient for this data set was 0.78 ($p < 0.0001$), suggesting significant colocalization of Ca_V 1.2 and BK_{Ca} channels in this cell. Analysis of cytofluorograms of similar images from 4 animals (7 cells analyzed) all showed significant correlations (Pearson's correlation coefficient = 0.76 ± 0.03; mean ± SE, $p < 0.05$ for each).

example, proximity ligation assays have been effectively used to study receptor protein localization in enzymatically isolated cremaster arteriolar SMCs,[51] demonstrating the feasibility of this approach in arteriolar SMCs.

3.4.4 Contraction of Single Smooth Muscle Cells

Agonist-induced contraction can also be studied using enzymatically isolated arteriolar SMCs. After isolation, a solution of SMCs is placed in a suitable chamber on the stage of an inverted microscope, the cells are imaged and agonists applied to the surface of single SMCs using pressure-ejection from an agonist-filled micropipette as shown in Figure 3.8. Because the SMCs are not tethered via the extracellular matrix and adjacent cells, we have found that they contract in an all-or-none fashion (Figure 3.8), with cells displaying a distinct threshold for the concentration of agonist that is required to produce contraction. This allows population concentration–contraction relationships to be

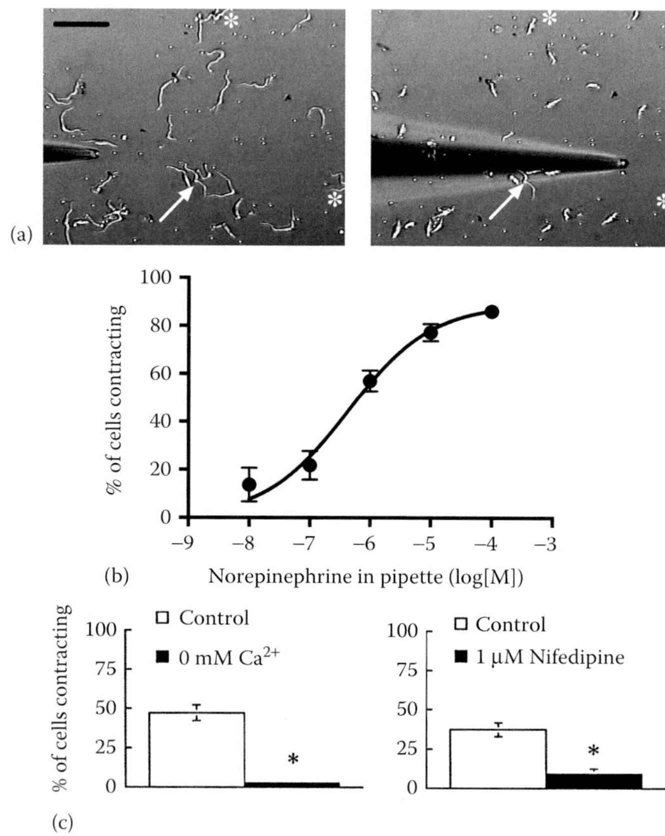

FIGURE 3.8 Contraction assays with isolated arteriolar SMCs. Panel (a) shows images of isolated arteriolar SMCs before (left panel) and after (right panel) pipette application of 10 μM norepinephrine. The white arrow points to a cell that did not contract. The white asterisks label two cells that contracted but were blown from the field of view of the camera. Scale bar = 100 μm. Panel (b) is reproduced from Jackson et al. and shows data from experiments similar to that shown in Panel (a) using various concentrations of norepinephrine in the pipette. The E_{max} is 89 ± 8% of cells contracting and the log EC_{50} is −6.4 ± 0.2. The *n*-values refer to the number of isolates studied at each concentration of norepinephrine. Panel (c) shows results of contraction assay experiments from Jackson et al., demonstrating that exposure of cells to solutions containing nominally 0 Ca^{2+} (left panel) or 1 μM nifedipine (right panel) inhibits contraction of cells to pipette application of 1 μM norepinephrine. * = $p < 0.05$, $n = 6$ cell isolates. (Reproduced from Jackson, W.F. et al., *Microcirculation*, 4, 35–50, 1997. With permission.)

constructed as shown in Figure 3.8b. We have used these single SMC contraction assays to show that, for example, norepinephrine-induced contraction of cremasteric arteriolar SMCs is inhibited by removal of extracellular Ca^{2+} or treatment with nifedipine, demonstrating the essential function of L-type VGCCs in norepinephrine-induced contraction of these SMCs.[33]

3.5 SUMMARY

The ability to isolate SMCs from defined arterioles in the microcirculation has allowed us to study expression and localization of channels that conduct Ca^{2+}, currents through Ca^{2+} channels, global and local Ca^{2+} signals, and contraction of these cells. Application of these approaches has demonstrated that arteriolar SMCs have different properties than SMCs isolated from upstream resistance arteries. Understanding the molecular basis for this regional heterogeneity in expression and function may provide new targets to treat vascular pathologies in hypertension, diabetes, obesity and aging.

3.6 CONFLICT OF INTEREST

None

ACKNOWLEDGMENTS

Supported by HL32469, HL086483 and P01 HL070687. Special thanks to Dr. Kenneth D. Cohen (Figures 3.4 and 3.5), Dr. Erika M. Boerman (Figures 3.6 and 3.7) and Erika Lange (Figures 3.6 and 3.7) for generating the data depicted in the noted figures.

REFERENCES

1. Davis, MJ, Hill, MA and Kuo, L. Local regulation of microvascular perfusion. *Compr Physiol.* 2011: 161–284.
2. Laughlin, MH, Davis, MJ, Secher, NH, van Lieshout, JJ, Arce-Esquivel, AA, Simmons GH, Bender, SB, et al. Peripheral circulation. *Compr Physiol.* 2012; 2: 321–447.
3. Baez, S. An open cremaster muscle preparation for the study of blood vessels by in vivo microscopy. *Microvasc Res.* 1973;5: 384–394.
4. Jackson, WF. Regional differences in mechanism of action of oxygen on hamster arterioles. *Am J Physiol.* 1993;265: H599–H603.
5. Gorczynski, RJ, and Duling, BR. Role of oxygen in arteriolar functional vasodilation in hamster striated muscle. *Am J Physiol.* 1978;235: H505–H515.
6. Segal, SS, and Duling, BR. Communication between feed arteries and microvessels in hamster striated-muscle: Segmental vascular-responses are functionally coordinated. *Circ Res.* 1986;59: 283–290.
7. Saito, Y, Eraslan, A, Lockard, V, and Hester, RL. Role of venular endothelium in control of arteriolar diameter during functional hyperemia. *Am J Physiol Heart Circ Physiol.* 1994;267: H1227–H1231.
8. Meininger, GA, and Faber, JE. Adrenergic facilitation of myogenic response in skeletal muscle arterioles. *Am J Physiol.* 1991;260: H1424–H1432.
9. Brekke, JF, Jackson, WF, and Segal, SS. Arteriolar smooth muscle Ca^{2+} dynamics during blood flow control in hamster cheek pouch. *J Appl Physiol.* 2006;101: 307–315.

10. Ikeoka, K, and Faber, JE. ANG II reverses selective inhibition of alpha 2-adrenoceptor sensitivity after in vitro isolation of arterioles. *Am J Physiol.* 1993;265: H1988–H1995.
11. Meininger, GA, Zawieja, DC, Falcone, JC, Hill, MA, and Davey, JP. Calcium measurement in isolated arterioles during myogenic and agonist stimulation. *Am J Physiol.* 1991;261: H950–H959.
12. Tateishi, J, and Faber, JE. Inhibition of arteriole alpha 2- but not alpha 1-adrenoceptor constriction by acidosis and hypoxia in vitro. *Am J Physiol.* 1995;268: H2068–H2076.
13. Messina, EJ, Sun, D, Koller, A, Wolin, MS, and Kaley, G. Role of endothelium-derived prostaglandins in hypoxia-elicited arteriolar dilation in rat skeletal muscle. *Circ Res.* 1992;71: 790–796.
14. Cipolla, MJ, Sweet, J, Chan, SL, Tavares, MJ, Gokina, N, and Brayden, JE. Increased pressure-induced tone in rat parenchymal arterioles vs. middle cerebral arteries: Role of ion channels and calcium sensitivity. *J Appl Physiol.* 2014;117: 53–59.
15. Westcott, EB, and Jackson, WF. Heterogeneous function of ryanodine receptors, but not IP3 receptors, in hamster cremaster muscle feed arteries and arterioles. *Am J Physiol Heart Circ Physiol.* 2011;300: H1616–H1630.
16. Westcott, EB, Goodwin, EL, Segal, SS, and Jackson, WF. Function and expression of ryanodine receptors and inositol 1,4,5-trisphosphate receptors in smooth muscle cells of murine feed arteries and arterioles. *J Physiol.* 2012;590: 1849–1869.
17. Pires, PW, Jackson, WF, and Dorrance, AM. Regulation of myogenic tone and structure of parenchymal arterioles by hypertension and the mineralocorticoid receptor. *Am J Physiol Heart Circ Physiol.* 2015;309: H127–H136.
18. Matin, N, Fisher, C, Jackson, WF, and Dorrance, AM. Bilateral common carotid artery stenosis in normotensive rats impairs endothelium-dependent dilation of parenchymal arterioles. *Am J Physiol Heart Circ Physiol.* 2016;310: H1321–H1329.
19. Diaz-Otero, JM, Garver, H, Fink, GD, Jackson, WF, and Dorrance, AM. Aging is associated with changes to the biomechanical properties of the posterior cerebral artery and parenchymal arterioles. *Am J Physiol Heart Circ Physiol.* 2016;310: H365–H375.
20. Hirst, GD, and Edwards, FR. Sympathetic neuroeffector transmission in arteries and arterioles. *Physiol Rev.* 1989;69: 546–604.
21. Klieber, HG, and Daut, J. A glibenclamide sensitive potassium conductance in terminal arterioles isolated from guinea pig heart. *Cardiovasc Res.* 1994;28: 823–830.
22. Little, TL, Beyer, EC, and Duling, BR. Connexin 43 and connexin 40 gap junctional proteins are present in arteriolar smooth muscle and endothelium in vivo. *Am J Physiol.* 1995;268: H729–H739.
23. Little, TL, Xia, J, and Duling, BR. Dye tracers define differential endothelial and smooth muscle coupling patterns within the arteriolar wall. *Circ Res.* 1995;76: 498–504.
24. Quayle, JM, Bonev, AD, Brayden, JE, and Nelson, MT. Calcitonin gene-related peptide activated ATP-sensitive K^+ currents in rabbit arterial smooth muscle via protein kinase A. *J Physiol.* 1994;475: 9–13.
25. Seidel, MF, Simard, JM, Hunter, SF, and Campbell, GA. Isolation of arteriolar microvessels and culture of smooth muscle cells from cerebral cortex of guinea pig. *Cell Tissue Res.* 1991;265: 579–587.
26. Dubey, RK, Roy, A, and Overbeck, HW. Culture of renal arteriolar smooth muscle cells: Mitogenic responses to angiotensin II. *Circ Res.* 1992;71: 1143–1152.
27. Imig, JD, and Anderson, GL. Small artery resistance increases during the development of renal hypertension. *Hypertension.* 1991;17: 317–322.
28. Stekiel, WJ, Contney, SJ, and Lombard, JH. Sympathetic neural control of vascular muscle in reduced renal mass hypertension. *Hypertension.* 1991;17: 1185–1191.
29. Bohlen, HG. The microcirculation in hypertension. *J Hypertens Suppl.* 1989;7: S117–S124.

30. Stekiel, WJ, Contney, SJ, and Rusch, NJ. Altered b-receptor control of in situ membrane potential in hypertensive rats. *Hypertension*. 1993;21: 1005–1009.
31. Falcone, JC, Granger, HJ, and Meininger, GA. Enhanced myogenic activation in skeletal muscle arterioles from spontaneously hypertensive rats. *Am J Physiol Heart Circ Physiol*. 1993;265: H1847–H1855.
32. Hernandez, I, Cowley, AW, Jr., Lombard, JH, and Greene, AS. Salt intake and angiotensin II alter microvessel density in the cremaster muscle of normal rats. *Am J Physiol Heart Circ Physiol*. 1992;263: H664–H667.
33. Jackson, WF, Huebner, JM, and Rusch, NJ. Enzymatic isolation and characterization of single vascular smooth muscle cells from cremasteric arterioles. *Microcirculation*. 1997;4: 35–50.
34. Jackson, WF, and Blair, KL. Characterization and function of Ca(2+)-activated K^+ channels in arteriolar muscle cells. *Am J Physiol*. 1998;274: H27–H34.
35. Jackson, WF. Hypoxia does not activate ATP-sensitive K^+ channels in arteriolar muscle cells. *Microcirculation*. 2000;7: 137–145.
36. Cohen, KD, and Jackson, WF. Hypoxia inhibits contraction but not calcium channel currents or changes in intracellular calcium in arteriolar muscle cells. *Microcirculation*. 2003;10: 133–141.
37. Hayoz, S, Bradley, V, Boerman, EM, Nourian, Z, Segal, SS, and Jackson, WF. Aging increases capacitance and spontaneous transient outward current amplitude of smooth muscle cells from murine superior epigastric arteries. *Am J Physiol Heart Circ Physiol*. 2014;306: H1512–H1524.
38. Duling, BR, Gore, RW, Dacey, RG, Jr., and Damon, DN. Methods for isolation, cannulation, and in vitro study of single microvessels. *Am J Physiol*. 1981;241: H108–H116.
39. Bagher, P, and Segal, SS. The mouse cremaster muscle preparation for intravital imaging of the microcirculation. *J Vis Exp*. 2011; 52: e2874.
40. Hungerford, JE, Sessa, WC, and Segal, SS. Vasomotor control in arterioles of the mouse cremaster muscle. *FASEB J*. 2000;14: 197–207.
41. Burns, WR, Cohen, KD, and Jackson, WF. K^+-induced dilation of hamster cremasteric arterioles involves both the Na^+/K^+-ATPase and inward-rectifier K^+ channels. *Microcirculation*. 2004;11: 279–293.
42. Jackson, WF, Boerman, EM, Lange, EJ, Lundback, SS, and Cohen, KD. Smooth muscle alpha1D-adrenoceptors mediate phenylephrine-induced vasoconstriction and increases in endothelial cell Ca^{2+} in hamster cremaster arterioles. *Br J Pharmacol*. 2008;155: 514–524.
43. Hill, MA, Nourian, Z, Ho, IL, Clifford, PS, Martinez-Lemus, L, and Meininger, GA. Small artery elastin distribution and architecture: Focus on three dimensional organization. *Microcirculation*. 2016; 23: 614–620.
44. Jackson, WF. Arteriolar oxygen reactivity: Where is the sensor and what is the mechanism of action? *J Physiol*. 2016;594: 5055–5077.
45. Nelson, MT, Cheng, H, Rubart, M, Santana, LF, Bonev, AD, Knot, HJ, and Lederer, WJ. Relaxation of arterial smooth muscle by calcium sparks. *Science*. 1995;270: 633–637.
46. Yang, Y, Murphy, TV, Ella, SR, Grayson, TH, Haddock, R, Hwang, YT, Braun, AP, et al. Heterogeneity in function of small artery smooth muscle BKCa: Involvement of the beta1-subunit. *J Physiol*. 2009;587: 3025–3044.
47. Liang, GH, Xi, Q, Leffler, CW, and Jaggar, JH. Hydrogen sulfide activates Ca(2+) sparks to induce cerebral arteriole dilatation. *J Physiol*. 2012;590: 2709–2720.
48. Dabertrand, F, Nelson, MT, and Brayden, JE. Acidosis dilates brain parenchymal arterioles by conversion of calcium waves to sparks to activate BK channels/novelty and significance. *Circ Res*. 2012;110: 285–294.

49. Fredriksson, S, Gullberg, M, Jarvius, J, Olsson, C, Pietras, K, Gustafsdottir, SM, Ostman, A, and Landegren, U. Protein detection using proximity-dependent DNA ligation assays. *Nat Biotechnol*. 2002;20: 473–477.
50. Adebiyi, A, Zhao, G, Narayanan, D, Thomas-Gatewood, CM, Bannister, JP, and Jaggar, JH. Isoform-selective physical coupling of TRPC3 channels to IP3 receptors in smooth muscle cells regulates arterial contractility. *Circ Res*. 2010;106: 1603–1612.
51. Hong, K, Li, M, Nourian, Z, Meininger, GA, and Hill, MA. Angiotensin II Type 1 receptor mechanoactivation involves RGS5 (Regulator of G Protein Signaling 5) in skeletal muscle arteries: Impaired trafficking of RGS5 in hypertension. *Hypertension*. 2017; 70: 1264–1272.

4 Ion Channel Trafficking in Arterial Smooth Muscle Cells and Functional Significance

M. Dennis Leo and Jonathan H. Jaggar

CONTENTS

4.1 INTRODUCTION

Many different ion channels control arterial smooth muscle cell contractility, as is described in other articles in this series of reviews.[1–3] The cellular current (I) generated by a membrane ion channel population is determined by the number of channels (N), their single-channel open probability (P_o) and single-channel current (i), such that $I = N.P_o.i$. Physiological and pathological stimuli and their downstream signaling processes can regulate ion channel function through transcriptional modulation and post-translational modification of protein subunits. Previous studies have primarily investigated mechanisms that regulate the P_o of plasma membrane-resident ion channels and the functional consequences of such modulation in arterial smooth muscle cells. Although the cellular current generated by ion channels is equally dependent upon N and P_o, it has only recently been demonstrated that physiological and pathological stimuli control the number of surface channel proteins in arterial smooth muscle cells. This review will summarize studies that have identified trafficking pathways of ion channels and their functional significance in arterial smooth muscle cells. At the end of this chapter, we also provide a protocol that will allow readers to perform surface biotinylation of intact arteries to measure protein distribution.

4.2 L-TYPE VOLTAGE-GATED Ca^{2+} ($Ca_V1.2$) CHANNELS

Voltage-gated Ca^{2+} channels couple membrane potential to Ca^{2+} influx in arterial smooth muscle. Voltage-gated Ca^{2+} channels are complexes composed of multiple subunits, including pore-forming α_{1C} ($Ca_V1.2$), auxiliary $\alpha_2\delta$, and β subunits (Figure 4.1). $Ca_V1.2$ is encoded by the CACNA1C gene, which has 55 exons in humans and 52 exons in rodents and undergoes alternative splicing to produce several variants that exhibit unique biophysical properties [1,2]. The principal variant of $Ca_V1.2$ that is expressed in rat arterial smooth muscle cells contains a novel N-terminus encoded by exon 1c ($Ca_V1.2e1c$), with a minority of protein the exon 1b variant [3]. $Ca_V1.2$ channels are high-voltage-activated and sensitive to several inhibitors, including dihydropyridines, benzothiazepines and phenylalkylamines, which are effective vasodilators [4–6].

Post-translational modification cleaves ~300 amino acids of the distal C-terminus of $Ca_V1.2$ channels in several cell types, including arterial smooth muscle cells,

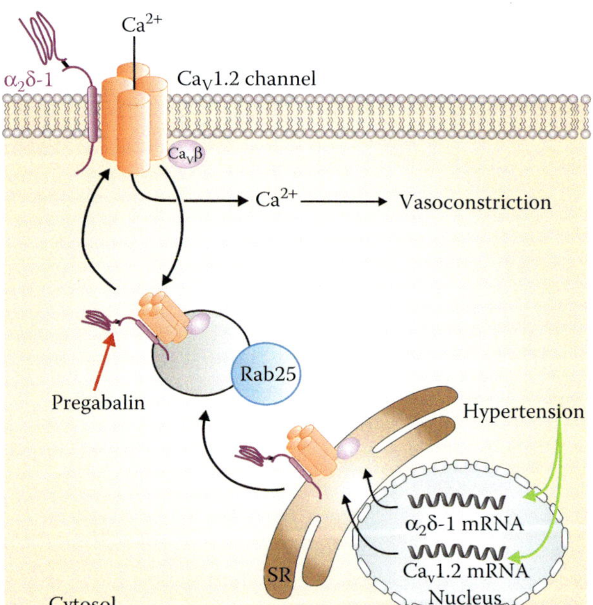

FIGURE 4.1 $Ca_V1.2$ trafficking in arterial smooth muscle cells. Rab25 promotes anterograde trafficking of $Ca_V1.2$ subunits. $\alpha_2\delta$-1 is required for functional surface expression of $Ca_V1.2$ subunits. The $\alpha_2\delta$-1 ligand, pregabalin, inhibits trafficking of $Ca_V1.2$, leading to vasodilation. In genetic hypertension, $\alpha_2\delta$-1 and $Ca_V1.2$ expression are both elevated, which increases surface trafficking of $Ca_V1.2$ channels, leading to vasoconstriction. Green arrow indicates activation. Red arrow indicates inhibition.

releasing a C-terminus fragment (CCT) [7,8]. Thus, $Ca_V1.2$ exists as both truncated ~190 kDa and full-length ~240 kDa proteins in many cell types [7–10]. Neuronal and skeletal muscle L-type Ca^{2+}channels are cleaved by calpain and recombinant cardiac $Ca_V1.2$ channels by chymotrypsin, but mechanisms of cleavage in arterial smooth muscle are unclear [11–13]. CCT contains a proline-rich domain that assists its re-association with truncated $Ca_V1.2$ protein [7,14]. The cleaved CCT acts as a bimodal vasodilator in cerebral arteries [8]. CCT re-association with $Ca_V1.2$ right-shifts voltage-activation, thereby reducing channel activity [8,14]. The CCT fragment also inhibits $Ca_V1.2$ transcription, leading to a reduction in total protein in arterial smooth muscle cells [7,8]. The combined effects of the shift in voltage-dependence and reduction in expression decrease $Ca_V1.2$ current density and inhibit pressure- and depolarization-induced vasoconstriction [8]. Mitochondria also modulate $Ca_V1.2$ gene expression in arterial smooth muscle cells [15]. Endothelin-1 (ET-1), a vasoconstrictor, activates IP_3R-mediated sarcoplasmic reticulum (SR) Ca^{2+} release, which elevates the Ca^{2+} concentration within spatially-localized mitochondria, stimulating reactive oxygen species ROS production [15]. ET-1-induced mitochondrial ROS activate the transcription factor NF-κB, which increases $Ca_V1.2$ transcription, leading to an increase in $Ca_V1.2$ current density and vasoconstriction [15]. This study demonstrated that mitochondria control $Ca_V1.2$ channel expression in arterial smooth muscle cells to modulate arterial contractility.

Biotinylation of intact arteries indicates that $Ca_V1.2$ α subunits are primarily plasma membrane-localized, with only a small fraction of total protein located intracellularly, in arterial smooth muscle cells [8–10,16]. The surface abundance of $Ca_V1.2$ in cerebral artery smooth muscle cells is controlled by Rab25, a small GTPase associated with apical recycling endosomes (Figure 4.1) [17]. In contrast, Rab11A or Rab4A do not regulate $Ca_V1.2$ surface protein or cellular distribution [17]. Acute knockdown of Rab25 reduced both plasma membrane and total $Ca_V1.2$ protein [17]. In the absence of Rab25-mediated anterograde trafficking, $Ca_V1.2$ protein was targeted for degradation through a mechanism that involved both proteasomal and lysosomal pathways, which appeared to act in series rather than in parallel [17]. The reduction in surface $Ca_V1.2$ protein decreased $Ca_V1.2$ current density but not the voltage dependence of current activation or inactivation [17]. This led to a decrease in both depolarization and pressure-induced vasoconstriction [17]. These findings indicated that Rab25 is essential for surface trafficking of $Ca_V1.2$ channels in arterial smooth muscle cells and identified Rab25 as novel target to induce vasodilation.

Auxiliary subunits regulate the surface expression and biophysical properties of recombinant $Ca_V1.2$ channels [18]. Four different $\alpha_2\delta$ subunits ($\alpha_2\delta$-1-4) exist that arise from alternative splicing of the CACNA2D1 gene [19,20]. After post-translational cleavage, the glycosylated extracellular α_2 and the smaller membrane-spanning δ subunit re-associate to form a functional protein [19–21]. $\alpha_2\delta$-1 is the only isoform that is expressed in arterial smooth muscle cells, where it is essential for $Ca_V1.2$ channel surface trafficking (Figure 4.1) [9]. In cerebral artery smooth muscle cells, $\alpha_2\delta$-1 surface traffics the $Ca_V1.2$e1c splice variant more effectively than $Ca_V1.2$e1b [10]. Both pregabalin (Lyrica™), an $\alpha_2\delta$-1/2 ligand used in the treatment of neuropathic pain, and $\alpha_2\delta$-1 knockdown reduced the surface abundance of plasma membrane $\alpha_2\delta$-1 and $Ca_V1.2$ α_1 subunits in arterial smooth muscle cells (Figure 4.1) [9]. $\alpha_2\delta$-1 knockdown inhibited both pressure-and depolarization-induced vasoconstriction [9]. Pregabalin caused vasodilation primarily by inhibiting $Ca_V1.2$ anterograde trafficking, but also by acting as a weak pore blocker of surface-resident $Ca_V1.2$ channels [9].

Hypertension is associated with a shift in the regulation of $Ca_V1.2$ channels by $\alpha_2\delta$-1 in arterial smooth muscle [16]. An upregulation of both protein transcription and translation increased $\alpha_2\delta$-1 and $Ca_V1.2$ protein in arteries of spontaneously hypertensive rats (Figure 4.1) [16]. $\alpha_2\delta$-1 increased more at the cell surface than did $Ca_V1.2\alpha_1$, causing a shift in the multisubunit composition of channels during hypertension [16]. The net result was an increase in $Ca_V1.2$ current density and a larger non-inactivating current, with no change in current voltage-dependence in arterial smooth muscle cells [16]. Pregabalin reduced surface $\alpha_2\delta$-1 and $Ca_V1.2\alpha_1$ more in hypertensive arteries than in control rat arteries, essentially normalizing surface levels of these proteins [16]. Pregabalin also normalized the subunit ratio of $\alpha_2\delta$-1 to $Ca_V1.2\alpha_1$ in hypertensive rat arteries. The pregabalin-induced changes in surface $Ca_V1.2$ channels reduced $Ca_V1.2$ currents and normalized the amount of non-inactivating current, suggesting that the hypertension-associated increase in $\alpha_2\delta$-1 surface protein was responsible for current modification. Pregabalin was a more effective vasodilator of hypertensive arteries than control arteries and effectively reversed the pathological vasoconstriction [16]. These studies suggested that smooth muscle cell $\alpha_2\delta$-1 could represent a novel target for antihypertensive drugs.

Four different genes encode $Ca_V\beta$ subunits (β1–β4), with each further subject to splice variation [22]. β subunits bind to the cytoplasmic linker connecting the first two transmembrane domains of $Ca_V1.2$ channels and can regulate properties including surface expression, voltage-sensitivity and interaction with proteins and kinases [22]. β3 subunit binding to $Ca_V1.2$ protein weakens the endoplasmic reticulum (ER) retention signal, leading to an increase in surface trafficking [23]. $Ca_V1.2\alpha$ total and surface protein were both lower, voltage-dependent Ca^{2+} current density was smaller and inactivation occurred more slowly in aortic smooth muscle cells of global $\beta3^{-/-}$ mice [24]. Angiotensin II infusion increased $Ca_V1.2$ protein in mesenteric arteries of control mice and elevated blood pressure and these effects were attenuated in global $\beta3^{-/-}$ mice [25]. In a mouse model of genetic hypertension, channel composition shifted from $\alpha_1c/\alpha_2\delta/\beta_3$ to $\alpha_1c/\alpha_2\delta/\beta_2$ in mesenteric artery smooth muscle cells, which decreased whole-cell $Ca_V1.2$ currents [26]. Collectively, these studies defined unique mechanisms by which $Ca_V1.2$ channels control arterial contractility and identified novel proteins that could be targeted to modulate $Ca_V1.2$ channel activity and arterial contractility.

4.3 VOLTAGE-DEPENDENT K^+ (K_V) CHANNELS

The K_V channel family contains ~40 different members classified into 12 subtypes (K_V1–12) [27]. K_V channels are six transmembrane domain proteins, with S4 the voltage sensor and S5–S6 forming the channel pore [27]. Depolarization stimulates the S4 segment to translocate to the extracellular surface of the plasma membrane, elevating P_o [28,29]. Several family members have been described to be expressed in vascular smooth muscle cells, including subtypes of K_V1–K_V4, K_V7, K_V9, and K_V11 [30–34]. K_V channels can form either homo- or hetero-tetramers, including in arterial smooth muscle cells, leading to diverse K^+ current phenotypes [35]. Activation of K_V channels causes membrane hyperpolarization in arterial smooth muscle cells, leading to vasodilation [36].

Studies have measured K_V channel isoform expression in either arterial smooth muscle cells or intact arteries, but few have performed quantitative analysis to determine the relative expression of channel subtypes expressed specifically in non-cultured, contractile smooth muscle cells. A recent study showed that in contractile rat mesenteric artery smooth muscle cells, $K_V1.5$ channels were the predominant isoform, accounting for ~60% of K_V channel transcript, with $K_V2.1$ and $K_V2.2$ each ~15% and total message for $K_V1.2$, 7.1 and 7.4 accounting for the remaining 10% [37]. Arterial biotinylation indicated that ~48% of total $K_V1.5$ and ~81% of total $K_V2.1$ protein was located at the cell surface in mesenteric arteries [37]. Membrane potential modulated $K_V1.5$ surface abundance by regulating $Ca_V1.2$ channel activity in arterial smooth muscle cells (Figure 4.2) [37]. Intravascular pressure and physiological membrane depolarization increased surface $K_V1.5$ protein by inhibiting both lysosomal and proteasomal degradation of internalized protein [37]. In contrast, pressure or membrane potential did not modulate $K_V2.1$ surface protein, indicating isoform specificity of this Ca^{2+}-dependent signaling pathway [37]. Explanations for the isoform-specific control of surface K_V channel proteins may involve amino acid sequence dissimilarity, differences in regulation by auxiliary subunits, or modulation by signal messengers. This study demonstrated that pressure-induced membrane depolarization increases $K_V1.5$ currents

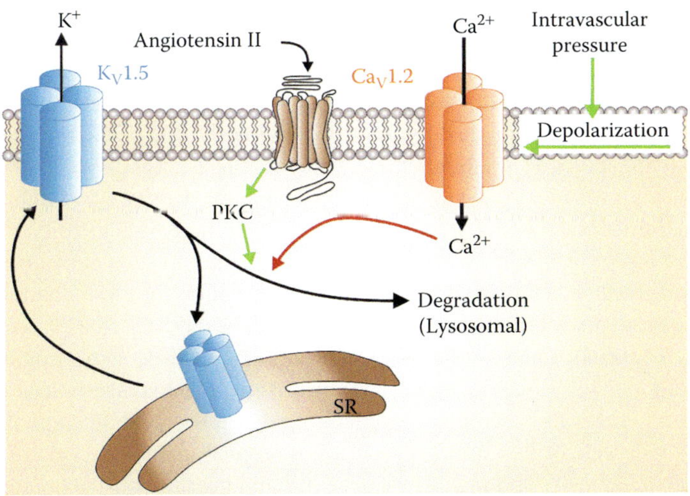

FIGURE 4.2 $K_V1.5$ trafficking in arterial smooth muscle cells. $K_V1.5$ continuously recycles to and from the plasma membrane. Pressure-induced membrane depolarization increases $[Ca^{2+}]_i$, which inhibits $K_V1.5$ degradation, allowing protein to return to the plasma membrane. Angiotensin II activates protein kinase C, which stimulates $K_V1.5$ internalization and degradation, leading to a decrease in surface channels and vasoconstriction. Green arrow indicates activation. Red arrow indicates inhibition.

not only by directly stimulating channel P_O, but also by increasing the number (N) of channels in the plasma membrane in arterial smooth muscle cells. The pressure-induced increase in both channel N and P_O opposes vasoconstriction.

Angiotensin II, a vasoconstrictor, binds to AT_1 receptors, which are coupled to $G_{q/11}$ proteins that stimulate phospholipase C in vascular smooth muscle cells [38]. Angiotensin II reduced surface $K_V1.5$ protein in mesenteric artery smooth muscle cells [39]. Data suggested that angiotensin II activates protein kinase C (PKC), which stimulates the degradation of internalized $K_V1.5$ channels, thereby reducing the amount of protein that returns to the plasma membrane (Figure 4.2). The decrease in $K_V1.5$ surface protein reduces $K_V1.5$ current density in smooth muscle cells and $K_V1.5$ function in pressurized arteries [39]. In contrast, angiotensin II did not alter surface or total $K_V2.1$ protein, further suggesting that diverse mechanisms control the surface expression of different K_V channel subtypes to control K_V currents and vascular contractility [39]. These studies also indicated that membrane potential and angiotensin II control $K_V1.5$ degradation through distinct mechanisms: membrane depolarization acting via $Ca_V1.2$ channels and angiotensin II through PKC (Figure 4.2). Membrane depolarization leads to a nanomolar increase in global $[Ca^{2+}]_i$, suggesting that a Ca^{2+}-sensing protein with an affinity in that concentration range, such as Ca^{2+}/calmodulin, inhibits $K_V1.5$ degradation [39]. Three isoforms of PKC are present in vascular smooth muscle cells: protein kinase Cα and protein kinase Cβ are DAG- and Ca^{2+}-dependent, and protein kinase Cε is stimulated by DAG [40]. Angiotensin II-mediated protein kinase Cε activation reduces whole-cell KV currents in arterial smooth muscle cells and protein kinase Cε inhibition reduces angiotensin II-induced vasoconstriction in mesenteric artery rings [41]. Thus, angiotensin II-induced $K_V1.5$ degradation may occur due to the activation of Ca^{2+}-independent protein kinase Cε. Future studies will be required to unravel the intracellular signaling mechanisms by which an increase in $[Ca^{2+}]_i$ inhibits $K_V1.5$ degradation and PKC activation stimulates $K_V1.5$ degradation to control contractility.

Serotonin stimulates $K_V1.5$ channel endocytosis in pulmonary artery smooth muscle cells [42]. $K_V1.5$ colocalized with caveolin-1 and 5-HT_{2A} receptors, suggesting that a compartmentalized signaling pathway may regulate channel internalization [42]. $K_V1.2$ has a high affinity binding site for postsynaptic density-95 (PSD95), a scaffolding protein, whereas $K_V1.5$ interacts with PSD95 via its low affinity binding site [43]. PSD95 knockdown reduced both K_V1 channel expression and currents in cerebral artery smooth muscle cells, which reduced vasoregulation by these channels [43]. A peptide sequence mimicking the $K_V1.2$ PSD95 binding motif decreased smooth muscle cell $K_V1.5$ currents and stimulated depolarization and vasoconstriction, suggesting that an interaction with PSD95 is essential for function [44]. Pial arteriole smooth muscle cells from (Tg)Notch3^{R169C} mice, a genetic model of cerebral autosomal dominant arteriopathy with subcortical infarcts and leukoencephalopathy (CADASIL) [45], had larger K_V current density and pial arterioles exhibited greater constriction to K_V channel inhibition than controls. Data suggested that pressure-induced cerebral vasoconstriction is attenuated in CADASIL due to an increase in the number of surface K_V1 channels in smooth muscle cells [45]. Oxyhemoglobin levels increase after subarachnoid hemorrhage (SAH) leading to cerebral vasospasm [46]. In rabbit cerebral arteries, oxyhemoglobin-induced tyrosine kinase activation decreased functional $K_V1.5$ expression that may contribute to SAH-induced cerebral vasoconstriction [47].

These studies show that physiological and pathological stimuli control surface K_V channels and that surface levels of different K_V channel subtypes can be distinctly regulated in arterial smooth muscle cells to control arterial contractility.

4.4 LARGE-CONDUCTANCE Ca^{2+}-ACTIVATED POTASSIUM (BK_{Ca}, BK, *SLO1*, $K_{Ca}1.1$) CHANNEL α AND β1 SUBUNITS

BK channels are formed from the homotetrameric assembly of seven transmembrane domain (S0–S6) pore-forming α subunits [48]. Amino acids in transmembrane segments S1–S4 confer channel voltage-sensitivity, whereas S5–S6 form the pore [49]. BK channels have a large single-channel conductance and are voltage-and Ca^{2+}-dependent [48,50]. The pore-forming BKα subunit is encoded by a single gene, KCNMA1 [51]. Alternative splicing gives rise to at least 10 different variants that exhibit variable degrees of surface expression [52,53]. A myometrial BK channel N-terminal variant with a 33-amino acid insert in the S1 transmembrane domain, SV1, contains an ER retention sequence that prevents it from trafficking to the plasma membrane [54]. The STRess axis regulated EXon (STREX) of BKα contains a 58 amino acid sequence insertion between RCK1 and RCK2 in the C-terminus [55]. STREX was first identified in chromaffin cells following hypophysectomy and occurs due to regulation by stress-axis hormones [55]. Protein kinase A (PKA) phosphorylation inhibits BK channels that contain the STREX variant but activates the BK channel ZERO variant [56]. In heterologous expression systems, the STREX variant demonstrated higher cell surface expression than the ZERO variant [57]. None, or only a small proportion, of BK channel mRNA contained the STREX insert in rat cerebral and cremaster arteries [57–59]. Virtually all BKα protein is plasma membrane-localized in cerebral and mesenteric artery smooth muscle cells [60,61]. BKα trafficked to the plasma membrane via a Rab4A-dependent pathway in cerebral artery smooth muscle cells, suggesting early endosomes shuttle the channel to the surface. (Figure 4.3) [61]. In contrast, BKα protein anterograde trafficking did not involve Rab11A or Rab11B in cerebral artery smooth muscle cells, indicating that recycling endosomes were not involved [62,63]. Angiotensin II stimulated PKC-dependent internalization and degradation of BKα in arterial smooth muscle cells (Figure 4.3) [61]. In contrast, nitric oxide (NO) did not alter surface BKα [60]. Angiotensin II-induced BKα internalization and degradation reduced BK currents in arterial myocytes, elevated myogenic tone and attenuated responses to BK channel modulators [61].

BK channel accessory subunits include the four β (β1-4) subunits and recently identified γ subunits, of which four (γ1-4) have been described [64]. Vascular smooth muscle cells express β1 subunits, which associate with BKα to increase channel apparent Ca^{2+}-sensitivity and slow channel activation and deactivation (Figure 4.4) [50]. In contrast to BKα channels, only a small fraction of total β1 protein is plasma membrane-resident, with most β1 located intracellularly in cerebral artery smooth muscle cells [60]. Experiments performed using Förster resonance energy transfer (FRET) and RNA interference (RNAi) indicated that intracellular β1 subunits are stored within Rab11A-positive recycling endosomes [60]. Activation of protein kinase G (PKG) by NO, an endothelium-derived

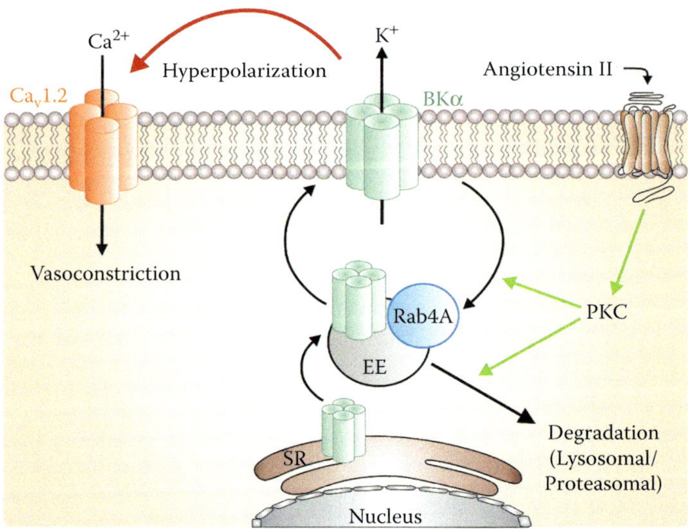

FIGURE 4.3 BKα subunit trafficking in arterial smooth muscle cells. BKα subunits are surface-trafficked by Rab4A-postive early endosomes (EE). Angiotensin II, a vasoconstrictor, stimulates protein kinase C (PKC), leading to BKα internalization and degradation. The decrease in surface BK channel protein reduces BK currents, leading to vasoconstriction. Green arrow indicates activation. Red arrow indicates inhibition.

vasodilator, stimulated rapid (seconds) surface trafficking of β1 subunits, which then associated with surface-resident BKα channels, increasing apparent Ca^{2+}-sensitivity and P_O [60]. This data was consistent with previous studies of recombinant channel proteins, which suggested that a recombinant BK channel tetramer can contain one to four β1 subunits and that the α:β1-tetramer ratio can shift BK channel voltage- and Ca^{2+}-dependence [48,65]. Endocytosis inhibitors increased surface β1 protein, suggesting that constant recycling of β1 to and from the cell surface occurs in arterial smooth muscle cells and that NO stimulates anterograde trafficking of β1 to increase surface protein [60]. In contrast to NO, angiotensin II did not alter surface levels of β1 subunits [61]. These studies provided the first evidence that native channel multisubunit composition is not rigid, but dynamic, and can be rapidly modulated by physiological vasoregulatory stimuli to control BK channel activity in arterial smooth muscle cells. Data also indicated that physiological stimuli rapidly modulate surface BK channel subunit composition to control activity and arterial contractility.

Intravascular pressure, by modulating membrane potential, is a major regulator of smooth muscle cell BK channel activity [5,31]. A recent study described a mechanism by which intravascular pressure and membrane potential control BK channel activity by modulating the plasma membrane abundance of auxiliary β1 subunits in arterial smooth muscle cells. Membrane depolarization stimulated voltage-dependent Ca^{2+} channels, leading to Ca^{2+} influx, which activated Rho-associated protein kinase (ROCK1 and ROCK2). ROCK phosphorylated Rab11A, stimulating anterograde trafficking of intracellular β1 subunits to the plasma membrane. These β1 subunits associated with plasma membrane-resident BK channels increasing their apparent Ca^{2+}-sensitivity, leading to channel activation and vasodilation. Similarly to NO, membrane depolarization rapidly stimulated trafficking and plasma membrane insertion of β1 subunits. Evidence suggested that depolarization and NO mobilize the same intracellular β1 subunit pool, but via distinct mechanisms: NO via PKG activation and membrane depolarization through Ca^{2+}-dependent ROCK activation. When combined with previous studies that measured the activity of surface-resident BK channels in excised patches, these data demonstrated that membrane depolarization activates surface BK channels both directly and indirectly by stimulating an increase in surface β1 subunits in arterial smooth muscle cells [66]. BK channel γ subunits are single-span, leucine-rich repeat containing (LRRC) proteins. LRRC26 is expressed in arterial smooth muscle cells and elevates BK channel voltage-sensitivity, leading to channel activation and vasodilation (Figure 4.4) [67,68]. γ1 knockdown using siRNA did not alter surface or total BKα or β1 subunit protein, suggesting that γ1 does not regulate BK channel trafficking in arterial smooth muscle cells.

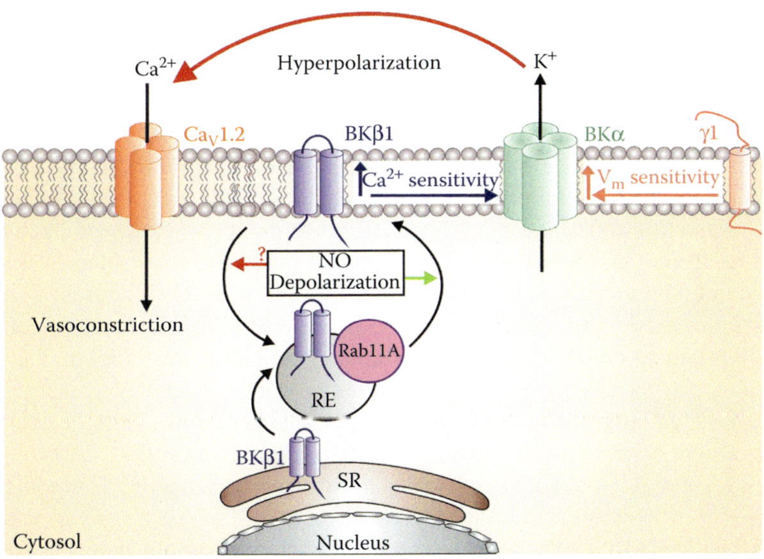

FIGURE 4.4 Auxiliary β1 subunit surface trafficking and γ1 subunits control BK channel activity. Rab11A-positive recycling endosomes (RE) continuously recycle surface β1 subunits, with the vast majority of protein located intracellular in resting arterial smooth muscle cells. Nitric oxide (NO) or membrane depolarization stimulate rapid (<1 min) anterograde trafficking of Rab11A-positive RE, which increases surface β1 protein. These β1 subunits associate with surface-resident BKα channels, increasing apparent Ca^{2+}-sensitivity. The increase in BK channel activity leads to membrane hyperpolarization and vasodilation. γ1 subunits are primarily located on the smooth muscle cell plasma, where they increase BK channel voltage-sensitivity, leading to vasodilation. SR: sarcoplasmic reticulum. Green arrow indicates activation. Red arrow indicates inhibition.

In summary, studies show that distinct mechanisms regulate the trafficking of different BK channel subunits to control surface BK channel activity and arterial contractility.

4.5 TRANSIENT RECEPTOR POTENTIAL CHANNELS

The transient receptor potential (TRP) family of non-selective cation channels comprises 28 distinct proteins in mammals [69]. TRP proteins are subdivided into six subfamilies, designated as TRPC (canonical), TRPV (vanilloid), TRPM (melastatin), TRPA (ankyrin), TRPP (polycystin), and TRPML (mucolipin). All TRP proteins have six transmembrane domains with both the N- and C-termini located intracellularly. Arterial smooth muscle cells express TRP channels from several different families [69]. TRP channel trafficking has primarily been investigated using recombinant proteins and non-vascular cells. Here, we discuss the few studies that have measured TRP cellular distribution and trafficking regulation in arterial smooth muscle cells.

Seven TRPC (TRPC1-7) channels have been identified, with several members expressed in vascular smooth muscle cells, including TRPC1, TRPC3 and TRPC6 [69]. Only ~30% of TRPC3 channel total protein was located in the plasma membrane in cerebral artery smooth muscle cells, whereas almost 100% was surface resident in mesenteric artery smooth muscle cells [70,71]. Thus, TRPC3 distribution may differ depending on the anatomical origin of the vasculature. TRPC3 channels were located in very close spatial proximity to SR type 1 IP_3 receptors (IP_3R1) [70]. Endothelin-1 increased intracellular IP_3, stimulating physical coupling of the IP_3R1 N-terminus to the TRPC3 channel C-terminus calmodulin and IP_3R binding (CIRB) domain, leading to I_{Cat} activation and vasoconstriction [70]. Caveolin-1 compartmentalized plasma membrane TRPC3 and SR IP_3R1 in close spatial proximity to enable physical coupling in cerebral artery smooth muscle cells [72]. Surface TRPC3 protein was ~two-fold higher in mesenteric artery smooth muscle cells of spontaneous hypertensive rats (SHR) than in controls, although cellular distribution was similar. These data indicated that hypertension is associated with an increase in TRPC3 protein and that stimulation of anterograde trafficking and/or inhibition of retrograde trafficking elevate the amount of protein at the surface [71]. This increase in surface TRPC3 protein enhanced ET-1 and IP_3-induced physical coupling to IP_3R1 in arterial smooth muscle cells, leading to vasoconstriction in hypertensive rats [71].

Both TRPM4 and TRPM8 are expressed in arterial smooth muscle cells [69]. Intravascular pressure and PKC stimulate TRPM4 channels, leading to Na^+ influx, membrane depolarization and vasoconstriction [73]. Biotinylation indicated that ~64% of TRPM4 protein is located in the plasma membrane of cerebral arteries [73]. Protein kinase C activation triggered surface-trafficking of TRPM4, which was associated with a corresponding decrease in intracellular protein [73]. This study suggested that intravascular pressure stimulates PKC, leading to surface trafficking of TRPM4 and smooth muscle cell depolarization and vasoconstriction [73].

TRPP1 (also termed PKD2 and previously TRPP2) channels are expressed in cerebral artery smooth muscle cells where protein is located primarily in the plasma membrane [74,75]. TRPP1 channel activation contributes to swelling-activated cation currents (I_{Cat}) in smooth muscle cells and pressure-induced vasoconstriction in rat cerebral arteries [74]. In contrast, a different study showed that siRNA-mediated TRPP1 knockdown increased stretch-induced I_{Cat} in murine mesenteric artery smooth muscle cells [75]. The authors proposed that TRPP1 channels inhibit stretch-activated I_{Cat} in arterial smooth muscle cells [75]. These studies suggest that TRPP1 channels may perform opposing functions in arterial smooth muscle cells of different species or vascular beds.

4.6 CONCLUSION

Many different surface ion channels regulate arterial smooth muscle cell contractility. Previous studies have focused on identifying mechanisms that control the activity of plasma membrane ion channels. Recent studies have discovered that physiological and pathological stimuli also regulate the number of surface ion channel proteins. Importantly, the modulation of surface channel number regulates arterial contractility. However, much remains unclear. Future studies should uncover novel ion channel trafficking pathways, regulatory mechanisms and physiological and pathological significance in arterial smooth muscle cells.

4.A APPENDIX—ARTERIAL SURFACE BIOTINYLATION PROTOCOL

4.A.1 Reagents and Equipment

1. Physiological saline solution (PSS)
2. EZ-Link Sulfo-NHS-LC-LC-Biotin (Catalog number: 21338, ThermoFisher Scientific, Waltham, MA)
3. EZ-Link Maleimide-PEG2-Biotin (Catalog number: 21901BID, Thermo Fisher Scientific, Waltham, MA)
4. Pierce™ Monomeric Avidin UltraLink™ Resin (Catalog number: 53146, ThermoFisher Scientific, Waltham, MA)
5. Low-retention micro centrifuge tubes
6. Rotating platform
7. Refrigerated centrifuge

4.A.2 Method

1. Dissect and clean arteries of any connective or adipose tissue.
2. Incubate arteries in PSS containing 1 mg/ml EZ-Link Sulfo-NHS-LC-LC-Biotin and 1 mg/ml EZ-Link Maleimide-PEG2-Biotin reagents for 1 h.
3. Centrifuge at 6000 rpm for 2 min at 4°C. Remove the biotin solution using a 1 ml pipette, carefully avoiding the arterial segments.
4. Quench the reaction by adding 1 ml of 100 mmol/L glycine in phosphate-buffered saline PBS (PBS, pH 7.4). Centrifuge at 6000 rpm for 2 min and carefully remove the supernatant. Repeat this step twice.

5. Wash the arteries twice using 1 ml PBS (pH 7.4). Centrifuge at 6000 rpm for 2 min after each wash to suspend the arteries and aid in removing the supernatant. Remove as much of the PBS as possible after the second wash.
6. Re-suspend the arterial segments in lysis buffer.
7. Cut the arteries into small pieces using dissection scissors and homogenize.
8. Centrifuge at 9000 rpm for 5 min at 4°C. Carefully collect the supernatant into a clean micro centrifuge tube. This is the *total protein* fraction.
9. Calculate protein concentration using a standard method.
10. Pipette protein lysate (30–50 μg total protein should be sufficient) into a low-retention 0.5 ml micro centrifuge tube. Based on the amount of protein present in the lysate, add an excess volume of avidin-bead slurry to the lysate to avoid the beads becoming protein-saturated. Bead binding capacity is ~1.2 mg/ml, so if the lysate contains 50 μg of protein, use 45–50 μl of bead slurry for the pulldown. Cap the tube and leave on a rotator at room temperature for 1 h.
11. Centrifuge the lysate-slurry mix at 12,000 rpm for 1 min at 4°C. Collect the supernatant into a fresh tube, taking care not to disturb the bead pellet at the bottom of the tube. Save the supernatant as this contains the non-biotinylated *intracellular proteins*. Add 4x Laemmli buffer to the intracellular fraction to denature proteins.
12. Wash the beads three times with 400 μl PBS (pH 7.4), centrifuging at 12,000 rpm for 1 min at 4°C after each wash step. Take care to remove the supernatant without disturbing the bead pellet.
13. Add 1x Laemmli buffer to the beads. Vortex for 1–2 sec and centrifuge at 12,000 rpm for 10 sec at 4°C to suspend the beads.
14. Heat-treat the tubes containing the beads and the intracellular fraction at 100°C for 3 min. Cool the tubes on ice for 5 min.
15. Centrifuge the tube containing the beads at 12,000 rpm at 4°C for 2 min. Carefully remove the supernatant into a fresh tube. This is the biotinylated surface fraction.
16. Surface and intracellular protein samples can be run on a standard sodium dodecyl sulfate polyacrylamide gel electrophoresis (SDS-PAGE) gel. Run the entire volume of the intracellular and membrane fractions as contiguous lanes on a gel. A fraction of the total protein lysate, if retained, can also be run in a separate lane on the same gel for reference. Prior to loading samples into wells, the volume of each fraction should be equalized using 1x Laemmli buffer. If the volume of the samples exceeds the capacity of the gel wells, a stacking gel can be used.
17. Follow standard Western blotting procedures.
18. Measure the intensity of the intracellular and surface bands and express as a percentage.
19. The precision of the procedure can be determined by re-probing the blot for a protein that should only be present in the intracellular fraction (e.g., Cytochrome C).

REFERENCES

1. Tang, Z. Z., M. C. Liang, S. Lu, D. Yu, C. Y. Yu, D. T. Yue, and T. W. Soong. 2004. Transcript scanning reveals novel and extensive splice variations in human L-type voltage-gated calcium channel, $Ca_V1.2$ α_1 subunit. *J Biol Chem* 279(43): 44335–44343.
2. Soldatov, N. M. 1994. Genomic structure of human L-type Ca^{2+} channel. *Genomics* 22(1): 77–87. doi:10.1006/geno.1994.1347.
3. Cheng, X., J. Liu, M. Asuncion-Chin, E. Blaskova, J. P. Bannister, A. M. Dopico, and J. H. Jaggar. 2007. A novel $Ca_V1.2$ N terminus expressed in smooth muscle cells of resistance size arteries modifies channel regulation by auxiliary subunits. *J Biol Chem* 282(40): 29211–29221.
4. Harder, D. R. 1984. Pressure-dependent membrane depolarization in cat middle cerebral artery. *Circ Res* 55(2): 197–202.
5. Knot, H. J., and M. T. Nelson. 1998. Regulation of arterial diameter and wall $[Ca^{2+}]$ in cerebral arteries of rat by membrane potential and intravascular pressure. *J Physiol* 508(Pt 1): 199–209.
6. Hofmann, F., M. Biel, and V. Flockerzi. 1994. Molecular basis for Ca^{2+} channel diversity. *Annu Rev Neurosci* 17: 399–418.
7. Gomez-Ospina, N., F. Tsuruta, O. Barreto-Chang, L. Hu, and R. Dolmetsch. 2006. The C terminus of the L-type voltage-gated calcium channel Ca(V)1.2 encodes a transcription factor. *Cell* 127(3): 591–606. doi:10.1016/j.cell.2006.10.017.
8. Bannister, J. P., M. D. Leo, D. Narayanan, W. Jangsangthong, A. Nair, K. W. Evanson, J. Pachuau, K. S. Gabrick, F. A. Boop, and J. H. Jaggar. 2013. The voltage-dependent L-type Ca^{2+} ($Ca_V1.2$) channel C-terminus fragment is a bi-modal vasodilator. *J Physiol* 591(12): 2987–2998. doi:10.1113/jphysiol.2013.251926.
9. Bannister, J. P., A. Adebiyi, G. Zhao, D. Narayanan, C. M. Thomas, J. Y. Feng, and J. H. Jaggar. 2009. Smooth muscle cell $\alpha_2\delta$-1 subunits are essential for vasoregulation by $Ca_V1.2$ channels. *Circ Res* 105(10): 948–955.
10. Bannister, J. P., C. M. Thomas-Gatewood, Z. P. Neeb, A. Adebiyi, X. Cheng, and J. H. Jaggar. 2011. $Ca_V1.2$ channel N-terminal splice variants modulate functional surface expression in resistance size artery smooth muscle cells. *J Biol Chem* 286(17): 15058–15066.
11. Gerhardstein, B. L., T. Gao, M. Bunemann, T. S. Puri, A. Adair, H. Ma, and M. M. Hosey. 2000. Proteolytic processing of the C terminus of the a_{1C} subunit of L-type calcium channels and the role of a proline-rich domain in membrane tethering of proteolytic fragments. *J Biol Chem* 275(12): 8556–8563.
12. De Jongh, K. S., A. A. Colvin, K. K. Wang, and W. A. Catterall. 1994. Differential proteolysis of the full-length form of the L-type calcium channel alpha 1 subunit by calpain. *J Neurochem* 63(4): 1558–1564.
13. Hell, J. W., R. E. Westenbroek, L. J. Breeze, K. K. Wang, C. Chavkin, and W. A. Catterall. 1996. N-methyl-D-aspartate receptor-induced proteolytic conversion of postsynaptic class C L-type calcium channels in hippocampal neurons. *Proc Natl Acad Sci USA* 93(8): 3362–3367.
14. Hulme, J. T., V. Yarov-Yarovoy, T. W. Lin, T. Scheuer, and W. A. Catterall. 2006. Autoinhibitory control of the $Ca_V1.2$ channel by its proteolytically processed distal C-terminal domain. *J Physiol* 576(Pt 1): 87–102. doi:10.1113/jphysiol.2006.111799.
15. Narayanan, D., Q. Xi, L. M. Pfeffer, and J. H. Jaggar. 2010. Mitochondria control functional $Ca_V1.2$ expression in smooth muscle cells of cerebral arteries. *Circ Res* 107(5): 631–641.

16. Bannister, J. P., S. Bulley, D. Narayanan, C. Thomas-Gatewood, P. Luzny, J. Pachuau, and J. H. Jaggar. 2012. Transcriptional upregulation of $\alpha_2\delta$-1 elevates arterial smooth muscle cell voltage-dependent Ca^{2+} channel surface expression and cerebrovascular constriction in genetic hypertension. *Hypertension* 60(4): 1006–1015. doi:10.1161/hypertensionaha.112.199661.
17. Bannister, J. P., S. Bulley, M. D. Leo, M. W. Kidd, and J. H. Jaggar. 2016. Rab25 influences functional $Ca_V1.2$ channel surface expression in arterial smooth muscle cells. *Am J Physiol Cell Physiol* 310(11): C885–C893. doi:10.1152/ajpcell.00345.2015.
18. Hofmann, F., V. Flockerzi, S. Kahl, and J. W. Wegener. 2014. L-type $Ca_V1.2$ calcium channels: From in vitro findings to in vivo function. *Physiol Rev* 94(1): 303–326. doi:10.1152/physrev.00016.2013.
19. Davies, A., J. Hendrich, A. T. Van Minh, J. Wratten, L. Douglas, and A. C. Dolphin. 2007. Functional biology of the $\alpha_2\delta$ subunits of voltage-gated calcium channels. *Trends Pharmacol Sci* 28(5): 220–228.
20. Cole, R. L., S. M. Lechner, M. E. Williams, P. Prodanovich, L. Bleicher, M. A. Varney, and G. Gu. 2005. Differential distribution of voltage-gated calcium channel alpha-2 delta ($\alpha_2\delta$) subunit mRNA-containing cells in the rat central nervous system and the dorsal root ganglia. *J Comp Neurol* 491(3): 246–269.
21. Andrade, A., A. Sandoval, N. Oviedo, M. De Waard, D. Elias, and R. Felix. 2007. Proteolytic cleavage of the voltage-gated Ca^{2+} channel $\alpha_2\delta$ subunit: Structural and functional features. *Eur J Neurosci* 25(6): 1705–1710.
22. Buraei, Z., and J. Yang. 2013. Structure and function of the beta subunit of voltage-gated Ca^{2+} channels. *Biochim Biophys Acta* 1828(7): 1530–1540. doi:10.1016/j.bbamem.2012.08.028.
23. Bichet, D., V. Cornet, S. Geib, E. Carlier, S. Volsen, T. Hoshi, Y. Mori, and M. De. Waard 2000. The I-II loop of the Ca^{2+} channel α_1 subunit contains an endoplasmic reticulum retention signal antagonized by the β subunit. *Neuron* 25(1): 177–190.
24. Murakami, M., H. Yamamura, T. Suzuki, M. G. Kang, S. Ohya, A. Murakami, I. Miyoshi et al. 2003. Modified cardiovascular L-type channels in mice lacking the voltage-dependent Ca^{2+} channel β3 subunit. *J Biol Chem* 278(44): 43261–43267. doi:10.1074/jbc.M211380200.
25. Kharade, S. V., S. K. Sonkusare, A. K. Srivastava, K. M. Thakali, T. W. Fletcher, S. W. Rhee, and N. J. Rusch. 2013. The β3 subunit contributes to vascular calcium channel upregulation and hypertension in angiotensin II-infused C57BL/6 mice. *Hypertension* 61(1): 137–142. doi:10.1161/hypertensionaha.112.197863.
26. Tajada, S., P. Cidad, O. Colinas, L. F. Santana, J. R. Lopez-Lopez, and M. T. Perez-Garcia. 2013. Down-regulation of $Ca_V1.2$ channels during hypertension: How fewer $Ca_V1.2$ channels allow more Ca^{2+} into hypertensive arterial smooth muscle. *J Physiol* 591(24): 6175–6191. doi:10.1113/jphysiol.2013.265751.
27. Gutman, G. A., K. G. Chandy, S. Grissmer, M. Lazdunski, D. McKinnon, L. A. Pardo, G. A. Robertson et al. 2005. International Union of Pharmacology. LIII. Nomenclature and molecular relationships of voltage-gated potassium channels. *Pharmacol Rev* 57(4): 473–508. doi:10.1124/pr.57.4.10.
28. Jiang, Y., A. Lee, J. Chen, V. Ruta, M. Cadene, B. T. Chait, and R. MacKinnon. 2003. X-ray structure of a voltage-dependent K^+ channel. *Nature* 423(6935): 33–41. doi:10.1038/nature01580.
29. Ghosh, S., D. A. Nunziato, and G. S. Pitt. 2006. KCNQ1 assembly and function is blocked by long-QT syndrome mutations that disrupt interaction with calmodulin. *Circ Res* 98(8): 1048–1054. doi:10.1161/01.RES.0000218863.44140.f2.
30. Lu, Y., S. T. Hanna, G. Tang, and R. Wang. 2002. Contributions of $K_V1.2$, $K_V1.5$ and $K_V2.1$ subunits to the native delayed rectifier K^+ current in rat mesenteric artery smooth muscle cells. *Life Sci* 71(12): 1465–1473.

31. Nelson, M. T., and J. M. Quayle. 1995. Physiological roles and properties of potassium channels in arterial smooth muscle. *Am J Physiol* 268(4 Pt 1): C799–C822.
32. Yeung, S. Y., V. Pucovsky, J. D. Moffatt, L. Saldanha, M. Schwake, S. Ohya, and I. A. Greenwood. 2007. Molecular expression and pharmacological identification of a role for K(v)7 channels in murine vascular reactivity. *Br J Pharmacol* 151(6): 758–770. doi:10.1038/sj.bjp.0707284.
33. Miguel-Velado, E., A. Moreno-Dominguez, O. Colinas, P. Cidad, M. Heras, M. T. Perez-Garcia, and J. R. Lopez-Lopez. 2005. Contribution of K_V channels to phenotypic remodeling of human uterine artery smooth muscle cells. *Circ Res* 97(12): 1280–1287. doi:10.1161/01.res.0000194322.91255.13.
34. Vandenberg, J. I., M. D. Perry, M. J. Perrin, S. A. Mann, Y. Ke, and A. P. Hill. 2012. hERG K^+ channels: Structure, function, and clinical significance. *Physiol Rev* 92(3): 1393–1478.
35. Ko, E. A., W. S. Park, A. L. Firth, N. Kim, J. X. Yuan, and J. Han. 2010. Pathophysiology of voltage-gated K^+ channels in vascular smooth muscle cells: Modulation by protein kinases. *Prog Biophys Mol Biol* 103(1): 95–101. doi:10.1016/j.pbiomolbio.2009.10.001.
36. Korovkina, V. P., and S. K. England. 2002. Molecular diversity of vascular potassium channel isoforms. *Clin Exp Pharmacol Physiol* 29(4): 317–323.
37. Kidd, M. W., M. D. Leo, J. P. Bannister, and J. H. Jaggar. 2015. Intravascular pressure enhances the abundance of functional $K_V1.5$ channels at the surface of arterial smooth muscle cells. *Sci Signal* 8(390): ra83. doi:10.1126/scisignal.aac5128.
38. Berk, B. C., and M. A. Corson. 1997. Angiotensin II signal transduction in vascular smooth muscle: Role of tyrosine kinases. *Circ Res* 80(5): 607–616.
39. Kidd, M. W., S. Bulley, and J. H. Jaggar. 2016. Angiotensin II reduces the surface abundance of $K_V1.5$ channels in arterial myocytes to stimulate vasoconstriction. *J Physiol* doi:10.1113/jp272893.
40. Newton, A. C. 1995. Protein kinase C: Structure, function, and regulation. *J Biol Chem* 270(48): 28495–28498.
41. Rainbow, R. D., R. I. Norman, D. E. Everitt, J. L. Brignell, N. W. Davies, and N. B. Standen. 2009. Endothelin-I and angiotensin II inhibit arterial voltage-gated K^+ channels through different protein kinase C isoenzymes. *Cardiovasc Res* 83(3): 493–500. doi:10.1093/cvr/cvp143.
42. Cogolludo, A., L. Moreno, F. Lodi, G. Frazziano, L. Cobeno, J. Tamargo, and F. Perez-Vizcaino. 2006. Serotonin inhibits voltage-gated K^+ currents in pulmonary artery smooth muscle cells: Role of 5-HT_{2A} receptors, caveolin-1, and $K_V1.5$ channel internalization. *Circ Res* 98(7): 931–938. doi:10.1161/01.RES.0000216858.04599.e1.
43. Joseph, B. K., K. M. Thakali, A. R. Pathan, E. Kang, N. J. Rusch, and S. W. Rhee. 2011. Postsynaptic density-95 scaffolding of Shaker-type K^+ channels in smooth muscle cells regulates the diameter of cerebral arteries. *J Physiol* 589 (Pt 21): 5143–5152. doi:10.1113/jphysiol.2011.213843.
44. Moore, C. L., P. L. Nelson, N. K. Parelkar, N. J. Rusch, and S. W. Rhee. 2014. Protein kinase A-phosphorylated K_V1 channels in PSD95 signaling complex contribute to the resting membrane potential and diameter of cerebral arteries. *Circ Res* 114(8): 1258–1267. doi:10.1161/circresaha.114.303167.
45. Dabertrand, F., C. Kroigaard, A. D. Bonev, E. Cognat, T. Dalsgaard, V. Domenga-Denier, D. C. Hill-Eubanks, J. E. Brayden, A. Joutel, and M. T. Nelson. 2015. Potassium channelopathy-like defect underlies early-stage cerebrovascular dysfunction in a genetic model of small vessel disease. *Proc Natl Acad Sci USA* 112(7): E796–E805. doi:10.1073/pnas.1420765112.
46. Dietrich, H. H., and R. G. Dacey. 2000. Molecular keys to the problems of cerebral vasospasm. *Neurosurgery* 46(3): 517–530.

47. Ishiguro, M., A. D. Morielli, K. Zvarova, B. I. Tranmer, P. L. Penar, and G. C. Wellman. 2006. Oxyhemoglobin-induced suppression of voltage-dependent K^+ channels in cerebral arteries by enhanced tyrosine kinase activity. *Circ Res* 99(11): 1252–1260. doi:10.1161/01.RES.0000250821.32324.e1.
48. Wu, R. S., and S. O. Marx. 2010. The BK potassium channel in the vascular smooth muscle and kidney: α- and β-subunits. *Kidney Int* 78(10): 963–974. doi:10.1038/ki.2010.325.
49. Tao, X., R. K. Hite, and R. MacKinnon. 2017. Cryo-EM structure of the open high-conductance Ca^{2+}-activated K^+ channel. *Nature* 541(7635): 46–51. doi:10.1038/nature20608.
50. Torres, Y. P., S. T. Granados, and R. Latorre. 2014. Pharmacological consequences of the coexpression of BK channel α and auxiliary β subunits. *Front Physiol* 5: 383. doi:10.3389/fphys.2014.00383.
51. Butler, A., S. Tsunoda, D. P. McCobb, A. Wei, and L. Salkoff. 1993. mSlo, a complex mouse gene encoding "maxi" calcium-activated potassium channels. *Science* 261(5118): 221–224.
52. Poulsen, A. N., H. Wulf, A. Hay-Schmidt, I. Jansen-Olesen, J. Olesen, and D. A. Klaerke. 2009. Differential expression of BK channel isoforms and β-subunits in rat neuro-vascular tissues. *Biochim Biophys Acta* 1788(2): 380–389. doi:10.1016/j.bbamem.2008.10.001.
53. Kyle, B. D., and A. P. Braun. 2014. The regulation of BK channel activity by pre- and post-translational modifications. *Front Physiol* 5: 316. doi:10.3389/fphys.2014.00316.
54. Zarei, M. M., N. Zhu, A. Alioua, M. Eghbali, E. Stefani, and L. Toro. 2001. A novel MaxiK splice variant exhibits dominant-negative properties for surface expression. *J Biol Chem* 276(19): 16232–16239. doi:10.1074/jbc.M008852200.
55. Xie, J., and D. P. McCobb. 1998. Control of alternative splicing of potassium channels by stress hormones. *Science* 280(5362): 443–446.
56. Tian, L., R. R. Duncan, M. S. Hammond, L. S. Coghill, H. Wen, R. Rusinova, A. G. Clark, I. B. Levitan, and M. J. Shipston. 2001. Alternative splicing switches potassium channel sensitivity to protein phosphorylation. *J Biol Chem* 276(11): 7717–7720. doi:10.1074/jbc.C000741200.
57. Nourian, Z., M. Li, M. D. Leo, J. H. Jaggar, A. P. Braun, and M. A. Hill. 2014. Large conductance Ca^{2+}-activated K^+ channel (BK_{Ca}) α-subunit splice variants in resistance arteries from rat cerebral and skeletal muscle vasculature. *PLoS One* 9(6): e98863. doi:10.1371/journal.pone.0098863.
58. Jaggar, J. H., A. Li, H. Parfenova, J. Liu, E. S. Umstot, A. M. Dopico, and C. W. Leffler. 2005. Heme is a carbon monoxide receptor for large-conductance Ca^{2+}-activated K^+ channels. *Circ Res* 97(8): 805–812.
59. Zhao, G., A. Adebiyi, Q. Xi, and J. H. Jaggar. 2007. Hypoxia reduces K_{Ca} channel activity by inducing Ca^{2+} spark uncoupling in cerebral artery smooth muscle cells. *Am J Physiol Cell Physiol* 292(6): C2122–C2128.
60. Leo, M. D., J. P. Bannister, D. Narayanan, A. Nair, J. E. Grubbs, K. S. Gabrick, F. A. Boop, and J. H. Jaggar. 2014. Dynamic regulation of beta1subunit trafficking controls vascular contractility. *Proc Natl Acad Sci USA* 111(6): 2361–2366. doi:10.1073/pnas.1317527111.
61. Leo, M. D., S. Bulley, J. P. Bannister, K. P. Kuruvilla, D. Narayanan, and J. H. Jaggar. 2015. Angiotensin II stimulates internalization and degradation of arterial myocyte plasma membrane BK channels to induce vasoconstriction. *Am J Physiol Cell Physiol* 309(6): C392–C402. doi:10.1152/ajpcell.00127.2015.
62. Sonnichsen, B., R. S. De, E. Nielsen, J. Rietdorf, and M. Zerial. 2000. Distinct membrane domains on endosomes in the recycling pathway visualized by multicolor imaging of Rab4, Rab5, and Rab11. *J Cell Biol* 149(4): 901–914.

63. Zerial, M., and H. McBride. 2001. Rab proteins as membrane organizers. *Nat Rev Mol Cell Biol* 2(2): 107–117.
64. Yan, J., and R. W. Aldrich. 2012. BK potassium channel modulation by leucine-rich repeat-containing proteins. *Proc Natl Acad Sci USA* 109(20): 7917–7922. doi:10.1073/pnas.1205435109.
65. Knaus, H. G., M. Garcia-Calvo, G. J. Kaczorowski, and M. L. Garcia. 1994. Subunit composition of the high conductance calcium-activated potassium channel from smooth muscle, a representative of the mSlo and slowpoke family of potassium channels. *J Biol Chem* 269(6): 3921–3924.
66. Leo, M. D., X. Zhai, K. P. Kuruvilla, S. Bulley, F. A. Boop, and J. H. Jaggar. 2017. Membrane depolarization activates BK channels through ROCK-mediated β1 subunit surface trafficking to limit vasoconstriction. *Sci Signal* 10 (478): pii: eaah5417.
67. Evanson, K. W., J. P. Bannister, M. D. Leo, and J. H. Jaggar. 2014. LRRC26 is a functional BK channel auxiliary gamma subunit in arterial smooth muscle cells. *Circ Res* 115(4): 423–431. doi:10.1161/circresaha.115.303407.
68. Yan, J., and R. W. Aldrich. 2010. LRRC26 auxiliary protein allows BK channel activation at resting voltage without calcium. *Nature* 466(7305): 513–536. doi:10.1038/nature09162.
69. Earley, S., and J. E. Brayden. 2015. Transient receptor potential channels in the vasculature. *Physiol Rev* 95(2): 645–690. doi:10.1152/physrev.00026.2014.
70. Adebiyi, A., G. Zhao, D. Narayanan, C. M. Thomas-Gatewood, J. P. Bannister, and J. H. Jaggar. 2010. Isoform-selective physical coupling of TRPC3 channels to IP_3 receptors in smooth muscle cells regulates arterial contractility. *Circ Res* 106(10): 1603–1612. doi:10.1161/circresaha.110.216804.
71. Adebiyi, A., C. M. Thomas-Gatewood, M. D. Leo, M. W. Kidd, Z. P. Neeb, and J. H. Jaggar. 2012. An elevation in physical coupling of type 1 inositol 1,4,5-trisphosphate (IP_3) receptors to transient receptor potential 3 (TRPC3) channels constricts mesenteric arteries in genetic hypertension. *Hypertension* 60(5): 1213–1219. doi:10.1161/hypertensionaha.112.198820.
72. Adebiyi, A., D. Narayanan, and J. H. Jaggar. 2011. Caveolin-1 assembles type 1 inositol 1,4,5-trisphosphate receptors and canonical transient receptor potential 3 channels into a functional signaling complex in arterial smooth muscle cells. *J Biol Chem* 286(6): 4341–4348.
73. Crnich, R., G. C. Amberg, M. D. Leo, A. L. Gonzales, M. M. Tamkun, J. H. Jaggar, and S. Earley. 2010. Vasoconstriction resulting from dynamic membrane trafficking of TRPM4 in vascular smooth muscle cells. *Am J Physiol Cell Physiol* 299(3). C682–C694.
74. Narayanan, D., S. Bulley, M. D. Leo, S. K. Burris, K. S. Gabrick, F. A. Boop, and J. H. Jaggar. 2013. Smooth muscle cell transient receptor potential polycystin-2 (TRPP2) channels contribute to the myogenic response in cerebral arteries. *J Physiol* 591(20): 5031–5046. doi:10.1113/jphysiol.2013.258319.
75. Sharif-Naeini, R., J. H. Folgering, D. Bichet, F. Duprat, I. Lauritzen, M. Arhatte, M. Jodar et al. 2009. Polycystin-1 and -2 dosage regulates pressure sensing. *Cell* 139(3): 587–596. doi: 10.1016/j.cell.2009.08.045.

5 Total Internal Reflection Fluorescence Microscopy in Vascular Smooth Muscle

Madeline Nieves-Cintrón, Sendoa Tajada, L. Fernando Santana, and Manuel F. Navedo

CONTENTS

5.1 INTRODUCTION

Total internal reflection fluorescence (TIRF) is a microscopy technique that relies on the evanescent field generated when light waves are totally internally reflected at the coverslip-buffer interface.[1,2] This distinct illumination mode allows the selective excitation of fluorophores and subsequent visualization of fluorescence signals in a restricted region of a sample near the glass-media interface. The evanescent field generated has excitation qualities similar to the original light source and only excites fluorophores within ~100 nm of the coverslip. The limited penetration of the evanescent field decreases out-of-focus fluorescent signals, thus enabling the study of cellular events at membrane/submembrane regions with high signal-to-noise ratio. Indeed, a critical advantage of TIRF microscopy (TIRFM) over other microscopy approaches is its ability to acquire images of membrane/submembrane regions with high resolution and contrast. This has led to a flurry of studies that examine mechanisms mediating protein trafficking, and calcium (Ca^{2+}) signals in membrane/submembrane regions.[3–6]

The elemental second messenger Ca^{2+} plays an important role in diverse signaling processes in excitable cells, including vascular smooth muscle. It is increasingly recognized that spatial and temporal regulation of Ca^{2+} signals provides the means for controlling signaling specificity and excitability in vascular smooth muscle.[7] Indeed, the proper integration of multiple Ca^{2+} signaling events that influence intracellular Ca^{2+} ($[Ca^{2+}]_i$) in vascular smooth muscle is required to arrive at the appropriate physiological response. This is achieved through steep Ca^{2+} ion gradients, which are maintained between the cytosol and the extracellular milieu, as well as intracellular organelles through active transport. Different classes of Ca^{2+}-permeable channels at the plasma membrane and sarcoplasmic reticulum (SR) contribute to the movement of Ca^{2+} ions down their gradient, thus affecting changes in intracellular Ca^{2+} and further fine-tuning Ca^{2+} signal specificity and cell excitability.[7] Whereas Ca^{2+} signals produced by Ca^{2+}-permeable channels located in the SR have been extensively examined using confocal and/or epifluorescence microscopy, those mediated by Ca^{2+}-permeable channels at the plasma membrane are less understood due to technical limitations with these techniques. The development of faster and brighter Ca^{2+} indicators, together with the advent of TIRFM, has facilitated the study of plasmalemmal Ca^{2+}-permeable channels and their contribution to Ca^{2+} signals and cellular function.

The Ca^{2+} signals produced by the opening of plasmalemmal Ca^{2+}-permeable channels are known as *sparklets*. In this chapter, we provide an overview of the basic concepts behind TIRFM, and its application to the study of the function and regulation of plasmalemmal Ca^{2+}-permeable channels in vascular smooth muscle. This approach could be easily applicable to the study of Ca^{2+}-permeable channels in other cells.

5.2 TOTAL INTERNAL REFLECTION FLUORESCENCE MICROSCOPY

In this section, key aspects of TIRFM are summarized, but more in-depth reviews about the physics behind TIRFM can be found in the published literature.[1,2,6] TIRFM utilizes an evanescent wave to selectively excite fluorophores in regions of a sample directly adjacent to the glass coverslip-buffer interface (e.g., plasma membrane of

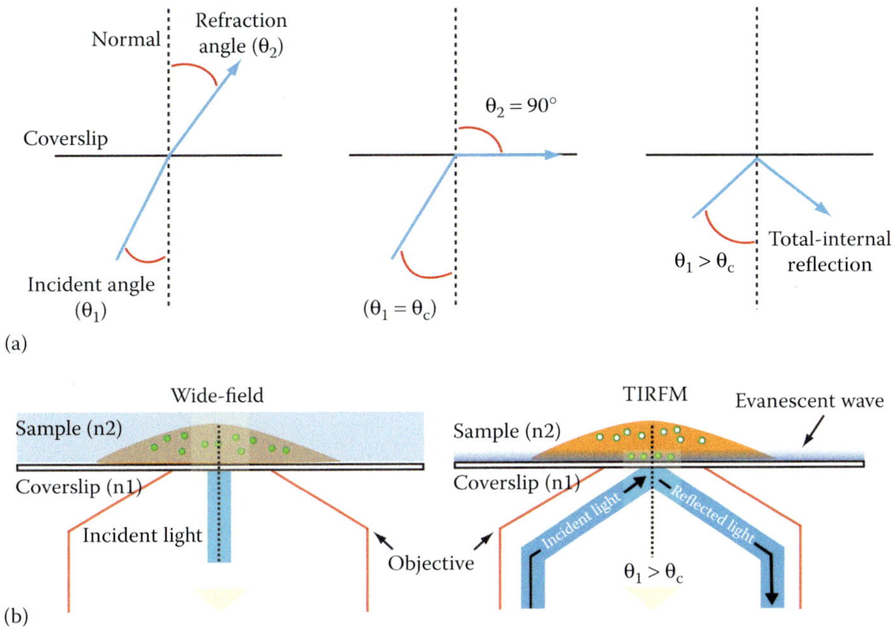

FIGURE 5.1 Diagram of a basic TIRFM system. (a) Schematic representation of the bending of incident light as it propagates through medium of different refractive indexes and (b) diagrams of specimen illumination in wide-field and TIRFM. Please see text for details.

adherent cells) (Figure 5.1). Such arrangement allows for unparalleled signal-to-noise ratio, where one can visualize the fluorophores located in the near-membrane space, with little noise being emitted from fluorophores in other parts of the cell. The principles at the heart of TIRFM are based on the laws of refraction of light and properties of the refractive media.[2] The refractive index (n) of a material describes how light propagates through that material compared to how it will travel in a vacuum. It is defined as $n = c/v$, where c is the speed of light in a vacuum and v is the velocity of light in the material. When light travels through mediums of different refractive indexes, the light rays will bend in function of the angle of incidence in accordance to Snell's law:

$$n_1 \sin\theta_1 = n_2 \sin\theta_2$$

where:

- n_1 is the medium with higher refractive index
- n_2 is the lower refractive index medium
- θ_1 is the angle of incidence with respect to the normal
- θ_2 is the angle of refracted light at the interface into the medium of lower refractive index

For example, vascular smooth muscle cells in aqueous media are allowed to adhere to the glass coverslip, which represents the media of high refractive index (n_1), whereas the aqueous buffer is the media with lower refractive index (n_2). The angle of incidence determines the angle of refraction (i.e., if the angle of incidence increases, so does the angle of refraction). When the angle of incidence θ_1 reaches a critical value θ_c (i.e., $\theta_1 = \theta_c$), the refraction angle $\theta_2 = 90°$ degrees, and the refracted light travels along the surface of the coverslip. According to Snell's law, the critical angle θ_c is given by

$$\theta_c = \sin^{-1}\left(\frac{n_2}{n_1}\right)$$

When the angle of incidence is greater than the critical angle, light is no longer refracted, but totally reflected (Figure 5.1a). The transition from refraction to reflection is a continuous process in which light goes from predominantly refracted with minimal reflection to total internal reflection back into the coverslip when the angle of incidence exceeds the critical angle. This phenomenon of total internal reflection leads to the generation of an electromagnetic field at the interface, which is known as the evanescent wave or evanescent field. This evanescent wave has excitation properties similar to those of the light source (Figure 5.1b). The evanescent wave propagates along the surface and its intensity decays exponentially with increasing distance from the interface, as described by the following equation:

$$I_z = I_0 \exp\left(\frac{-Z}{d}\right)$$

where:

I_z is the intensity at a perpendicular distance z from the interface
I_0 is the intensity of the evanescent field at the interface (i.e., $z = 0$)
d is the penetration depth

The penetration depth is the perpendicular distance from the interface at which the excitation intensity is 1/e of I_0. In other words, it is the distance at which the intensity of the evanescent field excitation decays 37% of its intensity at the interface. (In depth discussion of I_0 and how it relates to the incident beam angle and polarization are described elsewhere[1,2]). The wavelength (λ) of the excitation light, incidence angle of excitation (θ), and the refractive indices of the glass coverslip and the sample (n_1 and n_2, respectively) determine the penetration depth according to the following equation:

$$d = \frac{\lambda_0}{4\pi\sqrt{n_1^2 \sin\theta - n_2^2}}$$

Therefore, a lower penetration depth can be achieved with larger angles of incidence. In most TIRFM systems, the evanescent wave extends ~70–200 nm into the sample. This characteristic of the evanescent wave is the key to the axial resolution achieved with TIRFM (Figure 5.1). In practice, the penetration depth can be adjusted by changing the angle of incidence, since the excitation wavelength is usually limited by available fluorophores, and the refractive indices of the cell are out of the investigator's control. This is an important consideration when setting up a TIRFM system, particularly an objective-based system, since the objective numerical aperture (NA) directly influences the angles of incidence (see below).

TIRFM relies on the ability to introduce light at angles exceeding a critical angle. Initially, the required angle of illumination was achieved by using a prism to direct excitation light to the appropriate angle of incidence. Prism-based TIRFM has many advantages, including the cost effectiveness, generation of a purer evanescent wave due to minimal scattering of light and ease of customization. A drawback, however, is that it requires the glass coverslip to be optically coupled to the prism using liquid substances like glycerol, which complicates the set-up of each experiment and limits the potential use in high throughput applications. In addition, positioning of the prism in some configurations may limit access to the specimen, thus rendering manipulations cumbersome. This is an important consideration for studies using TIRFM to examine ion channel function, as it is recommended that TIRFM be applied in combination with whole-cell patch clamp (see Section 5.3.2).[8]

Objective-based TIRFM, in which excitation light is delivered through the objective, is also available. An advantage of objective-based TIRFM is that it does not require optical coupling between the coverslip and the prism, so it is easier to use. Other advantages include the ability to easily change angle of incidence and wavelength. Achieving angles larger than the critical angle requires an objective with high NA. The NA of the objective limits the range of possible angles of incidence. The minimal NA necessary for TIRFM is 1.4, but higher values (1.45–1.49) are desirable since they allow broader ranges of illumination angles and hence penetration depths.[6]

5.3 MEASURING Ca^{2+}-PERMEABLE CHANNEL ACTIVITY IN THE PLASMA MEMBRANE OF VASCULAR SMOOTH MUSCLE CELLS WITH TOTAL INTERNAL REFLECTION FLUORESCENCE MICROSCOPY

A particular innovative use of TIRFM has been on the recording of Ca^{2+} signals produced by the opening of Ca^{2+}-permeable channels at the plasma membrane of a cell, including vascular smooth muscle cells.[4,9–15] These Ca^{2+} signals are known as Ca^{2+} sparklets, a term first coined to define Ca^{2+} influx produced by the opening of voltage-gated Ca^{2+} channels in cardiac cells.[16] Since then, optical recording of Ca^{2+} influx produced by other Ca^{2+}-permeable channels and pores has been described, thus making the use of the term sparklet generic.[4,9–13,17–22] To remove ambiguity, the term *sparklet* is usually preceded by the name of the underlying channel (e.g., L-type $Ca_V1.2$ sparklet).

The optical recording of Ca^{2+} influx through Ca^{2+}-permeable channels is important because our ability to visualize micro and/or nanodomain Ca^{2+} signals at the plasma membrane in space and time is essential for fully understanding cell function under physiological and pathological conditions. The biophysical properties of Ca^{2+}-permeable channels at the plasma membrane of any cell have been mainly examined using the patch-clamp technique. Although powerful, this approach has some limitations in that it doesn't provide information about the localization and/or organization of functional Ca^{2+}-permeable channels. The combined use of patch-clamp with TIRFM has helped overcome these limitations and has revealed new features associated with channel activity and gating modality, which has improved our understanding of the function of plasmalemmal Ca^{2+}-permeable channels. Much of this new information has been obtained by optically recording the activity of plasmalemmal Ca^{2+}-permeable channels with TIRFM in vascular smooth muscle cells.[8,12,14,15] In this section we describe a protocol for isolation of vascular smooth muscle cells, which is critical for acquisition of high quality data, and technical details related to the use of TIRFM to study submembrane Ca^{2+} signals in these cells.

5.3.1 Vascular Smooth Muscle Cell Isolation

The quality of vascular smooth muscle cells is critical for successful recording of sparklets using TIRFM. Single vascular smooth muscle cells can be obtained by enzymatic digestion of freshly dissected arteries from different vascular beds. Dissociation protocols are provided for two well-established preparations: cerebral arteries and mesenteric arteries (third and fourth order). For cerebral vascular smooth muscle isolation, arteries are dissected at 4°C in magnesium physiological saline solution (Mg^{2+}PSS) buffer composed of (in mM): 140 NaCl, 5 KCl, 2 $MgCl_2$, 10 D-glucose, and 10 HEPES, pH 7.4 with NaOH. Clusters of arteries, which are floating freely and ideally settled at the bottom of a Wheaton V-vial, are digested in Mg^{2+}PSS buffer supplemented with papain (1 mg/mL; Worthington LS003119) and dithiothreitol (1 mg/mL; Sigma D9163) in a water bath at 37°C for 7 minutes, then incubated in Mg^{2+}PSS buffer supplemented with collagenase type H (0.3 mg/mL; Sigma C8051) and collagenase type F (0.7 mg/mL; Sigma C7926) at 37°C for 7 minutes. For mesenteric vascular smooth muscle cells isolation, clusters of freely floating arteries that settled at the bottom of a Wheaton V-vial are incubated in Mg^{2+}PSS buffer supplemented with papain (1 mg/mL) and dithiothreitol (1 mg/mL) in a water bath at 37°C for 10 minutes. After 10 minutes, the papain enzyme mix is taken out and replaced with Mg^{2+}PSS buffer supplemented with 1.77 mg ml collagenase type H (Sigma; C8051), Elastase (0.5 mg/ml; Worthington LS002292), and 1 mg/ml Soybean Trypsin Inhibitor (1 mg/ml; Sigma T9003). Immediately after the enzyme solution is removed, arteries are washed three times in ice-cold Mg^{2+}PSS buffer, being careful to not disturb the cluster of vessels. After the last wash, arteries are allowed to rest in ice-cold Mg^{2+}PSS buffer for 30 minutes before smooth muscle dissociation. In our experience, clusters of arteries that remain floating closer to the air-solution interface yield poor-quality cells after dissociation. Glass pipettes of decreasing diameters are used to triturate arteries to obtain single vascular smooth muscle cells by gently pipetting the digested arteries in solution up and down. Isolated cells are maintained in ice-cold Mg^{2+}PSS buffer until used. Figure 5.2 shows examples of high-quality and poor-quality cells for recording of sparklets using TIRFM.

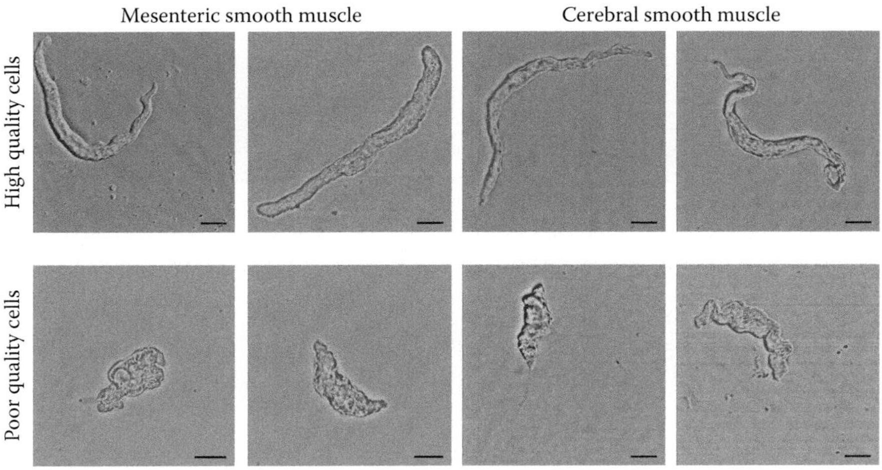

FIGURE 5.2 DIC image showing high quality (upper panels) or poor quality (lower panels) freshly dissociated mouse mesenteric and cerebral vascular smooth muscle cells for sparklet experiments using TIRFM. High quality smooth muscle cells exhibit relaxed, spindle-shaped and smoothed morphology, and attach very well to the coverslip. The resultant footprint of the TIRFM image may vary from that observed in the DIC as it is determined by cell area within the evanescent wave field. Poor quality cells are opaque and have a (semi-) contracted morphology that may be due to over digestion during the dissociation process. Scale bars = 10 μm.

5.3.2 Optical Recording of Plasmalemmal Ca^{2+}-Permeable Channels in Vascular Smooth Muscle Using Total Internal Reflection Fluorescence Microscopy

TIRFM, in combination with patch clamp electrophysiology, has been employed to optically record the activity of several plasmalemmal Ca^{2+}-permeable channels, including L-type $Ca_V1.2$ channels and transient receptor potential vanilloid 4 (TRPV4) channels, in vascular smooth muscle cells (Figure 5.3).[12,13] TIRFM has also been used to optically record TRPV4 channel activity in airway smooth muscle cells and L-type Ca^{2+} channel activity in smooth muscle from urinary bladder and portal veins.[20,23,24]

In a typical experiment, isolated smooth muscle cells are allowed to attach to a glass coverslip of a perfusion chamber mounted on an inverted microscope coupled to a TIRF module. After cell adhesion, cells are washed by continuous superfusion and patched in the whole-cell patch-clamp configuration. This accomplishes two main objectives: (1) control of membrane potential, which facilitates manipulation of the open probability of a channel and the driving force for Ca^{2+} influx, and (2) a conduit for delivery of substances (e.g., indicators, peptides, non-fluorescent Ca^{2+} buffers) through the patch pipette and into the cell. To eliminate Ca^{2+} release from the SR and prevent contamination of plasmalemmal Ca^{2+} signals with those produced by Ca^{2+} release from intracellular stores (e.g., Ca^{2+} sparks), cells are kept in physiological saline solution supplemented with the sarco/endoplasmic reticulum Ca^{2+}-ATPase (SERCA) inhibitor thapsigargin (1 μM), which prevents SR loading.

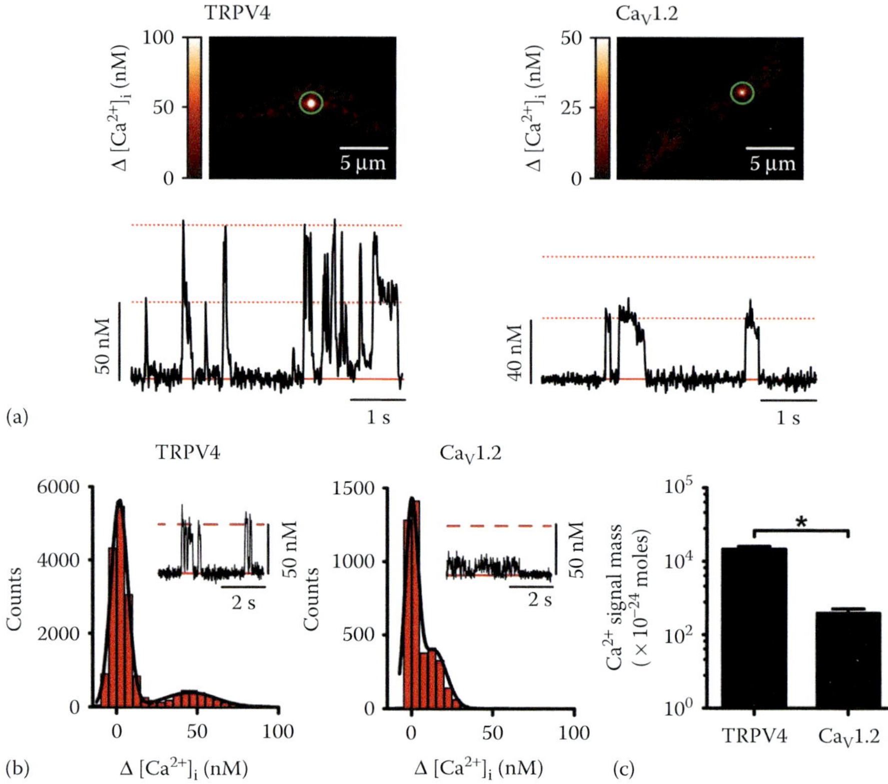

FIGURE 5.3 Examples of L-type Ca$_V$1.2 and TRPV4 sparklets in vascular smooth muscle. (a) TIRFM images of TRPV4 and L-type Ca$_V$1.2 sparklets recorded from freshly isolated vascular smooth muscle. TRPV4 sparklets were recorded with 2 mM extracellular Ca^{2+}, whereas L-type Ca$_V$1.2 sparklets were recorded with 20 mM extracellular Ca^{2+} in order to increase the driving force for Ca^{2+} and the amplitude of the quantal event. The traces below each image show the time-course of [Ca^{2+}]$_i$ in the site highlighted by the green circle in the images. Dotted lines represent the quantal level. (b) The amplitude of the signal mass of quantal TRPV4 and L-type Ca$_V$1.2 sparklets under physiological extracellular Ca^{2+} conditions. The insets show representative TRPV4 and L-type Ca$_V$1.2 sparklet recordings obtained with 2 mM extracellular Ca^{2+}. The solid line represents basal [Ca^{2+}]$_i$, whereas the dotted line marks the amplitude of quantal TRPV4 sparklets. Note that the quantal amplitude of L-type Ca$_V$1.2 sparklets significantly smaller than that of TRPV4 sparklets with 2 mM extracellular Ca^{2+}. The all-points histograms were generated from TRPV4 and L-type Ca$_V$1.2 sparklets recorded from vascular smooth muscle. The solid lines on the histograms represent best fits to data for TRPV4 (q = 48 nM) and L-type Ca$_V$1.2 (q = 18 nM) sparklets, and (c) the bar plot reveals that the signal mass of TRPV4 sparklets is larger than that of L-type Ca$_V$1.2 sparklets under similar experimental conditions. (Adapted from Mercado, J. et al., *J. Gen. Physiol.*, 143, 559–575, 2014. With permission.)

Once the whole-cell configuration is established, cells are loaded through the patch pipette with a fluorescent Ca^{2+} indicator, such as Fluo-5F (200 μM) and an excess of a non-fluorescent Ca^{2+} buffer, like EGTA (10 mM). Fluorescent indicators such as Fluo-5F are relatively fast binding Ca^{2+} dyes, whereas EGTA binds Ca^{2+} at a slower rate. The objective of this combination of indicator and EGTA is to facilitate that the faster Ca^{2+} indicator binds to Ca^{2+} first, thus producing a fluorescent signal. The duration of this signal will be limited as the slower but high affinity Ca^{2+} chelator EGTA buffers Ca^{2+} away. This maneuver will restrict the fluorescent signal to the Ca^{2+} entry site.[25] Images are acquired using through-the-lens TIRFM built around an inverted microscope equipped with 60x oil-immersion lens with appropriate NA (numerical aperture = 1.45), as described earlier. The microscope should be coupled to an ultra-fast (e.g., 100–500 Hz), highly sensitive (e.g., >90% quantum efficiency) electron multiplying charged-coupled device (EMCCD) camera, which is capable of detecting photons emitted from a single fluorescent molecule and recording two-dimensional images of dim fluorescent events at high speed. The use of this type of camera facilitates the recording of small, rare Ca^{2+} signal events, as well as allowing the imaging of relatively large areas of the cell. The very thin optical section afforded by the evanescent field during TIRFM (Figure 5.1b) extends ~100 nM from the surface of the coverslip, hence allowing the recording of small Ca^{2+} signals (e.g., sparklets) from submembrane regions, which are produced by the opening of Ca^{2+}-permeable channels, with high signal-to-noise ratio and adequate temporal resolution (Figure 5.3 for e.g., L-type $Ca_V1.2$ and TRPV4 sparklets).

5.4 SPARKLET ANALYSIS

To rigorously detect and analyze Ca^{2+} sparklet events, several automatic detection algorithms have been developed.[13,26,27] These algorithms facilitate analysis and prevent biases. An example of such an algorithm, developed by the Santana group, is the detection of Ca^{2+} signals as sparklets based on amplitude, and spatial and temporal criteria.[8,13,26] To be considered a sparklet, an elevation in Ca^{2+} needs to have an amplitude equal to or greater than the mean basal $[Ca^{2+}]_i$ plus three times its standard deviation. A sparklet site is then defined by a grid of 3 × 3 continuous pixels with $[Ca^{2+}]_i$ values at or above the amplitude threshold. Fluorescence signals can then be converted to concentration units to analyze sparklet events (after background subtraction) using the F_{max} equation as described by Maravall:

$$[Ca^{2+}] = K_d \frac{F/F_{max} - 1/R_f}{1 - F/F_{max}}$$

where:

- F is the fluorescence
- F_{max} is the indicator fluorescence intensity in the presence of saturating Ca^{2+} concentrations
- K_d is the indicator dissociation constant
- R is the indicator F_{max}/F_{min}
- F_{min} is the fluorescence intensity of the indicator in Ca^{2+} free solution

K_d and R_f are properties of the indicator and, as such, are not influenced by factors that may vary among cells like resting [Ca^{2+}], for example. F_{max} should be obtained at the end of each experiment by exposing the cells to 20 mM Ca^{2+} external solution supplemented with the Ca^{2+} ionophore ionomycin (10 μM). This approach facilitates calibration of fluorescence signals within and between cells, which is critical for comparison during different experimental conditions.

After the signal is transformed to [Ca^{2+}] units, and given that the time course of Ca^{2+} sparklet events is comparable to that of single-channel current recordings, open probability analysis analogous to that used for single-channel currents can be applied to determine sparklet activity.[28] Thus, Ca^{2+} sparklet activity can be expressed as nPs for each sparklet site, where n is the number of quantal levels, and Ps is the probability that a Ca^{2+} sparklet site is active. Amplitude histograms can also be generated from these data and fitted with a Gaussian function such that the quantal nature of the signal can be revealed, if any. Poisson distribution analysis can also be implemented to assess the randomness of the spatial distribution of Ca^{2+} sparklet sites.

5.5 DIVERSE Ca^{2+} SPARKLET EVENTS IN VASCULAR SMOOTH MUSCLE

The most extensive characterization of sparklet events in vascular smooth muscle has been performed for L-type $Ca_V1.2$ sparklets and TRPV4 sparklets.[12,13,26] L-type $Ca_V1.2$ channels and TRPV4 channels play important roles in smooth muscle contraction and relaxation pathways, respectively.[29,30] The identification of L-type $Ca_V1.2$ channels and TRPV4 channels as the molecular entities underlying L-type $Ca_V1.2$ sparklets and TRPV4 sparklets, respectively, is based on multiple pharmacological, biophysical and molecular biology criteria for each channel.[12,13,26] For instance, L-type $Ca_V1.2$ sparklets can be activated by the L-type Ca^{2+} channel agonist Bay-K 8644 and inhibited by the L-type Ca^{2+} channel antagonist nifedipine, whereas TRPV4 sparklets are activated by the specific TRPV4 agonist GSK1016790A and are inhibited by the specific TRPV4 inhibitor HC-067047, but not by nifedipine.[12,13,26] It is worth noting that additional highly potent and very selective TRPV4 inhibitors are now available (e.g., GSK293874). Importantly, simultaneous recording of Ca^{2+} signals and underlying currents indicate that L-type $Ca_V1.2$ sparklets and TRPV4 sparklets are associated with an inward current with appropriate biophysical properties for each particular channel.[12,13,26] Furthermore, sparklets in tsA-201 cells expressing L-type $Ca_V1.2$ channels or TRPV4 channels reproduce all the basic features of native L-type $Ca_V1.2$ sparklets or TRPV4 sparklets, respectively.[12,13,26] This rigorous characterization should be extended to the study of sparklet events produced by any other plasmalemmal Ca^{2+}-permeable channel when using TIRFM.

L-type $Ca_V1.2$ and/or TRPV4 sparklets may influence [Ca^{2+}]$_i$, which may have important implications for vascular smooth muscle excitability. The contribution of sparklet events to [Ca^{2+}]$_i$ can be defined by the *signal mass* method developed by Zou and colleagues.[12,28,31] For this analysis, simultaneous electrical and optical recording of events is recommended. Total fluorescence (F_{total}) produced by a sparklet event is obtained from raw images by summing the indicator-associated fluorescence from all the pixels in an area of the image sufficiently large as to encompass the entire fluorescence signal produced by a sparklet site. To determine the change in fluorescence (ΔF_{total}), the total fluorescence before the channel open is subtracted from the fluorescence at each time point during an opening. The signal mass is calculated by determining the peak of the integral of ΔF_{total} trace over time ($\int \Delta F_{total} dt$) for each Ca^{2+} sparklet. The charge influx (ΔQ) can be obtained by integrating the Ca^{2+}current underlying each sparklet. The signal mass of a Ca^{2+} sparklet and ΔQ have a linear relationship,[12,28] and the slope of this correlation can be used to convert Ca^{2+} sparklet signal mass into ΔQ values. The ΔQ values can then be used to provide an estimate of the potential impact of Ca^{2+} sparklets on $[Ca^{2+}]_i$ using the following equation:

$$\Delta[\mathrm{Ca}^{2+}]_i = \frac{\Delta Q_{Ca}}{2FVBC}$$

where:

ΔQ_{Ca} is the change in Ca^{2+} influx (as calculated from signal mass analysis)
F is the Faraday constant
V refers to the accessible cytosolic volume
BC is the buffering capacity of the cell

Using this approach, L-type $Ca_V1.2$ sparklets were found to have an impact on local and global $[Ca^{2+}]_i$.[28] Interestingly, the signal mass of TRPV4 sparklets is ~100 times larger than that for L-type $Ca_V1.2$ sparklets, suggesting that total Ca^{2+} flux is higher through TRPV4 channels than L-type $Ca_V1.2$ channels in vascular smooth muscle (Figure 5.3b).[12] Considering the strong link between an increase in $[Ca^{2+}]_i$ and vascular smooth muscle contraction, it is thus remarkable that TRPV4 channel activity is associated with relaxation pathways.[30] Note, however, that the frequency of TRPV4 sparklet events is significantly lower compared to that of L-type $Ca_V1.2$ sparklets, thus allowing L-type $Ca_V1.2$ channels to conceivably contribute to global changes in $[Ca^{2+}]_i$.[12] Furthermore, the higher Ca^{2+} flux through TRPV4 channels may strengthen the functional coupling between these channels and its intended target (i.e., ryanodine receptors in the SR leading to activation of BK_{Ca} channels) to contribute to relaxation of vascular smooth muscle.[12,30] These observations highlight the complex nature of local Ca^{2+} signals in modulating global $[Ca^{2+}]_i$ and vascular smooth muscle excitability.[32]

5.6 UNIQUE FEATURES OF Ca^{2+} SPARKLETS IN VASCULAR SMOOTH MUSCLE REVEALED BY TOTAL INTERNAL REFLECTION FLUORESCENCE MICROSCOPY

The ability to optically record submembrane Ca^{2+} signals (i.e., sparklets) produced by the opening of plasmalemmal Ca^{2+}-permeable channels that are critical for vascular smooth muscle excitability using TIRFM has revealed important features regarding the differences in gating modality, and distinct regulation of activity and spatial organization of functional channels.[8,14,15]

5.6.1 Heterogeneous Sparklet Activity

The first major finding revealed by TIRFM is that both L-type $Ca_V1.2$ sparklets and TRPV4 sparklet sites are restricted to particular membrane regions.[12,13,26] These results are particularly striking considering the broad expression of L-type $Ca_V1.2$ channels and TRPV4 channels throughout the plasmalemma of vascular smooth muscle cells, as well as the stochastic nature of the channels' activity.[12,13,33,34] Indeed, Poisson analysis suggests that the spatial distribution of both L-type $Ca_V1.2$ sparklets and TRPV4 sparklet sites is non-stochastic, being higher at specific sites within the plasmalemma of vascular smooth muscle cells.[12,13] In addition, it was found that some L-type $Ca_V1.2$ sparklets showed stochastic and sporadic activity, whereas other sites have nearly continuous Ca^{2+} influx (e.g., persistent L-type $Ca_V1.2$ sparklets), thus suggesting that their activity is heterogeneous.[13,17,26] Multiple approaches have been used to determine the mechanisms responsible for regional variation in persistent Ca^{2+} sparklet activity in vascular smooth muscle. Extensive reviews on the subject can be found elsewhere.[8,14,15] Briefly, a working model has been proposed in which heterogeneous L-type $Ca_V1.2$ channel activity is determined by regional activity of protein kinases and phosphatases acting on neighboring channels to modulate their gating properties.[13,17,26,35,36] The scaffolding protein A kinase anchoring protein 150 (AKAP150) facilitates the formation of these functional units by bringing together protein kinases and phosphatases to the vicinity of the L-type $Ca_V1.2$ channel.[35–38] This local control model has profound functional implications considering that a surge in the frequency of persistent L-type $Ca_V1.2$ sparklets may result in an increase in Ca^{2+} influx, leading to vascular smooth muscle contraction. Indeed, persistent L-type $Ca_V1.2$ sparklet activity is increased in vascular smooth muscle during angiotensin II-induced hypertension and in animal models of diabetes via mechanisms that engage distinct kinase signaling pathways.[35,36,39] A similar AKAP150-driven arrangement has been found for TRPV4 channels in vascular smooth muscle, but its relevance during pathological conditions remains to be elucidated.[12]

5.6.2 Coupled Gating of Ion Channels in Vascular Smooth Muscle

Another unique feature revealed by TIRFM is that small clusters of L-type $Ca_V1.2$ and TRPV4 channels in vascular smooth muscle can open and close in unison (e.g., coupled gating).[12,37,40,41] Consistent with this coupled gating model, closer examination of the amplitude distribution of L-type $Ca_V1.2$ sparklets and TRPV4 sparklets, as well as the respective underlying Ca^{2+} currents, has demonstrated that the frequency of multichannel openings exceeds the probability of simultaneous openings of independently gated channels. A complementary approach to study coupled events is by quantifying coupling behavior based on a Markov chain model.[42] This model can be applied to Ca^{2+} sparklet and/or Ca^{2+} current traces to yield a coupling coefficient or coupling strength of a particular site. In this model, the activity of an ion channel can be modeled as a first-order, discrete Markov chain, and the transition matrix can be estimated from the sparklet and/or current records and their corresponding channel opening time courses using a built-in

Hidden Markov parameter estimation function. The estimated probability matrix can be modeled as a partially coupled Markov chain with three free parameters: κ is the coupling coefficient (or coupling strength) and ρ and ζ are the channel open-to-open and close-to-close probabilities, respectively. The value of κ can range from 0 for channels gating independently to 1 for channels exhibiting complete synchronization of openings. Implementation of this analysis has confirmed that while the majority of sparklet events produced by L-type $Ca_V1.2$ channels and TRPV4 channels result from the independent gating of channels, there is a subpopulation of channels that can undergo coupled gating.[37,40,41,43]

Coupled gating of Ca^{2+}-permeable channels, such as L-type $Ca_V1.2$ channels and TRPV4 channels, has important implications for Ca^{2+} signaling in vascular smooth muscle. Indeed, this gating modality may result in Ca^{2+} signaling amplification by oligomerization of Ca^{2+}-permeable channels, which may contribute to regulate vascular smooth muscle excitability. Consistent with this, the frequency of L-type $Ca_V1.2$ channels with coupled events is significantly elevated in vascular smooth muscle during hypertension and in diabetes, which could enhance arterial tone during these pathological conditions.[35,37] Representative TRPV4 coupled events can be observed in Figure 5.3a. The identification of coupled events within Ca^{2+} sparklets exemplifies the usefulness of complementary approaches (e.g., optics, electrophysiology, mathematical modeling) in creating a better comprehensive picture of physiological and pathological phenomena.

5.7 ADDITIONAL APPLICATIONS OF TOTAL INTERNAL REFLECTION FLUORESCENCE MICROSCOPY IN VASCULAR SMOOTH MUSCLE

Because of its intrinsic properties, TIRFM has been extensively used to visualize submembrane protein trafficking in living cells, including vascular smooth muscle. This application requires tagging of the protein of interest with a fluorescent tag, such as the enhanced green fluorescent protein (eGFP), and subsequent expression of the construct of interest in the desired cells (e.g., vascular smooth muscle). Cells are allowed to settle and attach to the coverslip, after which, images are collected using the TIRFM with appropriate laser excitation before and after a stimulus. Subsequently, fluorescent intensity is measured in each frame and plotted against time. Using a comparable approach, it was found that dynamic membrane trafficking of the transient receptor potential melastatin 4 (TRPM4) channel in vascular smooth muscle cells contributed to vascular reactivity through a mechanism that requires protein kinase C (PKC) δ activity.[44] Another study found that in cultured A7r5 cells, hydrogen peroxide stimulated the trafficking of transient receptor potential canonical 6 (TRPC6) channels.[45] However, care should be taken with interpretation of these results and their extrapolation to effects on vascular reactivity, as many of these experiments are typically performed in cultured, non-contractile cells due to technical limitations of transfection/infection of native cells. Thus, these trafficking experiments should be complemented with additional approaches, such as surface biotinylation assay using native vascular smooth muscle.[46]

The Amberg Group developed another application that employs the TIRFM technique for the simultaneous recording of L-type $Ca_V1.2$ sparklets and the generation of endogenous reactive oxygen species (ROS).[47] Using the ROS indicator DCF and the Ca^{2+} indicator fluo 5F, this group demonstrated that the generation of ROS puncta localize and preceded the activation of persistent L-type $Ca_V1.2$ sparklets, which contributed to regulation of vascular reactivity during angiotensin II signaling.[47–49] Altogether, these data highlight the power of TIRFM in uncovering key mechanistic information in different processes.

5.8 CONCLUDING REMARKS

TIRFM is a powerful technique that has enabled the examination of numerous processes at the membrane/submembrane region with high spatiotemporal resolution. TIRFM has deepened our understanding of submembrane Ca^{2+} signals by facilitating the study of Ca^{2+} microdomains mediated by plasmalemmal Ca^{2+}-permeable channels. Accordingly, the use of TIRFM in combination with conventional electrophysiology has revealed unique features about the organization, function, and regulation of plasmalemmal L-type $Ca_V1.2$ channels and TRPV4 channels in vascular smooth muscle during physiological and pathological conditions. TIRFM can also be used to unmask distinctive elementary Ca^{2+} signals produced by other Ca^{2+}-permeable channels, as well as their trafficking patterns and potentially unique functional modulation by kinases, phosphatases, and signaling molecules in vascular smooth muscle. This will be critical to gain a better appreciation of how these processes shape vascular smooth muscle function and arterial reactivity in physiological and pathological conditions. Finally, it will be important to apply this technique to the study of Ca^{2+}-permeable channels and signaling proteins in native vascular smooth muscle cells from human subjects. This approach may reveal new information about the organization, function, regulation, and trafficking of ion channels and signaling proteins specific to humans.

ACKNOWLEDGMENTS

We thank Dr. Arsalan Syed and Miss Maria Paz Prada for critically reading previous versions of this chapter. This work was supported by National Institutes of Health grants 1R01HL098200, 1R01HL121059, 1R01HL085686, and 1R01HL085870.

REFERENCES

1. Axelrod, D. Selective imaging of surface fluorescence with very high aperture microscope objectives. *J Biomed Opt*. 2001;6:6–13.
2. Axelrod, D, Thompson, NL, Burghardt, TP. Total internal inflection fluorescent microscopy. *J Microsc*. 1983;129 Pt 1:19–28.
3. Nystoriak, MA, Nieves-Cintron, M, Navedo, MF. Capturing single L-type Ca(2+) channel function with optics. *Biochim Biophys Acta*. 2013;1833:1657–1664.
4. Demuro, A, Parker, I. Imaging single-channel calcium microdomains. *Cell Cal*. 2006;40:413–422.
5. Mattheyses, AL, Simon, SM, Rappoport, JZ. Imaging with total internal reflection fluorescence microscopy for the cell biologist. *J Cell Sci*. 2010;123:3621–3628.

6. Fish, KN. Total internal reflection fluorescence (TIRF) microscopy. *Curr Protoc Cytom*. 2009;Chapter 12:Unit12–18.
7. Amberg, GC, Navedo, MF. Calcium dynamics in vascular smooth muscle. *Microcirculation*. 2013;20:281–289.
8. Santana, LF, Navedo, MF. Molecular and biophysical mechanisms of Ca^{2+} sparklets in smooth muscle. *J Mol Cell Cardiol*. 2009;47:436–444.
9. Demuro, A, Parker, I. Imaging the activity and localization of single voltage-gated Ca^{2+} channels by total internal reflection fluorescence microscopy. *Biophys J*. 2004;86:3250–3259.
10. Demuro, A, Parker, I. Optical single-channel recording: Imaging Ca^{2+} flux through individual ion channels with high temporal and spatial resolution. *J Biomed Opt*. 2005;10:11002.
11. Demuro, A, Parker, I. Optical patch-clamping: Single-channel recording by imaging Ca^{2+} flux through individual muscle acetylcholine receptor channels. *J Gen Physiol*. 2005;126:179–192.
12. Mercado, J, Baylie, R, Navedo, MF, Yuan, C, Scott, JD, Nelson, MT, Brayden, JE, Santana, LF. Local control of TRPV4 channels by AKAP150-targeted pkc in arterial smooth muscle. *J Gen Physiol*. 2014;143:559–575.
13. Navedo, MF, Amberg, G, Votaw, SV, Santana, LF. Constitutively active L-type Ca^{2+} channels. *Proc Natl Acad Sci USA*. 2005;102:11112–11117.
14. Navedo, MF, Amberg, GC. Local regulation of L-type Ca(2+) channel sparklets in arterial smooth muscle. *Microcirculation*. 2013;20:290–298.
15. Navedo, MF, Santana, LF. Cav1.2 sparklets in heart and vascular smooth muscle. *J Mol Cell Cardiol*. 2013;58:67–76.
16. Wang, SQ, Song, LS, Lakatta, EG, Cheng, H. Ca^{2+} signalling between single L-type Ca^{2+} channels and ryanodine receptors in heart cells. *Nature*. 2001;410:592–596.
17. Navedo, MF, Amberg, GC, Westenbroek, RE, Sinnegger-Brauns, MJ, Catterall, WA, Striessnig, J, Santana, LF. Cav1.3 channels produce persistent calcium sparklets, but $Ca_v1.2$ channels are responsible for sparklets in mouse arterial smooth muscle. *Am J Physiol Heart Circ Physiol*. 2007;293:H1359–H1370.
18. Sonkusare, SK, Bonev, AD, Ledoux, J, Liedtke, W, Kotlikoff, MI, Heppner, TJ, Hill-Eubanks, DC, Nelson, MT. Elementary Ca^{2+} signals through endothelial TRPV4 channels regulate vascular function. *Science*. 2012;336:597–601.
19. Sullivan, MN, Gonzales, AL, Pires, PW, Bruhl, A, Leo, MD, Li, W, Oulidi, A, et al. Localized TRPA1 channel Ca^{2+} signals stimulated by reactive oxygen species promote cerebral artery dilation. *Sci Signal*. 2015;8:ra2.
20. Zhao, L, Sullivan, MN, Chase, M, Gonzales, AL, Earley, S. Calcineurin/nuclear factor of activated t cells-coupled vanilloid transient receptor potential channel 4 Ca^{2+} sparklets stimulate airway smooth muscle cell proliferation. *Amer J Respir Cell Mol Biol*. 2014;50:1064–1075.
21. Demuro, A, Smith, M, Parker, I. Single-channel Ca(2+) imaging implicates Aβ1-42 amyloid pores in Alzheimer's disease pathology. *J Cell Biol*. 2011;195:515–524.
22. Ullah, G, Demuro, A, Parker, I, Pearson, JE. Analyzing and modeling the kinetics of amyloid beta pores associated with Alzheimer's disease pathology. *PLoS One*. 2015;10:e0137357.
23. Sidaway, P, Teramoto, N. L-type Ca^{2+} channel sparklets revealed by TIRF microscopy in mouse urinary bladder smooth muscle. *PLoS One*. 2014;9:e93803.
24. McCarron, JG, Olson, ML, Currie, S, Wright, AJ, Anderson, KI, Girkin, JM. Elevations of intracellular calcium reflect normal voltage-dependent behavior, and not constitutive activity, of voltage-dependent calcium channels in gastrointestinal and vascular smooth muscle. *J Gen Physiol*. 2009;133:439–457.
25. Zenisek, D, Davila, V, Wan, L, Almers, W. Imaging calcium entry sites and ribbon structures in two presynaptic cells. *J Neurosci*. 2003;23:2538–2548.

26. Navedo, MF, Amberg, GC, Nieves, M, Molkentin, JD, Santana, LF. Mechanisms underlying heterogeneous Ca^{2+} sparklet activity in arterial smooth muscle. *J Gen Physiol.* 2006;127:611–622.
27. Francis, M, Qian, X, Charbel, C, Ledoux, J, Parker, JC, Taylor, MS. Automated region of interest analysis of dynamic Ca(2)+ signals in image sequences. *Am J Physiol Cell Physiol.* 2012;303:C236–C243.
28. Amberg, GC, Navedo, MF, Nieves-Cintrón, M, Molkentin, JD, Santana, LF. Calcium sparklets regulate local and global calcium in murine arterial smooth muscle. *J Physiol.* 2007;579:187–201.
29. Knot, HJ, Nelson, MT. Regulation of arterial diameter and wall $[Ca^{2+}]$ in cerebral arteries of rat by membrane potential and intravascular pressure. *J Physiol.* 1998;508:199–209.
30. Earley, S, Heppner, TJ, Nelson, MT, Brayden, JE. TRPV4 forms a novel Ca^{2+} signaling complex with ryanodine receptors and BKCa channels. *Circ Res.* 2005;97:1270–1279.
31. Zou, H, Lifshitz, LM, Tuft, RA, Fogarty, KE, Singer, JJ. Using total fluorescence increase (signal mass) to determine the Ca^{2+} current underlying localized Ca^{2+} events. *J Gen Physiol.* 2004;124:259–272.
32. Earley, S. Smooth muscle cell Ca(2)(+): Think locally, act globally. *Microcirculation.* 2013;20:279–280.
33. Owsianik, G, Talavera, K, Voets, T, Nilius, B. Permeation and selectivity of TRP channels. *Annu Rev Physiol.* 2006;68:685–717.
34. Dai, S, Hall, DD, Hell, JW. Supramolecular assemblies and localized regulation of voltage-gated ion channels. *Physiol Rev.* 2009;89:411–452.
35. Navedo, MF, Takeda, Y, Nieves-Cintron, M, Molkentin, JD, Santana, LF. Elevated Ca^{2+} sparklet activity during acute hyperglycemia and diabetes in cerebral arterial smooth muscle cells. *Am J Physiol. Cell Physiol.* 2010;298:C211–C220.
36. Nystoriak, MA, Nieves-Cintron, M, Patriarchi, T, Buonarati, OR, Prada, MP, Morotti, S, Grandi, E et al. Ser1928 phosphorylation by PKA stimulates L-type Ca^{2+} channel Cav1.2 and vasoconstriction during acute hyperglycemia and diabetes. *Sci Signal.* 2017;10.
37. Navedo, MF, Cheng, EP, Yuan, C, Votaw, S, Molkentin, JD, Scott, JD, Santana, LF. Increased coupled gating of L-type Ca^{2+} channels during hypertension and Timothy syndrome. *Circul Res.* 2010;106:748–756.
38. Navedo, MF, Nieves-Cintron, M, Amberg, GC, Yuan, C, Votaw, VS, Lederer, WJ, McKnight, GS, Santana, LF. AKAP150 is required for stuttering persistent Ca^{2+} sparklets and angiotensin II-induced hypertension. *Circ Res.* 2008;102:e1–e11.
39. Nieves-Cintron, M, Amberg, GC, Navedo, MF, Molkentin, JD, Santana, LF. The control of Ca^{2+} influx and NFATC3 signaling in arterial smooth muscle during hypertension. *Proc Natl Acad Sci USA.* 2008;105:15623–15628.
40. Dixon, RE, Moreno, CM, Yuan, C, Opitz-Araya, X, Binder MD, Navedo, MF, Santana, LF. Graded $Ca^{(2)(+)}$/calmodulin-dependent coupling of voltage-gated Cav1.2 channels. *eLife.* 2015;4.
41. Dixon, RE, Yuan, C, Cheng, EP, Navedo, M, Santana, LF. Calcium signaling amplification by oligomerization of L-type Cav1.2 channels. *Proc Natl Acad Sci USA.* 2012;109:1749–1754.
42. Chung, SH, Kennedy, RA. Coupled Markov chain model: Characterization of membrane channel currents with multiple conductance sublevels as partially coupled elementary pores. *Math Biosci.* 1996;133:111–137.
43. Cheng, EP, Yuan, C, Navedo, MF, Dixon, RE, Nieves-Cintron, M, Scott, JD, Santana, LF. Restoration of normal L-type Ca^{2+} channel function during Timothy syndrome by ablation of an anchoring protein. *Circul Res.* 2011;109:255–261.
44. Crnich, R, Amberg, GC, Leo, MD, Gonzales, AL, Tamkun, MM, Jaggar, JH, Earley, S. Vasoconstriction resulting from dynamic membrane trafficking of TRPM4 in vascular smooth muscle cells. *Am J Physiol. Cell Physiol.* 2010;299:C682–C694.

45. Ding, Y, Winters, A, Ding, M, Graham, S, Akopova, I, Muallem, S, Wang, Y, et al. Reactive oxygen species-mediated TRPC6 protein activation in vascular myocytes, a mechanism for vasoconstrictor-regulated vascular tone. *J Biol Chem.* 2011;286:31799–31809.
46. Bannister, JP, Adebiyi, A, Zhao, G, Narayanan, D, Thomas, CM, Feng, JY, Jaggar, JH. Smooth muscle cell $\alpha_2\beta$-1 subunits are essential for vasoregulation by Cav1.2 channels. *Circul Res.* 2009;105:948–955.
47. Amberg, GC, Earley, S, Glapa, SA. Local regulation of arterial L-type calcium channels by reactive oxygen species. *Circul Res.* 2010;107:1002–1010.
48. Chaplin, NL, Amberg, GC. Hydrogen peroxide mediates oxidant-dependent stimulation of arterial smooth muscle l-type calcium channels. *Am J Physiol. Cell Physiol.* 2012;302:C1382–C1393.
49. Chaplin, NL, Nieves-Cintron, M, Fresquez, AM, Navedo, MF, Amberg, GC. Arterial smooth muscle mitochondria amplify hydrogen peroxide microdomains functionally coupled to L-type calcium channels. *Circ Res.* 2015;117:1013–1023.

6 T-Type Ca^{2+} Channels in Vascular Smooth Muscle

Properties, Regulation, and Function

Osama F. Harraz, Ahmed M. Hashad, and Donald G. Welsh

CONTENTS

6.1 OVERVIEW

The mammalian cardiovascular system comprises a muscular pump and a distribution network of arteries, veins, and capillaries. Within this integrated system, it is the resistance arteries and their inherent ability to change diameter that determines the magnitude and distribution of tissue blood flow [1–3]. Under dynamic conditions, arterial tone is regulated by multiple stimuli including changes in tissue metabolism [4,5], neural activity [6,7], blood flow [8], and intravascular pressure [9,10]. These vasoactive stimuli influence arterial diameter by regulating the phosphorylation state of the myosin regulatory light chain. This phosphorylation step is controlled by myosin light chain kinase, a key enzyme whose activity is guided by global changes in cytosolic Ca^{2+} concentration ([Ca^{2+}]) [3,11,12].

Voltage-gated Ca^{2+} channels principally set Ca^{2+} influx and thus cytosolic [Ca^{2+}], a fundamental relationship best observed through the study of the myogenic response. The myogenic response refers to the inherent ability of resistance arteries to respond to increases and decreases in intraluminal pressure by constricting or dilating, respectively [9–11]. This phenomenon, first described a century ago [13], is intrinsic to vascular smooth muscle and is essential for optimizing capillary pressure and blood flow delivery to critical organs including the brain, heart, and kidneys [3,11]. The myogenic response is triggered by depolarization, a rightward shift in steady state membrane potential (V_M). This depolarization grades the opening of voltage-gated Ca^{2+} channels, elevates cytosolic [Ca^{2+}] and evokes myosin light chain phosphorylation [10,14,15]. L-type Ca^{2+} channel activation dominates vascular smooth muscle contraction and the myogenic response [10], and those channels are thus therapeutic targets for a number of antihypertensive medications. Intriguingly, more recent work has noted that T-type Ca^{2+} channels also contribute to the myogenic response in a secondary manner [16–22] and these integral membrane proteins will be the primary focus of this brief review.

6.2 VOLTAGE-GATED Ca^{2+} CHANNELS

In excitable cells (e.g., smooth muscle cells, cardiac myocytes, and neurons), voltage-gated Ca^{2+} channels predominate as the main influx route for extracellular Ca^{2+} in response to a depolarization [23,24]. As their name implies, changes in membrane potential determine the channel activity. A depolarizing stimulus, sensed by the channel's voltage sensor, evokes a conformational change in the channel structure, shifting the channel from the closed to the open state (*activation*). Channels experiencing sustained depolarization *inactivate* (open to inactive) in a voltage-dependent manner. Repolarization of the membrane *deactivates* open voltage-gated Ca^{2+} channels (open to closed) and makes them available for future activation. Ten genes encode for the mammalian family of voltage-gated Ca^{2+} channels and their products are divided into three subfamilies that include: (1) $Ca_V1.x$ (L-type); (2) $Ca_V2.x$ (P/Q-, N-, and R-types); and (3) $Ca_V3.x$ (T-type) channels [25]. The subtypes can be further classified with respect to their thresholds of activation into being high-voltage activated (HVA) and low-voltage activated (LVA) channels [24]. Members of the HVA classification include the $Ca_V1.x$ and $Ca_V2.x$ subtypes, while LVA channels are restricted to the $Ca_V3.x$ subfamily [24,25].

Structurally, a voltage-gated Ca^{2+} channel consists of a single pore-forming α_1 subunit that is sufficient to facilitate Ca^{2+} influx. Although not universal among the ten different voltage-gated Ca^{2+} channels, the α_1 subunit can co-associate with auxiliary subunits (β, $\alpha_2\delta$, and γ; Figure 6.1) to form a heteromultimeric complex in which accessory subunits optimize surface expression, gating kinetics, and channel regulation [24]. The α_1 subunit encompasses four domains (I, II, III, and IV), each composed of six transmembrane segments (S1–S6). The four domains are connected into a single polypeptide by intracellular

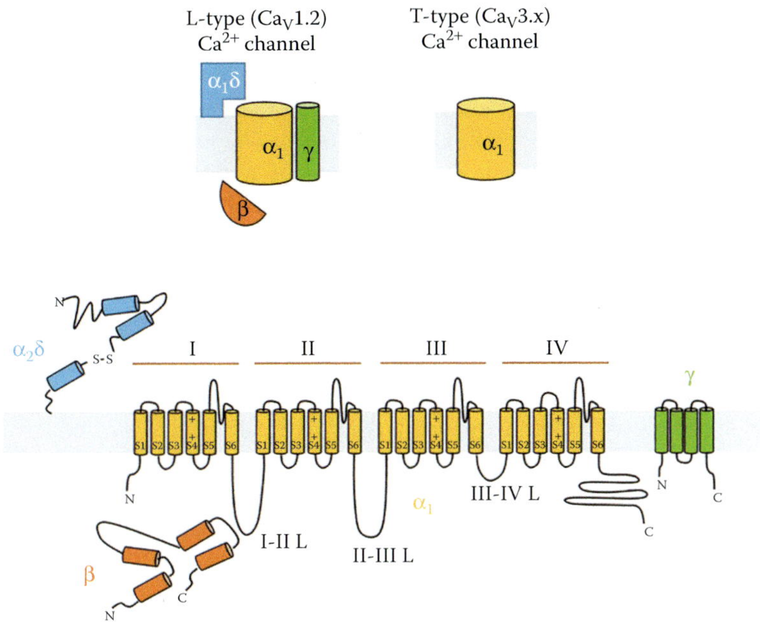

FIGURE 6.1 The subunit composition and topology of voltage-gated Ca^{2+} channels. High-voltage activated (HVA) channels (e.g., L-type) are heteromultimers of the pore-forming α_1 subunit and auxiliary subunits (β, $\alpha_2\delta$, and γ). Low-voltage activated (LVA) channels (T-type) are α_1 monomers. The α_1 subunit is composed of four domains (I-IV), and each domain comprises six transmembrane segments (S1–S6) and a re-entrant loop (green). The S4 segments (tagged +) in each domain represent the voltage sensor and contain positively charged amino acids. The selectivity filter is located between segment 5 and segment 6 in all domains. The domains are connected by intracellular linker loops (I-II L, II-III L, and III-IV L). The N- and C-termini are both intracellular.

linker loops (I II, II III, and III-IV); these linkers harbor sequence motifs implicated in auxiliary subunit association and post-translational modifications [26]. The C- and N-termini are intracellular and responsible for conferring key properties such as Ca^{2+}-dependent inactivation and protein kinase regulation [24,26–28]. The accessory β subunit is also intracellular and links to the I-II loop while the γ subunit is trans-membranous [24]. The $\alpha_2\delta$ subunit is extracellular and anchored to the plasma membrane via its δ component; it is thought to influence the expression and function of voltage-gated Ca^{2+} channels including the L-type Ca^{2+} channel expressed in vascular smooth muscle [29,30].

6.2.1 L-Type Ca^{2+} Channels in Arterial Smooth Muscle

L-type Ca^{2+} channels play a critical role in mediating Ca^{2+} flux that drives the myogenic response; the application of selective L-type dihydropyridine blockers (e.g., nifedipine) is known to attenuate pressure-induced increases in cytosolic [Ca^{2+}] and tone development in a variety of vessel beds [10,31]. Ca_V1.2 is the primary L-type channel involved and expression has been long confirmed in smooth muscle through mRNA and protein measurements, along with direct electrophysiological measurements [32]. The robust ability of L-type Ca^{2+} channel blockers to moderate pressure-induced tone has fostered the view that $Ca_V1.2$ is the sole Ca^{2+} channel subunit expressed in arterial smooth muscle [10,33]. While certainly a key conductance that enables extracellular Ca^{2+} influx, recent work has begun to reassess the breadth of this assertion given growing documentation of a dihydropyridine-insensitive T-type current in isolated smooth muscle cells [16,18,34].

6.2.2 T-Type Ca^{2+} Channels in Arterial Smooth Muscle

An increasing number of studies have detected expression of T-type Ca^{2+} channels in vascular smooth muscle isolated from cerebral, mesenteric, renal, coronary, pulmonary, ear and skeletal muscle arteries [18,35–40]. In contrast to L-type, T-type channels are LVA, with their activation and inactivation profiles shifted leftward to hyperpolarized voltages. The "T" denotes the *transient* nature of channel opening and they are further noted for rapid inactivation and small single-channel conductance (~1 pS in physiological $[Ca^{2+}]_o$; ~5 pS in 10 mM $[Ca^{2+}]_o$; and 6–8 pS in 110 mM $[Ca^{2+}]_o$ or $[Ba^{2+}]_o$) [24,41,42]. T-type channels are composed of a single monomeric α_1 subunit and there is to date no conclusive evidence of the $Ca_V3.x$ subunit co-associating with auxiliary proteins in native systems [24,41]. T-type Ca^{2+} channels are classified into three subtypes: $Ca_V3.1$ (α_{1G}), $Ca_V3.2$ (α_{1H}), and $Ca_V3.3$ (α_{1I}), encoded by the genes *CACNA1G*, *CACNA1H*, and *CACNA1I*, respectively. The former two subtypes ($Ca_V3.1$ and $Ca_V3.2$) are the main T-type channels expressed in cerebral and peripheral rodent resistance arteries [18,34,43,44]. Recent work has, however, noted a unique substitution in human cerebral arteries where $Ca_V3.3$ replaces $Ca_V3.1$ [21]. The $Ca_V3.3$ isoform has slower kinetics and displays relative depolarized shifts compared to $Ca_V3.1$, properties that may better suit T-type function in driving the myogenic response in human arteries.

6.2.2.1 Electrophysiological Delineation of Arterial T-Type Ca^{2+} Channels

The ensemble voltage-gated Ca^{2+} channel current in arterial smooth muscle is composed of L- and T-type subcomponents. The preferred electrophysiological means of delineating the $Ca_V3.x$ current is to assess the inward current once the L-type conductance has been eliminated with dihydropyridines [18,34] (Figure 6.2). The dihydropyridine of choice is nifedipine as it blocks the vascular $Ca_V1.2$ splice variant at nanomolar concentrations (IC_{50} ~ 10 nM) [45] with minimal impact on T-type channels [19,46]. The current component blocked by nifedipine (*nifedipine-sensitive*) is predominated by L-type $Ca_V1.2$ current. A residual voltage-dependent current is often observable in the presence of nifedipine (*nifedipine-insensitive*)—a current predominated by $Ca_V3.x$ channels—particularly in cells uncompromised by the dissociation procedure [16–18,34]. This residual component represents ~20% of the total current as measured in solutions using 10 mM Ba^{2+} as the charge carrier and using voltage-step protocols where Ba^{2+} currents reach a maximum amplitude at membrane potential between −15 and 15 mV. Such observations should not be interpreted to suggest that the L-type currents are four times greater than T-type, because $Ca_V1.2$ permeability to Ba^{2+} is 2–3 times greater than Ca^{2+}, while the two

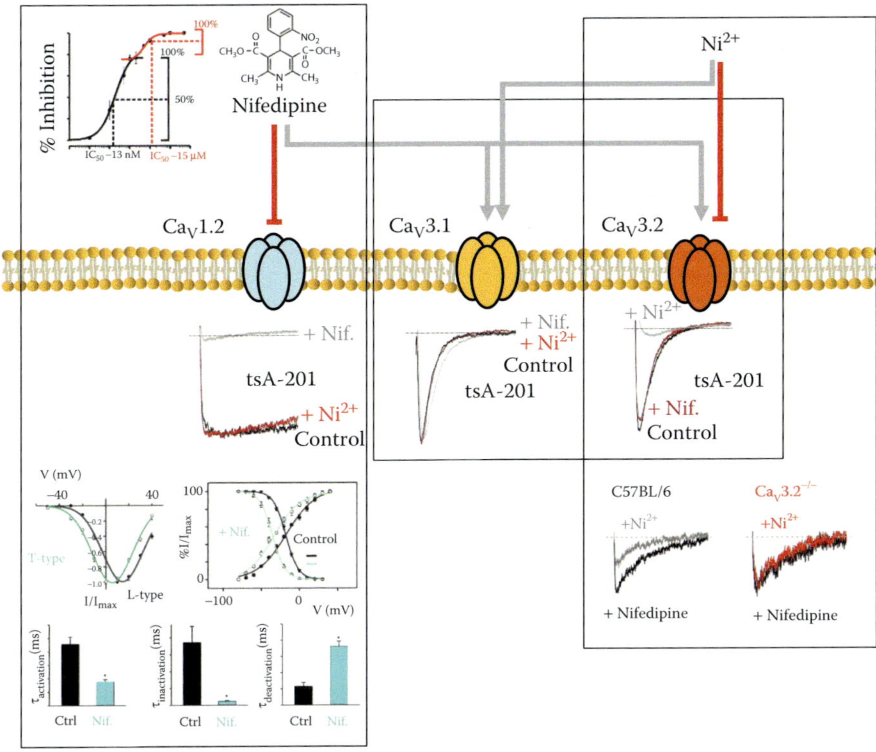

FIGURE 6.2 Delineation of arterial T-type Ca^{2+} channels. A schematic illustrates the approaches undertaken to separate the T-type conductance in rodent vascular smooth muscle. First, nifedipine dose response curve showed that 200 nM nifedipine should fully block $Ca_V1.2$ but not $Ca_V3.1$, or $Ca_V3.2$ currents in native smooth muscle cells; recordings in tsA-201 cells coincided with this perspective. Voltage dependence and kinetic analyses confirmed that nifedipine-sensitive and -insensitive currents were dominated by $Ca_V1.2$ and $Ca_V3.1$/$Ca_V3.2$, respectively. To delineate between $Ca_V3.1$ and $Ca_V3.2$ current, Ni^{2+} (50 μM) was used and it selectively abolished $Ca_V3.2$ current as denoted in tsA-201 cells and mouse mesenteric artery smooth muscle cells. Observations were originally published in Harraz et al. (2014a & 2015b).

divalent cations equally permeate T-type pores [41,47]. To determine whether the nifedipine-insensitive component is truly representative of T-type channel activity, investigators have relied on three strategies, the first focused on documenting the hyperpolarized shift in the current-voltage relationship and activation/inactivation profiles, consistent with low-voltage activation profile [18,34]. Second, a subset of studies have employed deeper kinetic analyses to determine whether the nifedipine-insensitive component activates/inactivates faster while deactivating slower than the nifedipine-sensitive (L-type) component [48,49]. Lastly, studies have also employed pharmacology to ascertain whether T-type blockers including mibefradil, NNC 55-0396, efonidipine, and kurtoxin abolished the residual current [34,48].

In considering the separation of T-type Ca^{2+} channels from L-type, three issues should be carefully acknowledged. First, nifedipine can suppress T-type channels if applied at concentrations in the micromolar range, ~100–1000 times above the IC_{50} for the arterial smooth muscle L-type Ca^{2+} channel [46]. Second, many T-type channel inhibitors (e.g., mibefradil, NNC 55-0396) lack target specificity and affect other conductances including, but not limited to, L-type Ca^{2+} channels and voltage-dependent K^+ channels [50–53]. This off targeting is evident in arterial smooth muscle [18,48,54], so careful protocol design is essential when applying T-type blockers in a functional setting where vessels are actively generating tone [34]. It is deemed necessary to first block co-expressed $Ca_V1.2$ currents with pharmacological tools not interfering with T-type current before applying the less selective $Ca_V3.x$ blockers [48]. Lastly, the vascular T-type current appears to display a depolarized voltage profile compared to other tissues [41], leading some to coin the term *high voltage activated* T-type currents [16]. In this regard, recent vascular studies have isolated a unique splice variant of $Ca_V3.1$ with depolarized voltage dependence [55,56]. Notwithstanding, the rightward shift may also reflect the experimental conditions and the permeant ionic species used in the experiment. Compared to physiological levels of $[Ca^{2+}]_o$ (i.e., 1.8 mM), high $[Ba^{2+}]_o$ evokes a screening effect and profound rightward shift in the voltage dependence of activation/inactivation for voltage-gated Ca^{2+} currents [43].

6.2.2.2 $Ca_V3.x$ Subtype Separation in Arterial Smooth Muscle

While an increasing number of studies have reported a T-type current in arterial smooth muscle, only a handful have attempted to separate among the two constitutive subtypes (i.e., $Ca_V3.1$ and $Ca_V3.2$) [19,48]. Separation is typically approached by exploiting the channels' differential sensitivity to Ni^{2+}; the IC_{50} for $Ca_V3.2$ ranges between 10 and 20 μM, whereas a 20-fold higher concentration is required to achieve 50% block of $Ca_V3.1$ in heterologous expression and native systems [19,57–59]. Using 50 μM Ni^{2+}, a concentration confirmed to selectively block $Ca_V3.2$ but not $Ca_V1.2$/$Ca_V3.1$ in TsA-201 cells, Harraz and colleagues have repeatedly shown that ~50% of the nifedipine-insensitive inward current in arterial smooth muscle is suppressed by this divalent cation [19,48] (Figure 6.2). This Ni^{2+}-induced suppression is clearly present in vascular smooth muscle from rat, mouse, and human cerebral arteries, consistent with the literature and over-expressions systems, but notably absent in cells from mesenteric arteries harvested from $Ca_V3.2^{-/-}$ knockout mice [19,21,44]. It should be noted that the availability of animal models in which $Ca_V3.1$ or $Ca_V3.2$ are genetically ablated [37,60] has made it possible for vascular investigations to explore these $Ca_V3.x$ isoform contribution to vascular function [20,44,61]. The ability to separate among the two sub-components has been important to the investigative field as it has enabled researchers to methodically examine the regulatory and functional properties of each conductance with clarity [22,49].

6.2.3 Regulation of Arterial T-Type Ca^{2+} Channels

Investigative interest in vascular $Ca_V3.x$ channels has largely focused on documenting their presence at the mRNA and protein level and providing robust electrophysiological evidence of a T-type channel fingerprint [18,34]. Emerging studies have begun to ascertain whether one or both $Ca_V3.x$ subtypes are a target of regulation and whether such modification impacts arterial tone development [43,49]. In sharp contrast to the scarcity of evidence in vascular tissues, work on neuronal and secretory tissues has richly reported that T-type channels are regulatory targets of hormones (e.g., angiotensin II) and neurotransmitters (e.g., serotonin, cholinergic, or adrenergic transmitters) [62–65]. These endogenous molecules are agonists that act on G-protein coupled receptors and their key downstream targets include: (1) protein kinase A (PKA), G (PKG), and C (PKC); (2) calmodulin dependent kinase II; and (3) enzymes linked to the generation of reactive oxygen species (ROS) [41,64,65].

6.2.3.1 Protein Kinase Regulation of Arterial T-Type Ca^{2+} Channels

Ion channels in arterial smooth muscle are known regulatory targets of second messenger/kinase signaling cascades, including those tied to PKA and PKG [33,66] G-protein coupled receptors activate PKA by stimulating adenylyl cyclase (AC) and initiating the production of cAMP; nitric oxide mobilizes soluble guanylyl cyclase (sGC) driving cGMP production and PKG activity. Both protein kinases are robustly linked to vasodilatory responses and in this regard, it follows that channels governing extracellular Ca^{2+} influx are likely targets of regulation [67,68]. In fact, a phosphorylation map of $Ca_V3.2$ channels in the brain has been developed and revealed that this isoform harbors many potential phosphorylation sites for different serine-threonine kinases, which influences channel gating [69].

Building upon this logic, Harraz et al. [43] began to examine the relationship between PKA and arterial T-type channels by activating the β-adrenergic receptor and its downstream PKA linked signaling pathway (Figure 6.3). Findings clearly demonstrated that β-adrenergic receptor activation (isoproterenol), stimulation of adenylyl cyclase (forskolin), or the use of a cell-permeable derivative of cAMP

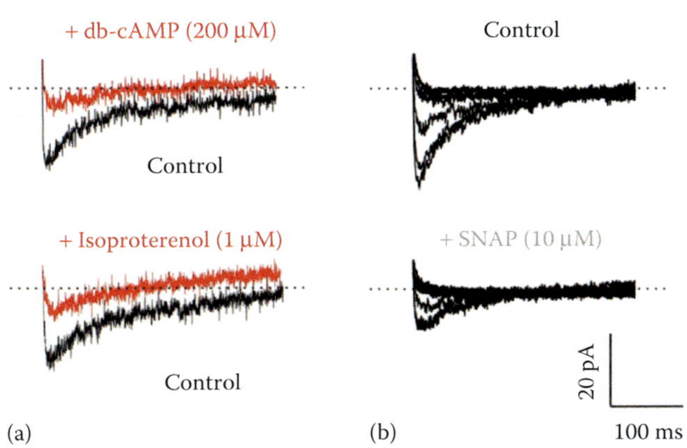

FIGURE 6.3 Regulation of arterial T-type Ca^{2+} channels. (a) Representative traces highlighting the suppressive effects of PKA activation on the T-type current (test voltage step −10 mV). PKA was activated indirectly through the β-adrenergic receptor (isoproterenol) or directly via db-cAMP, (b) representative traces highlighting the suppressive effects of PKG activation on the T-type current (voltage steps range between −50 and +10 mV). PKG was activated through the addition of SNAP, a nitric oxide donor. (From Harraz, O.F. and Welsh, D.G., *J. Cell Sci.*, 126, 2944–2954, 2013a; Harraz, O.F. et al., *Am. J. Physiol. Heart Circ. Physiol.*, 306, H279–H285, 2014b.)

inhibited T-type currents and evoked a leftward shift in steady-state inactivation. Inhibition of PKA had no effect on basal $Ca_V3.x$ currents but was clearly able to mask PKA-mediated suppression [43]. T-type suppression was absent in the presence of Ni^{2+}, suggestive of $Ca_V3.2$ being the primary target of PKA [43]. The selective targeting of $Ca_V3.2$ but not $Ca_V3.1$ is consistent with the expression of the 25bc splice variant of $Ca_V3.1$ in arterial smooth muscle [56], a variant known to lack a strong consensus site for PKA phosphorylation [70,71]. PKG signaling is a second regulatory pathway involved in arterial vasodilation and its activation by sGC and cGMP production also suppressed T-type currents [49] (Figure 6.3). Inhibitors of PKG or sGC masked current suppression, without effect on the basal $Ca_V3.x$ activity, a finding consistent with a regulatory model in which NO inhibits vascular T-type channels through the prototypic PKG signaling cascade and not through S-nitrosylation of putative extracellular cysteine residues [49]. Recent vascular studies have alternatively suggested that NO might regulate T-type activity by modulating superoxide production [54,72,73].

6.2.3.2 Reactive Oxygen Species and the Regulation of Arterial T-Type Ca^{2+} Channels

In vascular smooth muscle, ROS are generated through sources that include, but are not limited to, NADPH oxidase and mitochondria [74]. These sources modify key ion channels involved in membrane potential regulation and/or extracellular Ca^{2+} influx ($Ca_V1.2$), impacting tone development in both healthy and diseased tissues [75,76]. Howitt and colleagues first suggested that T-type channels were a defined target of ROS regulation, the result of which enhanced arterial tone development [73]. This perspective grew from functional measurements in which the T-type current was argued to increase in accordance with the oxidative stress that accompanies the activation of NADPH oxidase [73]. While direct electrophysiological evidence was absent, recent patch clamp observations have confirmed that T-type channels are a selective ROS target. In particular, Hashad et al. [77] has shown that angiotensin II (a vasoconstrictor peptide) suppresses the T-type current through augmenting the production of ROS via NADPH oxidase 1. This regulatory relationship is perhaps unsurprising in retrospect, as the same investigation has been noted by proximity ligation assay, a close association between $Ca_V3.2$ and NADPH oxidase 1 in vascular smooth muscle cells [77]. The impact of this targeted inhibition is discussed in the subsequent sections.

6.2.4 Arterial T-Type Ca^{2+} Channels and Vascular Tone Control

Ca^{2+} influx through T-type channels has attracted considerable interest in regard to arterial function and the contractile state of vascular smooth muscle [18,34]. In theory, Ca^{2+} movement through T-type channels could impact tone development through one of two general mechanisms [48] (Figure 6.4). First, extracellular influx, like that of L-type channels, could *directly* elevate the cytosolic Ca^{2+} pool augmenting myosin light chain phosphorylation [3,11,12]. Alternatively, influx through T-type channels could work *discretely and indirectly* targeting Ca^{2+} sensitive conductances that first alter membrane potential and then cytosolic [Ca^{2+}] through gating of the L-type channel. Each perspective will be discussed briefly in the following.

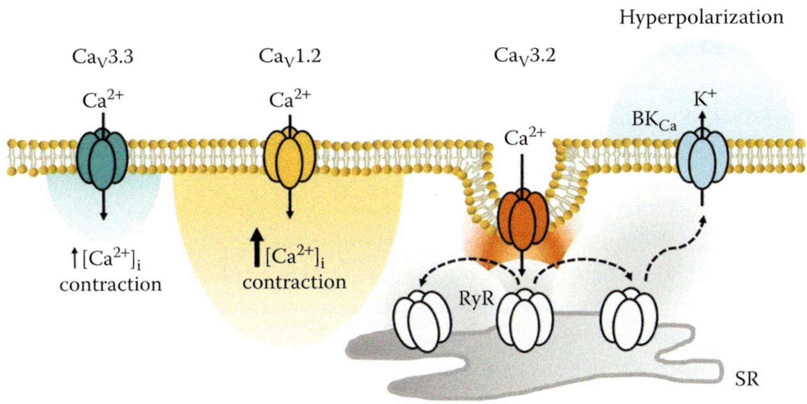

FIGURE 6.4 $Ca_V1.2/Ca_V3.3$ constrict and $Ca_V3.2$ dilates human cerebral arteries. A schematic depicts the roles of different voltage-gated Ca^{2+} channels in human cerebral arterial smooth muscle. $Ca_V1.2$ predominates as an entry route for Ca^{2+} to elicit a global rise in $[Ca^{2+}]_i$ and subsequently arterial smooth muscle contraction. $Ca_V3.3$ acts as an influx route that modestly contributes to smooth muscle contraction at relatively hyperpolarized membrane potentials. On the other hand, $Ca_V3.2$ channel triggers BK_{Ca}-mediated currents, possibly after the activation of RyR on the sarcoplasmic reticulum (SR) and Ca^{2+} spark generation, to induce V_M hyperpolarization and arterial relaxation.

6.2.4.1 Direct Effect of T-Type Ca^{2+} Influx on Tone Development

$Ca_V1.2$ is the key pore-forming subunit of arterial L-type channels and upon depolarization, its graded activation plays a central role in setting cytosolic $[Ca^{2+}]$ [10]. Although L-type channel activation is important to the myogenic response, it is clear that channel blockade does not fully eliminate all arterial responsiveness [34], suggesting that there are additional Ca^{2+} influx pathways driving cytosolic $[Ca^{2+}]$, which could include a contributory role for T-type channels. Work directed to this question in renal arteries has generated contradictory conclusions, some reporting that $Ca_V3.x$ activation elevates cytosolic $[Ca^{2+}]$ and drives a strong constriction [39,78] whereas others have found no direct role in either afferent/efferent arterioles [79]. Initial observations in the cerebral circulation have been similarly mixed and required careful re-evaluation in light of the selectivity concerns with T-type blockers like mibefradil [50–52].

Recent studies have instituted a more systematic approach to determine whether T-type channels are directly linked to arterial tone development [19,20,34]. In this regard, $Ca_V3.x$ expression was first confirmed in the vessels of interest while patch clamp electrophysiology was used to isolate a dihydropyridine-insensitive current with T-type like properties [19]. Functional experiments were carefully designed to monitor tone, prior to and following the sequential addition of L- and then T-type blockers to address off-targeting concerns. Irrespective of whether the work was performed in rat, mouse, and or human arteries, investigations have consistently observed a modest role for T-type channels in myogenic tone development, particularly at low pressures where V_M rests near −55 mV. Under these conditions, T-type Ca^{2+} channels will dominate extracellular Ca^{2+} influx given the hyperpolarized shift in their voltage dependence profile [18–21,34,41,48]. In contrast, L-type Ca^{2+} channel activation plays a more significant role at higher pressures when cerebral arteries depolarize to approximately −35 mV [10,34]. These observations together detail how T- and L-type Ca^{2+} channels can complement one another to elevate global cytosolic [Ca^{2+}] over a full range of physiological membrane potentials.

Rodent arterial smooth muscle expresses two subtypes of T-type Ca^{2+} channels ($Ca_V3.1$ and $Ca_V3.2$), and it was initially thought that both subtypes facilitate arterial constriction [18,34]. Recent reports, however, using vessels from $Ca_V3.1^{-/-}$ mice, indicate that this subtype is specifically responsible for enabling myogenic constriction at lower intravascular pressures [20]. $Ca_V3.2$ appears, in contrast, to be indirectly tied to a feedback response that antagonizes myogenic tone [19].

6.2.4.2 Indirect Effects of T-Type Ca^{2+} Influx on Tone Development

The idea that T-type Ca^{2+} influx in arterial smooth muscle might act in a *discrete and indirect* manner began with the peculiar observations of Campbell and colleagues. These investigators developed a $Ca_V3.2$ knockout mouse model and paradoxically observed augmented arterial constriction and a compromised ability of coronary arteries to dilate [37]. They subsequently hypothesized that $Ca_V3.2$ channels might be a part of a functional microdomain with the large conductance Ca^{2+}-activated K^+ channels (BK_{Ca}). In this scenario, localized Ca^{2+} influx via $Ca_V3.2$ would gate BK_{Ca} and augment outward K^+ current facilitating hyperpolarization [37]. While BK_{Ca} activation can indeed hyperpolarize resistance arteries, this K^+ conductance is more commonly viewed as a feedback conductance that moderates pressure-induced constriction [80], a perspective that we will discuss shortly.

To limit arterial constriction, vascular smooth muscle is encoded with a range of K^+ ion channels able to provide hyperpolarizing feedback in response to K^+ efflux. These include at least three specific subtypes of voltage-gated K^+ (K_V) channels (K_V1, K_V2, and K_V7), and the large conductance Ca^{2+}-activated K^+ channel (BK_{Ca}), all of which are present in a range of different vascular beds [81]. To activate smooth muscle BK_{Ca}, cytosolic [Ca^{2+}] must rise discretely to micromolar concentrations and this is achieved through the generation of Ca^{2+} sparks, sarcoplasmic reticulum (SR)-driven events dependent on ryanodine receptor (RyR) gating [80,82]. The opening of RyR is an integrated process, partially reliant on extracellular Ca^{2+} entry triggering the RyR cytosolic Ca^{2+} sensor [80]. The identity of this entry channel is uncertain, although past studies have alluded to candidates including TRPV4 or L-type channels [83–85]. While both are plausible candidates, their intrinsic properties are notably inconsistent with a triggering role. TRPV4 channels display voltage-independent properties [83], yet Ca^{2+} spark generation is graded in a voltage-dependent manner [86,87]. L-type channels exhibit Ca^{2+}-dependent inactivation; thus, if positioned in a diffusion-restricted microdomain, high [Ca^{2+}] would induce strong inactivation, impinging its ability to activate the RyR cytosolic gate [27]. As T-type $Ca_V3.x$ channels are voltage-gated, free of Ca^{2+}-dependent inactivation, and display a voltage window that overlaps with physiological membrane potentials, this conductance appears best suited for microdomain localization and to function as a trigger for Ca^{2+} spark generation [43].

It was in this context that investigators began to explore the vasomotor effect of Ni^{2+} on myogenically sensitive arteries from rat, mouse, and humans [19,21,44]. Irrespective of the species, this pharmacological maneuver has consistently induced arterial depolarization and constriction in myogenically active arteries, observations consistent with $Ca_V3.2$ driving a negative-feedback response dilating arteries [19]. To link Ca^{2+} influx ($Ca_V3.2$) to RyR triggering, Ca^{2+} sparks generation and the activation of BK_{Ca}, recent studies have employed a deep multi-layered approach beginning with structural analysis in rat cerebral arterial smooth muscle cells, noting that $Ca_V3.2$ localizes to caveolae in close proximity to RyR [19]. This was followed by computational and experimental work confirming the ability of $Ca_V3.2$ to trigger RyR and generate Ca^{2+} sparks through a Ca^{2+}-induced Ca^{2+} release mechanism [19]. Finally, perforated patch clamp electrophysiology confirmed that Ca^{2+} sparks triggered by $Ca_V3.2$ activate BK_{Ca} channels, eliciting membrane hyperpolarization and arterial relaxation [19]. In murine epigastric arteries, a single study argued that $Ca_V3.2$ channels do not modulate BK_{Ca} activity [88]. Critical caveats include the lack of evidence for the existence or absence of functional $Ca_V3.2$ currents; furthermore, BK_{Ca} currents were measured at membrane potentials where $Ca_V3.2$ channels, if present, would be almost completely inactivated and unable to drive negative feedback. Additionally, whole artery work in the same study did not reveal the expected vasoconstriction in response to BK_{Ca} inhibition [88]. Nevertheless, while findings in cerebral arteries were clear [19], interpretations critically hinged on the presumed selectivity of Ni^{2+} and there was additionally a lack of corroborative observations outside the cerebral circulation. In addressing these key concerns, Harraz et al. [44] employed wild-type and knockout $Ca_V3.2^{-/-}$ mice to test whether a similar functional axis (i.e., $Ca_V3.2$-RyR-BK_{Ca}) existed in mesenteric arteries. Consistent with an indirect role for $Ca_V3.2$ in tone regulation, genetic ablation of $Ca_V3.2$ was accompanied with enhanced myogenic tone and a lack of Ni^{2+} sensitivity [44,61]. A role for RyR and BK_{Ca} in moderating constriction was further evident as Ca^{2+} spark and BK_{Ca} activation was diminished with $Ca_V3.2$ inhibition in wild-type but not knockout mice [44].

6.2.5 Summary and Final Considerations

Concerted research over the past decade has delineated T-type channels and started to define their regulation and functional impact in arterial smooth muscle. Patch clamp electrophysiology along with carefully crafted pharmacology have successfully isolated the T-type current and separated it from the L-type current [43]. Arterial T-type channels appear to be a key regulatory target for signaling pathways linked to PKA, PKG, and NADPH oxidase [43,49,77]. Experimental evidence indicates that the two distinct Ca_V3 subtypes are unique in facilitating or antagonizing tone development by (1) directly elevating cytosolic [Ca^{2+}] or (2) indirectly activating Ca^{2+} sensitive conductance linked to membrane potential regulation [19]. The clear identification of T-type Ca^{2+} channels in arterial smooth muscle opens new research avenues as it pertains to the regulation and function of vascular tissue. What, for example, is the full extent to which $Ca_V3.x$ channels are regulated by vasoactive peptides? How is selective subtype targeting achieved and can key mechanistic pathways, proteins, and phosphorylation sites be identified? Further, there is considerable value in identifying the splice variants present in vascular tissue and documenting their key electrophysiological properties, akin to the preliminary work by Kuo and colleagues (2014). With careful consideration of mechanisms and experimental strategies, further work will continue to unlock new insights as to how T-type channels govern vascular tone development.

DISCLOSURES

The authors have no significant disclosures.

ACKNOWLEDGMENTS

This work was supported by an operating grant from the Canadian Institute of Health Research (RFN-106459; DG Welsh). D.G. Welsh is the Rorabeck Chair in Molecular Neuroscience and Vascular Biology at the University of Western Ontario. O.F. Harraz and A. Hashad were supported by scholarships from Alberta Innovates Health Solutions (AIHS) and the Canadian Institute of Health Research (Vanier).

REFERENCES

1. Segal, SS, and Duling, BR. 1986. Flow control among microvessels coordinated by intercellular conduction. *Science* 234: 868–870.
2. Duling, BR, Hogan, RD, Langille, BL, Lelkes, P, Segal, SS, Vatner, SF, Weigelt, H, and Young, MA. 1987. Vasomotor control: Functional hyperemia and beyond. *Fed Proc* 46: 251–263.
3. Cole, WC, and Welsh, DG. 2011. Role of myosin light chain kinase and myosin light chain phosphatase in the resistance arterial myogenic response to intravascular pressure. *Arch Biochem Biophys* 510: 160–173.
4. Harder, DR, Alkayed, NJ, Lange, AR, Gebremedhin, D, and Roman, RJ. 1998. Functional hyperemia in the brain: Hypothesis for astrocyte-derived vasodilator metabolites. *Stroke* 29: 229–234.
5. Filosa, JA, Bonev, AD, Straub, SV, Meredith, AL, Wilkerson, MK, Aldrich, RW, and Nelson, MT. 2006. Local potassium signaling couples neuronal activity to vasodilation in the brain. *Nat Neurosci* 9: 1397–1403.
6. Brayden, JE, and Bevan, JA. 1985. Neurogenic muscarinic vasodilation in the cat. An example of endothelial cell-independent cholinergic relaxation. *Circ Res* 56: 205–211.
7. Si, ML, and Lee, TJ. 2002. Alpha7-nicotinic acetylcholine receptors on cerebral perivascular sympathetic nerves mediate choline-induced nitrergic neurogenic vasodilation. *Circ Res* 91: 62–69.
8. Garcia-Roldan, JL, and Bevan, JA. 1990. Flow-induced constriction and dilation of cerebral resistance arteries. *Circ Res* 66: 1445–1448.
9. Harder, DR, Gilbert, R, and Lombard, JH. 1987. Vascular muscle cell depolarization and activation in renal arteries on elevation of transmural pressure. *Am J Physiol Renal Physiol* 253: F778–F781.
10. Knot, HJ, and Nelson, MT. 1998. Regulation of arterial diameter and wall [Ca^{2+}] in cerebral arteries of rat by membrane potential and intravascular pressure. *J Physiol* 508: 199–209.
11. Davis, MJ. 1993. Myogenic response gradient in an arteriolar network. *Am J Physiol Heart Circ Physiol* 264: H2168–H2179.
12. Gallagher, PJ, Herring, BP, and Stull, JT. 1997. Myosin light chain kinases. *J Muscle Res Cell Motil* 18: 1–16.

13. Bayliss, WM. 1902. On the local reactions of the arterial wall to changes of internal pressure. *J Physiol* 28: 220–231.
14. Welsh, DG, Nelson, MT, Eckman, DM, and Brayden, JE. 2000. Swelling-activated cation channels mediate depolarization of rat cerebrovascular smooth muscle by hyposmolarity and intravascular pressure. *J Physiol* 527: 139–148.
15. Welsh, DG, Morielli, AD, Nelson, MT, and Brayden, JE. 2002. Transient receptor potential channels regulate myogenic tone of resistance arteries. *Circ Res* 90: 248–250.
16. Morita, H, Cousins, H, Onoue, H, Ito, Y, and Inoue, R. 1999. Predominant distribution of nifedipine-insensitive, high voltage-activated Ca^{2+} channels in the terminal mesenteric artery of guinea pig. *Circ Res* 85: 596–605.
17. Morita, H, Shi, J, Ito, Y, and Inoue, R. 2002. T-channel-like pharmacological properties of high voltage-activated, nifedipine-insensitive Ca^{2+} currents in the rat terminal mesenteric artery. *Br J Pharmacol* 137: 467–476.
18. Kuo, IY, Ellis, A, Seymour, VA, Sandow, SL, and Hill, CE. 2010. Dihydropyridine-insensitive calcium currents contribute to function of small cerebral arteries. *J Cereb Blood Flow Metab* 30: 1226–1239.
19. Harraz, OF, Abd El-Rahman, RR, Bigdely-Shamloo, K, Wilson, SM, Brett, SE, Romero, M, Gonzales, AL, et al. 2014a. $Ca_V3.2$ channels and the induction of negative feedback in cerebral arteries. *Circ Res* 115: 650–661.
20. Bjorling, K, Morita, H, Olsen, MF, Prodan, A, Hansen, PB, Lory, P, Holstein-Rathlou, NH, and Jensen, LJ. 2013. Myogenic tone is impaired at low arterial pressure in mice deficient in the low-voltage-activated $Ca_V3.1$ T-type Ca^{2+} channel. *Acta Physiol* 207: 709–720.
21. Harraz, OF, Visser, F, Brett, SE, Goldman, D, Zechariah, A, Hashad, AM, Menon, BK, Watson, T, Starreveld, Y, and Welsh, DG. 2015a. $Ca_V1.2/Ca_V3.x$ channels mediate divergent vasomotor responses in human cerebral arteries. *J Gen Physiol* 145: 405–418.
22. Hashad, AM, Mazumdar, N, Romero, M, Nygren, A, Bigdely-Shamloo, K, Harraz, OF, Puglisi, JL, Vigmond, EJ, Wilson, SM, and Welsh, DG. 2017. Interplay among distinct Ca^{2+} conductances drives Ca^{2+} sparks/spontaneous transient outward currents in rat cerebral arteries. *J Physiol* 595: 1111–1126.
23. Clapham, DE. 2007. Calcium signalling. *Cell* 131: 1047–1058.
24. Catterall, WA. 2011. Voltage-gated calcium channels. *Cold Spring Harb Perspect Biol* 3: a003947.
25. Catterall, WA, Perez-Reyes, E, Snutch, TP, and Striessnig, J. 2005. International Union of Pharmacology. XLVIII. Nomenclature and structure-function relationships of voltage-gated calcium channels. *Pharmacol Rev* 57: 411–425.
26. Zamponi, GW, Striessnig, J, Koschak, A, and Dolphin, AC. 2015. The physiology, pathology, and pharmacology of voltage-gated calcium channels and their future therapeutic potential. *Pharmacol Rev* 67: 821–870.
27. Peterson, BZ, DeMaria, CD, Adelman, JP, and Yue, DT. 1999. Calmodulin is the Ca^{2+} sensor for Ca^{2+}-dependent inactivation of L-type calcium channels. *Neuron* 22: 549–558.
28. Hulme, JT, Lin, TW, Westenbroek, RE, Scheuer, T, and Catterall, WA. 2003. Beta-adrenergic regulation requires direct anchoring of PKA to cardiac $Ca_V1.2$ channels via a leucine zipper interaction with A kinase-anchoring protein 15. *Proc Natl Acad Sci USA* 100: 13093–13098.
29. Wakamori, M, Mikala, G, and Mori, Y. 1999. Auxiliary subunits operate as a molecular switch in determining gating behaviour of the unitary N-type Ca^{2+} channel current in Xenopus oocytes. *J Physiol* 517: 659–672.
30. Bannister, JP, Bulley, S, Narayanan, D, Thomas-Gatewood, C, Luzny, P, Pachuau, J, and Jaggar, JH. 2012. Transcriptional upregulation of $\alpha_2\delta$-1 elevates arterial smooth muscle cell voltage-dependent Ca^{2+} channel surface expression and cerebrovascular constriction in genetic hypertension. *Hypertension* 60: 1006–1015.

31. Mikkelsen, E, and Andersson, KE. 1978. Contractile effects of prostaglandin F2α on isolated human peripheral arteries and veins. *Acta Pharmacol Toxicol (Copenh)* 43: 398–404.
32. Moosmang, S, Schulla, V, Welling, A, Feil, R, Feil, S, Wegener, JW, Hofmann, F, and Klugbauer, N. 2003. Dominant role of smooth muscle L-type calcium channel $Ca_V1.2$ for blood pressure regulation. *EMBO J* 22: 6027–6034.
33. Wellman, GC, Cartin, L, Eckman, DM, Stevenson, AS, Saundry, CM, Lederer, WJ, and Nelson, MT. 2001. Membrane depolarization, elevated Ca^{2+} entry, and gene expression in cerebral arteries of hypertensive rats. *Am J Physiol Heart Circ Physiol* 281: H2559–H2567.
34. Abd El-Rahman, RR, Harraz, OF, Brett, SE, Anfinogenova, Y, Mufti, RE, Goldman, D, and Welsh, DG. 2013. Identification of L- and T-type Ca^{2+} channels in rat cerebral arteries: Role in myogenic tone development. *Am J Physiol Heart Circ Physiol* 304: H58–H71.
35. Benham, CD and Tsien, RW. 1988. Noradrenaline modulation of calcium channels in single smooth muscle cells from rabbit ear artery. *J Physiol* 404:767–784.
36. Van Bavel, E, Sorop, O, Andreasen, D, Pfaffendorf, M, and Jensen, BL. 2002. Role of T-type calcium channels in myogenic tone of skeletal muscle resistance arteries. *Am J Physiol Heart Circ Physiol* 283: H2239–H2243.
37. Chen, CC, Lamping, KG, Nuno, DW, Barresi, R, Prouty, SJ, Lavoie, JL, Cribbs, LL, et al. Abnormal coronary function in mice deficient in alpha1H T-type Ca^{2+} channels. *Science* 302: 1416–1418.
38. Jensen, LJ, Salomonsson, M, Jensen, BL, and Holstein-Rathlou, NH. 2004. Depolarization induced calcium influx in rat mesenteric small arterioles is mediated exclusively via mibefradil-sensitive calcium channels. *Br J Pharmacol* 142: 709–718.
39. Poulsen, CB, Al-Mashhadi, RH, Cribbs, LL, Skott, O, and Hansen, PB. 2011. T-type voltage gated calcium channels regulate the tone of mouse efferent arterioles. *Kidney Int* 79: 443–451.
40. Chevalier, M, Gilbert, G, Roux, E, Lory, P, Marthan, R, Savineau, JP, and Quignard, JF. 2014. T type calcium channels are involved in hypoxic pulmonary hypertension. *Cardiovasc Res* 103: 597–606.
41. Perez-Reyes, E. 2003. Molecular physiology of low-voltage-activated T-type calcium channels. *Physiol Rev* 83: 117–161.
42. Talavera, K, and Nilius, B. 2006. Biophysics and structure-function relationship of T-type Ca^{2+} channels. *Cell Calcium* 40: 97–114.
43. Harraz, OF, and Welsh, DG. 2013a. Protein kinase A regulation of T-type Ca^{2+} channels in rat cerebral arterial smooth muscle. *J Cell Sci* 126: 2944–2954.
44. Harraz, OF, Brett, SE, Zechariah, A, Romero, M, Puglisi, JL, Wilson, SM, and Welsh, DG. 2015b. Genetic ablation of $Ca_V3.2$ channels enhances the arterial myogenic response by modulating the RyR-BKCa axis. *Arterioscler Thromb Vasc Biol* 35: 1843–1851.
45. Liao, P, Yu, D, Li, G, Yong, TF, Soon, JL, Chua, YL, and Soong, TW. 2007. A smooth muscle $Ca_V1.2$ calcium channel splice variant underlies hyperpolarized window current and enhanced state-dependent inhibition by nifedipine. *J Biol Chem* 282: 35133–35142.
46. Akaike, N, Kanaide, H, Kuga, T, Nakamura, M, Sadoshima, J, and Tomoike, H. 1989. Low voltage-activated calcium current in rat aorta smooth muscle cells in primary culture. *J Physiol* 416: 141–160.
47. Li, Z, Wang, X, Gao, G, Qu, D, Yu, B, Huang, C, Elmslie, KS, and Peterson, BZ. 2010. A single amino acid change in $Ca_V1.2$ channels eliminates the permeation and gating differences between Ca^{2+} and Ba^{2+}. *J Membr Biol* 233: 23–33.
48. Harraz, OF, and Welsh, DG. 2013b. T-type Ca^{2+} Channels in cerebral arteries: Approaches, hypotheses and speculation. *Microcirculation* 20: 299–306.

49. Harraz, OF, Brett, SE, and Welsh, DG. 2014b. Nitric oxide suppresses vascular voltage-gated T-type Ca^{2+} channels through cGMP/PKG signaling. *Am J Physiol Heart Circ Physiol* 306: H279–H285.
50. Nilius, B, Prenen, J, Kamouchi, M, Viana F, Voets, T, and Droogmans, G. 1997. Inhibition by mibefradil, a novel calcium channel antagonist of Ca^{2+}- and volume-activated Cl^- channels in macrovascular endothelial cells. *Br J Pharmacol* 121: 547–555.
51. Perchenet, L, and Clement-Chomienne, O. 2000. Characterization of mibefradil block of the human heart delayed rectifier $hK_V1.5$. *J Pharmacol Exp Ther* 295: 771–778.
52. Koh, SD, Monaghan, K, Ro, S, Mason, HS, Kenyon, JL, and Sanders, KM. 2001. Novel voltage dependent non-selective cation conductance in murine colonic myocytes. *J Physiol* 533: 341–355.
53. Hong da, H, Yang, D, Choi, IW, Son, YK, Jung, WK, Kim, DJ, Han J, Na, SH, and Park, WS. 2012. The T-type Ca^{2+} channel inhibitor mibefradil inhibits voltage-dependent K^+ channels in rabbit coronary arterial smooth muscle cells. *J Pharmacol Sci* 120: 196–205.
54. Howitt, L, Kuo, IY, Ellis, A, Chaston, DJ, Shin, HS, Hansen, PB, and Hill, CE. 2013. Chronic deficit in nitric oxide elicits oxidative stress and augments T-type calcium-channel contribution to vascular tone of rodent arteries and arterioles. *Cardiovasc Res* 98: 449–457.
55. Chemin, J, Monteil, A, Bourinet, E, Nargeot, J, and Lory, P. 2001. Alternatively spliced alpha(1G) ($Ca_V3.1$) intracellular loops promote specific T-type Ca^{2+} channel gating properties. *Biophys J* 80: 1238–1250.
56. Kuo, IY, Howitt, L, Sandow, SL, McFarlane, A, Hansen, PB, and Hill, CE. 2014. Role of T-type channels in vasomotor function: Team player or chameleon? *Pflugers Arch* 466: 767–779.
57. Lee, JH, Gomora, JC, Cribbs, LL, and Perez-Reyes, E. 1999. Nickel block of three cloned T type calcium channels: Low concentrations selectively block alpha1H. *Biophys J* 77: 3034–3042.
58. Niwa, N, Yasui, K, Opthof, T, Takemura, H, Shimizu, A, Horiba, M, Lee, JK, Honjo, H, Kamiya, K, and Kodama, I. 2004. $Ca_V3.2$ subunit underlies the functional T-type Ca^{2+} channel in murine hearts during the embryonic period. *Am J Physiol Heart Circ Physiol* 286: H2257–H2263.
59. Autret, L, Mechaly, I, Scamps, F, Valmier, J, Lory, P, and Desmadryl, G. 2005. The involvement of $Ca_V3.2$/alpha1H T type calcium channels in excitability of mouse embryonic primary vestibular neurons. *J Physiol* 567: 67–78.
60. Kim, D, Song, I, Keum, S, Lee, T, Jeong, MJ, Kim, SS, McEnery, MW, and Shin, HS. 2001. Lack of the burst firing of thalamocortical relay neurons and resistance to absence seizures in mice lacking alpha(1G) T-type Ca(2+) channels. *Neuron* 31: 35–45.
61. Mikkelsen, MF, Björling, K, and Jensen, LJ. 2016. Age-dependent impact of Ca_V 3.2 T-type calcium channel deletion on myogenic tone and flow-mediated vasodilatation in small arteries. *J Physiol* 594: 5881–5898.
62. Berger, AJ, and Takahashi, T. 1990. Serotonin enhances a low-voltage-activated calcium current in rat spinal motoneurons. *J Neurosci* 10: 1922–1928.
63. Fraser, DD, and MacVicar, BA. 1991. Low-threshold transient calcium current in rat hippocampal lacunosum-moleculare interneurons: Kinetics and modulation by neurotransmitters. *J Neurosci* 11: 2812–2820.
64. Yunker, AM, and McEnery, MW. 2003. Low-voltage-activated ("T-Type") calcium channels in review. *J Bioenerg Biomembr* 35: 533–575.
65. Iftinca, MC, and Zamponi, GW. 2009. Regulation of neuronal T-type calcium channels. *Trends Pharmacol Sci* 30: 32–40.

66. Nelson, MT, Patlak, JB, Worley, JF, and Standen, NB. 1990. Calcium channels, potassium channels, and voltage dependence of arterial smooth muscle tone. *Am J Physiol* 259:C3–C18.
67. Scott, JD. 1991. Cyclic nucleotide-dependent protein kinases. *Pharmacol Ther* 50: 123–145.
68. Pelligrino, DA, and Wang, Q. 1998. Cyclic nucleotide crosstalk and the regulation of cerebral vasodilation. *Prog Neurobiol* 56: 1–18.
69. Blesneac, I, Chemin, J, Bidaud, I, Huc-Brandt, S, Vandermoere, F, and Lory, P. 2015. Phosphorylation of the Cav3.2 T-type calcium channel directly regulates its gating properties. *Proc Natl Acad Sci USA* 112: 13705–13710.
70. Emerick, MC, Stein, R, Kunze, R, McNulty, MM, Regan, MR, Hanck, DA, and Agnew, WS. 2006. Profiling the array of $Ca_V3.1$ variants from the human T-type calcium channel gene CACNA1G: alternative structures, developmental expression, and biophysical variations. *Proteins* 64: 320–342.
71. Zhong, X, Liu, JR, Kyle, JW, Hanck, DA, and Agnew, WS. 2006. A profile of alternative RNA splicing and transcript variation of CACNA1H, a human T-channel gene candidate for idiopathic generalized epilepsies. *Hum Mol Genet* 15: 1497–1512.
72. Selemidis, S, Dusting, GJ, Peshavariya, H, Kemp-Harper, BK, and Drummond, GR. 2007. Nitric oxide suppresses NADPH oxidase-dependent superoxide production by S-nitrosylation in human endothelial cells. *Cardiovasc Res* 75: 349–358.
73. Howitt, L, Matthaei, KI, Drummond, GR, and Hill, CE. 2015. Nox1 upregulates the function of vascular T-type calcium channels following chronic nitric oxide deficit. *Pflugers Arch* 467: 727–735.
74. Nguyen Dinh Cat, A, Montezano, AC, Burger, D, and Touyz, RM. 2013. Angiotensin II, NADPH oxidase, and redox signaling in the vasculature. *Antioxid Redox Signal* 19: 1110–1120.
75. Narayanan, D, Xi Q, Pfeffer, LM, and Jaggar, JH. 2010. Mitochondria control functional $Ca_V1.2$ expression in smooth muscle cells of cerebral arteries. *Circ Res* 107: 631–641.
76. Chaplin, NL, Nieves-Cintrón, M, Fresquez, AM, Navedo, MF, and Amberg, GC. 2015. Arterial smooth muscle mitochondria amplify hydrogen peroxide microdomains functionally coupled to L-type calcium channels. *Circ Res* 117: 1013–1023.
77. Hashad, AM, Sancho, M, and Welsh, DG. 2016. Angiotensin II inhibits T-type Ca^{2+} channels in cerebral arterial smooth muscle cells. *FASEB J* 30: 945.12.
78. Feng, MG, and Navar, LG. 2004. Angiotensin II-mediated constriction of afferent and efferent arterioles involves T-type Ca^{2+} channel activation. *Am J Nephrol* 24: 641–648.
79. Smirnov, SV, Loutzenhiser, K, and Loutzenhiser, R. 2013. Voltage-activated Ca^{2+} channels in rat renal afferent and efferent myocytes: No evidence for the T-type Ca^{2+} current. *Cardiovasc Res* 97: 293–301.
80. Nelson, MT, Cheng, H, Rubart, M, Santana, LF, Bonev, AD, Knot, HJ, and Lederer, WJ. 1995. Relaxation of arterial smooth muscle by calcium sparks. *Science* 270: 633–637.
81. Nelson, MT, and Quayle, JM. 1995. Physiological roles and properties of potassium channels in arterial smooth muscle. *Am J Physiol Cell Physiol* 268: C799–C822.
82. Ledoux, J, Werner, ME, Brayden, JE, and Nelson, MT. 2006. Calcium-activated potassium channels and the regulation of vascular tone. *Physiology* 21: 69–78.
83. Earley, S, Heppner, TJ, Nelson, MT, and Brayden, JE. 2005. TRPV4 forms a novel Ca^{2+} signaling complex with ryanodine receptors and BK_{Ca} channels. *Circ Res* 97: 1270–1279.
84. Essin, K, Welling, A, Hofmann, F, Luft, FC, Gollasch, M, and Moosmang, S. 2007. Indirect coupling between $Ca_V1.2$ channels and ryanodine receptors to generate Ca^{2+} sparks in murine arterial smooth muscle cells. *J Physiol* 584: 205–219.

85. Takeda, Y, Nystoriak, MA, Nieves-Cintron, M, Santana, LF, and Navedo, MF. 2011. Relationship between Ca^{2+} sparklets and sarcoplasmic reticulum Ca^{2+} load and release in rat cerebral arterial smooth muscle. *Am J Physiol Heart Circ Physiol* 301: H2285–H2294.
86. Jaggar, JH, Stevenson, AS, and Nelson, MT. 1998. Voltage dependence of Ca^{2+} sparks in intact cerebral arteries. *Am J Physiol Cell Physiol* 274: C1755–C1761.
87. Perez, GJ, Bonev, AD, Patlak, JB, and Nelson, MT. 1999. Functional coupling of ryanodine receptors to KCa channels in smooth muscle cells from rat cerebral arteries. *J Gen Physiol* 113: 229–238.
88. Mullan, B, Pettis, J, and Jackson, WF. 2017. T-type voltage-gated Ca2+ channels do not contribute to the negative feedback regulation of myogenic tone in murine superior epigastric arteries. *Pharmacol Res Perspect* 5: e00320.

7 Calcium Sensitization in Smooth Muscle Involving Regulation of Myosin Light Chain Phosphatase Activity

Justin A. MacDonald, Michael P. Walsh, and William C. Cole

CONTENTS

7.1 INTRODUCTION

Smooth muscle tissues display a remarkable diversity of contractile behavior, varying from low-amplitude, sustained alterations in tone in resistance arteries, to large-amplitude propagating waves of contraction in the gastrointestinal tract in peristalsis and uterus in labor, and intermittent relaxation from sustained constriction in the sphincters of the gut and urogenital tract. These behavioral differences are largely the result of varied patterns of membrane electrical activity (e.g., graded depolarization *versus* action potentials) and associated temporal changes in voltage-dependent Ca^{2+} influx and cytosolic free Ca^{2+} concentration ($[Ca^{2+}]_i$). However, the extent and time course of force development by smooth muscle is also regulated by cellular signaling pathways activated by receptor occupancy and other stimuli that modulate the sensitivity of the contractile apparatus to $[Ca^{2+}]_i$.

Actin-activated myosin Mg^{2+}-ATPase activity, cross-bridge cycling, and force generation in smooth muscle are dependent on a thick filament regulatory mechanism, involving phosphorylation of myosin regulatory light chains (LC_{20}) by the Ca^{2+}-calmodulin (CaM)-dependent enzyme, myosin light chain kinase (MLCK), and dephosphorylation of LC_{20} by myosin light chain phosphatase (MLCP). Ca^{2+}-CaM-dependent regulation of MLCK was initially thought to be solely responsible for determining the onset, strength, and duration of contraction. However, we now appreciate that the activity of MLCP is also regulated by intricate cellular signaling mechanisms to modulate the magnitude and time course of contraction. The current paradigm of smooth muscle contraction holds that force generation is under dual regulation by MLCK and MLCP activities; excitatory stimuli can cause contraction by elevating $[Ca^{2+}]_i$ and activating MLCK, and/or by suppressing MLCP activity without the requirement of an additional increase in $[Ca^{2+}]_i$. In the latter case, MLCP inhibition alters the balance of MLCK and MLCP activities to favor MLCK-dependent LC_{20} phosphorylation, resulting in a leftward shift in the force-$[Ca^{2+}]_i$ relation, and the phenomenon of Ca^{2+} sensitization of the contractile apparatus.

This chapter reviews our current understanding of the molecular basis of Ca^{2+} sensitization in smooth muscle. The well-established major signaling pathways and mechanisms controlling MLCP activity are covered with an emphasis on emerging details concerning the regulation of RhoA signaling by Rho guanine nucleotide exchange factors (RhoGEFs), novel kinases identified to phosphorylate MLCP, and emerging pharmacologic tools for suppression of kinases suspected to regulate MLCP activity. Space limitations preclude a discussion of differences in contractile protein expression, MLCP regulation by multi-site phosphorylation of its targeting subunit, Ca^{2+} desensitization mechanisms, or the contribution of dynamic alterations in actin cytoskeleton structure that also influence the Ca^{2+} sensitivity of smooth muscle contraction. Interested readers are referred to recent review articles for further information on these topics [1–7].

7.2 HISTORICAL PERSPECTIVE AND CLASSIC MECHANISMS OF Ca^{2+} SENSITIZATION

That the regulation of smooth muscle contraction differs from the strict depolarization- and Ca^{2+}-dependent excitation-contraction coupling of striated skeletal and cardiac muscle was evident by the 1960s, but consensus concerning the mechanism responsible did not emerge until the late 1980s. The presence of a distinct regulatory mechanism(s) was initially indicated by the ability of excitatory agonists to increase force development in depolarized smooth muscle tissues maintained *in vitro* in solutions with elevated extracellular potassium concentration [8], and the presence of greater maximal force in response to agonist treatment compared to complete depolarization with extracellular potassium [9]. These observations were subsequently reinforced by experiments using intracellular fluorescent Ca^{2+} indicators that showed that potassium depolarization caused a greater sustained increase in $[Ca^{2+}]_i$, but less force than excitatory agonists [10–15]. These findings implied that receptor signaling could cause contraction by a mechanism independent of alterations in membrane potential or $[Ca^{2+}]_i$. Several explanations were advanced, including that "some stimuli may also release a potentiator which augments the contractile force developed at any given free calcium level" [9] by "an increase in the sensitivity of the contractile apparatus to Ca^{2+} [16]."

Direct evidence of Ca^{2+} sensitization of the contractile apparatus, and identification of the key mechanisms responsible for MLCP inhibition, was initially obtained by examining agonist-evoked contraction in smooth muscle tissues treated with membrane permeabilizing agents. This approach permitted precise control of $[Ca^{2+}]_i$, as well as the introduction of signaling intermediates, activators, and inhibitors into the intracellular compartment. Force generation in response to excitatory agonists at a constant, sub-maximal $[Ca^{2+}]_i$ was found to be accompanied by: (1) increased levels of LC_{20} phosphorylation [17–19]; (2) G protein-dependent cellular signaling requiring the presence of GTP, that was mimicked by GTP[γS] and inhibited by GDP[βS] [14,17–21]; (3) RhoA signaling that activated Rho-associated coiled-coil kinase (ROCK) [22–24]; (4) a decline in MLCP activity [18,19] due to (5) ROCK-mediated phosphorylation of the myosin-targeting subunit 1 (MYPT1) of MLCP [25–28], and/or (6) diacylglycerol (DAG)- and Ca^{2+}-dependent protein kinase C (PKC)-mediated phosphorylation of the 17 kDa PKC-potentiated protein phosphatase 1 inhibitor protein (CPI-17) that binds to the catalytic subunit and thereby inhibits MLCP activity [29].

Figure 7.1 illustrates the initial mechanisms identified to evoke Ca^{2+} sensitization in smooth muscle. It provides a useful framework to appreciate the fundamentals of Ca^{2+} sensitization but does not provide sufficient detail to account for the widely recognized receptor-, tissue- and time-dependent variability in the contribution of MLCP inhibition to force generation by smooth muscle [14,30–32]. CPI-17-dependent sensitization is rapid in onset, paralleling the kinetics of the rise in $[Ca^{2+}]_i$ evoked by agonist challenge, whereas the ROCK-mediated

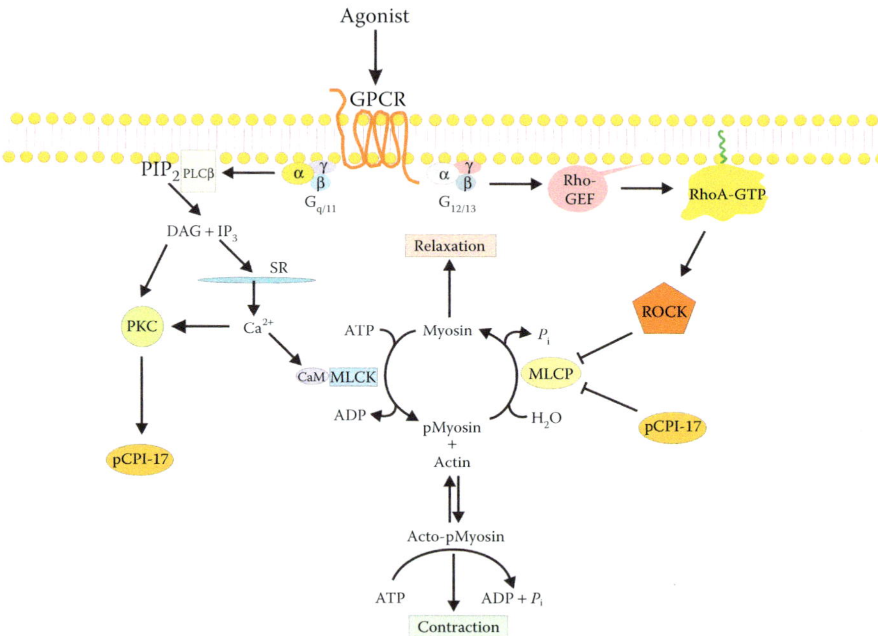

FIGURE 7.1 Initial mechanisms identified to evoke Ca^{2+}-induced contraction and Ca^{2+} sensitization of smooth muscle contraction. A variety of contractile stimuli acting via G protein-coupled receptors (GPCRs) activate the heterotrimeric G proteins $G_{q/11}$, which in turn activate phospholipase C-β (PLCβ) to generate inositol 1,4,5-trisphosphate (IP_3) and DAG by the hydrolysis of membrane phosphatidylinositol 4,5-bisphosphate (PIP_2). IP_3 binds to IP_3 receptors (Ca^{2+} channels) in the sarcoplasmic reticulum (SR) membrane to release Ca^{2+} to the cytosol. Ca^{2+} binds to CaM, which activates MLCK, leading to phosphorylation of the LC_{20} of myosin II at Ser-19, activation of cross-bridge cycling and actomyosin MgATPase activity, and contraction driven by the energy derived from the hydrolysis of ATP. In addition, Ca^{2+} in combination with the other product of PIP_2 hydrolysis, DAG, activates PKC. PKC phosphorylates the 17 kDa cytosolic phosphatase inhibitor, CPI-17, which thereby becomes a potent inhibitor of MLCP. This increases the MLCK: MLCP activity ratio, resulting in increased LC_{20} phosphorylation and force. Finally, several contractile agonists are also coupled to members of the $G_{12/13}$ family of heterotrimeric G proteins, which sequentially activate RhoGEFs and the small GTPase RhoA. Activated RhoA translocates to the plasma membrane, into which it inserts via its exposed geranylgeranyl moiety and activates the ROCK. Activated ROCK phosphorylates the myosin targeting subunit of MLCP (MYPT1) at Thr-696 and Thr-853 with a reduction in phosphatase activity, increased MLCK: MLCP activity ratio and contraction. Since the latter pathway is Ca^{2+}-independent, the result is Ca^{2+} sensitization of smooth muscle contraction. Phosphorylated forms of CPI-17 and myosin are denoted pCPI-17 and pMyosin, respectively, here and in Figure 7.2.

phosphorylation of MYPT1 is slower; this is indicative of two temporally-distinct components of Ca^{2+} sensitization mediated by fast PKC- and slow RhoA-mediated signaling [33]. Readers are directed to the recent article by Masumi Eto and Toshio Kitazawa for a comprehensive review of CPI-17-induced Ca^{2+} sensitization in smooth muscle [34].

7.3 G PROTEIN-COUPLED RECEPTOR SIGNALING AND Ca^{2+} SENSITIZATION

G protein-coupled receptors (GPCRs) coupled to $G\alpha_{q/11}$ and $G\alpha_{12/13}$, respectively, were initially implicated in Ca^{2+}-dependent and Ca^{2+}-independent mechanisms of contraction in smooth muscle according to the presence in other cell types of: (1) $G_{q/11}$-coupled phospholipase C (PLC)-β activity that regulates $[Ca^{2+}]_i$ and MLCK activity via IP_3-dependent release of Ca^{2+} from internal stores, as well as DAG-mediated activation of nonselective cation channels, membrane depolarization, and voltage-dependent Ca^{2+} influx; and (2) $G_{12/13}$-mediated activation of RhoA and ROCK signaling, MLCP inhibition, and Ca^{2+} sensitization [30,35–37]. However, it is now apparent that activation of $G_{12/13}$ or $G_{q/11}$ can stimulate Rho/ROCK activity, so Ca^{2+} sensitization due to stimulation of $G_{q/11}$-coupled GPCRs signaling is not limited to the PKC-CPI-17-mediated mechanism.

Evidence of $G_{q/11}$- and $G_{12/13}$-dependent signaling following GPCR activation was obtained by Gohla et al. [31] using cultured aortic smooth muscle cells, and the hydrolysis-resistant GTP analogue GTP-azidoanilide to permit photolabeling and subsequent identification of the receptor-activated G proteins by immunoprecipitation. Endothelin-1, vasopressin, and angiotensin II receptors were found to couple to both $G_{q/11}$ and $G_{12/13}$, and cell contraction evoked by endothelin-1 involved $G_{q/11}$ and PLC-β–mediated Ca^{2+}-dependent activation of MLCK, and $G_{12/13}$-mediated, MLCP inhibition via RhoA-ROCK signaling [31]. Evidence of differential signaling via $G_{q/11}$ and/or $G_{12/13}$ proteins was obtained by screening additional agonists using aortic tissues from mice with conditional, smooth muscle-specific $G_{q/11}$- or $G_{12/13}$-deficiency [32]. $G_{q/11}$-knock-out abolished contractions evoked by phenylephrine or angiotensin II, but caused partial inhibition of responses to serotonin, endothelin-1, vasopressin, and the thromboxane A_2 mimetic U46619. In contrast, $G_{12/13}$-deficiency did not affect responses to phenylephrine, serotonin, or vasopressin, whereas contractions to angiotensin II, vasopressin, U46619, and endothelin-1 were suppressed to varied extents. These data indicate that excitatory agonists evoke contraction in smooth muscle by activating $G_{q/11}$ (e.g., phenylephrine via α_1-adenoceptors), $G_{12/13}$ (e.g., U46619 via thromboxane A_2 TP receptors), or both heterotrimeric G proteins. Thus, differential signaling through $G_{q/11}$ and $G_{12/13}$ is a potential cause of variability in the contribution of Ca^{2+} sensitization to force generation in smooth muscle; for example, CPI-17-mediated Ca^{2+} sensitization is stimulated by $G_{q/11}$-, but not by $G_{12/13}$-coupled receptors. However, the presence of additional differences in signaling is indicated by the varied amplitude of the residual responses to the different agonists in the knock-out tissues [32]. As noted earlier, the RhoA/ROCK pathway of Ca^{2+} sensitization was historically associated with $G_{12/13}$-coupled receptor activation but is now recognized to be evoked by $G_{q/11}$-coupled receptors as well [38,39]. Characterization of the molecular basis of $G_{q/11}$ and $G_{12/13}$ regulation of RhoA activity is an important area of emerging knowledge regarding the regulation of Ca^{2+} sensitization in smooth muscle as it may illuminate novel targets that can be exploited to provide greater selective control compared to ROCK inhibition.

7.3.1 Regulation of G Protein-Coupled Receptor Signaling by G Protein-Coupled Receptor Kinases and Regulators of G Protein Signaling

Acute and chronic GPCR signaling are regulated by mechanisms that attenuate receptor responsiveness, referred to as desensitization. For example, G protein-coupled receptor kinases (GRKs) phosphorylate GPCRs, which facilitates the binding of β-arrestins that uncouple the receptors from G protein activation to diminish subsequent cellular signaling [40], as well as evoking phosphorylation-independent mechanisms that attenuate signaling [41]. GRK regulation has been demonstrated for $G_{q/11}$-coupled receptors, but regulation of $G_{12/13}$-coupled receptors has not been described. Endothelin-1, phenylephrine, and angiotensin II $G_{q/11}$-mediated signaling in vascular smooth muscle were shown to be sensitive to GRK2 activity, and reduced levels of GRK2 expression were associated with hypertension in humans and animal models [42–44]. Mice in which GRK2 expression was reduced by GRK2-shRNA knockdown demonstrated reduced $G_{q/11}$-coupled receptor desensitization, increased levels of $[Ca^{2+}]_i$, greater contractile responses to angiotensin II and phenylephrine, and elevated blood pressure [44,45]. Differential desensitization by GRKs can selectively inhibit $G_{q/11}$-coupled GPCRs while leaving $G_{12/13}$-dependent signaling-mediated Ca^{2+} sensitization intact, even when evoked by a single excitatory agonist that couples to both receptors.

Regulators of G protein signaling (RGS) are GTPase-activating proteins (GAPs) that enhance the intrinsic GTPase activity of G protein α-subunits and terminate signaling through $G_{i/o}$ and $G_{q/11}$, but not $G_{12/13}$ or G_S (RhoGEFs with RGS domains possess GAP activity for $G_{12/13}$ proteins; see the following). Two RGS proteins are expressed in vascular smooth muscle cells (VSMCs), RGS-2 and RGS-5, and appear to contribute to the down-regulation of Ca^{2+} sensitization in response to endothelium-derived nitric oxide signaling [46–48]. Knock-out of the $G_{q/11}$-specific inhibitor RGS-2 in mice was associated with enhanced angiotensin II, thrombin, and α-adrenoceptor signaling, leading to increases in $[Ca^{2+}]_i$ and MYPT1 phosphorylation in cultured aortic myocytes, greater force development by aortic and mesenteric tissues, and elevated blood pressure (mean arterial pressure of ~135 *versus* 85 mmHg) [49–52].

RGS-5 appears to have a similar function as RGS-2, but results from initial knock-out mice were confusing with hyper-, normo- or hypotension detected, possibly due to knock-out strategy, genetic background, and/or method of blood pressure assessment [53]. Holobotovskyy et al. [54] showed that RGS-5 knock-out is associated with profound hypertension and an increase in peak contractile response of mesenteric arterial smooth muscle to angiotensin II, mediated by a greater elevation in $[Ca^{2+}]_i$, and increased phosphorylation of MYPT1 and LC_{20} in femoral arterial VSMCs. Interestingly, ROCK-selective inhibitors (HA1077, fasudil hydrochloride, or Y27632) reversed angiotensin II-induced hypertension in RGS-5 knock-out mice, leading the authors to suggest that $G_{12/13}$ could be affected, or that $G_{q/11}$-mediated mechanisms of RhoA-ROCK activation were enhanced in the absence of RGS-5 [54].

7.3.2 Regulation of RhoA Signaling in Ca^{2+} Sensitization

Variability in the activation of Ca^{2+} sensitization by excitatory agonists and other stimuli in smooth muscle may derive from differences in the regulation of RhoA activity (Figure 7.2). RhoA is a member of the Rho family of small GTPases that also includes Rac1 and Cdc42 [55]. RhoGTPases act as off-on switches in intracellular signaling cascades, cycling between an inactive, cytoplasmic GDP-bound state, and an activated, membrane-associated GTP-bound state. The intrinsic rates of GDP dissociation and GTP hydrolysis of GTPases are slow, so switching between off and on conformations is facilitated by three interacting molecules: RhoGEFs, GAPs, and guanine nucleotide dissociation inhibitors (GDIs).

RhoGEFs are triggered by activated Gα-subunits, which stimulate their guanine nucleotide exchange activity, facilitating the replacement of GDP with GTP on RhoA. RhoGAPs terminate RhoA signaling by enhancing intrinsic GTPase activity of the Rho protein. GDIs bind inactive GTPases at the membrane, and act to sequester/chaperone GDP-bound RhoGTPase to the cytoplasm by shielding its hydrophobic prenylated C-terminus [56–66]. Heterogeneity in G protein-evoked RhoA signaling is provided by variability in each step of the regulation of RhoA through the expression of structurally and, therefore, functionally distinct RhoGEFs, RhoGAPs and GDIs. The presence of unique combinations of structural domains in each of these proteins confers specificity to their interactions with different regulatory and scaffolding molecules, including second messengers, kinases, and lipids [56–66].

The RhoGEFs that specifically interact with RhoGTPases are encoded by two gene families: the Dibble (DBL) family is the largest with more than 70 known members, including several containing a regulator of G protein-signaling homology (RGS) domains that interact with RhoA, Rac and/or Cdc42, and dedicator of cytokinesis (DOC) homology domain-containing family with 11 members that interact with Rac and/or Cdc42, but not RhoA [59,67]. DBL proteins are characterized by the presence of the DBL homology catalytic domain that is specifically responsible for the nucleotide exchange activity [59,67], a pleckstrin homology (PH) interaction domain that is required for full GEF activity and binding to other proteins to determine subcellular localization [68], the RGS domain that permits their interaction and activation by Gα-subunits (i.e., RhoGEFs are GAPs for G protein GTPase activity) and, in some cases, PSD-95/Disc-large/ZO-1 (PDZ) homology domains that mediate interactions with cellular membranes, coiled-coil oligomerization domains, proline-rich, acidic autoinhibitory domains, and actin-binding motifs, many of which contribute to spatio-temporal control of GEF function [69].

$G_{12/13}$-coupled GPCRs specifically signal through four different RGS-containing RhoGEFs: p115-RhoGEF, PDZ-RhoGEF, leukemia-associated RhoGEF (LARG), and lymphoid blast crisis (Lbc)-RhoGEF [68,70]. G_{13}, but not G_{12}, was reported to promote GEF activity of p115-RhoGEF and PDZ-RhoGEF, providing an additional level of specificity to $G_{12/13}$-coupled GPCR signaling [71]; and $G_{q/11}$-coupled GPCRs were found to stimulate RhoA activity by interacting with p63-RhoGEF that in turn activates RhoA/ROCK signaling [39,72–74]. Momotani et al. [39] analyzed the pattern of RhoGEF transcript expression in mouse portal vein smooth muscle and detected the presence of mRNAs for all four RGS-RhoGEFs, confirming earlier

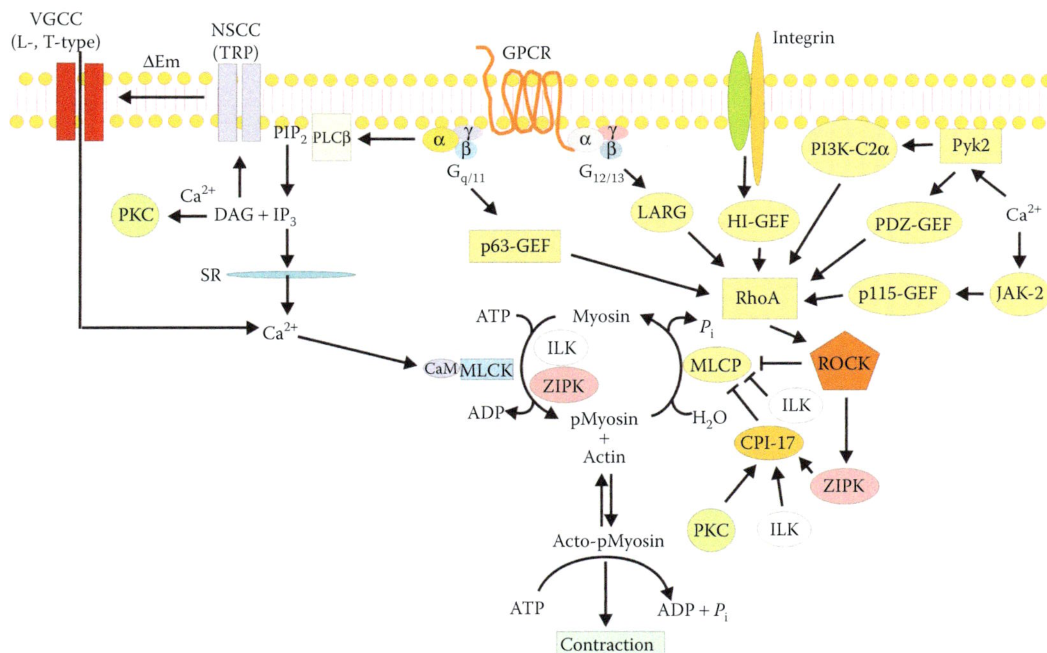

FIGURE 7.2 Contemporary signaling pathways for regulation of Ca^{2+} sensitization of smooth muscle contraction. More details are now available regarding the membrane depolarization (ΔEm)-induced entry of extracellular Ca^{2+} triggered by opening of non-selective cation channels (NSCC) of the TRP family, leading to the activation of L- and T-type voltage-gated Ca^{2+} channels (VGCC). IP_3-induced Ca^{2+} release from the SR contributes to the increase in $[Ca^{2+}]_i$ leading to activation of MLCK, PKC, and the Ca^{2+}-dependent tyrosine kinase Pyk2. Activation of a variety of RhoGEFs, coupled to $G_{q/11}$, $G_{12/13}$, and integrins or downstream of Ca^{2+} or the C2α isoform of PI 3-kinase (PI3K-C2α), leads to RhoA and ROCK activation and Ca^{2+} sensitization of contraction. ZIPK, which is a substrate of ROCK, directly phosphorylates myosin LC_{20} (at Ser-19 and Thr-18) and ILK, which also phosphorylates LC_{20} at both sites, phosphorylates both MYPT1 and CPI-17 at phosphatase inhibitory sites, thereby enhancing Ca^{2+} sensitization.

reports (LARG-RhoGEF [75]; PDZ-RhoGEF [76,77]; p115-RhoGEF [78]), as well as RhoGEFs for Rac and Cdc42, and p63-RhoGEF that was the most abundant transcript detected [39,74].

The expression of multiple RhoGEF transcripts suggests that different GPCRs may specifically recruit one or more RhoGEFs to evoke RhoA/ROCK signaling and Ca^{2+} sensitization with distinct temporal characteristics [39,79,80]. For example, angiotensin II- and endothelin-1-evoked contractions of aortic smooth muscle were reduced by LARG-RhoGEF deficiency [32,80], by suppressing p63-RhoGEF expression [39,74] or, respectively, by reducing either p115-RhoGEF [38], or LARG-RhoGEF and PDZ-RhoGEF [80]. In contrast, phenylephrine-evoked contractions were reduced by p63-RhoGEF knock-down, but unaffected by LARG-RhoGEF ablation [32], and U46619-evoked Ca^{2+} sensitization was reduced by LARG-RhoGEF and PDZ-RhoGEF deficiency [80], but resistant to p63-RhoGEF [39] and p115-RhoGEF suppression [38]. The pathway characterized for p115-RhoGEF was suggested to involve Ca^{2+}-dependent activation of Janus kinase-2 and subsequent stimulation of GEF activity by tyrosine phosphorylation, rather than a direct G protein mediated activation [38].

Some smooth muscle tissues exhibit an increase in sustained force generation following KCl-induced depolarization that is due in part to Ca^{2+} sensitization via RhoA/ROCK-mediated phosphorylation of MYPT1 [81–84]. The RhoA/ROCK signaling and MYPT1 phosphorylation was independent of GPCR and G protein activation but suppressed by the tyrosine kinase inhibitor genistein [85]. Ca^{2+}-dependent activation of proline-rich tyrosine kinase PYK2 was identified as the cause of RhoA activation and translocation to the membrane in response to membrane depolarization [86,87]. PYK2 is a non-receptor tyrosine kinase in the focal adhesion kinase (FAK) family that participates in a variety of cell functions activated by integrins, chemokine and cytokine immune receptors, growth factor receptors, and various GPCRs [88]. The mechanism of RhoA activation by PYK2 was not determined; however, Ying et al. [89] suggested that tyrosine phosphorylation of PDZ-RhoGEF by PYK2 was responsible for RhoA activation in response to elevated $[Ca^{2+}]_i$ in cultured rat aortic cells. Alternatively, phosphoinositide 3-kinase class II α-isoform (PI3K-C2α) was found to be required for Ca^{2+}-dependent activation of RhoA/ROCK and MYPT1 phosphorylation in aortic smooth muscle [90,91], raising the possibility of RhoA stimulation due to phosphatidylinositol 3,4,5-trisphosphate produced by PI3K-C2α [87].

The emerging view of RhoA regulation in smooth muscle cells is that it is dependent on a complex, sometimes redundant pattern of signaling through different RhoGEFs [79], including RhoGEFs such as GEF-H1 and LARG that couple mechanical forces exerted on cells through integrins to evoke contraction [92,93], as may occur in the myogenic response of resistance arterial VSMCs [94]. However, it is important to recognize that this understanding is based on the use of a limited selection of different smooth muscle tissues and GPCR agonists. Additionally, non-GPCR mechanisms of RhoA stimulation that are also mediated by RhoGEF signaling such as integrins, and regulation of RhoGEF activity by mechanisms in addition to G-proteins, including phosphorylation by second messengers such as phosphoinositides, cAMP, and PKC, have not been addressed in detail. These mechanisms may provide further sophistication to the signaling networks controlling the extent of Ca^{2+} sensitization in smooth muscle via control of RhoA activity.

An additional consideration that warrants further attention is the contribution of scaffolding proteins that provide for spatial organization of the elements of the signaling pathways causing MLCP inhibition and Ca^{2+} sensitization. For example, a yeast two-hybrid screen was employed to detect proteins that interact with RhoA and MLCP in an attempt to identify proteins that regulate the interaction of RhoA-ROCK with MLCP [95]. Surks et al. [95] identified the expression in cultured human aortic and coronary arterial smooth muscle cells of a RhoA-interacting protein possessing 88% sequence identity with mouse p116-RIP3 RhoA-binding protein that suppresses neuronal cell rounding and neurite retraction due to RhoA signaling [96], which they referred to as myosin phosphatase–Rho interacting protein MP-RIP (also known as M-RIP and p116RIP; [97,98]). MP-RIP is a cytoskeletal scaffolding protein that binds directly to both RhoA and MYPT1 [95], targeting MLCP to the contractile apparatus to dephosphorylate LC_{20} [98–101]. The binding site for MP-RIP is within the leucine zipper region of MYPT1 at its C-terminus [95]. MP-RIP silencing prevented RhoA/ROCK-dependent phosphorylation of MYPT1, but not activation of ROCK by RhoA [101]; MP-RIP does not bind ROCK, leaving the issue of how ROCK is targeted to MLCP unresolved. MP-RIP expression was recently shown to be higher in the phasic smooth muscle of gastric antrum than tonic smooth muscle of the fundus [102], implying that the differential contributions of Ca^{2+} sensitization in phasic and tonic smooth muscles may depend in part on distinct expression profiles of scaffolding proteins that organize signaling pathways for MLCP inhibition.

7.3.3 Rho-Associated Coiled-Coil Kinase-Mediated Regulation of Ca^{2+} Sensitization

In the active state, RhoGTPases interact with and activate multiple (>30) effector proteins, including in the case of RhoA, the ~158 kDa serine/threonine kinase, ROCK [103–105]. The interaction of RhoA with ROCK is thought to disrupt the intramolecular autoinhibition of ROCK's catalytic site, resulting in kinase activation. Specifically, ROCK has an N-terminal kinase domain, a central coiled-coil domain, and a C-terminal PH domain that acts as an intramolecular autoinhibitory fold in the absence of RhoA-GTP. Binding of RhoA-GTP to the Rho-binding domain just proximal to the PH domain relieves the autoinhibition and activates ROCK activity [106]. Two isoforms of ROCK, ROCK1 (ROCKβ) and ROCK2 (ROCKα), sharing 65% sequence identity, are expressed in smooth muscle [107]. The coiled-coil region of ROCK2 between residues 354–775 appears to mediate the interaction with MYPT1, specifically within its coiled-coil domain between residues 704 and 764 [107]. This interaction with MYPT1 may be specific to ROCK2, due to differences in the coiled-coil domain in ROCK1 and ROCK2 that share only 58% sequence identity.

The ROCK signaling nexus is a pivotal regulator of cellular functions, including motility and adhesion, membrane permeability, differentiation, proliferation, and apoptosis. As a result, ROCK has been the subject of intense focus by structural biologists and medicinal chemists, and there have been significant advances in the development of potent and selective inhibitors [108,109]. In the context of smooth muscle contraction, several small molecule inhibitors of ROCK have been accepted as pharmacological tools to examine the role of the kinase in Ca^{2+} sensitization in health and disease. These have traditionally included HA-1077 (fasudil), hydroxyfasudil, Y-27632, and H-1152. The molecules are effective inhibitors of both ROCK isoforms and are commonly used at relatively high concentrations in organ bath or cell culture experiments where significant off-target effects may exist [110]. There are to date only limited reports of the application of next-generation ROCK inhibitors in the study of smooth muscle contraction and Ca^{2+} sensitivity [111–113]. Such molecules include GSK-269962A (increased potency [113]), SLX-2119 (selective inhibition of ROCK2 with >200-fold selectivity over ROCK1 [112]) and SB-7720770-B (increased potency [113]). The smooth muscle research community should be encouraged to utilize these molecules in future studies.

7.4 REGULATION OF MYOSIN LIGHT CHAIN PHOSPHATASE ACTIVITY BY MYOSIN TARGETING SUBUNIT 1 PHOSPHORYLATION

MLCP is a complex of the type 1 protein serine/threonine phosphatase catalytic subunit (HGNC: PPP1C, δ isoform; denoted PP1cδ), the ~130-kDa myosin phosphatase targeting subunit (HGNC: protein phosphatase 1 regulatory subunit 12A, PPP1R12A; denoted MYPT1) and a 20 kDa (M20) regulatory subunit with an unresolved biological function. Several groups identified MYPT1 as a regulatory subunit of the MLCP holoenzyme that provided specific targeting of the PP1cδ catalytic activity toward myosin [114–117]. MYPT1 possesses two well-defined PP1c-binding elements, the so-called *RVxF* and *MyPhoNE* motifs, in its N-terminal region (Figure 7.3) [118]. A crystal structure of PP1cδ in complex with the N-terminal region of MYPT1 revealed that the binding of MYPT1 reshapes the catalytic cleft of PP1c to provide increased myosin substrate specificity [119].

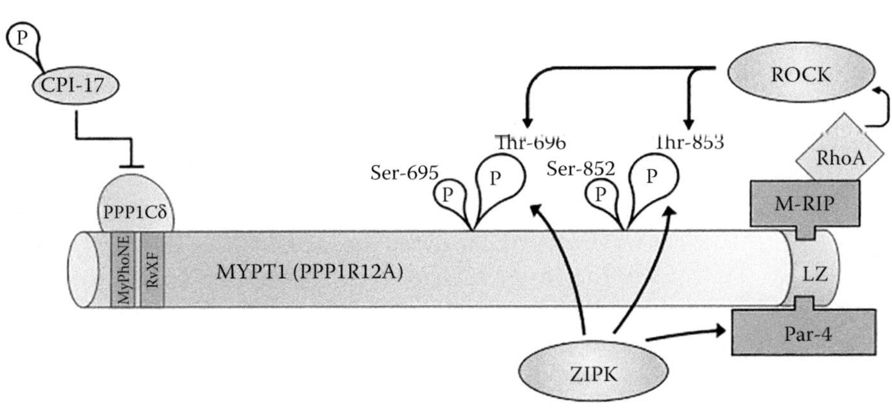

FIGURE 7.3 Mechanisms of MLCP regulation. The phosphorylation status of two dominant inhibitory sites (Thr-696 and Thr-853) on the myosin phosphatase-targeting subunit (MYPT1) primarily involves integration of ROCK and ZIPK signaling. The prostate-apoptosis response (Par)-4 protein and the myosin phosphatase-RhoA-interacting protein (M-RIP) provide additional signal regulation. Moreover, the C-kinase phosphatase inhibitor of 17-kDa (CPI-17), when phosphorylated, acts as a potent inhibitor of myosin phosphatase.

As indicated earlier, MLCP was initially thought to be unregulated, but it is now appreciated that the enzymatic activity of the holoenzyme is controlled directly by phosphorylation of MYPT1 at multiple sites [7,120–122] and/or indirectly by phosphorylation of the MLCP inhibitor protein, CPI-17 [34,123,124]. An examination of the MYPT1 sequence suggests an abundance of potential phosphorylation sites. Indeed, many directed biochemical and proteomic studies have defined multiple MYPT1 phosphorylation events (reviewed recently in [7]). While greater than 100 unique MYPT1 phosphorylation events have been described in the literature, the two best-characterized phosphorylation sites on MYPT1 are Thr-696 and Thr-853 (numbering for the human isoform is used throughout this chapter).

Trinkle-Mulcahy et al. [25] were the first to demonstrate that phosphorylation of MYPT1 suppressed MLCP activity and suggested that phosphorylation and dephosphorylation of the MYPT1 subunit was the regulatory mechanism for modulating MLCP enzymatic activity *in situ*. The native MLCP holoenzyme was demonstrated to possess an associated protein kinase that phosphorylated MYPT1; however, the kinase responsible was not identified at the time of publication [125]. A dominant phosphorylation site for the associated kinase was identified to be Thr-696. Moreover, the MLCP activity was progressively inhibited as the extent of phosphorylation increased, but the oligomeric state of the MLCP holoenzyme was not affected. The endogenous MLCP-associated kinase was later identified through purification and Edman peptide sequencing to be zipper-interacting protein kinase (ZIPK) [126]. ZIPK was shown to associate with MYPT1 and phosphorylate the inhibitory Thr-696 site in intact smooth muscle preparations.

Coincident with these early studies of MLCP regulation, Kimura and colleagues demonstrated that ROCK-dependent phosphorylation of MYPT1 was associated with a decrease in MLCP activity toward LC_{20} [26]. These investigators also demonstrated that the GTP-bound, active form of RhoA specifically interacted with the C-terminal domain of MYPT1. Additional independent investigations provided confirmation of the ability of ROCK to phosphorylate the MYPT1 protein and inhibit MLCP activity [28]. Two major sites of ROCK-dependent phosphorylation on MYPT1 were identified with biochemical analyses in concert with site-directed mutagenesis; that is, Thr-696 and Thr-853 [27]. Phosphospecific antibodies to pThr-696 and pThr-853 were developed and confirmed *in situ* phosphorylation of these inhibitory sites in cells following activation of RhoA [27,127]. The phosphorylation of either Thr-696 or Thr-853 by ROCK was demonstrated to suppress MLCP activity *in vitro* [128]. The rates of ROCK-dependent phosphorylation of the two sites were shown to be similar, and the efficacy of inhibition following phosphorylation was equivalent [128]. While still incompletely understood, the biochemical mechanism for Thr-696 inhibition of MLCP activity is thought to involve autoinhibition [129]. Mapping of the active site of MLCP shows a charge and sequence complementarity around Thr-696 and Thr-853 of MYPT1 that resembles those around Ser-19 of LC_{20}. It was proposed that phosphorylation at Thr-696 produces a potent autoinhibitory site, interacting with the active site of MLCP to induce an inactive form of the holoenzyme. Phosphorylation of Thr-853 may also generate an autoinhibited form of MLCP [129] or result in release of the phosphatase holoenzyme from the myosin filament [130,131].

There is still some debate among researchers regarding the functional effect of MYPT1 Thr-696 and Thr-853 phosphorylations and their impact upon the activity of the MLCP holoenzyme. *In vitro* biochemical assessments by Khasnis and colleagues [132] suggest that phosphorylation of Thr-696 provides direct inhibition of MLCP activity, whereas phosphorylation of Thr-853 does not alter enzymatic activity. Although previous studies associated the phosphorylation of Thr-853 with enzymatic inhibition of MLCP *in vitro* [128], investigations also suggest the site provides a second mechanism for the inhibition of MLCP activity by promoting the dissociation of the MLCP holoenzyme from myosin and intracellular relocalization to the plasma membrane or nucleus [130,131]. These findings were recapitulated *in vivo* with the generation of MYPT1 knock-in mice (e.g., heterozygous T696A/–, T696A/+ and homozygous T853A/T853A mice [133,134]).

Many studies with various tissues and stimuli demonstrate that phosphorylation of MYPT1 at Thr-696 and/or Thr-853 occurs in concert with Ca^{2+} sensitization of smooth muscle contraction (reviewed in [135,136]). High basal levels of Thr-696 phosphorylation are generally identified in smooth muscle tissues, with only modest increases in response to contractile stimuli. In most cases, there is little if any basal Thr-853 phosphorylation; however, Thr-853 phosphorylation has been found to increase dramatically in response to G protein activation. It is the phosphorylation of Thr-853 that is usually associated with ROCK-dependent MLCP inhibition and sensitization of force development. However, additional kinases can phosphorylate Thr-853 *in vitro* [7]. Significant pathological impact of mutational changes in the MYPT1 gene is predicted from high-quality exome (protein-coding region) DNA sequence data for 60,706 individuals of diverse ancestries generated as part of the Exome Aggregation Consortium (ExAC; exac.broadinstitute.org) [137]. The MYPT1 gene has more synonymous and fewer missense variants in the global sequence data than expected, so that the MYPT1 gene is predicted to be extremely intolerant of loss-of-function mutations. Global knockout of MYPT1 is embryonic-lethal in mice, but MYPT1 smooth muscle-specific knockouts driven by the α-actin promoter (SMA-Cre) are viable and demonstrate altered contractile properties of gastrointestinal smooth muscle [138,139], and enhanced vasoconstrictor responses to agonists and elevated mean arterial blood pressure [140].

While intolerant missense mutations of the Thr-696 and Thr-853 residues are not apparent within the ExAC sequence data, abnormal MYPT1 phosphorylation has been observed in experimental models of disease. In animal models of diabetes, basal levels of myogenic tone, LC_{20}, and Thr-853 phosphorylation were elevated in resistance arterioles of pre-diabetic Goto-Kakizaki rats, whereas loss of pressure-dependent myogenic constriction in diabetic Goto-Kakizaki animals was associated with an absence of transmural pressure-dependent MYPT1 Thr-853 phosphorylation [141]. In models of erectile dysfunction, higher Thr-696 phosphorylation was observed in penile tissues following cavernous nerve injury [142]. MYPT1 Thr-696 phosphorylation is elevated in smooth muscle from rat models of chronic hypoxia-induced pulmonary hypertension [143,144]. In resistance vessels isolated from the spontaneously-hypertensive rat, Thr-853 phosphorylation of MYPT1 was elevated at the hypertensive, but not the prehypertensive stage, as compared with normotensive controls [145]. In a model of diabetic gastroparesis, lower basal levels of Thr-696 and

Thr-853 MYPT1 phosphorylations were associated with impaired gastric antrum smooth muscle function [146]. Ca^{2+} sensitivity and MYPT1 phosphorylation (both Thr-696 and Thr-853) of intestinal smooth muscle were also decreased in a rodent model of intestinal edema following resuscitation fluid administration and mesenteric venous hypertension [147].

Studies of MYPT1 gene knockout in mice support its role in the regulation of MLCP activity and vascular smooth muscle contraction, but do not support an important role for MYPT1 phosphorylation by ROCK and PKG in the regulation of contraction and relaxation, respectively. Global knockout of MYPT1 in the mouse results in embryonic lethality [148], whereas smooth muscle-specific knockout has no significant effect on development or birth, and the mice survive to adulthood with only modest phenotypic changes [138]. This dramatic difference is most likely due to expression of MYPT1 in non-muscle as well as smooth muscle cells, its involvement in fundamental cellular processes such as cell migration and cell division, and its interactions with proteins other than myosin [121,122]. The lack of an obvious phenotype in the smooth muscle-specific knockout mouse indicates that MYPT1 is not essential for smooth muscle function, but MYPT1-deficient intestinal smooth muscle demonstrated a significant increase in force during the sustained phase of contraction, with reduced rates of contraction and relaxation and slower shortening velocity, and in the Ca^{2+} sensitivity of force development [138]. These findings are consistent with activation of PP1c-mediated LC_{20} dephosphorylation by MYPT1. Detailed analysis of smooth muscle-specific MYPT1 deletion revealed enhanced LC_{20} phosphorylation, contractility, and blood pressure in vascular smooth muscle [140]. Importantly, the authors demonstrated that MYPT1 deletion had no effect on cardiac and renal functions or vascular structure, indicating that the increase in blood pressure could not be ascribed to such effects. Specifically, MYPT1 deletion resulted in increased LC_{20} phosphorylation and force in isolated mesenteric arteries in response to membrane depolarization with KCl, and several contractile agonists acting via GPCRs (norepinephrine, endothelin-1, angiotensin II, vasopressin, and U46619). As was observed in control tissues, KCl- and norepinephrine-induced contractions of MYPT1-knockout vessels were inhibited by ROCK and PKC inhibitors and were associated with suppression of CPI-17 phosphorylation. Likewise, acetylcholine- and sodium nitroprusside-induced relaxation of KCl- or norepinephrine-induced contractions were comparable in control and MYPT1-knockout vessels [140], indicating that PKG-mediated MYPT1 phosphorylation does not contribute significantly to the relaxation response, which may be accounted for by the NO/cGMP/PKG pathway leading to a reduction in $[Ca^{2+}]_i$ [149].

The amount of PP1cδ in the bladder of the MYPT1 knockout mice was reduced by ~70% compared to wild-type; nevertheless, MLCP activity was comparable in wild-type and MYPT1-knockout bladder [150]. The authors attributed this to reduced MLCP activity in wild-type tissue due to constitutive phosphorylation of MYPT1 at Thr-696 and Thr-853. MYPT1 had previously been shown to be constitutively phosphorylated at these sites in unstimulated rabbit urethral smooth muscle, and membrane depolarization and α_1-adrenoceptor activation failed to increase phosphorylation at either site [151].

Recent studies on bladder smooth muscle from embryonic and postnatal knock-in mice with mutations T696A or T853A led to the conclusion that phosphorylation at Thr-696, but not Thr-853, mediates force maintenance through MLCP inhibition and increased LC_{20} phosphorylation [133]. The MYPT1-T696A mutation was found to inhibit sustained force, as well as LC_{20} phosphorylation, while the T853A mutation had no effect on maximal force development, little effect on force maintenance, and no contribution to agonist-induced (and Thr-853 phosphorylation-dependent) contractile responses. Furthermore, the authors demonstrated that force maintenance by Thr-696 phosphorylation was independent of ROCK activation. These results are consistent with the observations of Khasnis et al. [132], who developed a procedure to co-express PP1cδ and MYPT1 in mammalian cells and thereby reconstitute MYPT1 retaining the properties of the native holoenzyme. They found that Thr-696, but not Thr-853, phosphorylation inhibited MLCP activity. Thus, phosphorylation of Thr-696 may titrate the basal activity of cellular MLCP and affect the temporal phosphorylation at Thr-853 that is synchronized with myosin targeting [132].

While some of the apparent discrepancies regarding the effects of MYPT1 phosphorylation may be due to tissue-specific differences (particularly phasic compared to tonic smooth muscles), other possibilities include the use of pharmacological compared to physiological contractile stimuli and the method of application of the stimulus. Most studies that support a role for agonist-mediated phosphorylation of MYPT1 in Ca^{2+} sensitization of force have utilized bath application of relatively high concentrations of agonists. Electric field stimulation of mouse bladder smooth muscle, which releases the neurotransmitter acetylcholine to induce contraction via cholinergic receptors, was used as a physiological stimulus and found to increase LC_{20} phosphorylation and force but had no effect on the constitutive levels of MYPT1 phosphorylation at Thr-696 or Thr-853, in contrast to bath application of acetylcholine [152]. Similar results were reported for gastric fundus [153] and ileal smooth muscles [139]. Specifically, conditional knockout of MYPT1 or the knock-in mutation T853A had no effect on the contractile response to electric field stimulation in adult ileal smooth muscle [139]. It is possible that these neurophysiological responses differ in phasic and tonic smooth muscles given that the myogenic response in resistance arteries from the cerebral and skeletal muscle circulations involves increases in MYPT1 phosphorylation at Thr-853 [154,155]. As noted by Gao et al. [139], it would be interesting to investigate these responses in MYPT1-knockout and MYPT1-T853A mice.

7.5 ALTERNATIVE KINASE REGULATION OF MYOSIN LIGHT CHAIN PHOSPHATASE ACTIVITY

7.5.1 Integrin-Linked Kinase

Integrin-linked kinase (ILK) has been implicated in the regulation of MLCP activity in smooth muscle. Originally identified by yeast two-hybrid screening as a binding partner to the cytoplasmic domain of β-integrins [156], ILK is recognized to play a central role in integrin-mediated signal transduction [157–160]. Through higher level protein–protein interactions with the particularly-interesting new cysteine-histidine-rich (PINCH) and parvin proteins, ILK links integrins, and other components of focal adhesions (e.g., growth factor receptor kinase and GTPase signaling pathways) to the actin cytoskeleton [159]. Accordingly, the PINCH-ILK-PARVIN signaling complex regulates focal adhesion assembly, extracellular matrix (ECM) adhesions, cytoskeleton organization, and signaling to modulate cell migration, spreading, and survival. ILK is a 59-kDa Ser/Thr protein kinase with N-terminal ankyrin repeats and a C-terminal kinase domain (HGNC: 6040; UniProtKB: Q13418; ILK_HUMAN). In the context of smooth muscle contraction, the first observation of ILK involvement was reported by Walsh and colleagues [161]. The investigators identified two distinct pools of ILK within chicken gizzard, which included a Triton X-100-insoluble component from myofilament preparations that phosphorylated myosin in a Ca^{2+}-independent manner. ILK was also shown to account, at least in part, for the phosphorylation of myosin and contraction that occurred in isolated myofilaments without a change in $[Ca^{2+}]_i$. Additional *in vitro* studies revealed that ILK was associated with the MLCP holoenzyme and provided evidence that ILK could phosphorylate MYPT1 at the inhibitory Thr-696 residue [162], phosphorylate the MLCP inhibitors CPI-17 and PHI-1 [163], and diphosphorylate LC_{20} at Ser-19 and Thr-18 [164]. Taken together, these findings support a role for ILK in the contractile response to agonists that induce the inhibition of MLCP in vascular smooth muscle. There has been long-standing debate regarding the kinase activity of ILK, with some reports classifying the protein as a pseudokinase [157,160], and others demonstrating direct phosphorylation capabilities toward various substrates [158,165]. To aid in the assessment of the functional activity of ILK, several novel small molecule inhibitors have been developed in recent years. Examples of ILK inhibitors include QLT0267 [166], OSU-T315 [167], and KP392 [168]. QLT0267 was demonstrated to alter F-actin architecture in studies of breast cancer cells. The OSU-T315 compound was described to possess specificity for ILK activity toward various substrates including LC_{20} [167]. Regrettably, there are few if any reports of the application of ILK small molecule inhibitors in the study of VSM contraction. Future investigations are required to examine the potential of ILK small molecule inhibitors to modulate MLCP activity and smooth muscle Ca^{2+} sensitization.

7.5.2 Zipper-Interacting Protein Kinase

Zipper-interacting protein kinase (ZIPK), also known as death-associated protein kinase (DAPK)-3 (HGNC: 2676; UnitProtKB: O43293; DAPK3_HUMAN), is a 52 kDa Ser/Thr protein kinase linked to the regulation of a number of cell processes, including cell motility, smooth muscle contraction, and programmed cell death (i.e., apoptosis and autophagy) [169–171]. ZIPK possesses an N-terminal kinase domain, a central autoinhibitory region, a nuclear localization sequence, and a C-terminal leucine zipper motif. The protein is a member of a larger family of protein kinases known as the death-associated protein kinases (DAPKs). The kinase domain is most similar to other DAPK proteins (e.g., DAPK1) but also shares significant sequence and structural conservation with calmodulin-dependent kinase (CaMK) family members such as MLCK [172]. In contrast to MLCK and the other DAPKs, ZIPK does not have a CaM-binding domain, and its activity is regulated independently of $[Ca^{2+}]_i$. In addition to phosphorylating LC_{20} at both Ser-19 and Thr-18 in a Ca^{2+}-independent manner [173,174], ZIPK can regulate the activity of MLCP *in vitro* by associating with, and phosphorylating, its targeting subunit MYPT1 [126,175] and CPI-17. In this context, ZIPK was described as the major MYPT1-associated kinase in aortic smooth muscle cells [176]. These targets are thought to be the primary mediators of ZIPK influence on contraction; however, additional ZIPK regulators have been revealed recently. Notably, the prostate apoptosis response-4 (Par-4) protein was suggested to act as a scaffold for ZIPK [177]. While the upstream activating pathways for ZIPK are not completely resolved, evidence suggests that the kinase can be activated in vascular smooth muscle in response to a variety of external stimuli, including phenylephrine, angiotensin II, endothelin-1, U46619, and KCl-evoked membrane depolarization [175]. These contractile responses suggest that ZIPK can signal downstream of GPCRs that act through both $G_{12/13}$ and $G_{q/11}$.

Small resistance arteries (i.e., <250 μm diameter) regulate total peripheral resistance through intrinsic tone development originating within VSMCs in response to the mechanical stimulus of transmural pressure and circumferential wall-stress (i.e., the myogenic response) [178,179]. Alterations in $[Ca^{2+}]_i$ cannot fully explain the myogenic response, as concurrent measurement of $[Ca^{2+}]_i$ and vessel diameter has shown that contractions can continue to gain strength despite minimal further increase in $[Ca^{2+}]_i$ [180]. Thus, Ca^{2+} sensitization mechanisms are proposed to supply a critical regulatory capacity to the myogenic process [181]. Given the dominant role for ZIPK identified in VSM Ca^{2+} sensitization and emerging evidence for the up-regulation of ZIPK in models of hypertension and type 2 diabetes that are associated with alterations in myogenic contraction, it is possible that changes in ZIPK expression and functional capacity contribute to amplify the myogenic response of resistance arteries and promote vascular myopathies. In this regard, specific small molecule modulators of ZIPK may provide a novel strategy for pharmacological intervention [182].

The first reported small molecule inhibitor of ZIPK/DAPK was based on an aminopyridazine chemical structure [183]. Although potency was in the micromolar range, a single injection of this inhibitor attenuated brain tissue damage and neuronal biomarker loss when administered to animal models of ischemic brain injury. These therapeutic effects were validated in oxygen glucose deprivation and middle cerebral artery occlusion models. Another class of ZIPK inhibitors developed on an oxo-β-carboline backbone display nanomolar potencies [184]; however, these compounds have significant off-target actions against a large number of protein kinases, including CaMKII, CK1α, PKN2, PKA, and ROCK2. Another class of DAPK1/ZIPK inhibitors was established on a benzylidene oxazolone backbone [185,186]. These molecules display nanomolar potencies toward DAPK1 and ZIPK; however, a significant off-target action of these oxazolone inhibitors was demonstrated against another contractile kinase, namely ROCK2 [111]. This finding places the conclusions of studies of oxazolone compounds in animal models of hypertension into doubt, as the concurrent inhibition of ROCK2 during treatment could provide an equally rational explanation for some outcomes [187]. Finally, a group of thiol-substituted pyrazolo[3,4-d]pyrimidinones were identified with a fluorescence-linked enzyme chemoproteomic strategy (FLECS), an expansion of proteome mining in which inhibitors of a fluorescently-tagged ZIPK were screened against a background of the entire purinome [188]. The lead molecule, HS-38, does not possess any of the off-target liabilities of other published DAPK inhibitors, but does display inhibitory potential toward the proto-oncogene proviral integration site for moloney murine leukemia virus (PIM) kinases. Assessments of HS-38 effects on VSMCs show that the compound can attenuate LC_{20} phosphorylation and force development in isolated arterial strips [175,188]. It is important to note that structure-activity relationship explorations have not provided selectivity filters for DAPK1 *versus* DAPK3/ZIPK. So, all molecules available for the inhibition of ZIPK are also presumed to target DAPK1.

7.5.3 Mitogen-Activated Protein Kinase

Mitogen-activated protein kinases (MAPKs) are well-known to participate in the regulation of smooth muscle contractility. For example, the extracellular-regulated kinase (ERK) pathway was shown to be involved in phenylephrine-induced contraction of uterine artery [189] and aorta [190], and in the angiotensin II-induced contraction of lower esophageal sphincter [191,192]. The p38-MAPK pathway contributed to angiotensin II- and endothelin-induced contraction of aortic smooth muscle cells [193,194] and noradrenaline-induced contraction of mesenteric artery [195]. While these reports have revealed a role for MAPKs in the contractile function of tonic (e.g., vascular) smooth muscles, additional data indicate that MAPKs are also important contributors to gastrointestinal smooth muscle contractility [196–202]. This is important because, with intestinal inflammation, it is thought that smooth muscle cells undergo a phenotypic alteration whereby normal rhythmic contractions are supplanted by sustained Ca^{2+}-independent contractions that persist long after the mucosal response to injury has subsided. Thus, the study of underlying mechanisms for the regulation of intestinal smooth muscle Ca^{2+} sensitization will be important to understand the basis of loss of functional intestinal efficiency that characterizes the inflammatory bowel diseases and other intestinal motility disorders.

Recent evidence implicates the contribution of MAPKs to the regulation of MLCP activity during Ca^{2+} sensitization of ileal smooth muscle. For example, blockade of MAPKs resulted in accelerated ML-7 (MLCK inhibitor)-dependent relaxation with proportional reductions in LC_{20} phosphorylation; MLCP activity was increased after treatment with MAPK inhibitors but phosphorylation of MYPT1 and CPI-17, which were increased during carbachol-evoked contraction, were unaffected by treatment with MAPK inhibitors [200]. Neither ERK nor p38MAPK is known to directly phosphorylate MYPT1 (at Thr-696 or Thr-853); however, Matsumura and colleagues [203,204] have reported that MYPT1 could be phosphorylated at multiple sites (Ser-432, Ser-473, and Ser-601) during mitotic events by the proline-directed cdc2 kinase. Furthermore, they suggested that MAPKs could also phosphorylate these sites, although data were not included in the publication. The phosphorylation of MYPT1 by MAPKs might not affect activity but could modulate the association of other signaling proteins with MLCP. This situation is thought to account for the antagonism observed between MYPT1 phosphorylation (Ser-473) and the activity of polo-like kinase (PLK1) during cell division, specifically that phosphorylation of MYPT1 by cdc2 kinase creates a binding motif for PLK1 that mediates

MLCP-PLK1 complex formation and prohibits downstream signaling [204]. Also, it is possible that MAPK-dependent phosphorylation of MYPT1 in intestinal smooth muscle may antagonize ZIPK signaling. ZIPK interacts with MYPT1, and it is possible that MLCP controls ZIPK activity by reducing phosphorylation at an activation site, or by prohibiting its interaction with MLCP. Alternatively, the phosphorylation of MYPT1 by MAPKs may promote the association of an additional phosphatase activity to prevent accumulation of Thr-696/Thr-853 phosphorylation. Resolving these issues will be aided by development of novel ERK1/2 and p38MAPK inhibitor compounds [205–207] that can be employed to assess the function of these kinases in the abnormal regulation of Ca^{2+} sensitization in gastrointestinal smooth muscle motility disorders [199,208].

7.6 CONCLUSION

Our understanding of the cellular signaling pathways and mechanisms responsible for evoking Ca^{2+} sensitization of smooth muscle contraction has advanced considerably in recent years. However, knowledge is still incomplete, many gaps still exist in the signaling pathways depicted in Figure 7.2, and significant fundamental issues remain to be resolved. For example, based largely on the work of Zhu, Stull, and colleagues [133,139], and of Khasnis et al. [132], the physiological relevance of MYPT1 phosphorylation at Thr-696 and Thr-853 has been challenged. The possibility of involvement of additional kinases in the phosphorylation of LC_{20} (at Ser-19 or both Ser-19, and Thr-18), MYPT1 at Thr-696, Thr-853 and other regulatory sites, and CPI-17 at Thr-38 and other regulatory sites exists. For example, inhibition of p90 ribosomal S6 kinase (RSK) has recently been shown to reduce the Ca^{2+} sensitivity of force development and to decrease the phosphorylation of LC_{20} at Ser-19 and of MYPT1 at Thr-853 in α-toxin-permeabilized pulmonary arterial smooth muscle [209]. Finally, even the physiological relevance of RhoA-mediated activation of ROCK has been questioned recently [210].

ACKNOWLEDGMENTS

The authors acknowledge grant support from the Canadian Institutes of Health Research (MOP-97988 to WCC & MPW and MOP-133543 to JAM).

REFERENCES

1. Gunst, S. J., and W. Zhang. 2008. Actin cytoskeletal dynamics in smooth muscle: A new paradigm for the regulation of smooth muscle contraction. *Am J Physiol Cell Physiol* 295: C576–C587.
2. Frei, E., M. Huster, P. Smital, J. Schlossmann, F. Hofmann, and J. W. Wegener. 2009. Calcium-dependent and calcium-independent inhibition of contraction by cGMP/cGKI in intestinal smooth muscle. *Am J Physiol Gastrointest Liver Physiol* 297: G834–G839.
3. Zhou, G. P. 2011. The structural determinations of the leucine zipper coiled-coil domains of the cGMP-dependent protein kinase Iα and its interaction with the myosin binding subunit of the myosin light chain phosphatase. *Protein Pept Lett* 18: 966–978.

4. Walsh, M. P., and W. C. Cole. 2013. The role of actin filament dynamics in the myogenic response of cerebral resistance arteries. *J Cereb Blood Flow Metab* 33: 1–12.
5. Dippold, R. P., and S. A. Fisher. 2014. Myosin phosphatase isoforms as determinants of smooth muscle contractile function and calcium sensitivity of force production. *Microcirculation* 21: 239–248.
6. Reho, J. J., X. Zheng, and S. A. Fisher. 2014. Smooth muscle contractile diversity in the control of regional circulations. *Am J Physiol Heart Circ Physiol* 306: H163–H172.
7. Walsh, M. P., and J. A. MacDonald. 2017. Regulation of smooth muscle myosin light chain phosphatase by multi-site phosphorylation of the myosin targeting subunit, MYPT1. *Cardiovasc Hematolog Dis* in press. doi:10.2174/1871529x18666180326120638.
8. Evans, D. H., H. O. Schild, and S. Thesleff. 1958. Effects of drugs on depolarized plain muscle. *J Physiol* 143: 474–485.
9. Somlyo, A. V., and A. P. Somlyo. 1968. Electromechanical and pharmacomechanical coupling in vascular smooth muscle. *J Pharmacol Exp Ther* 159: 129–145.
10. Morgan, J. P., and K. G. Morgan. 1982. Vascular smooth muscle: The first recorded Ca^{2+} transients. *Pflugers Arch* 395: 75–77.
11. DeFeo, T. T., and K. G. Morgan. 1985. Calcium-force relationships as detected with aequorin in two different vascular smooth muscles of the ferret. *J Physiol* 369: 269–282.
12. Bradley, A. B., and K. G. Morgan. 1987. Alterations in cytoplasmic calcium sensitivity during porcine coronary artery contractions as detected by aequorin. *J Physiol* 385: 437–448.
13. Takuwa, Y., and H. Rasmussen. 1987. Measurement of cytoplasmic free Ca^{2+} concentration in rabbit aorta using the photoprotein, aequorin. Effect of atrial natriuretic peptide on agonist-induced Ca^{2+} signal generation. *J Clin Invest* 80: 248–257.
14. Himpens, B., T. Kitazawa, and A. P. Somlyo. 1990. Agonist-dependent modulation of Ca^{2+} sensitivity in rabbit pulmonary artery smooth muscle. *Pflugers Arch* 417: 21–28.
15. Rembold, C. M. 1990. Modulation of the [Ca^{2+}] sensitivity of myosin phosphorylation in intact swine arterial smooth muscle. *J Physiol* 429: 77–94.
16. Morgan, J. P., and K. G. Morgan. 1984. Stimulus-specific patterns of intracellular calcium levels in smooth muscle of ferret portal vein. *J Physiol* 351: 155–167.
17. Fujiwara, T., T. Itoh, Y. Kubota, and H. Kuriyama. 1989. Effects of guanosine nucleotides on skinned smooth muscle tissue of the rabbit mesenteric artery. *J Physiol* 408: 535–547.
18. Kitazawa, T., B. D. Gaylinn, G. H. Denney, and A. P. Somlyo. 1991. G-protein-mediated Ca^{2+} sensitization of smooth muscle contraction through myosin light chain phosphorylation. *J Biol Chem* 266: 1708–1715.
19. Kubota, Y., M. Nomura, K. E. Kamm, M. C. Mumby, and J. T. Stull. 1992. GTP gamma S-dependent regulation of smooth muscle contractile elements. *Am J Physiol Cell Physiol* 262: C405–C410.
20. Gong, M. C., K. Iizuka, G. Nixon, J. P. Browne, A. Hall, J. F. Eccleston, M. Sugai, S. Kobayashi, A. V. Somlyo, and A. P. Somlyo. 1996. Role of guanine nucleotide-binding proteins—ras-family or trimeric proteins or both—in Ca^{2+} sensitization of smooth muscle. *Proc Natl Acad Sci USA* 93: 1340–1345.
21. Otto, B., A. Steusloff, I. Just, K. Aktories, and G. Pfitzer. 1996. Role of Rho proteins in carbachol-induced contractions in intact and permeabilized guinea-pig intestinal smooth muscle. *J Physiol* 496: 317–329.
22. Gong, M. C., H. Fujihara, A. V. Somlyo, and A. P. Somlyo. 1997. Translocation of RhoA associated with Ca^{2+} sensitization of smooth muscle. *J Biol Chem* 272: 10704–10709.
23. Fujihara, H., L. A. Walker, M. C. Gong, E. Lemichez, P. Boquet, A. V. Somlyo, and A. P. Somlyo. 1997. Inhibition of RhoA translocation and calcium sensitization by in vivo ADP-ribosylation with the chimeric toxin DC3B. *Mol Biol Cell* 8: 2437–2447.

24. Fu, X., M. C. Gong, T. Jia, A. V. Somlyo, and A. P. Somlyo. 1998. The effects of the Rho-kinase inhibitor Y-27632 on arachidonic acid-, GTPgammaS-, and phorbol ester-induced Ca^{2+}-sensitization of smooth muscle. *FEBS Lett* 440: 183–187.
25. Trinkle-Mulcahy, L., K. Ichikawa, D. J. Hartshorne, M. J. Siegman, and T. M. Butler. 1995. Thiophosphorylation of the 130-kDa subunit is associated with a decreased activity of myosin light chain phosphatase in alpha-toxin-permeabilized smooth muscle. *J Biol Chem* 270: 18191–18194.
26. Kimura, K., M. Ito, M. Amano, K. Chihara, Y. Fukata, M. Nakafuku, B. Yamamori et al. 1996. Regulation of myosin phosphatase by Rho and Rho-associated kinase (Rho-kinase). *Science* 273: 245–248.
27. Feng, J., M. Ito, K. Ichikawa, N. Isaka, M. Nishikawa, D. J. Hartshorne, and T. Nakano. 1999, Inhibitory phosphorylation site for Rho-associated kinase on smooth muscle myosin phosphatase. *J Biol Chem* 274: 37385–37390.
28. Feng, J., M. Ito, Y. Kureishi, K. Ichikawa, M. Amano, N. Isaka, K. Okawa et al. 1999. Rho-associated kinase of chicken gizzard smooth muscle. *J Biol Chem* 274: 3744–3752.
29. Hayashi, Y., S. Senba, M. Yazawa, D. L. Brautigan, and M. Eto. 2001. Defining the structural determinants and a potential mechanism for inhibition of myosin phosphatase by the protein kinase C-potentiated inhibitor protein of 17 kDa. *J Biol Chem*. 276: 39858–39863.
30. Somlyo, A. P., and A. V. Somlyo. 1994. Signal transduction and regulation in smooth muscle. *Nature* 372: 231–236.
31. Gohla, A., G. Schultz, and S. Offermanns. 2000. Role for G(12)/G(13) in agonist-induced vascular smooth muscle cell contraction. *Circ Res* 87: 221–227.
32. Wirth, A., Z. Benyo, M. Lukasova, B. Leutgeb, N. Wettschureck, S. Gorbey, P. Orsy et al. 2008. G12–G13-LARG-mediated signalling in vascular smooth muscle is required for salt-induced hypertension. *Nat Med* 14: 64–68.
33. Dimopoulos, G. J., S. Semba, K. Kitazawa, M. Eto, and T. Kitazawa. 2007. Ca^{2+}-dependent rapid Ca^{2+} sensitization of contraction in arterial smooth muscle. *Circ Res* 100: 121–129.
34. Eto, M., and T. Kitazawa. 2017. Diversity and plasticity in signalling pathways that regulate smooth muscle responsiveness: Paradigms and paradoxes for the myosin phosphatase, the master regulator of smooth muscle contraction. *J Smooth Muscle Res* 53: 1–19.
35. Somlyo, A. P., and A. V. Somlyo. 2003. Ca^{2+} sensitivity of smooth muscle and non-muscle myosin II: Modulated by G proteins, kinases, and myosin phosphatase. *Physiol Rev* 83. 1325–1358.
36. Large, W. A., S. N. Saleh, and A. P. Albert. 2009. Role of phosphoinositol 4,5-bisphosphate and diacylglycerol in regulating native TRPC channel proteins in vascular smooth muscle. *Cell Calcium* 45: 574–582.
37. Albert, A. P., and W. A. Large. 2006. Signal transduction pathways and gating mechanisms of native TRP-like cation channels in vascular myocytes. *J Physiol* 570: 45–51.
38. Guilluy, C., J. Brégeon, G. Toumaniantz, M. Rolli-Derkinderen, K. Retailleau, L. Loufrani, D. Henrion et al. 2010. The Rho exchange factor Arhgef1 mediates the effects of angiotensin II on vascular tone and blood pressure. *Nat Med* 16: 183–190.
39. Momotani, K., M. V. Artamonov, D. Utepbergenov, U. Derewenda, Z. S. Derewenda, and A. V. Somlyo. 2011. p63RhoGEF couples Gα(q/11)-mediated signaling to Ca^{2+} sensitization of vascular smooth muscle contractility. *Circ Res* 109: 993–1002.
40. Zhang, J., S. S. Ferguson, L. S. Barak, M. J. Aber, B. Giros, R. J. Lefkowitz, and M. G. Caron. 1997. Molecular mechanisms of G protein-coupled receptor signaling: Role of G protein-coupled receptor kinases and arrestins in receptor desensitization and resensitization. *Recept Chan* 5: 193–199.

41. Ferguson, S. S. 2007. Phosphorylation-independent attenuation of GPCR signalling. *Trends Pharmacol Sci* 28: 173–179.
42. Brinks, H. L., and A. D. Eckhart. 2010. Regulation of GPCR signalling in hypertension. *Biochim Biophys Acta* 1802: 1268–1275.
43. Morris, G. E., C. P. Nelson, N. B. Standen, R. A. Challiss, and J. M. Willets. 2010. Endothelin signalling in arterial smooth muscle is tightly regulated by G protein-coupled receptor kinase 2. *Cardiovasc Res* 85: 424–433.
44. Tutunea-Fatan, E., F. A. Caetano, R. Gros, and S. S. Ferguson. 2015. GRK2 targeted knock-down results in spontaneous hypertension, and altered vascular GPCR signalling. *J Biol Chem* 290: 5141–5155.
45. Cohn, H. I., D. M. Harris, S. Pesant, M. Pfeiffer, R. H. Zhou, W. J. Koch, G. W. Dorn 2nd, and A. D. Eckhart. 2008. Inhibition of vascular smooth muscle G protein-coupled receptor kinase 2 enhances alpha1D-adrenergic receptor constriction. *Am J Physiol Heart Circ Physiol* 295: H1695–H1704.
46. Watson, N., M. E. Linder, K. M. Druey, J. H. Kehrl, and K. J. Blumer. 1996. RGS family members: GTPase-activation proteins for heterotrimeric G-protein alpha-subunits. *Nature* 383: 172–175.
47. Wieland, T., S. Lutz, and P. Chidiac. 2017. Regulators of G protein signalling: A spotlight on emerging functions in the cardiovascular system. *Curr Opin Pharmacol* 7: 201–207.
48. Osei-Owusu, P., and K. J. Blumer. 2015. Regulator of G protein signalling 2: A versatile regulator of vascular function. *Prog Mol Biol Transl Sci* 133: 77–92.
49. Heximer, S. P., N. Watson, M. E. Linder, K. J. Blumer, and J. R. Hepler. 1997. RGS-2/G0S8 is a selective inhibitor of Gqalpha function. *Proc Natl Acad Sci USA* 94: 14389–14393.
50. Tang, K. M., G. R. Wang, P. Lu, R. H. Karas, M. Aronovitz, S. P. Heximer, K. M. Kaltenbronn et al. 2003. Regulator of G-protein signalling-2 mediates vascular smooth muscle relaxation and blood pressure. *Nat Med* 9: 1506–1512.
51. Heximer, S. P., R. H. Knutsen, X. Sun, K. M. Kaltenbronn, M. H. Rhee, N. Peng, A. Oliveira-dos-Santos et al. 2003. Hypertension and prolonged vasoconstrictor signalling in RGS2-deficient mice. *J Clin Invest* 111: 445–452.
52. Sun, X., K. M. Kaltenbronn, T. H. Steinberg, and K. J. Blumer. 2005. RGS-2 is a mediator of nitric oxide action on blood pressure and vasoconstrictor signalling. *Mol Pharmacol* 67: 631–639.
53. Ganss, R. 2015. Keeping the balance right: Regulator of G protein signalling 5 in vascular physiology and pathology. *Prog Mol Biol Transl Sci* 133: 93–121.
54. Holobotovskyy, V., M. Manzur, M. Tare, J. Burchell, E. Bolitho, H. Viola, L. C. Hool, L. F. Arnolda, D. J. McKitrick, and R. Ganss. 2013. Regulator of G-protein signalling 5 controls blood pressure homeostasis and vessel wall remodeling. *Circ Res.* 112: 781–791.
55. Hall, A. 2015. Rho GTPases and the control of cell behaviour. *Biochem Soc Trans* 33: 891–895.
56. Cherfils, J., and P. Chardin. 1999. GEFs: Structural basis for their activation of small GTP-binding proteins. *Trends Biochem Sci* 24: 306–311.
57. Vetter, I. R., and A. Wittinghofer. 2001. The guanine nucleotide-binding switch in three dimensions. *Science* 294: 1299–1304.
58. Dvorsky, R., and M. R. Ahmadian. 2004. Always look on the bright site of Rho: Structural implications for a conserved intermolecular interface. *EMBO Rep* 5: 1130–1136.
59. Rossman, K. L., C. J. Der, and J. Sondek. 2005. GEF means go: Turning on RHO GTPases with guanine nucleotide-exchange factors. *Nat Rev Mol Cell Biol* 6: 167–180.

60. DerMardirossian, C., and G. M. Bokoch. 2005. GDIs: Central regulatory molecules in Rho GTPase activation. *Trends Cell Biol* 15: 356–363.
61. Bernards, A. 2006. Ras superfamily and interacting proteins database. *Methods Enzymol* 407: 1–9.
62. Bos, J. L., H. Rehmann, and A. Wittinghofer. 2007. GEFs and GAPs: Critical elements in the control of small G proteins. *Cell* 129: 865–877.
63. Tcherkezian, J., and N. Lamarche-Vane. 2007. Current knowledge of the large RhoGAP family of proteins. *Biol Cell* 99: 67–86.
64. Siehler, S. 2009. Regulation of RhoGEF proteins by G12/13-coupled receptors. *Br J Pharmacol* 158: 41–49.
65. Cherfils, J., and M. Zeghouf. 2013. Regulation of small GTPases by GEFs, GAPs, and GDIs. *Physiol Rev* 93: 269–309.
66. Miller, N. L., E. G. Kleinschmidt, and D. D. Schlaepfer. 2014. RhoGEFs in cell motility: Novel links between Rgnef and focal adhesion kinase. *Curr Mol Med* 14: 221–234.
67. Zheng, Y. 2001. DBL family guanine nucleotide exchange factors. *Trends Biochem Sci* 26: 724–732.
68. Fukuhara, S., H. Chikumi, and J. S. Gutkind. 2001. RGS-containing RhoGEFs: The missing link between transforming G proteins and Rho? *Oncogene* 20: 1661–1668.
69. Aittaleb, M., C. A. Boguth, and J. J. Tesmer. 2010. Structure and function of heterotrimeric G protein-regulated Rho guanine nucleotide exchange factors. *Mol Pharmacol* 77: 111–125.
70. Dutt, P., N. Nguyen, and D. Toksoz. 2004. Role of Lbc RhoGEF in $G\alpha_{12/13}$-induced signals to Rho GTPase. *Cell Signal* 16: 201–209.
71. Tanabe, S., B. Kreutz, N. Suzuki, and T. Kozasa. 2004. Regulation of RGS-RhoGEFs by $G\alpha_{12}$ and $G\alpha_{13}$ proteins. *Methods Enzymol* 390: 285–294.
72. Lutz, S., A. Freichel-Blomquist, Y. Yang, U. Rümenapp, K. H. Jakobs, M. Schmidt, and T. Wieland. 2005. The guanine nucleotide exchange factor p63RhoGEF, a specific link between Gq/11-coupled receptor signaling and RhoA. *J Biol Chem* 280: 11134–11139.
73. Toumaniantz, G., D. Ferland-McCollough, C. Cario-Toumaniantz, P. Pacaud, and G. Loirand. 2010. The Rho protein exchange factor Vav3 regulates vascular smooth muscle cell proliferation and migration. *Cardiovasc Res* 86: 131–140.
74. Wuertz, C. M., A. Lorincz, C. Vettel, M. A. Thomas, T. Wieland, and S. Lutz. 2010. p63RhoGEF: A key mediator of angiotensin II-dependent signalling and processes in vascular smooth muscle cells. *FASEB J* 24: 4865–4876.
75. Moriki, N., M. Ito, T. Seko, Y. Kureishi, R. Okamoto, T. Nakakuki, M. Kongo, N. Isaka, K. Kaibuchi, and T. Nakano. 2004. RhoA activation in vascular smooth muscle cells from stroke-prone spontaneously hypertensive rats. *Hypertens Res* 27: 263–270.
76. Derewenda, U., A. Oleksy, A. S. Stevenson, J. Korczynska, Z. Dauter, A. P. Somlyo, J. Otlewski, A. V. Somlyo, and Z. S. Derewenda. 2004. The crystal structure of RhoA in complex with the DH/PH fragment of PDZRhoGEF, an activator of the $Ca^{(2+)}$ sensitization pathway in smooth muscle. *Structure* 12: 1955–1965.
77. Hilgers, R. H., J. Todd, and R. C. Webb. 2007. Increased PDZ-RhoGEF/RhoA/Rho kinase signaling in small mesenteric arteries of angiotensin II-induced hypertensive rats. *J Hypertens* 25: 1687–1697.
78. Ying, Z., L. Jin, A. M. Dorrance, and R. C. Webb. 2004. Increased expression of mRNA for regulator of G protein signaling domain-containing Rho guanine nucleotide exchange factors in aorta from stroke-prone spontaneously hypertensive rats. *Am J Hypertens* 17: 981–985.
79. Momotani, K., and A. V. Somlyo. 2012. p63RhoGEF: A new switch for G(q)-mediated activation of smooth muscle. *Trends Cardiovasc Med* 22: 122–127.

80. Artamonov, M. V., K. Momotani, A. Stevenson, D. R. Trentham, U. Derewenda, Z. S. Derewenda, P. W. Read, J. S. Gutkind, and A. V. Somlyo. 2013. Agonist-induced Ca^{2+} sensitization in smooth muscle: Redundancy of Rho guanine nucleotide exchange factors (RhoGEFs) and response kinetics, a caged compound study. *J Biol Chem* 288: 34030–34040.
81. Yanagisawa, T., and Y. Okada. 1994. KCl depolarization increases Ca^{2+} sensitivity of contractile elements in coronary arterial smooth muscle. *Am J Physiol Heart Circ Physiol* 267: H614–H621.
82. Mita, M., H. Yanagihara, S. Hishinuma, M. Saito, and M. P. Walsh. 2002. Membrane depolarization-induced contraction of rat caudal arterial smooth muscle involves Rho-associated kinase. *Biochem J* 364: 431–440.
83. Sakamoto, K., M. Hori, M. Izumi, T. Oka, K. Kohama, H. Ozaki, and H. Karaki. 2003. Inhibition of high K^+-induced contraction by the ROCKs inhibitor Y-27632 in vascular smooth muscle: Possible involvement of ROCKs in a signal transduction pathway. *J Pharmacol Sci* 92: 56–69.
84. Ratz, P. H., K. M. Berg, N. H. Urban, and A. S. Miner. 2005. Regulation of smooth muscle calcium sensitivity: KCl as a calcium-sensitizing stimulus. *Am J Physiol Cell Physiol* 288: C769–C783.
85. Mita, M., H. Tanaka, H. Yanagihara, J. Nakagawa, S. Hishinuma, C. Sutherland, M. P. Walsh, and M. Shoji. 2013. Membrane depolarization-induced RhoA/Rho-associated kinase activation and sustained contraction of rat caudal arterial smooth muscle involves genistein-sensitive tyrosine phosphorylation. *J Smooth Muscle Res* 49: 26–45.
86. Mills, R. D., M. Mita, J. Nakagawa, M. Shoji, C. Sutherland, and M. P. Walsh. 2015. A role for the tyrosine kinase Pyk2 in depolarization-induced contraction of vascular smooth muscle. *J Biol Chem* 290: 8677–8692.
87. Mills, R. D., M. Mita, and M. P. Walsh. 2015. A role for the Ca($^{2+}$)-dependent tyrosine kinase Pyk2 in tonic depolarization-induced vascular smooth muscle contraction. *J Muscle Res Cell Motil* 36: 479–489.
88. Hall, J. E., W. Fu, and M. D. Schaller. 2011. Focal adhesion kinase: Exploring Fak structure to gain insight into function. *Int Rev Cell Mol Biol* 288: 185–225.
89. Ying, Z., F. R. Giachini, R. C. Tostes, and R. C. Webb. 2009. PYK2/PDZ-RhoGEF links Ca^{2+} signalling to RhoA. *Arterioscler Thromb Vasc Biol* 29: 1657–1663.
90. Wang, Y., K. Yoshioka, M. A. Azam, N. Takuwa, S. Sakurada, Y. Kayaba, N. Sugimoto et al. 2006. Class II phosphoinositide 3-kinase α-isoform regulates Rho, myosin phosphatase and contraction in vascular smooth muscle. *Biochem J* 394: 581–592.
91. Yoshioka, K., N. Sugimoto, N. Takuwa, and Y. Takuwa. 2007. Essential role for class II phosphoinositide 3-kinase α-isoform in Ca^{2+}-induced, Rho- and Rho kinase-dependent regulation of myosin phosphatase and contraction in isolated vascular smooth muscle cells. *Mol Pharmacol* 71: 912–920.
92. Guilluy, C., V. Swaminathan, R. Garcia-Mata, E. T. O'Brien, R. Superfine, and K. Burridge. 2011. The Rho GEFs LARG and GEF-H1 regulate the mechanical response to force on integrins. *Nat Cell Biol* 13: 722–727.
93. Schiller, H. B., M. R. Hermann, J. Polleux, T. Vignaud, S. Zanivan, C. C. Friedel, Z. Sun et al. 2013. β1- and αv-class integrins cooperate to regulate myosin II during rigidity sensing of fibronectin-based microenvironments. *Nat Cell Biol* 2013;15:625–636.
94. Colinas, O., A. Moreno-Domínguez, H. L. Zhu, E. J. Walsh, M. T. Pérez-García, M. P. Walsh, and W. C. Cole. 2015. α5-Integrin-mediated cellular signaling contributes to the myogenic response of cerebral resistance arteries. *Biochem Pharmacol* 97: 281–291.
95. Surks, H. K., C. T. Richards, and M. E. Mendelsohn. 2003. Myosin phosphatase-Rho interacting protein. A new member of the myosin phosphatase complex that directly binds RhoA. *J Biol Chem* 278: 51484–51493.

96. Gebbink, M. F., O. Kranenburg, M. Poland, F. P. van Horck, B. Houssa, and W. H. Moolenaar. 1997. Identification of a novel, putative Rho-specific GDP/GTP exchange factor and a RhoA-binding protein: Control of neuronal morphology. *J Cell Biol* 137: 1603–1613.
97. Mulder, J., A. Ariaens, D. van den Boomen, and W. H. Moolenaar. 2004. p116RIP targets myosin phosphatase to stress fibers of the actin cytoskeleton in cultured A7r5 cells and is essential for RhoA/ROCK-regulated neuritogenesis. *Mol Biol Cell* 15: 5516–5527.
98. Koga, Y., and M. Ikebe. 2005. p116RIP decreases myosin II phosphorylation by activating myosin light chain phosphatase and by inactivating RhoA. *J Biol Chem* 280: 4983–4991.
99. Vallenius, T., K. Vaahtomeri, B. Kovac, A. M. Osiceanu, M. Viljanen, and T. P. Mäkelä. 2011. An association between NUAK2 and MRIP reveals a novel mechanism for regulation of actin stress fibers. *J Cell Sci* 124: 384–393.
100. Surks, H. K., N. Riddick, and K. Ohtani. 2005. M-RIP targets myosin phosphatase to stress fibers to regulate myosin light chain phosphorylation in vascular smooth muscle cells. *J Biol Chem* 280: 42543–42551.
101. Riddick, N., K. Ohtani, and H. K. Surks. 2007. Targeting by myosin phosphatase-RhoA interacting protein mediates RhoA/ROCK regulation of myosin phosphatase. *J Cell Biochem* 103: 1158–1170.
102. Bhetwal, B. P., C. L. An, S. A. Fisher, and B. A. Perrino. 2011. Regulation of basal LC20 phosphorylation by MYPT1 and CPI-17 in murine gastric antrum, gastric fundus, and proximal colon smooth muscles. *Neurogastroenterol Motil* 23: e425–e436.
103. Leung, T., E. Manser, L. Tan, and L. Lim. 1995. A novel serine/threonine kinase binding the Ras-related RhoA GTPase which translocates the kinase to peripheral membranes. *J Biol Chem* 270: 29051–29054.
104. Amano, M., M. Ito, K. Kimura, Y. Fukata, K. Chihara, T. Nakano, Y. Matsuura, and K. Kaibuchi. 1996. Phosphorylation and activation of myosin by Rho-associated kinase (Rho-kinase). *J Biol Chem* 271: 20246–20249.
105. Bishop, A. L., and A. Hall. 2000. Rho GTPases and their effector proteins. *Biochem J* 348: 241–255.
106. Riento, K., and A. J. Ridley. 2003. Rocks: Multifunctional kinases in cell behaviour. *Nat Rev Mol Cell Biol* 4: 446–456.
107. Wang, Y., X. R. Zheng, N. Riddick, M. Bryden, W. Baur, X. Zhang, and H. K. Surks. 2009. ROCK isoform regulation of myosin phosphatase and contractility in vascular smooth muscle cells. *Circ Res* 104: 531–540.
108. Defert, O., and S. Boland. 2017. Rho kinase inhibitors: A patent review (2014–2016). *Expert Opin Ther Pat* 27: 1–9.
109. Guan, R., X. Xu, M. Chen, H. Hu, H. Ge, S. Wens, S. Zhou, and R. Pi. 2013. Advances in the studies of roles of Rho/Rho-kinase in diseases and the development of its inhibitors. *Eur J Med Chem* 70: 613–622.
110. Wirth, A. 2010. Rho kinase and hypertension. *Biochim Biophys Acta* 1802: 1276–1284.
111. Al-Ghabkari, A., J. T. Deng, P. C. McDonald, S. Dedhar, M. Alshehri, M. P. Walsh, and J. A. MacDonald. 2016. A novel inhibitory effect of oxazol-5-one compounds on ROCKII signaling in human coronary artery vascular smooth muscle cells. *Sci Rep* 6: 32118.
112. Boerma, M., Q. Fu, J. Wang, D. S. Loose, A. Bartolozzi, J. L. Ellis, S. McGonigle et al. 2008. Comparative gene expression profiling in three primary human cell lines after treatment with a novel inhibitor of Rho kinase or atorvastatin. *Blood Coagul Fibrinolysis* 19: 709–718.
113. Doe, C., R. Bentley, D. J. Behm, R. Lafferty, R. Stavenger, D. Jung, M. Bamford et al. 2007. Novel Rho kinase inhibitors with anti-inflammatory and vasodilatory activities. *J Pharmacol Exp Ther* 320: 89–98.

114. Alessi, D., L. K. MacDougall, M. M. Sola, M. Ikebe, and P. Cohen. 1992. The control of protein phosphatase-1 by targetting subunits. The major myosin phosphatase in avian smooth muscle is a novel form of protein phosphatase-1. *Eur J Biochem* 210: 1023–1035.
115. Okubo, S., M. Ito, Y. Takashiba, K. Ichikawa, M. Miyahara, H. Shimizu, T. Konishi et al. 1994. A regulatory subunit of smooth muscle myosin bound phosphatase. *Biochem Biophys Res Commun* 200: 429–434.
116. Shimizu, H., M. Ito, M. Miyahara, K. Ichikawa, S. Okubo, T. Konishi, M. Naka et al. 1994. Characterization of the myosin-binding subunit of smooth muscle myosin phosphatase. *J Biol Chem* 269: 30407–30411.
117. Shirazi, A., K. Iizuka, P. Fadden, C. Mosse, A. P. Somlyo, A. V. Somlyo, and T. A. Haystead. 1994. Purification and characterization of the mammalian myosin light chain phosphatase holoenzyme. The differential effects of the holoenzyme and its subunits on smooth muscle. *J Biol Chem* 269: 31598–31606.
118. Hendrickx, A., M. Beullens, H. Ceulemans, T. Den Abt, A. Van Eynde, E. Nicolaescu, B. Lesage, and M. Bollen. 2009. Docking motif-guided mapping of the interactome of protein phosphatase-1. *Chem Biol* 16: 365–371.
119. Terrak, M., F. Kerff, K. Langsetmo, T. Tao, and R. Dominguez. 2004. Structural basis of protein phosphatase 1 regulation. *Nature* 429: 780–784.
120. Ito, M., T. Nakano, F. Erdődi, and D. J. Hartshorne. 2004. Myosin phosphatase: Structure, regulation and function. *Mol Cell Biochem* 259: 197–209.
121. Matsumura, F., and D. J. Hartshorne. 2008. Myosin phosphatase target subunit: Many roles in cell function. *Biochem Biophys Res Commun* 369: 149–156.
122. Grassie, M. E., L. D. Moffat, M. P. Walsh, and J. A. MacDonald. 2011 The myosin phosphatase targeting protein (MYPT) family: A regulated mechanism for achieving substrate specificity of the catalytic subunit of protein phosphatase type 1δ. *Arch Biochem Biophys* 510: 147–159.
123. Eto, M. 2009. Regulation of cellular protein phosphatase-1 (PP1) by phosphorylation of the CPI-17 family, C-kinase-activated PP1 inhibitors. *J Biol Chem* 284: 35273–35277.
124. Eto, M., and D. L. Brautigan. 2012. Endogenous inhibitor proteins that connect Ser/Thr kinases and phosphatases in cell signaling. *IUBMB Life* 64: 732–729.
125. Ichikawa, K., M. Ito, and D. J. Hartshorne. 1996. Phosphorylation of the large subunit of myosin phosphatase and inhibition of phosphatase activity. *J Biol Chem* 271: 4733–4740.
126. MacDonald, J. A., M. A. Borman, A. Murányi, A. V. Somlyo, D. J. Hartshorne, and T. A. Haystead. 2001. Identification of the endogenous smooth muscle myosin phosphatase-associated kinase. *Proc Natl Acad Sci USA* 98: 2419–2424.
127. Kawano, Y., Y. Fukata, N. Oshiro, M. Amano, T. Nakamura, M. Ito, F. Matsumura, M. Inagaki, and K. Kaibuchi. 1999. Phosphorylation of myosin-binding subunit (MBS) of myosin phosphatase by Rho-kinase in vivo. *J Cell Biol* 147: 1023–1038.
128. Murányi, A., D. Derkach, F. Erdődi, A. Kiss, M. Ito, and D. J. Hartshorne. 2005. Phosphorylation of Thr695 and Thr850 on the myosin phosphatase target subunit: Inhibitory effects and occurrence in A7r5 cells. *FEBS Lett* 579: 6611–6615.
129. Khromov, A., N. Choudhury, A. S. Stevenson, A. V. Somlyo, and M. Eto. 2009. Phosphorylation-dependent autoinhibition of myosin light chain phosphatase accounts for Ca^{2+} sensitization force of smooth muscle contraction. *J Biol Chem* 284: 21569–21579.
130. Velasco, G., C. Armstrong, N. Morrice, S. Frame, and P. Cohen. 2002. Phosphorylation of the regulatory subunit of smooth muscle protein phosphatase 1M at Thr850 induces its dissociation from myosin. *FEBS Lett* 52: 101–104.

131. Lontay, B., A. Kiss, P. Gergely, D. J. Hartshorne, and F. Erdődi. 2005. Okadaic acid induces phosphorylation and translocation of myosin phosphatase target subunit 1 influencing myosin phosphorylation, stress fiber assembly and cell migration in HepG2 cells. *Cell Sig* 17: 1265–1275.
132. Khasnis, M., A. Nakatomi, K. Gumpper, and M. Eto. 2014. Reconstituted human myosin light chain phosphatase reveals distinct roles of two inhibitory phosphorylation sites of the regulatory subunit, MYPT1. *Biochemistry* 53: 2701–2709.
133. Chen, C. P., X. Chen, Y. N. Qiao, P. Wang, W. Q. He, C. H. Zhang, W. Zhao et al. 2015. In vivo roles for myosin phosphatase targeting subunit-1 phosphorylation sites T694 and T852 in bladder smooth muscle contraction. *J Physiol* 593: 681–700.
134. MacDonald, J. A. 2015. A tale of two threonines: Myosin phosphatase inhibition and calcium sensitization of smooth muscle. *J Physiol* 593: 487–488.
135. Perrino, B. A. 2016. Calcium sensitization mechanisms in gastrointestinal smooth muscles. *J Neurogastroenterol Motil* 22: 213–225.
136. Butler, T., J. Paul, N. Europe-Finner, R. Smith, and E. C. Chan. 2013. Role of serine-threonine phosphoprotein phosphatases in smooth muscle contractility. *Am J Physiol Cell Physiol* 304: C485–C504.
137. Lek, M., K. J. Karczewski, E. V. Minikel, K. E. Samocha, E. Banks, T. Fennell, A. H. O'Donnell-Luria et al. 2016. Consortium exome aggregation. Analysis of protein-coding genetic variation in 60,706 humans. *Nature* 536: 285–291.
138. He, W. Q., Y. N. Qiao, Y. J. Peng, J. M. Zha, C. H. Zhang, C. Chen, C. P. Chen et al. 2013. Altered contractile phenotypes of intestinal smooth muscle in mice deficient in myosin phosphatase target subunit 1. *Gastroenterol* 144: 1456–1465.
139. Gao, N., A. N. Chang, W. He, C. P. Chen, Y. N. Qiao, M. Zhu, K. E. Kamm, and J. T. Stull. 2016. Physiological signalling to myosin phosphatase targeting subunit-1 phosphorylation in ileal smooth muscle. *J Physiol* 594: 3209–3925.
140. Qiao, Y. N., W. Q. He, C. P. Chen, C. H. Zhang, W. Zhao, P. Wang, L. Zhang et al. 2014. Myosin phosphatase target subunit 1 (MYPT1) regulates the contraction and relaxation of vascular smooth muscle and maintains blood pressure. *J Biol Chem* 289: 22512–22523.
141. Abd-Elrahman, K. S., O. Colinas, E. J. Walsh, H. L. Zhu, C. M. Campbell, M. P. Walsh, and W. C. Cole. 2017. Abnormal myosin phosphatase targeting subunit 1 phosphorylation and actin polymerization contribute to impaired myogenic regulation of cerebral arterial diameter in the type 2 diabetic Goto-Kakizaki rat. *J Cereb Blood Flow Metab* 37: 227–240.
142. Cho, M. C., K. Park, S. W. Kim, and J. S. Paick. 2015. Restoration of erectile function by suppression of corporal apoptosis, fibrosis and corporal veno-occlusive dysfunction with rho-kinase inhibitors in a rat model of cavernous nerve injury. *J Urol* 193: 1716–1723.
143. Guilluy, C., V. Sauzeau, M. Rolli-Derkinderen, P. Guerin, C. Sagan, P. Pacaud, and G. Loirand. 2005. Inhibition of RhoA/Rho kinase pathway is involved in the beneficial effect of sildenafil on pulmonary hypertension. *Br J Pharmacol* 146: 1010–1018.
144. Oka, M., N. Homma, L. Taraseviciene-Stewart, K. G. Morris, D. Kraskauskas, N. Burns, N. F. Voelkel, and I. F. McMurtry. 2007. Rho kinase-mediated vasoconstriction is important in severe occlusive pulmonary arterial hypertension in rats. *Circ Res* 100: 923–929.
145. Seok, Y. M., M. A. Azam, Y. Okamoto, A. Sato, K. Yoshioka, M. Maeda, I. Kim, and Y. Takuwa. 2010. Enhanced Ca^{2+}-dependent activation of phosphoinositide 3-kinase class IIalpha isoform-Rho axis in blood vessels of spontaneously hypertensive rats. *Hypertension* 56: 934–941.

146. Bhetwal, B. P., C. An, S. A. Baker, K. L. Lyon, and B. A. Perrino. 2013. Impaired contractile responses and altered expression and phosphorylation of $Ca(^{2+})$ sensitization proteins in gastric antrum smooth muscles from ob/ob mice. *J Muscle Res Cell Motil* 34: 137–149.
147. Chu, J., C. T. Miller, K. Kislitsyna, G. A. Laine, R. H. Stewart, C. S. Cox, and K. S. Uray. 2012. Decreased myosin phosphatase target subunit 1 (MYPT1) phosphorylation via attenuated rho kinase and zipper-interacting kinase activities in edematous intestinal smooth muscle. *Neurogastroenterol Motil* 24: 257–266.
148. Okamoto, R., M. Ito, N. Suzuki, M. Kongo, N. Moriki, H. Saito, H. Tsumura et al. 2005. The targeted disruption of the MYPT1 gene results in embryonic lethality. *Transgenic Res* 14: 337–340.
149. Hofmann, F. 2005. The biology of cGMP-dependent protein kinases. *J Biol Chem* 280: 1–4.
150. Tsai, M.-H., A. N. Chang, J. Huang, W. He, H. L. Sweeney, M. Zhu, K. E. Kamm, and J. T. Stull. 2014. Constitutive phosphorylation of myosin phosphatase targeting subunit-1 in smooth muscle. *J Physiol* 592: 3031–3051.
151. Walsh, M. P., K. Thornbury, W. C. Cole, G. Sergeant, M. Hollywood, and N. McHale. 2011. Rho-associated kinase plays a role in rabbit urethral smooth muscle contraction, but not via enhanced myosin light chain phosphorylation. *Am J Physiol Renal Physiol* 300: F73–F85.
152. Tsai, M. H., K. E. Kamm, and J. T. Stull. 2012. Signalling to contractile proteins by muscarinic and purinergic pathways in neutrally stimulated bladder smooth muscle. *J Physiol* 590: 5107–5121.
153. Bhetwal, B. P., K. M. Sanders, D. M. Trappanese, R. S. Moreland, and B. A. Perrino. 2013. Ca^{2+} sensitization pathways accessed by cholinergic neurotransmission in the murine gastric fundus. *J Physiol* 591: 2971–2986.
154. Johnson, R. P., A. F. El-Yazbi, K. Takeya, E. J. Walsh, M. P. Walsh, and W. C. Cole. 2009. Ca^{2+} sensitization via phosphorylation of myosin phosphatase targeting subunit at threonine-855 by Rho kinase contributes to the arterial myogenic response. *J Physiol* 587: 2537–2553.
155. Moreno-Domínguez, A., O. Colinas, A. El-Yazbi, E. J. Walsh, M. A. Hill, M. P. Walsh, and W. C. Cole. 2013. Ca^{2+} sensitization due to myosin light chain phosphatase inhibition and cytoskeletal reorganization in the myogenic response of skeletal muscle resistance arteries. *J Physiol* 591: 1235–1250.
156. Hannigan, G. E., C. Leung-Hagesteijn, L. Fitz-Gibbon, M. G. Coppolino, G. Radeva, J. Filmus, J. C. Bell, and S. Dedhar. 1996. Regulation of cell adhesion and anchorage-dependent growth by a new beta 1-integrin-linked protein kinase. *Nature* 379: 91–96.
157. Ghatak, S., J. Morgner, and S. A. Wickstrom. 2013. ILK: A pseudokinase with a unique function in the integrin-actin linkage. *Biochem Soc Trans* 41: 995–1001.
158. Hannigan, G. E., P. C. McDonald, M. P. Walsh, and S. Dedhar. 2011. Integrin-linked kinase: Not so "pseudo" after all. *Oncogene* 30: 4375–4385.
159. Kovalevich, J., B. Tracy, and D. Langford. 2011. PINCH: More than just an adaptor protein in cellular response. *J Cell Physiol* 226: 940–947.
160. Wickstrom, S. A., A. Lange, E. Montanez, and R. Fassler. 2010. The ILK/PINCH/parvin complex: The kinase is dead, long live the pseudokinase! *EMBO J* 29: 281–291.
161. Deng, J. T., J. E. Van Lierop, C. Sutherland, and M. P. Walsh. 2001. Ca^{2+}-independent smooth muscle contraction. A novel function for integrin-linked kinase. *J Biol Chem* 276: 16365–16373.
162. Murányi, A., J. A. MacDonald, J. T. Deng, D. P. Wilson, T. A. Haystead, M. P. Walsh, F. Erdődi, E. Kiss, Y. Wu, and D. J. Hartshorne. 2002. Phosphorylation of the myosin phosphatase target subunit by integrin-linked kinase. *Biochem J* 366: 211–216.

163. Deng, J. T., C. Sutherland, D. L. Brautigan, M. Eto, and M. P. Walsh. 2002. Phosphorylation of the myosin phosphatase inhibitors, CPI-17 and PHI-1, by integrin-linked kinase. *Biochem J* 367: 517–524.
164. Wilson, D. P., C. Sutherland, M. A. Borman, J. T. Deng, J. A. MacDonald, and M. P. Walsh. 2005. Integrin-linked kinase is responsible for Ca^{2+}-independent myosin diphosphorylation and contraction of vascular smooth muscle. *Biochem J* 392: 641–648.
165. Maydan, M., P. C. McDonald, J. Sanghera, J. Yan, C. Rallis, S. Pinchin, G. E. Hannigan et al. 2010. Integrin-linked kinase is a functional Mn2+-dependent protein kinase that regulates glycogen synthase kinase-3beta (GSK-3beta) phosphorylation. *PLoS One* 5: e12356.
166. Koul, D., R. Shen, S. Bergh, Y. Lu, J. E. de Groot, T. J. Liu, G. B. Mills, and W. K. Yung. 2005. Targeting integrin-linked kinase inhibits Akt signaling pathways and decreases tumor progression of human glioblastoma. *Mol Cancer Ther* 4: 1681–1688.
167. Lee, S. L., E. C. Hsu, C. C. Chou, H. C. Chuang, L. Y. Bai, S. K. Kulp, and C. S. Chen. 2011. Identification and characterization of a novel integrin-linked kinase inhibitor. *J Med Chem* 54: 6364–6374.
168. Persad, S., S. Attwell, V. Gray, M. Delcommenne, A. Troussard, J. Sanghera, and S. Dedhar. 2000. Inhibition of integrin-linked kinase (ILK) suppresses activation of protein kinase B/Akt and induces cell cycle arrest and apoptosis of PTEN-mutant prostate cancer cells. *Proc Natl Acad Sci USA* 97: 3207–3212.
169. Haystead, T. A. 2005. ZIP kinase, a key regulator of myosin protein phosphatase 1. *Cell Signal* 17: 1313–1322.
170. Ihara, E., and J. A. MacDonald. 2007. The regulation of smooth muscle contractility by zipper-interacting protein kinase. *Can J Physiol Pharmacol* 8: 79–87.
171. Shiloh, R., S. Bialik, and A. Kimchi. 2014. The DAPK family: A structure-function analysis. *Apoptosis* 19: 286–297.
172. Ihara, E., E. Edwards, M. A. Borman, D. P. Wilson, M. P. Walsh, and J. A. MacDonald. 2007. Inhibition of zipper-interacting protein kinase function in smooth muscle by a myosin light chain kinase pseudosubstrate peptide. *Am J Physiol Cell Physiol* 292: C1951–C1959.
173. Moffat, L. D., S. B. Brown, M. E. Grassie, A. Ulke-Lemee, L. M. Williamson, M. P. Walsh, and J. A. MacDonald. 2011. Chemical genetics of zipper-interacting protein kinase reveal myosin light chain as a bona fide substrate in permeabilized arterial smooth muscle. *J Biol Chem* 286: 36978–36991.
174. Borman, M. A., J. A. MacDonald, A. Murányi, D. J. Hartshorne, and T. A. Haystead. 2002. Smooth muscle myosin phosphatase-associated kinase induces Ca^{2+} sensitization via myosin phosphatase inhibition. *J Biol Chem* 277: 23441–23446.
175. MacDonald, J. A., C. Sutherland, D. A. Carlson, S. Bhaidani, A. Al-Ghabkari, K. Sward, T. A. Haystead, and M. P. Walsh. 2016. A small molecule pyrazolo[3,4-d] pyrimidinone inhibitor of zipper-interacting protein kinase suppresses calcium sensitization of vascular smooth muscle. *Mol Pharmacol* 89: 105–117.
176. Endo, A., H. K. Surks, S. Mochizuki, N. Mochizuki, and M. E. Mendelsohn. 2004. Identification and characterization of zipper-interacting protein kinase as the unique vascular smooth muscle myosin phosphatase-associated kinase. *J Biol Chem* 279: 42055–42061.
177. Vetterkind, S., E. Lee, E. Sundberg, R. H. Poythress, T. C. Tao, U. Preuss, and K. G. Morgan. 2010. Par-4: A new activator of myosin phosphatase. *Mol Biol Cell* 21: 1214–1224.
178. Hill, M. A., M. J. Davis, G. A. Meininger, S. J. Potocnik, and T. V. Murphy. 2006. Arteriolar myogenic signalling mechanisms: Implications for local vascular function. *Clin Hemorheol Microcirc* 34: 67–79.

179. Hill, M. A., H. Zou, S. J. Potocnik, G. A. Meininger, and M. J. Davis. 2001. Invited review: Arteriolar smooth muscle mechanotransduction: Ca(2+) signaling pathways underlying myogenic reactivity. *J Appl Physiol* 91: 973–983.
180. Osol, G., J. F. Brekke, K. McElroy-Yaggy, and N. I. Gokina. 2002. Myogenic tone, reactivity, and forced dilatation: A three-phase model of in vitro arterial myogenic behavior. *Am J Physiol Heart Circ Physiol* 283: H2260–H2267.
181. Cole, W. C., and D. G. Welsh. 2011. Role of myosin light chain kinase and myosin light chain phosphatase in the resistance arterial myogenic response to intravascular pressure. *Arch Biochem Biophys* 510: 160–173.
182. Hill, M. A., G. A. Meininger, M. J. Davis, and I. Laher. 2009. Therapeutic potential of pharmacologically targeting arteriolar myogenic tone. *Trends Pharmacol Sci* 30: 363–374.
183. Velentza, A. V., M. S. Wainwright, M. Zasadzki, S. Mirzoeva, A. M. Schumacher, J. Haiech, P. J. Focia, M. Egli, and D. M. Watterson. 2003. An aminopyridazine-based inhibitor of a pro-apoptotic protein kinase attenuates hypoxia-ischemia induced acute brain injury. *Bioorg Med Chem Lett* 13: 3465–3470.
184. Huber, K., L. Brault, O. Fedorov, C. Gasser, P. Filippakopoulos, A. N. Bullock, D. Fabbro et al. 2012. 7,8-dichloro-1-oxo-beta-carbolines as a versatile scaffold for the development of potent and selective kinase inhibitors with unusual binding modes. *J Med Chem* 55: 403–413.
185. Okamoto, M., K. Takayama, T. Shimizu, K. Ishida, O. Takahashi, and T. Furuya. 2009. Identification of death-associated protein kinases inhibitors using structure-based virtual screening. *J Med Chem* 52: 7323–7327.
186. Okamoto, M., K. Takayama, T. Shimizu, A. Muroya, and T. Furuya. 2010. Structure-activity relationship of novel DAPK inhibitors identified by structure-based virtual screening. *Bioorg Med Chem* 18: 2728–2734.
187. Usui, T., M. Okada, Y. Harand, and H. Yamawaki. 2012. Death-associated protein kinase 3 mediates vascular inflammation and development of hypertension in spontaneously hypertensive rats. *Hypertension* 60: 1031–1039.
188. Carlson, D. A., A. S. Franke, D. H. Weitzel, B. L. Speer, P. F. Hughes, L. Hagerty, C. N. Fortner et al. 2013. Fluorescence linked enzyme chemoproteomic strategy for discovery of a potent and selective DAPK1 and ZIPK inhibitor. *ACS Chem Biol* 8: 2715–2723.
189. Xiao, D., L. D. Longo, and L. Zhang. 2005. Alpha1-adrenoceptor-mediated phosphorylation of MYPT-1 and CPI-17 in the uterine artery: Role of ERK/PKC. *Am J Physiol Heart Circ Physiol* 288: H2828–H2835.
190. Dessy, C., I. Kim, C. L. Sougnez, R. Laporte, and K. G. Morgan. 1998. A role for MAP kinase in differentiated smooth muscle contraction evoked by alpha-adrenoceptor stimulation. *Am J Physiol Cell Physiol* 275: C1081–C1086.
191. Cao, W., U. D. Sohn, K. N. Bitar, J. Behar, P. Biancani, and K. M. Harnett. 2003. MAPK mediates PKC-dependent contraction of cat esophageal and lower esophageal sphincter circular smooth muscle. *Am J Physiol Gastrointest Liver Physiol* 285: G86–G95.
192. Puri, R. N., Y. P. Fan, and S. Rattan. 2002. Role of pp60(c-src) and p(44/42) MAPK in ANG II-induced contraction of rat tonic gastrointestinal smooth muscles. *Am J Physiol Gastrointest Liver Physiol* 283: G390–G399.
193. Lee, H. M., J. K. Won, J. Kim, H. J. Park, H. J. Kim, H. Y. Roh, S. H. Lee, C. K. Lee, and B. Kim. 2007. Endothelin-1 induces contraction via a Syk-mediated p38 mitogen-activated protein kinase pathway in rat aortic smooth muscle. *J Pharmacol Sci* 103: 427–433.
194. Meloche, S., J. Landry, J. Huot, F. Houle, F. Marceau, and E. Giasson. 2000. p38 MAP kinase pathway regulates angiotensin II-induced contraction of rat vascular smooth muscle. *Am J Physiol Heart Circ Physiol* 279: H741–H751.

195. Ohanian, J., P. Cunliffe, E. Ceppi, A. Alder, E. Heerkens, and V. Ohanian. 2001. Activation of p38 mitogen-activated protein kinases by endothelin and noradrenaline in small arteries, regulation by calcium influx and tyrosine kinases, and their role in contraction. *Arterioscler Thromb Vasc Biol* 21: 1921–1927.
196. Barona, I., D. S. Fagundes, S. Gonzalo, L. Grasa, M. P. Arruebo, M. A. Plaza, and M. D. Murillo. 2011. Role of TLR4 and MAPK in the local effect of LPS on intestinal contractility. *J Pharm Pharmacol* 63: 657–662.
197. Al-Shboul, O., A. D. Nalli, D. P. Kumar, R. Zhou, S. Mahavadi, J. F. Kuemmerle, J. R. Grider, and K. S. Murthy. 2014. Jun kinase-induced overexpression of leukemia-associated Rho GEF (LARG) mediates sustained hypercontraction of longitudinal smooth muscle in inflammation. *Am J Physiol Cell Physiol* 306: C1129–C1141.
198. Anderson Jr, C. D., D. M. Kendig, M. Al-Qudah, S. Mahavadi, K. S. Murthy, and J. D. Grider. 2014. Role of various kinases in muscarinic M3 receptor-mediated contraction of longitudinal muscle of rat colon. *J Smooth Muscle Res* 50: 103–119.
199. Ihara, E., P. L. Beck, M. Chappellaz, J. Wong, S. A. Medlicott, and J. A. MacDonald. 2009. Mitogen-activated protein kinase pathways contribute to hypercontractility and increased Ca^{2+} sensitization in murine experimental colitis. *Mol Pharmacol* 75: 1031–1041.
200. Ihara, E., Q. Yu, M. Chappellaz, and J. A. MacDonald. 2015. ERK and p38MAPK pathways regulate myosin light chain phosphatase and contribute to Ca^{2+} sensitization of intestinal smooth muscle contraction. *Neurogastroenterol Motil* 2: 135–146.
201. Ihara, E., L. Moffa, J. Ostrander, M. P. Walsh, and J. A. MacDonald. 2007. Characterization of protein kinase pathways responsible for Ca^{2+} sensitization in rat ileal longitudinal smooth muscle. *Am J Physiol Gastrointest Liver Physiol* 293: G699–G710.
202. Cao, W., U. D. Sohn, J. N. Bitar, J. Behar, P. Biancani, and K. M. Harnett. 2003. MAPK mediates PKC-dependent contraction of cat esophageal and lower esophageal sphincter circular smooth muscle. *Am J Physiol Gastrointest Liver Physiol* 285: G86–G95.
203. Totsukawa, G., Y. Yamakita, S. Yamashiro, H. Hosoya, D. J. Hartshorne, and F. Matsumura. 1999. Activation of myosin phosphatase targeting subunit by mitosis-specific phosphorylation. *J Cell Biol* 144: 735–744.
204. Yamashiro, S., Y. Yamakita, G. Totsukawa, H. Goto, K. Kaibuchi, M. Ito, D. J. Hartshorne, and F. Matsumura. 2008. Myosin phosphatase-targeting subunit 1 regulates mitosis by antagonizing polo-like kinase 1. *Dev Cell* 14: 787–797.
205. Dominguez, C., D. A. Powers, and N. Tamayo. 2005. p38 MAP kinase inhibitors: Many are made, but few are chosen. *Curr Opin Drug Discov Devel* 8: 421–430.
206. Jha, S., E. J. Morris, A. Aruza, M. S. Mansueto, G. K. Schroeder, J. Arbanas, D. McMasters et al. 2016. Dissecting therapeutic resistance to ERK inhibition. *Mol Cancer Ther* 15: 548–559.
207. Santarpia, L., S. M. Lippman, and A. K. El-Naggar. 2012. Targeting the MAPK-RAS-RAF signaling pathway in cancer therapy. *Expert Opin Ther Targets* 16: 103–119.
208. Ihara, E., H. Akiho, K. Nakamura, S. R. Turner, and J. A. MacDonald. 2011. MAPKs represent novel therapeutic targets for gastrointestinal motility disorders. *World J Gastrointest Pathophysiol* 2: 19–25.
209. Artamonov, M., K. Momotani, D. Utepbergenov, A. Franke, A. Khromov, Z. S. Derewenda, and A. V. Somlyo. 2003. The p90 Ribosomal S6 kinase (RSK) is a mediator of smooth muscle contractility. *PLoS One* 8(3): e58703.
210. Truebestein, L., D. J. Elsner, E. Fuchs, and T. A. Leonard. 2015. A molecular ruler regulates cytoskeletal remodelling by the Rho kinases. *Nat Commun* 6: 10029.

8 Approaches for Investigating the Functional Significance of Transient Receptor Potential Cation Channels in Smooth Muscle Cells

Harry A. T. Pritchard, Paulo W. Pires, and Scott Earley

CONTENTS

8.1 INTRODUCTION

The transient receptor potential (TRP) superfamily of cation channels regulate numerous functions and signal transduction pathways in excitable and non-excitable cells throughout the body.[1] The 28 genes encoding TRP protein subunits have been assigned to six subfamilies according to amino acid sequence homology. These subfamilies are: canonical (TRPC1-7), vanilloid (TRPV1-6), melastatin (TRPM1-8), ankyrin (TRPA1), mucolipin (TRPML1-3), and polycystin (TRPP1-3). TRP channel subunits are expressed as six-transmembrane domain proteins with intracellular amino and carboxy termini, and multiple TRP protein subunits are present in all cells. Four homologous TRP subunits assemble to form functional cation channels, and in some cases association of two or three different subunits can create heteromultimeric channels with distinctive properties.[2–7] Although TRP channels are commonly described as non-selective cation channels, in reality specific gradations in ionic selectivity among the various TRP channels have significant influence on physiological function, which will be discussed.

The goal of this chapter is to summarize experimental approaches and strategies useful for studying the impact of TRP channels on the regulation of smooth muscle cell (SMC) contractility and proliferation. The physiological and pathophysiological roles of TRP channels in SMCs have been extensively reviewed elsewhere.[8–10] Here, we will briefly review useful stratagems for identifying TRP channel expression patterns in SMCs and consider patch-clamp electrophysiological approaches for studying channel biophysics and regulation. The use of optical techniques to record the single-channel activity of Ca^{2+}-permeable TRP channels is also briefly discussed. Finally, we describe techniques used to establish the roles of TRP channels in the regulation of SMC contractility and proliferation.

8.2 TRP CHANNEL EXPRESSION

Multiple TRP channel subunits are present in all tissues, but specific expression profiles vary widely between cell types. Ascertaining where, when, and how much of a specific channel protein is expressed within the cells under investigation is an important first step in elucidating functional significance. This is typically accomplished through standard RNA and protein analysis methods (endpoint and quantitative RT-PCR, Western blot, immunolabeling, etc.), but obtaining sufficient quantities of untainted SMCs for analysis can be challenging. Tissues composed of SMCs, such as blood vessels, the gastrointestinal track, and the urinary bladder typically contain other types of cells, including endothelial cells, peripheral neurons, fibroblasts, and interstitial cells of Cajal. Consequently, protein and RNA extracted from whole tissues will consist of a mixture of these cell types. Cell culture may be appropriate for studying SMC proliferation, but such methods induce drastic phenotypic changes and are unsuitable for investigating control of contractility. Several strategies for cell isolation have been used to overcome this limitation: Following enzymatic dispersal, contractile SMCs identified using phase-contrast microscopy can be manually collected using a micropipette mounted on a micromanipulator. This approach is laborious but can be effective on a small scale.[11] An alternative method is to use magnetic beads coated with a primary antibody raised against a specific cell surface protein to either positively select or negatively exclude certain cells from a mixed population. The highest throughput can be achieved using fluorescence-activated cell sorting (FACS) flow cytometry to sort and isolate dispersed SMCs. Isolation can rely on antibody-based approaches to recognize cell-specific surface antigens or can employ genetically-encoded reporter mice selectively expressing a reporter protein, such as enhanced green fluorescent protein (eGFP), within SMCs.[12–14] Relatively large numbers of cells, limited only by the amount of available starting material, can be rapidly isolated for molecular analysis using this method. Individual cells can also be isolated with the FACS approach and analyzed using high-throughput single-cell genomic methods to probe for population variances among SMCs.[15] However, efficient isolation based on reporter gene methods is entirely dependent on the specificity of the promoter directing reporter expression. For example, the *acta2* promoter is often used to drive expression in SMCs but is also expressed by pericytes,[16] which are present in the cerebral microcirculation. Appropriate cell-type specificity controls are necessary to verify fidelity of isolation.

Genetic mouse models where the TRP channel of choice is tagged with a fluorescent reporter protein can be used to visualize cell type-specific TRP channel expression within complex tissues.[17] In addition, the development of transgenic mice expressing reporter proteins under the control of the promoter of a gene of interest provides a powerful method for identifying tissue-specific expression patterns. This method has been used to study the distribution of TRPV1 channels throughout the body[18] and to demonstrate expression of the channel within SMCs in specific segments of the cerebral vasculature.[19] Although standard transgenic methods for generating these reporter mice are laborious, expensive, and time consuming, broader application of this approach may be facilitated by emerging CRISPR/Cas9 technology.

8.3 TRP CHANNEL PHARMACOLOGY

Selective pharmacological activators and inhibitors are commonly used to study the properties and physiological function of all ion channels, including TRP channels. The study of TRPV4 channels has been accelerated by the availability of potent and highly selective agonists (GSK10106790A[20,21] and RN1747[22]) and antagonists (HC-067047,[23] RN1734,[22] and GSK2193874[24]). TRPV1 channels are selectively activated by low concentration of the dietary molecule capsaicin,[25] a substance found in hot peppers, and are selectively inhibited by capsazepine and other compounds (JNJ17203212,[26] A 784168,[27] and AMG 517[28]). A few other TRP channels are activated by dietary substances including TRPM8 (menthol from mint),[29,30] TRPV3 (carvacrol from oregano)[31] and TRPA1 (allicin from garlic and allyl isothiocyanate [AITC] from mustard oil).[32] However, none of these compounds are completely discriminating and can have off-target effects. Selective blockers of TRPA1 channels are available (HC-030031,[33] A 967079,[34] and AP18[35]) and can be used in combination with AITC to identify TRPA1 currents in native cells. ML-SA1[36] and MK6-83[37] selectively activate TRPML channels but do not distinguish between subfamily members. A newer compound, SN 2, has shown some specificity for activation of TRPML3 compared with TRPML1 and TRPML2.[38] In general, there are few pharmacological tools that can be used to study TRPC channels. For example, although the pyrole compound Pyr3 effectively blocks the activity of TRPC3 channels,[39] it has also been shown to inhibit store-operated Orai1 channels with similar potency.[40] However, the recently described compound Pico145 shows selectivity for TRPC1/4/5 channels at sub-nanomolar concentrations with no effect on other TRPC channels and other TRP channels.[41] Furthermore, this compound shows particular efficacies for different heteromultimeric channels, which could provide insight into TRPC expression patterns in native tissue. The cell-permeable hydroxytricyclic derivative 9-phenanthrol blocks TRPM4 but not TRPM5 channels[42,43] and has been used to study SMC contractility in cerebral arteries[44–46] and the urinary bladder,[47–49] but has reported off-target effects[50,51] which may limit utility. Few suitable options are currently widely available for the other TRP channels, but this area is rapidly developing. Various strategies are used to circumvent these limitations, including knockout and knockdown approaches and the use of blocking antibodies targeting extracellular epitopes of the pore region of the channel to block conductance.[52–55] Further detailed information on the pharmacology of TRP channels can be found elsewhere.[56]

8.4 TRP CHANNEL ELECTROPHYSIOLOGY

Ion channels were classically described by function rather than molecular identity.[57] The situation was reversed for the mammalian TRP channels, which were discovered and cloned on the basis of early bioinformatic analysis that discovered genes homologous to *Drosophila trp* and *trpl*.[58,59] In a series of groundbreaking studies,[25,30,60–76] the biophysical characteristics of recombinant channels were investigated by applying patch-clamp electrophysiology to cultured cells overexpressing cloned TRP genes. Application of the conventional whole-cell patch-clamp configuration over a range of command potentials was used to define the current-voltage (I/V) relationships and activation/deactivation kinetics. Single-channel current recordings obtained from *inside-out* isolated membrane patches were used to estimate unitary conductance, and ion substitution protocols provided information about relative permeability for various ionic species. Together, these characteristics provide an identifying fingerprint for the currents conducted by each channel when studied under ideal conditions, offering a starting point for recognizing similar currents in native cells. Current isolation in native cells is considerably more challenging compared with the overexpression approach, as currents are likely to be much lower in amplitude and are usually contaminated by currents conducted by other types of channels present in the cell. Manipulation of ionic gradients, in combination with available pharmacology and genetic knockdown/knockout approaches, are frequently used to isolate and identify specific currents in native SMCs.[77–80] Particular currents may be easier to isolate and recognize than others. For example, currents conducted by TRPV1[25] and TRPML family members,[81] respectively, show characteristic strong outward or inward rectification and are readily identifiable in native cells in conjunction with available selective pharmacology. In contrast, currents conducted by TRPM2 channels display linear rectification,[68] lack selective pharmacological activators and inhibitors, and are consequently difficult to conclusively distinguish from leak currents.

Although valuable for investigating biophysical properties of ion channels, the conventional whole-cell and inside-out patch-clamp configurations may not be suitable for studying aspects of endogenous regulation of TRP channel activity.

Intracellular components required for regulated activity, including intracellular [Ca^{2+}], endogenous Ca^{2+} buffers, membrane phospholipids such as phosphatidylinositol 4,5-bisphosphate (PtdIns(4,5)P_2), and adenosine triphosphate (ATP) are diluted after rupture of the plasma membrane. Use of the nystatin[82] or amphotericin B[83] perforated patch-clamp configuration may be more appropriate in these situations, as this arrangement provides voltage control of the plasma membrane, but Ca^{2+} ions and other critical cytosolic components are not dialyzed into the patching pipette. For example, application of perforated patch-clamp was necessary to elucidate endogenous Ca^{2+}-dependent pathways regulating TRPM4 channels in vascular SMC.[45,84,85] The cell attached patch-clamp configuration has also been successfully used to study TRPC channel regulation in SMCs. For example, this approach was used to show that TRPC1/5 channels require PtdIns(4,5)P_2 and protein kinase C (PKC) for activation,[54,86] whereas TRPC3/6/7 channels are directly activated by diacylglycerol (DAG; produced by PtdIns(4,5)P_2 hydrolysis) but inhibited by PtdIns(4,5)P_2.[88–92]

In general, currents conducted by TRP channels present in SMCs are composed of an inward flow of Na^+ and Ca^{2+} ions, an outward flow of K^+ ions, and flux of minor cations, such as Mg^{2+}, Zn^{2+}, and others. Particular TRP channels conduct specific combinations of cations that dictate ultimate cellular function. Insight into the composition of these complex currents is provided by ion substitution experiments designed to estimate the relative permeability (P) of the channel for different cations. For example, studies where extracellular Na^+ was substituted for the impermeant cation N-Methyl-D-glucamine (NMGD) revealed that TRPV5 and TRPV6 channels have a relative permeability for Ca^{2+} versus Na^+ ions (P_{Ca}:P_{Na}) of ~100:1,[67,93,94] indicating that currents conducted by these channels have a high Ca^{2+} fraction. Conversely, similar studies show that TRPM4 and TRPM5 are impermeable to Ca^{2+} and only carry monovalent (primarily Na^+ and K^+ ions) charges.[70,71,95,96] Most of the other TRP channels conduct a more complex combination of cations. For channels such as TRPC1 ($P_{Ca} = P_{Na}$), Na^+ is the major inward charge carrier at depolarized membrane potentials in physiological solution where extracellular [Na^+] is ~100-fold greater than [Ca^{2+}] and the fractional Ca^{2+} current is small. The relative K^+ permeability of the channel is also expected to influence the Ca^{2+} fractional current.[97] For example, if $P_K = P_{Na}$, the reversal potential of the monovalent portion of the current is 0 mV and will not affect Ca^{2+} permeation. However, if $P_{Na} > P_K$ the reversal potential will be positive, which will reduce the driving force for Ca^{2+} influx, whereas if $P_K > P_{Na}$ the driving force for Ca^{2+} influx will be increased. The influence of differential K^+ permeability may explain why the Ca^{2+} fraction of TRPV4 currents is greater when compared to TRPC6 currents, despite similar P_{Ca}:P_{Na}.[97]

8.5 OPTICAL RECORDING OF TRP CHANNEL SPARKLETS

The activity of single TRP channels that conduct currents with a high Ca^{2+} fraction can be optically recorded using total internal reflection fluorescence (TIRF) microscopy. TIRF microscopy relies on the generation of a low-energy evanescent illumination field when excitation light is completely reflected away from a glass surface supporting the sample. The intensity of the evanescent field decays exponentially with distance and only penetrates approximately ~100 nm into the specimen, enabling excitation of fluorophores only at or very near the plasma membrane. This maneuver enables excellent signal-to-noise ratio for signals within the evanescent field. Ca^{2+} signals, representing Ca^{2+} influx through single or clustered ion channels, are recorded using TIRF from cells loaded with Ca^{2+} indicators dyes such as Fluo-4-AM[98] or expressing genetically encoded Ca^{2+} biosensor proteins such as GCaMPs.[99] For historical reasons, these signals are referred to as sparklets[100] and are typically associated with a particular ion channel, that is, TRPV4 sparklets. TIRF offers a number of advantages over patch-clamp electrophysiology, including higher throughput and the ability to precisely determine the subcellular spatial distribution of active ion channels. Further benefits include the ability to precisely define the biophysical characteristics of the Ca^{2+} signals, including kinetics, amplitude, spatial spread, number of active sites in a field of view, and the number of individual channels involved in generating a sparklet. Ca^{2+} influx across the cell membrane can be directly visualized with high spatial resolution, identifying subcellular regions where the Ca^{2+} concentration is much greater than the bulk cytosol. This information is not provided by classic electrophysiology, where Ca^{2+} influx is assumed to be uniformly distributed throughout a cell, only impacting the global Ca^{2+} concentration. TIRF can also be combined with patch-clamp to simultaneously record sparklets and ionic currents, providing direct correlation between the electrophysiological and imaging data. TIRF is difficult to apply to intact tissues because specimens must be completely flat and tightly attached to the glass coverslip to be illuminated by the evanescent field. However, confocal microscopy has been used to record TRPV4 sparklets in the endothelium of intact arteries[101,102] and this method may also be useful for other cells with a relatively flat morphology. An excellent recent review of TRP channel Ca^{2+} signaling can be found elsewhere.[103]

The Santana laboratory pioneered the use of TIRF to record Ca^{2+} channel activity in SMCs in an elegant series of studies investigating the regulation of L-type voltage-gated Ca^{2+} channel (VGCC) sparklets.[104] In theory, sparklets can be recorded from any TRP channel conducting currents with a sufficiently large Ca^{2+} fraction. To date, TRPV1 sparklets have been recorded from dorsal root ganglion neurons,[105] and TRPV3, TRPV4, and TRPA1 sparklets have been recorded from the endothelium.[106–108] TIRF has also been used to study the regulation and function of TRPV4 sparklets in contractile SMCs from cerebral arteries[109] and to study the role of TRPV4 sparklets in the proliferation of airway SMCs.[110] Interestingly, the Ca^{2+} signal mass of a single TRPV4 sparklet in SMCs has been estimated to be 100-fold greater than that of an L-type Ca^{2+} channel sparklet.[109] Apart from a preliminary report of TRPV1 sparklets in middle meningeal artery SMCs,[111] no other studies have investigated TRP channel activity in SMCs using optical recording methods. The development of improved fluorescent indicators for other ions (i.e., Na^+) could enable the use of TIRF or confocal microscopy to study the single-channel activity of TRP channels present in SMCs that conduct mixed cationic currents with a low Ca^{2+} fraction.

8.6 TRP CHANNEL FUNCTION

8.6.1 TRP Channel Regulation of Smooth Muscle Cell Contractility

The principal function of SMCs is to contract or relax under appropriate conditions. SMCs contract when myofilaments are engaged as a result of increases in the global intracellular [Ca^{2+}] or in response to increased Ca^{2+} sensitivity of the appropriate regulatory pathways. Global cytosolic [Ca^{2+}] in SMCs is primarily controlled by the activity of VGCCs, which are in turn regulated by the membrane potential. Membrane depolarization supports Ca^{2+} influx and contraction whereas hyperpolarization favors Ca^{2+} efflux and relaxation. The resting membrane potential for SMCs under physiological conditions is normally between −80 and −40 mV, close to E_K (−88 mV) and significantly hyperpolarized compared to E_{Na} (+60 mV). Under these conditions, activation of a cation channel in SMCs where $P_{Na} \approx P_K$, such as TRPM4, generates an influx of Na^+, favoring membrane depolarization and contraction, as has been demonstrated for vascular and detrusor SMCs.[44–48,79,80,84,85,112–114] Direct influx of Ca^{2+} through Ca^{2+}-permeable TRP channels can potentially directly engage in contractile pathways to cause constriction, as has been demonstrated for TRPV1 channels in SMCs in specific elements of the cerebral vasculature.[19] Interestingly, Ca^{2+} influx through TRPV4 channels causes relaxation of SMCs by sensitizing type 2 ryanodine receptors (RyR2) on the sarcoplasmic reticulum (SR) to stimulate Ca^{2+}-induced Ca^{2+}-release.[109,115] This effect increases the frequency of spontaneous transient Ca^{2+} release events (Ca^{2+} sparks) that are functionally coupled with multiple large conductance Ca^{2+}-activated K^+ channels (BK) on the plasma membrane. Macroscopic outward K^+ currents generated by this mechanism cause membrane hyperpolarization and relaxation of SMCs.[78] TRP channel also act as receptor operated channels downstream of G-protein coupled receptors, acting as receptor-operated channels and participating in agonist-induced contractile responses.[116]

Contractility is typically assessed using isometric tension recordings of muscle strips or rings and pressure myography for isolated blood or lymphatic vessels. Selective pharmacology (when available) in combination with appropriate genetic knockout animals or knockdown approaches can be effectively used to conclusively elucidate the influence of channel activity on contractile function. A major limitation of constitutive global knockout models is the potential for compensatory upregulation of complementary genes to maintain function. For example, arteries from global TRPC6 knockout mice unexpectedly demonstrate enhanced reactivity to vasoconstrictor stimuli because of increased expression of TRPC3 channels.[117] Complications due to compensation can be mitigated through the use of inducible SMC-specific conditional knockouts, as demonstrated by a recent study showing that TRPP1[8] (also known as *PKD2* or TRPP2 by outdated nomenclature) is vital for myogenic constriction of mouse mesenteric arteries.[77] However, this approach is not without limitations. Conditional knockouts typically use the *cre-lox* system targeted with a SMC-specific promotor such as the myosin heavy chain or *acta2*, but the potential for off-target effects in other cell types must be considered. Furthermore, inducible knockout approaches often use injection of tamoxifen to activate gene silencing. Tamoxifen is an estrogen receptor antagonist[118] and has the potential to provoke off-target effects. For example, tamoxifen has been shown to interfere with phosphatidylinositol 4,5-bisphosphate-channel interactions,[119] and block Cl^- channels[120–122] and PKC activity,[123] all vital to SMC contractility. Pre-load, or the amount of baseline contractile tone, must be taken into account for all *ex vivo* myography experiments. SMC contractility is defined by the relative activities of myosin light chain kinase (MLCK) and myosin light chain phosphatase (MLCP), which can be dependent or independent of the intracellular Ca^{2+} concentration established by the activity of VGCCs. If, for example a TRPV4 agonist is applied to hyperpolarize the SMC plasma membrane,[78] no effect on contraction would be produced if the preload was established independent of VGCC activity. Appropriate controls for these confounding factors are necessary.

Short-term knockdown studies employing small interfering RNA (siRNA) performed using serum-free organ culture can be rapidly deployed to investigate channel function in contractile studies and are less vulnerable to physiological compensation. Emerging silencing morpholino oligonucleotide-based knockdown approaches are promising,[124] as this technique may generate consistent knockdown of protein expression with fewer off-target effects compared with siRNA.[125,126] The overall impact of SMC TRP channels *in vivo* can be best studied using conditional knockout approaches combined with functional assays, such as radiotelemetry recording of blood pressure and heart rate[127] or measurements of local blood flow using laser Doppler or multiphoton microscopy.[128]

8.6.2 TRP Channel Regulation of Smooth Muscle Cell Proliferation

Fully differentiated contractile SMCs can revert to a non-contractile proliferative state as a result of injury or disease. Proliferative SMC cells are phenotypically distinct from contractile SMCs and express a different set of ion channels, cell surface receptors, and other proteins to support distinct regulatory pathways. Contractile SMCs can enter the proliferative state under standard cell culture conditions, providing a convenient method for studying the mechanisms underlying these pathological changes. Further, cultured cells provide an abundant source of material for molecular studies. The knockout, knockdown, and pharmacological approaches discussed earlier are used to study the functional effects of specific TRP channels on SMC proliferation. Research into this area has shown that TRPC3[129] and TRPV4[110,130] are associated with airway SMC proliferation, which may be involved with the fixed component of asthma-induced airway obstruction.[131,132] Conversely, TRPM8 and TRPA1 activation has been shown to inhibit airway SMC proliferation.[133] TRPC1/4/6 channels are upregulated in proliferating pulmonary artery SMC[134–139] and TRPV1 has been shown to be important for hypoxia-induced proliferation of pulmonary artery SMCs associated with hypoxia-induced pulmonary artery remodeling.[140] TRPM7 has been shown to play a role in angiotensin II-induced SMC proliferation.[141] TRPM3 is present in both proliferative and contractile vascular SMCs,[142] but the functional significance of the channel in this context is uncertain.

8.7 CONCLUSIONS

In conclusion, the experimental approaches described here have been used to demonstrate important roles for many TRP channels in detecting diverse external stimuli to directly or indirectly initiate signal transduction pathways regulating SMC contractility and proliferation. Future studies utilizing new state-of-the art technologies, including super-resolution microscopy, optogenetics, and CRISPR/Cas9 gene editing, are expected to provide a deeper understanding of the roles of TRP channels in SMC function.

ACKNOWLEDGMENTS

This work was supported by grants from the National Heart, Lung, and Blood Institute (R01HL091905, R01HL137852 and R01HL139585 to SE) and the American Heart Association (15POST24720002 to PWP).

REFERENCES

1. Wu, LJ, Sweet, TB, and Clapham, DE. International Union of Basic and Clinical Pharmacology. LXXVI. Current progress in the mammalian TRP ion channel family. *Pharmacol Rev* 2010;62:381–404.
2. Ma, X, Qiu, S, Luo, J, Ma, Y, Ngai, CY, Shen, B, Wong, CO, Huang, Y, and Yao, X. Functional role of vanilloid transient receptor potential 4-canonical transient receptor potential 1 complex in flow-induced Ca^{2+} influx. *Arterioscler Thromb Vasc Biol* 2010;30:851–858.
3. Du, J, Ma, X, Shen, B, Huang, Y, Birnbaumer, L, and Yao, X. TRPV4, TRPC1, and TRPP2 assemble to form a flow-sensitive heteromeric channel. *FASEB J* 2014;28:4677–4685.
4. Strubing, C, Krapivinsky, G, Krapivinsky, L, and Clapham, DE. Formation of novel TRPC channels by complex subunit interactions in embryonic brain. *J Biol Chem* 2003;278:39014–39019.
5. Schaefer, M. Homo- and heteromeric assembly of TRP channel subunits. *Pflugers Arch* 2005;451:35–42.
6. Hofmann, T, Schaefer, M, Schultz, G, and Gudermann, T. Subunit composition of mammalian transient receptor potential channels in living cells. *Proc Natl Acad Sci USA* 2002;99:7461–7466.
7. Goel, M, Sinkins, WG, and Schilling, WP. Selective association of TRPC channel subunits in rat brain synaptosomes. *J Biol Chem* 2002;277:48303–48310.
8. Earley, S, and Brayden, JE. Transient receptor potential channels in the vasculature. *Physiol Rev* 2015;95:645–690.
9. Holzer, P. Transient receptor potential (TRP) channels as drug targets for diseases of the digestive system. *Pharmacol Ther* 2011;131:142–170.
10. Merrill, L, Gonzalez, EJ, Girard, BM, and Vizzard, MA. Receptors, channels, and signalling in the urothelial sensory system in the bladder. *Nat Rev Urol* 2016;13:193–204.
11. Harhun, MI, Szewczyk, K, Laux, H, Prestwich, SA, Gordienko, DV, Moss, RF, and Bolton, TB. Interstitial cells from rat middle cerebral artery belong to smooth muscle cell type. *J Cell Mol Med* 2009;13:4532–4539.
12. Cobine, CA, Sotherton, AG, Peri, LE, Sanders, KM, Ward, SM, and Keef, KD. Nitrergic neuromuscular transmission in the mouse internal anal sphincter is accomplished by multiple pathways and postjunctional effector cells. *Am J Physiol Gastroint Liver Physiol* 2014;307:G1057–G1072.

13. Peri, LE, Sanders, KM, and Mutafova-Yambolieva, VN. Differential expression of genes related to purinergic signaling in smooth muscle cells, PDGFRalpha-positive cells, and interstitial cells of Cajal in the murine colon. *Neurogastroenterol Motil* 2013;25:e609–e620.
14. Lee, H, Koh, BH, Peri, LE, Sanders, KM, and Koh, SD. Functional expression of SK channels in murine detrusor PDGFR+ cells. *J Physiol* 2013;591:503–513.
15. Macaulay, IC, Ponting, CP, and Voet, T. Single-cell multiomics: Multiple measurements from single cells. *Trends Genet* 2017;33:155–168.
16. Gonzales, AL, Longden, TA, Shui, B, Kotlikoff, MI, and Nelson, MT. Contractile pericytes determine the direction of blood flow at capillary bifurcations. *FASEB J* 2016;30:949.3.
17. Wyatt, A, Wartenberg, P, Candlish, M, Krasteva-Christ, G, Flockerzi, V, and Boehm, U. Genetic strategies to analyze primary TRP channel-expressing cells in mice. *Cell Calcium* 2017;67:91–104.
18. Cavanaugh, DJ, Chesler, AT, Jackson, AC, Sigal, YM, Yamanaka, H, Grant, R, O'Donnell, D et al. TRPV1 reporter mice reveal highly restricted brain distribution and functional expression in arteriolar smooth muscle cells. *J Neurosci* 2011;31:5067–5077.
19. Koide, M, Hughes, E, Syed, AU, Sonkusare, S, Manuelyan, I, Nelson, MT, and Wellman, GC. TRPV1-mediated Ca^{2+} influx in arterial smooth muscle. *J Gen Physiol* 2016;148:20a.
20. Thorneloe, KS, Sulpizio, AC, Lin, Z, Figueroa, DJ, Clouse, AK, McCafferty, GP, Chendrimada, TP et al. N-((1S)-1-{[4-((2S)-2-{[(2,4-dichlorophenyl)sulfonyl]amino}-3-hydroxypropanoyl)-1-piperazinyl]carbonyl}-3-methylbutyl)-1-benzothiophene-2-carboxamide (GSK1016790A), a novel and potent transient receptor potential vanilloid 4 channel agonist induces urinary bladder contraction and hyperactivity: Part I. *J Pharmacol Exp Ther* 2008;326:432–442.
21. Willette, RN, Bao, W, Nerurkar, S, Yue, TL, Doe, CP, Stankus, G, Turner, GH et al. Systemic activation of the transient receptor potential vanilloid subtype 4 channel causes endothelial failure and circulatory collapse: Part 2. *J Pharmacol Exp Ther* 2008;326:443–452.
22. Vincent, F, Acevedo, A, Nguyen, MT, Dourado, M, DeFalco, J, Gustafson, A, Spiro, P, Emerling, DE, Kelly, MG, and Duncton, MA. Identification and characterization of novel TRPV4 modulators. *Biochem Biophys Res Commun* 2009;389:490–494.
23. Everaerts, W, Zhen, X, Ghosh, D, Vriens, J, Gevaert, T, Gilbert, JP, Hayward, NJ et al. Inhibition of the cation channel TRPV4 improves bladder function in mice and rats with cyclophosphamide-induced cystitis. *Proc Natl Acad Sci USA* 2010;107:19084–19089.
24. Thorneloe, KS, Cheung, M, Bao, W, Alsaid, H, Lenhard, S, Jian, MY, Costell, M et al. An orally active TRPV4 channel blocker prevents and resolves pulmonary edema induced by heart failure. *Sci Transl Med* 2012;4:159ra148.
25. Caterina, MJ, Schumacher, MA, Tominaga, M, Rosen, TA, Levine, JD, and Julius, D. The capsaicin receptor: A heat-activated ion channel in the pain pathway. *Nature* 1997;389:816–824.
26. Swanson, DM, Dubin, AE, Shah, C, Nasser, N, Chang, L, Dax, SL, Jetter, M et al. Identification and biological evaluation of 4-(3-trifluoromethylpyridin-2-yl)piperazine-1-carboxylic acid (5-trifluoromethylpyridin-2-yl)amide, a high affinity TRPV1 (VR1) vanilloid receptor antagonist. *J Med Chem* 2005;48:1857–1872.
27. Cui, M, Honore, P, Zhong, C, Gauvin, D, Mikusa, J, Hernandez, G, Chandran, P et al. TRPV1 receptors in the CNS play a key role in broad-spectrum analgesia of TRPV1 antagonists. *J Neurosci* 2006;26:9385–9393.
28. Blum, CA, Caldwell, T, Zheng, X, Bakthavatchalam, R, Capitosti, S, Brielmann, H, De Lombaert, S et al. Discovery of novel 6,6-heterocycles as transient receptor potential vanilloid (TRPV1) antagonists. *J Med Chem* 2010;53:3330–3348.

29. McKemy DD, Neuhausser, WM, and Julius, D. Identification of a cold receptor reveals a general role for TRP channels in thermosensation. *Nature* 2002;416:52–58.
30. Peier, AM, Moqrich, A, Hergarden, AC, Reeve, AJ, Andersson, DA, Story, GM, Earley, TJ et al. A TRP channel that senses cold stimuli and menthol. *Cell* 2002;108:705–715.
31. Xu, H, Delling, M, Jun, JC, and Clapham, DE. Oregano, thyme and clove-derived flavors and skin sensitizers activate specific TRP channels. *Nat Neurosci* 2006;9:628–635.
32. Jordt, SE, Bautista, DM, Chuang, HH, McKemy, DD, Zygmunt, PM, Hogestatt, ED, Meng, ID, and Julius, D. Mustard oils and cannabinoids excite sensory nerve fibres through the TRP channel ANKTM1. *Nature* 2004;427:260–265.
33. McNamara, CR, Mandel-Brehm, J, Bautista, DM, Siemens, J, Deranian, KL, Zhao, M, Hayward, NJ et al. TRPA1 mediates formalin-induced pain. *Proc Natl Acad Sci USA* 2007;104:13525–13530.
34. Chen, J, Joshi, SK, DiDomenico, S, Perner, RJ, Mikusa, JP, Gauvin, DM, Segreti, JA et al. Selective blockade of TRPA1 channel attenuates pathological pain without altering noxious cold sensation or body temperature regulation. *Pain* 2011;152:1165–1172.
35. Petrus, M, Peier, AM, Bandell, M, Hwang, SW, Huynh, T, Olney, N, Jegla, T, and Patapoutian, A. A role of TRPA1 in mechanical hyperalgesia is revealed by pharmacological inhibition. *Mol Pain* 2007;3:40.
36. Shen, D, Wang, X, Li, X, Zhang, X, Yao, Z, Dibble, S, Dong, XP et al. Lipid storage disorders block lysosomal trafficking by inhibiting a TRP channel and lysosomal calcium release. *Nat Commun* 2012;3:731.
37. Chen, CC, Keller, M, Hess, M, Schiffmann, R, Urban, N, Wolfgardt, A, Schaefer, M et al. A small molecule restores function to TRPML1 mutant isoforms responsible for mucolipidosis type IV. *Nat Commun* 2014;5:4681.
38. Grimm, C, Jors, S, Saldanha, SA, Obukhov, AG, Pan, B, Oshima, K, Cuajungco, MP, Chase, P, Hodder, P, and Heller, S. Small molecule activators of TRPML3. *Chem Biol* 2010;17:135–148.
39. Kiyonaka, S, Kato, K, Nishida, M, Mio, K, Numaga, T, Sawaguchi, Y, Yoshida, T et al. Selective and direct inhibition of TRPC3 channels underlies biological activities of a pyrazole compound. *Proc Natl Acad Sci USA* 2009;106:5400–5405.
40. Schleifer, H, Doleschal, B, Lichtenegger, M, Oppenrieder, R, Derler, I, Frischauf, I, Glasnov, T, Kappe, C, Romanin, C, and Groschner, K. Novel pyrazole compounds for pharmacological discrimination between receptor-operated and store-operated Ca(2+) entry pathways. *Br J Pharmacol* 2012;167:1712–1722.
41. Rubaiy, HN, Ludlow, MJ, Henrot, M, Gaunt, HJ, Miteva, K, Cheung, SY, Tanahashi, Y et al. Picomolar, selective, and subtype-specific small-molecule inhibition of TRPC1/4/5 channels. *J Biol Chem* 2017;292:8158–8173.
42. Guinamard, R, Hof, T, and Del Negro, CA. The TRPM4 channel inhibitor 9-phenanthrol. *Br J Pharmacol* 2014;171:1600–1613.
43. Grand, T, Demion, M, Norez, C, Mettey, Y, Launay, P, Becq, F, Bois, P, and Guinamard, R. 9-phenanthrol inhibits human TRPM4 but not TRPM5 cationic channels. *Br J Pharmacol* 2008;153:1697–1705.
44. Gonzales, AL, Garcia, ZI, Amberg, GC, and Earley, S. Pharmacological inhibition of TRPM4 hyperpolarizes vascular smooth muscle. *Am J Physiol Cell Physiol* 2010;299:C1195–C1202.
45. Gonzales, AL, Amberg, GC, and Earley, S. Ca^{2+} release from the sarcoplasmic reticulum is required for sustained TRPM4 activity in cerebral artery smooth muscle cells. *Am J Physiol Cell Physiol* 2010;299:C279–C288.
46. Crnich, R, Amberg, GC, Leo, MD, Gonzales, AL, Tamkun, MM, Jaggar, JH, and Earley, S. Vasoconstriction resulting from dynamic membrane trafficking of TRPM4 in vascular smooth muscle cells. *Am J Physiol Cell Physiol* 2010;299:C682–C694.

47. Smith, AC, Parajuli, SP, Hristov, KL, Cheng, Q, Soder, RP, Afeli, SA, Earley, S, Xin, W, Malysz, J, and Petkov, GV. TRPM4 channel: A new player in urinary bladder smooth muscle function in rats. *Am J Physiol Renal Physiol* 2013;304:F918–F929.
48. Smith, AC, Hristov, KL, Cheng, Q, Xin, W, Parajuli, SP, Earley, S, Malysz, J, and Petkov, GV. Novel role for the transient potential receptor melastatin 4 channel in guinea pig detrusor smooth muscle physiology. *Am J Physiol Cell Physiol* 2013;304:C467–C477.
49. Parajuli, SP, Hristov, KL, Sullivan, MN, Xin, W, Smith, AC, Earley, S, Malysz, J, and Petkov, GV. Control of urinary bladder smooth muscle excitability by the TRPM4 channel modulator 9-phenanthrol. *Channels (Austin)* 2013;7:537–540.
50. Garland, CJ, Smirnov, SV, Bagher, P, Lim, CS, Huang, CY, Mitchell, R, Stanley, C, Pinkney, A, and Dora, KA. TRPM4 inhibitor 9-phenanthrol activates endothelial cell intermediate conductance calcium-activated potassium channels in rat isolated mesenteric artery. *Br J Pharmacol* 2015;172:1114–1123.
51. Burris, SK, Wang, Q, Bulley, S, Neeb, ZP, and Jaggar, JH. 9-Phenanthrol inhibits recombinant and arterial myocyte TMEM16A channels. *Br J Pharmacol* 2015;172:2459–2468.
52. Xu, SZ, Sukumar, P, Zeng, F, Li, J, Jairaman, A, English, A, Naylor, J, Ciurtin, C et al. TRPC channel activation by extracellular thioredoxin. *Nature* 2008;451:69–72.
53. Xu, SZ, and Beech, DJ. TRPC1 is a membrane-spanning subunit of store-operated Ca(2+) channels in native vascular smooth muscle cells. *Circ Res* 2001;88:84–87.
54. Shi, J, Ju, M, Abramowitz, J, Large, WA, Birnbaumer, L, and Albert, AP. TRPC1 proteins confer PKC and phosphoinositol activation on native heteromeric TRPC1/C5 channels in vascular smooth muscle: Comparative study of wild-type and TRPC1-/- mice. *FASEB J* 2012;26:409–419.
55. Chen, J, Crossland, RF, Noorani, MM, and Marrelli, SP. Inhibition of TRPC1/TRPC3 by PKG contributes to NO-mediated vasorelaxation. *Am J Physiol Heart Circ Physiol* 2009;297:H417–H424.
56. Bon, RS, and Beech, DJ. In pursuit of small molecule chemistry for calcium-permeable non-selective TRPC channels—mirage or pot of gold? *Br J Pharmacol* 2013;170:459–474.
57. Hille, B. *Ionic Channels of Excitable Membranes*. Sunderland, MA: Sinauer, 2001.
58. Wes, PD, Chevesich, J, Jeromin, A, Rosenberg, C, Stetten, G, and Montell, C. TRPC1, a human homolog of a Drosophila store-operated channel. *Proc Natl Acad Sci USA* 1995;92:9652–9656.
59. Zhu, X, Chu, PB, Peyton, M, and Birnbaumer, L. Molecular cloning of a widely expressed human homologue for the Drosophila trp gene. *FEBS Lett* 1995;373:193–198.
60. Vaca, L, Sinkins, WG, Hu, Y, Kunze, DL, and Schilling, WP. Activation of recombinant trp by thapsigargin in Sf9 insect cells. *Am J Physiol* 1994;267:C1501–C1505.
61. Xu, XZ, Li, HS, Guggino, WB, and Montell, C. Coassembly of TRP and TRPL produces a distinct store-operated conductance. *Cell* 1997;89:1155–1164.
62. Zitt, C, Obukhov, AG, Strubing, C, Zobel, A, Kalkbrenner, F, Luckhoff, A, and Schultz, G. Expression of TRPC3 in Chinese hamster ovary cells results in calcium-activated cation currents not related to store depletion. *J Cell Biol* 1997;138:1333–1341.
63. Jordt, SE, and Julius, D. Molecular basis for species-specific sensitivity to "hot" chili peppers. *Cell* 2002;108:421–430.
64. Smith, GD, Gunthorpe, MJ, Kelsell, RE, Hayes, PD, Reilly, P, Facer, P, Wright, JE et al. TRPV3 is a temperature-sensitive vanilloid receptor-like protein. *Nature* 2002;418:186–190.
65. Xu, H, Ramsey, IS, Kotecha, SA, Moran, MM, Chong, JA, Lawson, D, Ge, P et al. TRPV3 is a calcium-permeable temperature-sensitive cation channel. *Nature* 2002;418:181–186.
66. Guler, AD, Lee, H, Iida, T, Shimizu, I, Tominaga, M, and Caterina, M. Heat-evoked activation of the ion channel, TRPV4. *J Neurosci* 2002;22:6408–6414.

67. Yue, L, Peng, J-B, Hediger, MA, and Clapham, DE. CaT1 manifests the pore properties of the calcium-release-activated calcium channel. *Nature* 2001;410:705–709.
68. Perraud, AL, Fleig, A, Dunn, CA, Bagley, LA, Launay, P, Schmitz, C, Stokes, AJ et al. ADP-ribose gating of the calcium-permeable LTRPC2 channel revealed by Nudix motif homology. *Nature* 2001;411:595–599.
69. Grimm, C, Kraft, R, Sauerbruch, S, Schultz, G, and Harteneck, C. Molecular and functional characterization of the melastatin-related cation channel TRPM3. *J Biol Chem* 2003;278:21493–21501.
70. Launay, P, Fleig, A, Perraud, AL, Scharenberg, AM, Penner, R, and Kinet, JP. TRPM4 is a Ca^{2+}-activated nonselective cation channel mediating cell membrane depolarization. *Cell* 2002;109:397–407.
71. Hofmann, T, Chubanov, V, Gudermann, T, and Montell, C. TRPM5 is a voltage-modulated and Ca(2+)-activated monovalent selective cation channel. *Curr Biol* 2003;13:1153–1158.
72. Nadler, MJ, Hermosura, MC, Inabe, K, Perraud, AL, Zhu, Q, Stokes, AJ, Kurosaki, T et al. LTRPC7 is a Mg. ATP-regulated divalent cation channel required for cell viability. *Nature* 2001;411:590–595.
73. Runnels, LW, Yue, L, and Clapham, DE. The TRPM7 channel is inactivated by PIP(2) hydrolysis. *Nat Cell Biol* 2002;4:329–336.
74. Story, GM, Peier, AM, Reeve, AJ, Eid, SR, Mosbacher, J, Hricik, TR, Earley, TJ et al. ANKTM1, a TRP-like channel expressed in nociceptive neurons, is activated by cold temperatures. *Cell* 2003;112:819–829.
75. Hanaoka, K, Qian, F, Boletta, A, Bhunia, AK, Piontek, K, Tsiokas, L, Sukhatme, VP, Guggino, WB, and Germino, GG. Co-assembly of polycystin-1 and -2 produces unique cation-permeable currents. *Nature* 2000;408:990–994.
76. Birnbaumer, L, Zhu, X, Jiang, M, Boulay, G, Peyton, M, Vannier, B, Brown, D et al. On the molecular basis and regulation of cellular capacitative calcium entry: Roles for TRP proteins. *Proc Natl Acad Sci USA* 1996;93:15195–15202.
77. Narayanan, D, Bulley, S, Leo, MD, Burris, SK, Gabrick, KS, Boop, FA, and Jaggar, JH. Smooth muscle cell transient receptor potential polycystin-2 (TRPP2) channels contribute to the myogenic response in cerebral arteries. *J Physiol* 2013;591:5031–5046.
78. Earley, S, Heppner, TJ, Nelson, MT, and Brayden, JE. TRPV4 forms a novel Ca^{2+} signaling complex with ryanodine receptors and BKCa channels. *Circ Res* 2005;97:1270–1279.
79. Earley, S, Waldron, BJ, and Brayden, JE. Critical role for transient receptor potential channel TRPM4 in myogenic constriction of cerebral arteries. *Circ Res* 2004;95:922–929.
80. Earley, S, Straub, SV, and Brayden, JE. Protein kinase C regulates vascular myogenic tone through activation of TRPM4. *Am J Physiol Heart Circ Physiol* 2007;292:H2613–H2622.
81. Dong, XP, Cheng, X, Mills, E, Delling, M, Wang, F, Kurz, T, and Xu, H. The type IV mucolipidosis-associated protein TRPML1 is an endolysosomal iron release channel. *Nature* 2008;455:992–996.
82. Horn, R, and Marty, A. Muscarinic activation of ionic currents measured by a new whole-cell recording method. *J Gen Physiol* 1988;92:145–159.
83. Rae, J, Cooper, K, Gates, P, and Watsky, M. Low access resistance perforated patch recordings using amphotericin B. *J Neurosci Methods* 1991;37:15–26.
84. Gonzales, AL, Yang, Y, Sullivan, MN, Sanders, L, Dabertrand, F, Hill-Eubanks, DC, Nelson, MT, and Earley, S. A PLCgamma1-dependent, force-sensitive signaling network in the myogenic constriction of cerebral arteries. *Sci Signal* 2014;7:ra49.
85. Gonzales, AL, and Earley, S. Endogenous cytosolic Ca(2+) buffering is necessary for TRPM4 activity in cerebral artery smooth muscle cells. *Cell Calcium* 2012;51:82–93.

86. Shi, J, Birnbaumer, L, Large, WA, and Albert, AP. Myristoylated alanine-rich C kinase substrate coordinates native TRPC1 channel activation by phosphatidylinositol 4,5-bisphosphate and protein kinase C in vascular smooth muscle. *FASEB J* 2014;28:244–255.
87. Helliwell, RM, and Large, WA. Alpha 1-adrenoceptor activation of a non-selective cation current in rabbit portal vein by 1,2-diacyl-sn-glycerol. *J Physiol* 1997;499:417–428.
88. Hofmann, T, Obukhov, AG, Schaefer, M, Harteneck, C, Gudermann, T, and Schultz, G. Direct activation of human TRPC6 and TRPC3 channels by diacylglycerol. *Nature* 1999;397:259–263.
89. Albert, AP, and Large, WA. Synergism between inositol phosphates and diacylglycerol on native TRPC6-like channels in rabbit portal vein myocytes. *J Physiol* 2003;552:789–795.
90. Peppiatt-Wildman, CM, Albert, AP, Saleh, SN, and Large, WA. Endothelin-1 activates a Ca(2+)-permeable cation channel with TRPC3 and TRPC7 properties in rabbit coronary artery myocytes. *J Physiol* 2007;580:755–764.
91. Albert, AP, Pucovský, V, Prestwich, SA, and Large, WA. TRPC3 properties of a native constitutively active Ca(2+)-permeable cation channel in rabbit ear artery myocytes. *J Physiol* 2006;571:361–369.
92. Nilius, B, Vennekens, R, Prenen, J, Hoenderop, JGJ, Bindels, RJM, and Droogmans, G. Whole-cell and single channel monovalent cation currents through the novel rabbit epithelial Ca(2+) channel ECaC. *J Physiol* 2000;527:239–248.
93. Vennekens, R, Hoenderop, JG, Prenen, J, Stuiver, M, Willems, PH, Droogmans, G, Nilius, B, and Bindels, RJ. Permeation and gating properties of the novel epithelial Ca(2+) channel. *J Biol Chem* 2000;275:3963–3969.
94. Nilius, B, Prenen, J, Droogmans, G, Voets, T, Vennekens, R, Freichel, M, Wissenbach, U, and Flockerzi, V. Voltage dependence of the Ca^{2+}-activated cation channel TRPM4. *J Biol Chem* 2003;278:30813–30820.
95. Prawitt, D, Monteilh-Zoller, MK, Brixel, L, Spangenberg, C, Zabel, B, Fleig, A, and Penner, R. TRPM5 is a transient Ca^{2+}-activated cation channel responding to rapid changes in $[Ca^{2+}]i$. *Proc Natl Acad Sci USA* 2003;100:15166–15171.
96. Hill-Eubanks, DC, Gonzales, AL, Sonkusare, SK, and Nelson, MT. Vascular TRP channels: Performing under pressure and going with the flow. *Physiology* (*Bethesda*) 2014;29:343–360.
97. Gee, KR, Brown, KA, Chen, WN, Bishop-Stewart, J, Gray, D, and Johnson, I. Chemical and physiological characterization of fluo-4 Ca(2+)-indicator dyes. *Cell Calcium* 2000;27:97–106.
98. Ji, G, Feldman, ME, Deng, KY, Greene, KS, Wilson, J, Lee, JC, Johnston, RC et al. Ca^{2+}-sensing transgenic mice: Postsynaptic signaling in smooth muscle. *J Biol Chem* 2004;279:21461–21468.
99. Wang, SQ, Song, LS, Lakatta, EG, and Cheng, H. Ca^{2+} signalling between single L-type Ca^{2+} channels and ryanodine receptors in heart cells. *Nature* 2001;410:592–596.
100. Sonkusare, SK, Bonev, AD, Ledoux, J, Liedtke, W, Kotlikoff, MI, Heppner, TJ, Hill-Eubanks, DC, and Nelson, MT. Elementary Ca^{2+} signals through endothelial TRPV4 channels regulate vascular function. *Science* 2012;336:597–601.
101. Sonkusare, SK, Dalsgaard, T, Bonev, AD, Hill-Eubanks, DC, Kotlikoff, MI, Scott, JD, Santana, LF, and Nelson, MT. AKAP150-dependent cooperative TRPV4 channel gating is central to endothelium-dependent vasodilation and is disrupted in hypertension. *Sci Signal* 2014;7:ra66.
102. Mulier, M, Vriens, J, and Voets, T. TRP channel pores and local calcium signals. *Cell Calcium* 2017;66:19–24.
103. Navedo, MF, Amberg, GC, Votaw, VS, and Santana, LF. Constitutively active L-type Ca^{2+} channels. *Proc Natl Acad Sci USA* 2005;102:11112–11117.

104. Senning, EN, and Gordon, SE. Activity and Ca(2)(+) regulate the mobility of TRPV1 channels in the plasma membrane of sensory neurons. *Elife* 2015;4:e03819.
105. Pires, PW, Sullivan, MN, Pritchard, HA, Robinson, JJ, and Earley, S. Unitary TRPV3 channel Ca^{2+} influx events elicit endothelium-dependent dilation of cerebral parenchymal arterioles. *Am J Physiol Heart Circ Physiol* 2015;309:H2031–H2041.
106. Sullivan, MN, Gonzales, AL, Pires, PW, Bruhl, A, Leo, MD, Li, W, Oulidi, A et al. Localized TRPA1 channel Ca^{2+} signals stimulated by reactive oxygen species promote cerebral artery dilation. *Sci Signal* 2015;8:ra2.
107. Sullivan, MN, Francis, M, Pitts, NL, Taylor, MS, and Earley, S. Optical recording reveals novel properties of GSK1016790A-induced vanilloid transient receptor potential channel TRPV4 activity in primary human endothelial cells. *Mol Pharmacol* 2012;82:464–472.
108. Mercado, J, Baylie, R, Navedo, MF, Yuan, C, Scott JD, Nelson, MT, Brayden, JE, and Santana, LF. Local control of TRPV4 channels by AKAP150-targeted PKC in arterial smooth muscle. *J Gen Physiol.* 2014;143:559–575.
109. Zhao, L, Sullivan, MN, Chase, M, Gonzales, AL, and Earley, S. Calcineurin/nuclear factor of activated T cells-coupled vanilloid transient receptor potential channel 4 Ca^{2+} sparklets stimulate airway smooth muscle cell proliferation. *Am J Respir Cell Mol Biol* 2014;50:1064–1075.
110. Manuelyan, I, Syed, A, Koide, M, Shui, B, Sonkusare, S, Kotlikoff, M, Nelson, M, and Wellman, G. TRPV1-mediated Ca^{2+} influx and middle meningeal artery constriction. *FASEB J* 2015;29 (1 Supplement), 943.6.
111. Earley, S. TRPM4 channels in smooth muscle function. *Pflugers Arch* 2013;465:1223–1231.
112. Garcia, ZI, Bruhl, A, Gonzales, AL, and Earley, S. Basal protein kinase Cdelta activity is required for membrane localization and activity of TRPM4 channels in cerebral artery smooth muscle cells. *Channels (Austin)* 2011;5:210–214.
113. Earley, S. Vanilloid and melastatin transient receptor potential channels in vascular smooth muscle. *Microcirculation* 2010;17:237–249.
114. Earley, S, Heppner, TJ, Nelson, MT, and Brayden, JE. TRPV4 forms a novel Ca^{2+} signaling complex with ryanodine receptors and BKCa channels. *Circ Res* 2005;97:1270–1279.
115. Reading, SA, Earley, S, Waldron, BJ, Welsh, DG, and Brayden, JE. TRPC3 mediates pyrimidine receptor-induced depolarization of cerebral arteries. *Am J Physiol Heart Circ Physiol* 2005;288:H2055–2061.
116. Dietrich, A, Mederos, YSM, Gollasch, M, Gross, V, Storch, U, Dubrovska, G, Obst, M et al. Increased vascular smooth muscle contractility in TRPC6-/- mice. *Mol Cell Biol* 2005;25:6980–6989.
117. Kuiper, GG, Carlsson, B, Grandien, K, Enmark, E, Haggblad, J, Nilsson, S and Gustafsson, JA. Comparison of the ligand binding specificity and transcript tissue distribution of estrogen receptors alpha and beta. *Endocrinology* 1997;138:863–870.
118. Ponce-Balbuena, D, Lopez-Izquierdo, A, Ferrer, T, Rodriguez-Menchaca, AA, Arechiga-Figueroa, IA, and Sanchez-Chapula, JA. Tamoxifen inhibits inward rectifier K^+ 2.x family of inward rectifier channels by interfering with phosphatidylinositol 4,5-bisphosphate-channel interactions. *J Pharmacol Exp Ther* 2009;331:563–573.
119. Nilius, B, Sehrer, J, and Droogmans, G. Permeation properties and modulation of volume-activated Cl^--currents in human endothelial cells. *Br J Pharmacol* 1994;112:1049–1056.
120. Greenwood, IA, and Large, WA. Properties of a Cl^- current activated by cell swelling in rabbit portal vein vascular smooth muscle cells. *Am J Physiol* 1998;275:H1524–H1532.
121. Sones, WR, Davis, AJ, Leblanc, N, and Greenwood, IA. Cholesterol depletion alters amplitude and pharmacology of vascular calcium-activated chloride channels. *Cardiovasc Res* 2010;87:476–484.

122. O'Brian, CA, Liskamp, RM, Solomon, DH, and Weinstein, IB. Inhibition of protein kinase C by tamoxifen. *Cancer Res* 1985;45:2462–2465.
123. Jepps, TA, Carr, G, Lundegaard, PR, Olesen, SP, and Greenwood, IA. Fundamental role for the KCNE4 ancillary subunit in Kv7.4 regulation of arterial tone. *J Physiol* 2015;593:5325–5340.
124. Summerton, J. Morpholino antisense oligomers: The case for an RNase H-independent structural type. *Biochim Biophys Acta* 1999;1489:141–158.
125. Summerton, JE. Morpholino, siRNA, and S-DNA compared: Impact of structure and mechanism of action on off-target effects and sequence specificity. *Curr Top Med Chem* 2007;7:651–660.
126. Li, W, Peng, H, Mehaffey, EP, Kimball, CD, Grobe, JL, van Gool, JM, Sullivan, MN et al. Neuron-specific (pro)renin receptor knockout prevents the development of salt-sensitive hypertension. *Hypertension* 2014;63:316–323.
127. Longden, TA, Dabertrand, F, Koide, M, Gonzales, AL, Tykocki, NR, Brayden, JE, Hill-Eubanks, D, and Nelson, MT. Capillary K^+-sensing initiates retrograde hyperpolarization to increase local cerebral blood flow. *Nat Neurosci* 2017;20:717–726.
128. Xiao, JH, Zheng, YM, Liao, B, and Wang, YX. Functional role of canonical transient receptor potential 1 and canonical transient receptor potential 3 in normal and asthmatic airway smooth muscle cells. *Am J Respir Cell Mol Biol* 2010;43:17–25.
129. Jia, Y, Wang, X, Varty, L, Rizzo, CA, Yang, R, Correll, CC, Phelps, PT, Egan, RW, and Hey, JA. Functional TRPV4 channels are expressed in human airway smooth muscle cells. *Am J Physiol Lung Cell Mol Physiol* 2004;287:L272–L278.
130. White, TA, Xue, A, Chini, EN, Thompson, M, Sieck, GC, and Wylam, ME. Role of transient receptor potential C3 in TNF-α–enhanced calcium influx in human airway myocytes. *Am J Respir Cell Mol Biol* 2006;35:243–251.
131. Gombedza, F, Kondeti, V, Al-Azzam, N, Koppes, S, Duah, E, Patil, P, Hexter, M, Phillips D, Thodeti, CK, and Paruchuri, S. Mechanosensitive transient receptor potential vanilloid 4 regulates Dermatophagoides farinae-induced airway remodeling via 2 distinct pathways modulating matrix synthesis and degradation. *FASEB J* 2017;31:1556–1570.
132. Zhang, L, An, X, Wang, Q, and He, M. Activation of cold-sensitive channels TRPM8 and TRPA1 inhibits the proliferative airway smooth muscle cell phenotype. *Lung* 2016;194:595–603.
133. Golovina, VA, Platoshyn, O, Bailey, CL, Wang, J, Limsuwan, A, Sweeney, M, Rubin, LJ, and Yuan, JX. Upregulated TRP and enhanced capacitative Ca(2+) entry in human pulmonary artery myocytes during proliferation. *Am J Physiol Heart Circ Physiol.* 2001;280:H746–H755.
134. Sweeney, M, Yu, Y, Platoshyn, O, Zhang, S, McDaniel, SS, and Yuan, JX. Inhibition of endogenous TRP1 decreases capacitative Ca^{2+} entry and attenuates pulmonary artery smooth muscle cell proliferation. *Am J Physiol Lung Cell Mol Physiol* 2002;283:L144–L155.
135. Yu, Y, Fantozzi, I, Remillard, CV, Landsberg, JW, Kunichika, N, Platoshyn, O, Tigno, DD, Thistlethwaite, PA, Rubin, LJ, and Yuan, JX. Enhanced expression of transient receptor potential channels in idiopathic pulmonary arterial hypertension. *Proc Natl Acad Sci USA* 2004;101:13861–13866.
136. Yu, Y, Keller, SH, Remillard, CV, Safrina, O, Nicholson, A, Zhang, SL, Jiang, W et al. A functional single-nucleotide polymorphism in the TRPC6 gene promoter associated with idiopathic pulmonary arterial hypertension. *Circulation* 2009;119:2313–2322.
137. Yu, Y, Sweeney, M, Zhang, S, Platoshyn, O, Landsberg, J, Rothman, A, and Yuan, JX. PDGF stimulates pulmonary vascular smooth muscle cell proliferation by upregulating TRPC6 expression. *Am J Physiol Cell Physiol* 2003;284:C316–C330.

138. Zhang, S, Remillard, CV, Fantozzi, I, and Yuan, JX. ATP-induced mitogenesis is mediated by cyclic AMP response element-binding protein-enhanced TRPC4 expression and activity in human pulmonary artery smooth muscle cells. *Am J Physiol Cell Physiol* 2004;287:C1192–C1201.
139. Wang, YX, Wang, J, Wang, C, Liu, J, Shi, LP, Xu, M, and Wang, C. Functional expression of transient receptor potential vanilloid-related channels in chronically hypoxic human pulmonary arterial smooth muscle cells. *J Membr Biol* 2008;223:151–159.
140. Yang, M, Zhao, T, Lin, J, Ju, T, and Zhang, L. Inhibition of TRPM7 attenuates rat aortic smooth muscle cell proliferation induced by angiotensin II: Role of genistein. *J Cardiovasc Pharmacol* 2015;66:16–24.
141. Naylor, J, Li, J, Milligan, CJ, Zeng, F, Sukumar, P, Hou, B, Sedo, A et al. Pregnenolone sulphate- and cholesterol-regulated TRPM3 channels coupled to vascular smooth muscle secretion and contraction. *Circ Res* 2010;106:1507–1515.

9 Mitochondria Structure and Position in the Local Control of Calcium Signals in Smooth Muscle Cells

John G. McCarron, Christopher Saunter, Calum Wilson, John M. Girkin, and Susan Chalmers

CONTENTS

9.1 INTRODUCTION

Features of Ca^{2+} signals including the amplitude, duration, frequency and location are encoded by various physiological stimuli. These features of the signals are decoded by cells to selectively activate smooth muscle functions that include contraction and proliferation [1–3]. Central, therefore, to an appreciation of how smooth muscle is controlled is an understanding of the regulation of Ca^{2+}. In smooth muscle, Ca^{2+} signals arise from two major sources. The first is the extracellular space from which Ca^{2+} enters the cell via channels such as voltage-dependent Ca^{2+} channels, store-operated Ca^{2+} (SOC) channels and various members of the transient receptor potential channel family. The second major Ca^{2+} source is the internal Ca^{2+} store (sarcoplasmic reticulum; SR) [4–6]. The SR accumulates Ca^{2+} using sarco/endoplasmic reticulum Ca^{2+}-ATPases (SERCA) and Ca^{2+} is released from the SR via the ligand-gated channel/receptor complexes, the IP_3 receptor (IP_3R) and ryanodine receptor (RyR). Release of Ca^{2+} via IP_3R is activated by

IP_3 generated in response to many G-protein or tyrosine kinase-linked receptor activators including drugs [7,8]. RyR may be activated pharmacologically (e.g., caffeine), by Ca^{2+} influx from outside the cell in the process of Ca^{2+}-induced Ca^{2+} release (CICR), or when the stores Ca^{2+} content exceeds normal physiological values, that is in store overload [2,9–12].

Activation of either Ca^{2+} influx or Ca^{2+} release results in an increase of the cytoplasmic Ca^{2+} concentration ($[Ca^{2+}]_c$) from the resting value of ~100 nM to ~1 μM for many seconds throughout the cell, and transiently (e.g., 100 ms) to much higher values (e.g., 50 μM) in small parts of the cytoplasm close to sites of influx or Ca^{2+} release. These local Ca^{2+} signals begin with the opening of one or a few channels, allowing a large flux of the ion into the cytoplasm. Influx to the cytoplasm via voltage-dependent Ca^{2+} channels occurs at rates of ~0.6 million Ca^{2+} ions per second per channel (0.2 pA current). The influx generates a significant local concentration gradient near the channel in which $[Ca^{2+}]$ declines from ~10 μM to ~100 nM over a few hundred nanometers from the plasma membrane [2,13–17]. Voltage-dependent Ca^{2+} channel open time is brief (~1 ms) and the gradient dissipates rapidly with rates of change in the subplasma membrane space on the order of ~5000 μM s^{-1} [2] as compared to a much slower rate of ~0.5 μM s^{-1} in the bulk cytoplasm [2] after a global $[Ca^{2+}]$ rise [18,19]. The large difference in rate of decline in the subplasma membrane space and bulk cytoplasm arise because local changes are driven mostly by buffering and diffusion while the slower rate of decline in bulk cytoplasm is determined by pumps. High local $[Ca^{2+}]$ and the rapid rates of change near channels may target processes with rapid Ca^{2+} binding kinetics to selectively activate particular functions [20–23]. The high local $[Ca^{2+}]$ signals arising from influx also, in turn, may activate IP_3R or RyR to amplify the local signals or propagate through the cell as global signals with slower but more widespread effects [24–30]. The transition of signals from those involving single to multiple channels and from local to global Ca^{2+} increases creates a multitude of signals with various locations, magnitudes and time courses [31–34] so that various cellular biological responses may be selectively activated.

It is acknowledged that a major way that Ca^{2+} signaling specifically targets particular biological processes is by increases in concentration of the ion being selectively localized to certain regions of the cell (Figure 9.1) [36,37]. In native smooth muscle cells, mitochondria contribute to the localization of Ca^{2+} signals and to the modulation of the amplitude of Ca^{2+} signals [38–42]. Mitochondria regulate these local signals by the organelles' ability to take up and release the ion. Ca^{2+} uptake occurs through the mitochondrial Ca^{2+} uniporter while efflux is mediated by the mitochondrial Na^+/Ca^{2+} exchanger. Mitochondrial Ca^{2+} uptake and efflux may regulate cytoplasmic Ca^{2+} concentrations both directly and indirectly. Direct regulation occurs by alteration of bulk Ca^{2+} levels (Figures 9.2 and 9.3) [18,40,44–47]. Indirect regulation occurs as a result of mitochondrial influence on the activity of SR or plasma membrane Ca^{2+} channels. This chapter describes how the structure and positioning of mitochondria contribute to the control of Ca^{2+} signaling, including a previously unrecognized ability of the position of the organelles to increase local Ca^{2+} entry via voltage-dependent Ca^{2+} channels.

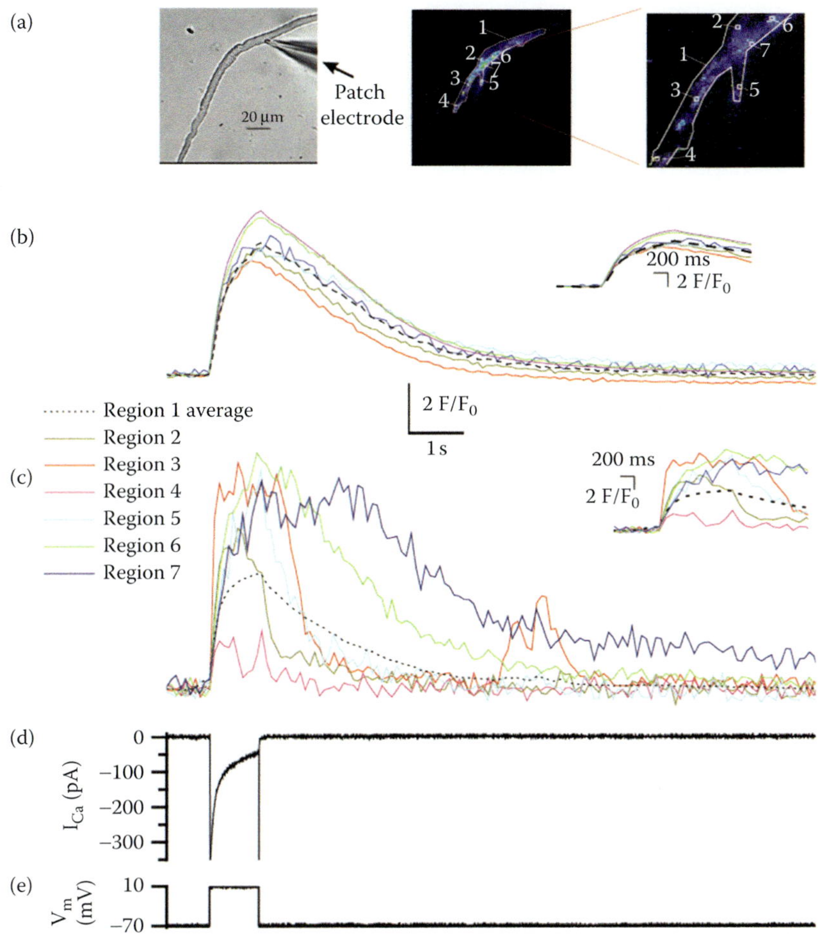

FIGURE 9.1 Simultaneous wide-field epi-fluorescence and total internal reflection fluorescence (TIRF) $[Ca^{2+}]$ measurements in a voltage-clamped smooth muscle. Depolarization (−70 to +10 mV; e), activated a voltage-dependent Ca^{2+} current (I_{Ca}; d) to evoke a rise in $[Ca^{2+}]$ in both the subplasma membrane space (c) and bulk cytoplasm (b). The rise in $[Ca^{2+}]$ that occurred in the subplasma membrane space (measured by TIRF) (c) was more rapid in onset than that seen in the bulk cytoplasm (measured by wide-field epi-fluorescence) (b). Changes in the fluorescence ratio with time (b and c) are each derived from precisely the same 2 × 2-pixel boxes (regions 1–6 in (a), middle and right (expanded) panel; drawn at a 3 × 3-pixel size to facilitate visualization) and from a larger region encompassing the entire TIRF region (region 7). Significantly, while the cytosolic $[Ca^{2+}]$ increase that occurred in the bulk cytoplasm (b) was approximately uniform and simultaneous throughout the cell, those in subplasma membrane space (c) had a wide range of amplitudes and various time courses. Insets in (b) and (c) show the rising phase of the transients on an expanded time base. Black dotted lines are the average responses. (Reproduced from McCarron, J.G. et al., *J. Gen. Physiol.*, 133, 439–457, 2009. With permission.)

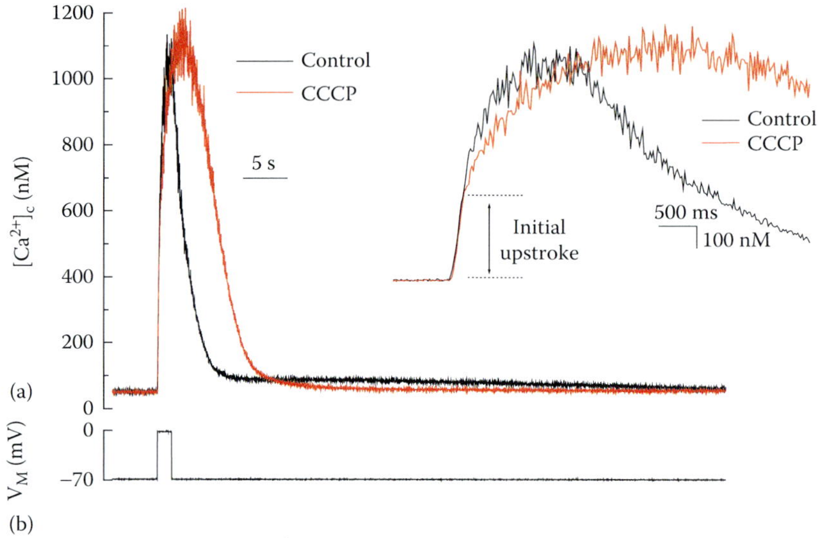

FIGURE 9.2 Preventing mitochondrial Ca^{2+} uptake does not alter the initial upstroke or peak cytosolic $[Ca^{2+}]$ after depolarization-evoked Ca^{2+} entry. Depolarization (−70 to 0 mV; b) of a voltage-clamped single smooth muscle cell activated a voltage-dependent Ca^{2+} current (not shown) and an increased cytosolic $[Ca^{2+}]$ (a). After carbonyl cyanide *m*-chlorophenyl hydrazone (CCCP; 5 μM; transients in red lines) neither the initial rate of cytosolic $[Ca^{2+}]$ rise (see inset) or peak cytosolic $[Ca^{2+}]$ achieved were significantly altered. The insets show the transients on an expanded time base. Cytosolic $[Ca^{2+}]$ was measured at a frequency of 50 Hz using the membrane-impermeable fura-2 (potassium salt, 50 μM) introduced into the cell from the patch pipette. (Reproduced from McCarron, J.G. et al., *Pflugers Arch.*, 464, 51–62, 2012. With permission.)

9.2 ROLE OF MITOCHONDRIA IN Ca^{2+} SIGNALING

Mitochondria are major hubs for cellular signaling. In smooth muscle, mitochondria control contractility, proliferation, and growth through regulation of cytoplasmic Ca^{2+} concentrations. A critical feature of mitochondria's ability to control Ca^{2+} signaling is the position and structure of the organelles within cells. Understanding of the precise relationship between Ca^{2+} signaling and mitochondrial structure and position in living, fully-differentiated, (native) cells is preliminary, which may be due to the uncertainty in mitochondrial structure in these cells. Most studies describing mitochondrial structure are derived from cells maintained in culture conditions. Yet, the structure and arrangement of mitochondria in cultured cells differ substantially from the organelles in fully differentiated smooth muscle cells [48–51]. In cultured cells the position of mitochondria contributes significantly to regulation of cytoplasmic Ca^{2+} signals and to the control of mitochondrial Ca^{2+} concentration ($[Ca^{2+}]_m$). For example, SOC increased the $[Ca^{2+}]_m$ in the cultured endothelial cell culture line ECV304, suggesting the organelles accumulate this Ca^{2+} influx. In ECV304 cells, 14% of mitochondria are positioned close to the plasma membrane, a position that will facilitate Ca^{2+} uptake. Less than 4% of mitochondria are close to the internal

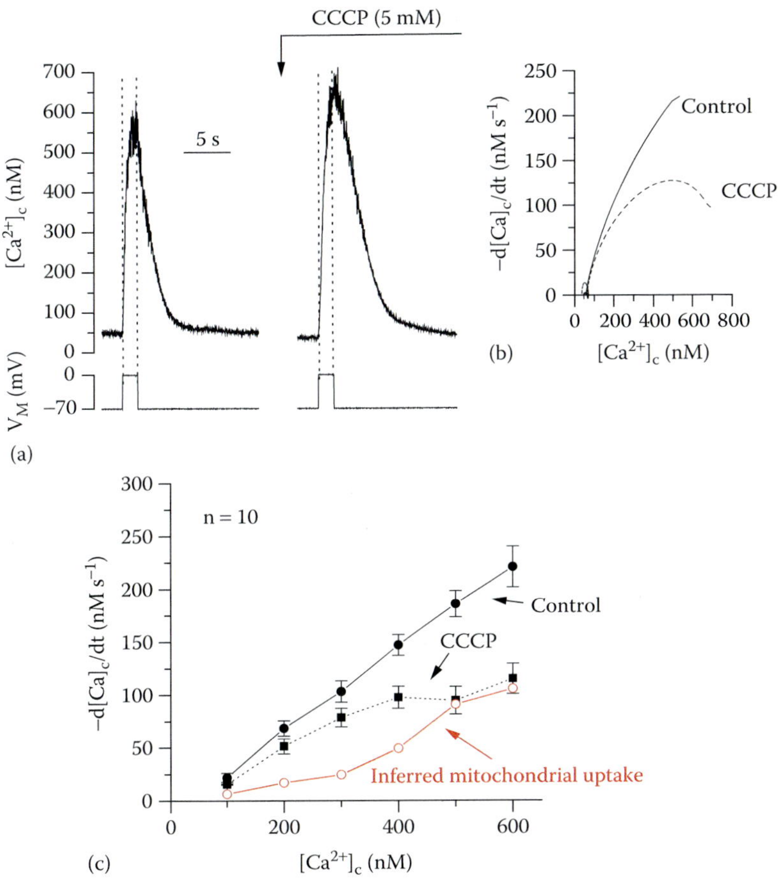

FIGURE 9.3 Mitochondria contribute to cytosolic $[Ca^{2+}]$ decline following voltage-dependent Ca^{2+} entry in smooth muscle. (a) Depolarization (−70 to 0 mV) activated a voltage-dependent Ca^{2+} current (not shown) and increased cytosolic $[Ca^{2+}]$. Carbonyl cyanide *m*-chlorophenyl hydrazone (CCCP; 5 μM) slowed the rate of decline of cytosolic $[Ca^{2+}]$ on repolarization compared with control. (b) The rate of decline of the two transients (a) is shown in (b). The derivative (b) was obtained from high-order polynomial fits to the declining phase of the transient and shows a significant slowing when mitochondria were prevented from accumulating Ca^{2+}. (c) A summary of the rates of decline for ten cells in the presence and absence of CCCP. The inferred mitochondrial contribution to the decline of cytosolic $[Ca^{2+}]$ (red line) was obtained by subtracting control rates from those seen in CCCP and shows that mitochondrial Ca^{2+} uptake occurred above 200 nM $[Ca^{2+}]_c$. (Reproduced from McCarron, J.G. et al., *J. Physiol.*, 516: 149–161, 1999. With permission.)

store and store release of Ca^{2+} did not increase $[Ca^{2+}]_m$. Conversely in HeLa cells, where 65% of mitochondria are close to the internal store and <6% are in close proximity to the plasma membrane, store release of Ca^{2+} caused a much greater increase in $[Ca^{2+}]_m$ than did SOC [52]. These observations suggest mitochondrial position is critical in determining the organelles' uptake of Ca^{2+}. In turn, the position of the mitochondria regulates SOC [53,54]. Forced relocation of mitochondria away from the cell periphery and towards the nucleus (by overexpressing dynamitin, a protein involved in mitochondrial movement) decreased SOC following store depletion in HeLa cells [55]. These results point to the position of mitochondria as being critical in regulation of Ca^{2+} signaling.

In native fully-differentiated cell types the distribution of mitochondria may also regulate Ca^{2+} signaling. For example, in fully-differentiated cardiac myocytes or neurons, mitochondria appear to be located particularly close to sites of initiation of Ca^{2+} signals from voltage-dependent Ca^{2+} channels on the plasma membrane [56]. At these sites mitochondria contribute to the clearance of Ca^{2+} from the mouth of the channel, limiting Ca^{2+}-dependent inactivation and thus prolonging channel open time [57,58] and facilitating Ca^{2+} entry. In smooth muscle, however, mitochondrial Ca^{2+} uptake does not appear to alter the kinetics of voltage-dependent Ca^{2+} entry (Figure 9.2). This observation suggests the gating of voltage-dependent Ca^{2+} channels is not altered by mitochondria. Notwithstanding the absence of a direct effect on voltage-dependent Ca^{2+} channels, mitochondria may modulate Ca^{2+} rises arising from voltage-dependent Ca^{2+} entry in smooth muscle by modulating Ca^{2+}-dependent feedback processes operative on other ion channels. Mitochondria appear to increase clearance of bulk cytoplasmic Ca^{2+} concentrations, and as a result decrease the activity of Ca^{2+}-activated chloride channels (Cl_{Ca}) [59] and increase the activity of Ca^{2+}-activated K^+ channels (K_{Ca}) [60]. Pharmacological inhibition of the ability of mitochondria to remove Ca^{2+} from the cytoplasm prolonged the activity of Cl_{Ca} in portal vein smooth muscle cells [59] and inhibited K_{Ca} currents in cerebral artery smooth muscle [60]. Modulation of Cl_{Ca} or K_{Ca} will alter the membrane potential to then regulate Ca^{2+} influx via voltage-dependent Ca^{2+} channels as a result.

By taking up Ca^{2+}, mitochondria modulate the time course and amplitude of Ca^{2+} signals to shape the resulting message [1]. For example, we have shown that the Ca^{2+} transient arising from activation of voltage-dependent Ca^{2+} channels has an accelerated rate of decline as a consequence of mitochondrial Ca^{2+} uptake; when uptake is inhibited, the rate of Ca^{2+} decline is substantially slowed (Figures 9.2 and 9.3). In the example shown in Figure 9.3, mitochondria modulate Ca^{2+} signals over the cytosolic [Ca^{2+}] range 600 nM–200 nM, which demonstrates that mitochondria have a high affinity for Ca^{2+} (in the sub-micromolar range; Figure 9.3). Mitochondrial Ca^{2+} uptake does not alter the rate, or extent, of Ca^{2+} *influx* via voltage-dependent Ca^{2+} channels [1] (the likely reasons for this are discussed later on). Some studies have shown that increased mitochondrial reactive oxygen species (ROS) production can alter voltage-dependent Ca^{2+} influx [61], however, this is likely to be due to ROS-dependent alterations in channel open time or conductance, rather than altered mitochondrial Ca^{2+} uptake during voltage-dependent Ca^{2+} influx.

Mitochondria also have the capacity to modulate Ca^{2+} signals that are ~2 orders of magnitude larger than the global Ca^{2+} transient arising from voltage-dependent Ca^{2+} channels and in the tens of micromolar range [62]. One notable example in smooth muscle is mitochondrial regulation of IP_3-mediated Ca^{2+} signals that arise from the activity of a single IP_3R cluster (*Ca^{2+} puffs*; Figure 9.4). When mitochondria are prevented from taking up Ca^{2+} (using uncouplers, complex I inhibitors or uniporter inhibitors), the upstroke of Ca^{2+} puffs is inhibited [42] (Figure 9.4). The duration of the upstroke of a puff is determined by the open time of IP_3R in the cluster.

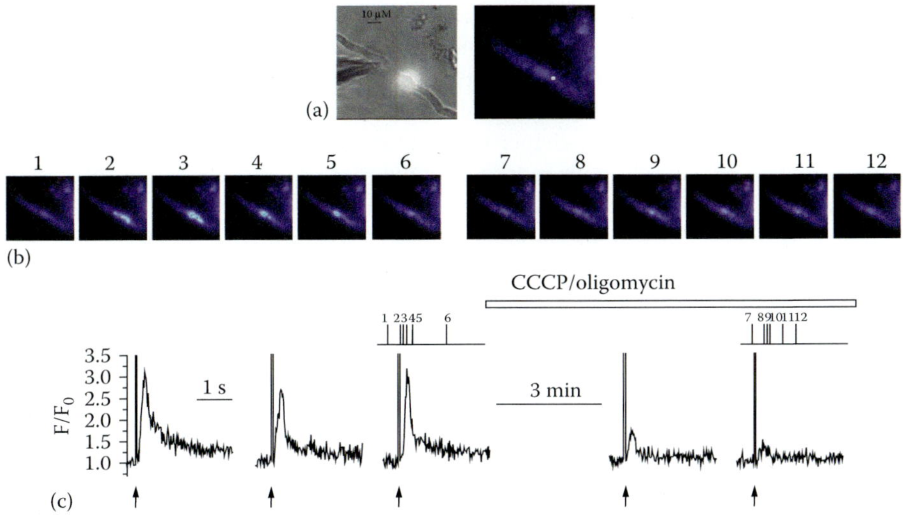

FIGURE 9.4 Preventing mitochondrial Ca^{2+} uptake by depolarizing $\Delta\Psi m$ with CCCP/oligomycin inhibits Ca^{2+} release from IP_3R clusters (Ca^{2+} puffs). At –70 mV, locally photolyzed caged $InsP_3$ (25 µM) (↑, C) in a ~20 µm diameter region (a; bright spot in left-hand panel, see also whole cell electrode, left side) evoked Ca^{2+} puffs in an EGTA (300 µM) buffered smooth muscle cell (b, c). Note: There are two individual Ca^{2+} puff sites in response to photorelease of $InsP_3$; one site releases Ca^{2+} just before the other site. Flash photolysis of $InsP_3$ every ~60 s generated approximately comparable cytosolic $[Ca^{2+}]$ increases. (c) Superfusion of carbonyl cyanide *m*-chlorophenyl hydrazone (CCCP) and oligomycin (1 µM and 6 µM, respectively) while continuing to photolyze $InsP_3$ at ~60 intervals, decreased the amplitude of $InsP_3$-mediated Ca^{2+} puffs (b, c). The cytosolic $[Ca^{2+}]$ images (b) are derived from the time points indicated by the corresponding numbers in (c). Cytosolic $[Ca^{2+}]_c$ changes in (b) are expressed by color; dark blue low and light blue high cytosolic $[Ca^{2+}]$. Measurements were made from a single 3 × 3-pixel box (a; right hand panel, white square). The large increase in fluorescence at time 0 is the flash artefact triggering photolysis of caged IP_3. (Reproduced from Olson, M.L. et al., *J. Biol. Chem.*, 285, 2040–2050, 2010. With permission.)

The observation [42] that mitochondria modulates the puff upstroke suggests that the organelle's uptake is fast enough to modulate Ca^{2+} concentration near IP_3R clusters during the channels open time, requiring mitochondria to have a very low affinity for Ca^{2+} (Ca^{2+} puffs >10 µM) and to be positioned close to the channel.

Electron microscopy studies of porcine tracheal smooth muscle cells suggest that 99% of mitochondria were within 30 nm of SR and 48% of the mitochondrial outer membrane was within 30 nm with the SR [63] with some mitochondria fully ensheathed by SR [63,64]. Given the proximity between sites of Ca^{2+} release and mitochondria, the question arises as to whether or not there is a direct intermolecular link between the two structures. While there are no high-resolution studies of this link in smooth muscle, electron tomographs [65] and transmission electron micrographs [66] in other cells show tethers that connect the endoplasmic reticulum (ER) to the mitochondrial outer membrane that may be significant in Ca^{2+} signaling. Indeed, immunocytochemical studies show that those regions of the ER in close proximity to mitochondria are also enriched with IP_3R to create *hotspots* for the transfer of Ca^{2+} from the ER to mitochondria [62]. Several molecular candidates for the tethers have been identified in cell types other than smooth muscle cells. The IP_3R itself has been shown to be tethered to the mitochondrial voltage-dependent anion channel (VDAC) via the chaperone glucose-regulated protein 75 (GRP75), an interaction that promotes mitochondrial accumulation of Ca^{2+} released by IP_3R [67]. Interestingly, both *in vitro* treatment of epithelial cells with high glucose and long-term hyperglycemia in a mouse model of diabetes reduced GRP75 levels, reducing ER-mitochondrial contact sites [68]. The protein mitofusin-2 is located on both mitochondrial and ER membranes and is proposed to form homodimeric tethers between the organelles [69] that contribute to Ca^{2+} signaling. Disrupting linkages between the SR and mitochondria by gene silencing of mitofusin-2 decreased mitochondrial Ca^{2+} uptake during IP_3-mediated $[Ca^{2+}]_c$ rises [65,69]. The multifunctional sorting protein PACS-2, which is localized to the ER and may play a role in apoptotic pathway initiation, has also been proposed to be important for maintaining close apposition of ER and mitochondria in the A7 astrocyte cell line [70]. siRNA silencing of PACS-2 caused mitochondrial fragmentation and separation from the ER, altering Ca^{2+} signaling by increasing IP_3-mediated Ca^{2+} release [70]. The sigma-1 receptor, a protein involved in cell stress responses and present on the ER at regions in close contact with mitochondria, may also couple the membranes of the two organelles [71]. The sigma-1 receptor may form a complex with another chaperone protein (BiP) to stabilize IP_3R at regions of ER-mitochondrial proximity. Depletion of ER Ca^{2+} dissociates the sigma-1 receptor from BiP, leading to prolonged Ca^{2+} signaling into mitochondria via IP_3Rs [71].

The question arises: how can mitochondria—by removing Ca^{2+} from the cytoplasm (i.e., lowering $[Ca^{2+}]$)—generate a larger cytoplasmic $[Ca^{2+}]$ rise? IP_3R are regulated by Ca^{2+}-dependent positive and negative feedback mechanisms [72,73]. By removing Ca^{2+}, mitochondria may limit a negative feedback inhibition of Ca^{2+} on IP_3R [42]. There are at least two types of Ca^{2+}-dependent negative feedback mechanisms that may deactivate smooth muscle IP_3R. In the first, a Ca^{2+}-dependent deactivation of IP_3R occurs at cytosolic $[Ca^{2+}]$ exceeding ~300 nM [73]. The onset is rapid and the deactivation is persistent and lasts for ~5 s after the cytosolic $[Ca^{2+}]$ increase has ended [74]. Another form of Ca^{2+}-dependent deactivation of IP_3R, once initiated by an increased cytosolic $[Ca^{2+}]$, persisted for tens of seconds after cytosolic $[Ca^{2+}]$ had regained resting values, that is, IP_3R became, at least partially, refractory [30,75]. Each of these processes would persistently restrict Ca^{2+} release via IP_3R. Mitochondrial Ca^{2+} uptake, by buffering the Ca^{2+} rise at IP_3R, presumably prevented a persistent deactivation of IP_3R that would otherwise limit the overall release of Ca^{2+} [30,75].

The control that mitochondria exert on Ca^{2+} puffs enables the organelle to exert particularly dramatic effects on the global Ca^{2+} events such as Ca^{2+} waves and oscillations. When mitochondria are prevented from taking up Ca^{2+}, waves and oscillations halt [42,45], that is, modulation of the time course of a Ca^{2+} rise by the organelle determines whether or not some Ca^{2+} signals occur at all. The position and structure of the organelles, together with nano-architectural features of SR-mitochondria junctions, is critical in determining precisely how mitochondria regulate local and global Ca^{2+} signals.

9.3 MITOCHONDRIA STRUCTURE IN NATIVE CELLS

The precise structure and position of mitochondria have been studied most extensively in cultured smooth muscle cells because of the relative ease that the organelles can be visualized in these cells. The thin axial depth of cultured cells and ability to transfect the cells to express unique, mitochondrially-targeted fluorophores result in exceptional signal to noise for imaging mitochondrial structure. In cultured cells, mitochondria exist in a wide range of sizes and shapes (Figure 9.5) and the organelle may change rapidly from solitary ovoid shapes to extensive branched networks and

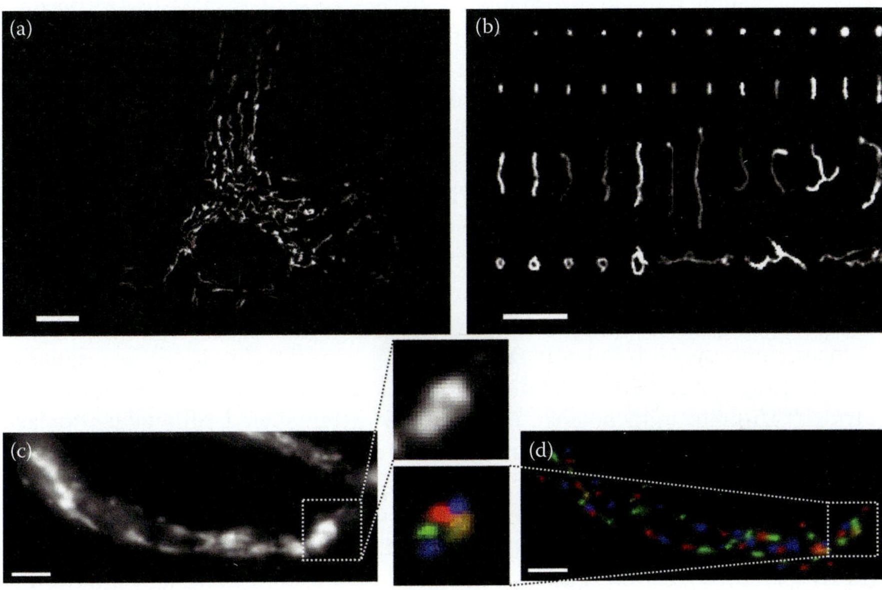

FIGURE 9.5 Mitochondrial structures in cultured and fully-differentiated smooth muscle cells. (a) A cultured single vascular smooth muscle cell showing the arrangement of mitochondria. The organelle is scattered through the cytoplasm and is arranged in various orientations. Mitochondria were labeled with the mitotracker green. (b) Example mitochondria showing the diverse phenotypes which include small spheres, swollen spheres, straight rods, twisted rods, branched rods, and loops. (a and b reproduced from McCarron, J. et al., *J. Vasc. Res.*, 5, 357–371, 2013. With permission.) (c) A native fully-differentiated smooth muscle cell from a cerebral resistance artery showing the arrangement of mitochondria as revealed by the fluorophore tetramethylrhodamine, ethyl ester (TMRE). The organelle is distributed throughout the cytoplasm and appears to be mainly organized parallel to the long axis of the cell. However, the significant axial depth of the cells (when compared to cultured smooth muscle) and multiple overlapping fluorescence point sources result in an optically-confused image with structures that are hard to interpret (d) FaLM permits the size, shape, and position of mitochondria to determined and shows the organelles comprise mainly spheres and short rods. The inset between c and d shows the same example region of the cell before and after application of FaLM. (c and d modified from Chalmers, S. et al., *Nat. Sci. Rep.*, 30, 2000–2013, 2015. With permission.) Scale bars = 10 μm.

even single continuous mitochondrial structure throughout the cell [62,76,77]. This almost continuous re-shaping creates a diversity of structures (Figure 9.5), presumably each with different physiological roles although the precise functions are not yet fully understood [51,78]. On the other hand, mitochondria within fully-differentiated live smooth muscle cells are difficult to visualize because of the significant axial cell depth. This depth results in multiple confused overlapping fluorescence point sources at different distances through the cell derived from various mitochondria. The overlapping fluorescence is difficult to interpret and may be single large organelles or multiple small mitochondria. Exclusive fluorophore targeting is also more difficult in native cells than in cultured cells. Whether or not the arrangements of mitochondria seen in cultured cells occur in fully-differentiated cells is uncertain.

Live cell imaging is required to appreciate the precise relationship between position and structure of mitochondria and the control of Ca^{2+} signaling in fully-differentiated cells. Confocal or other optical sectioning imaging methods do not provide sufficient resolution to determine detailed structure of the organelle in fully-differentiated cells. Detailed insights into the structure of individual mitochondria at fixed points in time have been provided by electron microscopy [EM; 79]. However, EM cannot easily resolve the entire cellular mitochondrial complement or provide information on dynamic changes or functional connectivity and is not compatible with Ca^{2+} imaging. EM [80] and super-resolution techniques [81–89] do not unambiguously reveal the extent of the electrically-continuous inner mitochondrial membrane and are less suited to simultaneously studying Ca^{2+} signaling in live cells, nor do they currently have the imaging speed frequently required.

We recently developed a technique to determine the structure of individual, electrically-discrete mitochondria in live, fully-differentiated cells using conventional fluorescence imaging. Our approach is rapid and quantifies the shape, size and location of mitochondria even when the organelles are densely clustered. The approach measures transient changes in the membrane potential that are unique to each mitochondrion to determine their individual shapes. Changes in membrane potential ($\Delta\Psi m$) arise from transient openings of the mitochondrial permeability transition pore [mPTP; 90,91] and can be visualized using a rapidly-repartitioning cationic fluorophore such as TMRE, as *flickers* of fluorescence [92,93]. These flickers demarcate individual, electrically-discrete mitochondria [92,94]. Drawing inspiration from PALM/BaLM super-resolution techniques [81,82], we used these flicker events to extract structural information of each organelle, facilitating a more detailed analysis of the live-cell fluorescence images. These mitochondrial flickers differ from the correlation of photo-activation, bleaching, or blinking in PALM/BaLM because the flickering objects are not single molecule emitters with a well-defined point-spread function but extended organelles of unknown shape. Nonetheless, the size, shape, and position of the mitochondria can be rapidly determined to the conventional resolution limit from a pixel-by-pixel spatio-temporal covariance of the derived flickering fluorescence signals in an approach that we call Flicker-assisted Localization Microscopy (FaLM) [49,50].

Using FaLM, we found that mitochondria in fully-differentiated resistance artery smooth muscle cells were distributed through the cytoplasm as solitary, spherical-, or short rod-shaped entities (Figure 9.5) [50]. We found mitochondria had a median area of 0.46 μm², a median length of 0.9 μm, and a median length:width ratio of 1.5. We also found the median density of mitochondria in resistance artery smooth muscle cells was 0.125 mitochondria per μm² and that mitochondria occupied 7.0% of the cell volume. Each mitochondrion had 2–5 immediate neighbors with ~2 μm between the organelle centers. The mean center distance of a mitochondria from the plasma membrane was ~2 μm [49]. Given a width of mitochondria of 0.6 μm, then, the edge distance of the organelle from the plasma membrane is 1.4 μm. The opening of a single voltage-dependent Ca^{2+} channel (0.2 pA current, mean open-time 1 ms) generates a high local [Ca^{2+}] that decreases steeply with distance from the plasma membrane as the ion diffuses into an increasing hemispheric volume. At a distance of 1.4 μm from the plasma membrane the [Ca^{2+}] arising from the opening of the channel is only ~5 nM above resting values (assuming a buffer power of 100). It is unlikely that this microdomain of Ca^{2+} will be significantly altered by mitochondria. Clearly, the local rise in [Ca^{2+}], and potential contribution of mitochondria, will increase if more than one channel opens or if mitochondria position is different as occurs in some other cell types.

9.4 MITOCHONDRIAL POSITION AND SIGNALING VIA VOLTAGE-DEPENDENT Ca^{2+} CHANNELS

Mitochondria situated immediately at sites of Ca^{2+} influx or release into the cytoplasm may form part of subcellular Ca^{2+}-signaling microdomains and have a privileged access to Ca^{2+} from these sites [38,42,44,62,95]. While, in smooth muscle, Ca^{2+} influx via voltage-dependent Ca^{2+} channels appears not to be tightly controlled at a local level by mitochondria (Figure 9.2) [44,45], nonetheless, the sites of mitochondria near the plasma membrane are associated with *larger* Ca^{2+} signals than elsewhere in the plasma membrane (Figure 9.6). Thus, a single voltage-clamped cerebral resistance artery smooth muscle cell, activated with a depolarizing pulse (−70 to 0 mV) responded with voltage-dependent Ca^{2+} entry and a rise in cytoplasmic Ca^{2+} concentration (Figure 9.6). In this same cell, mitochondrial position was determined (Figure 9.6) using the $\Delta\Psi$m-sensitive dye TMRE (see Section 9.3). Interestingly, the amplitude of the local rise in cytoplasmic Ca^{2+} concentration was largest at these sites close to mitochondria (Figure 9.6).

After the depolarization ended, at precisely the same sites, there was a substantial undershoot in the cytoplasmic Ca^{2+} concentration when compared to regions away from mitochondria (Figure 9.6). These results suggest that mitochondria may regulate (increase) voltage-dependent Ca^{2+} entry and aid the removal of Ca^{2+} from the cytoplasm positioned close to the plasma membrane.

The upstroke of the voltage-dependent Ca^{2+} transient is unaltered when mitochondria are prevented from taking up Ca^{2+} (Figure 9.2), which suggests that mitochondria are not regulating channel gating. The question arises as to how do mitochondria regulate voltage-dependent Ca^{2+} entry? One possibility is that mitochondria are not involved in the short-term regulation of Ca^{2+} entry but rather in the longer-term level of either expression or distribution of the voltage-dependent Ca^{2+} channel. IP_3-mediated Ca^{2+} release has been shown to induce mitochondrial-ROS generation,

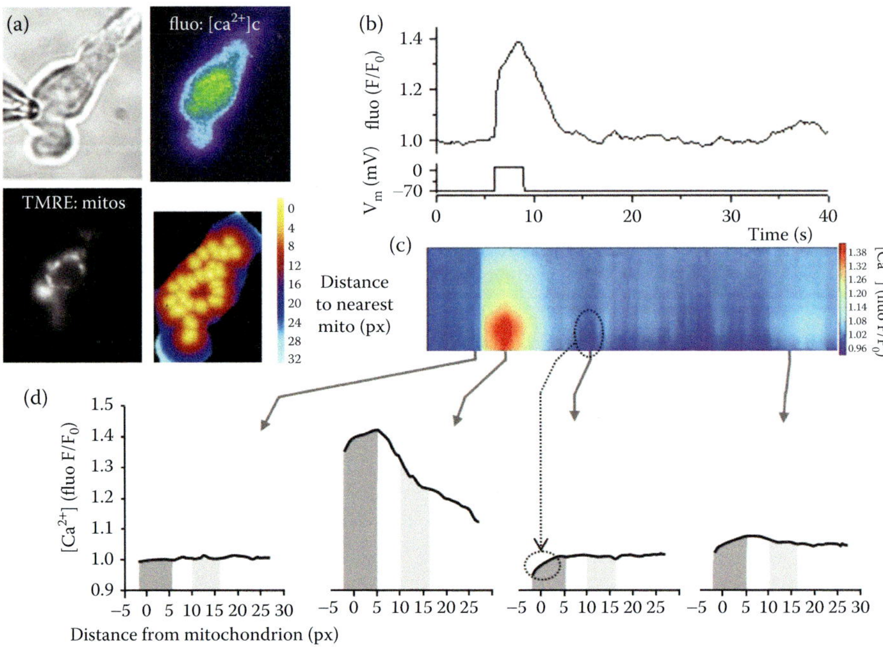

FIGURE 9.6 Mitochondrial position and the magnitude of [Ca^{2+}] changes arising from voltage-dependent Ca^{2+} entry. Depolarization (−70 to 0 mV; b), of a single voltage-clamped smooth muscle cell (a—upper left panel) activated a voltage-dependent Ca^{2+} current (not shown) and rise in [Ca^{2+}] (a—upper right panel, b, and c). The rise in [Ca^{2+}] declined after the depolarization ended. Mitochondrial position was determined at the same time by labeling the organelles with TMRE (a—lower left panel). The magnitude of the rise in [Ca^{2+}] throughout the cell (b and c) was compared on a pixel-by-pixel basis to the center of mass of each nearest mitochondrion (a—lower left and right panels). The rise in [Ca^{2+}] was largest close to a mitochondrion (c and d). In (c) the Ca^{2+} signal in each pixel is plotted as a function of time (x-axis; same scale as b) and distance to the nearest mitochondrion (left y-axis). The magnitude of the [Ca^{2+}] is colored—red high, blue low. After the depolarization had ended, (b) there was an undershoot in [Ca^{2+}] at those regions closest to a mitochondrion (c; shown as darker blue regions and indicated by the dotted lines). (d) Plots the pixel [Ca^{2+}] as a function of distance from the nearest mitochondrion (0 on the x axis is the mitochondrial position) at 4 time points indicated by the arrows from (c). The shaded regions highlight the differences in Ca^{2+} signals close (darker gray) and further (lighter gray) from a mitochondrion. Those regions that are closest to mitochondria experience the highest [Ca^{2+}] concentration after a depolarization and the greatest undershoot (indicated by the dotted lines) as [Ca^{2+}] returns to resting values.

activate the transcription factor NF-kappaB, and stimulate voltage-dependent Ca^{2+} channel transcription [96]. It is tempting to speculate that mitochondria may therefore control voltage-dependent Ca^{2+} channel expression following persistent entry via voltage-dependent Ca^{2+} channels. It is also tempting to speculate that the regional changes in Ca^{2+} (Figure 9.1) arising from voltage-dependent Ca^{2+} channel entry arise from mitochondrial control of the distribution of the channel. The localized decreases in Ca^{2+} near mitochondria also offer support to the proposed control mechanisms operating to regulate and maintain IP_3-mediated Ca^{2+} release (Figure 9.4).

ACKNOWLEDGMENTS

This work was funded by the Wellcome Trust (092292/Z/10/Z; 202924/Z/16/Z) and the British Heart Foundation (PG/16/54/32230; PG16/82/32439), whose support is gratefully acknowledged. CW is supported by a Sir Henry Wellcome Postdoctoral Research Fellowship (204682/Z/16/Z).

REFERENCES

1. Chalmers, S, Olson, ML, MacMillan, D, Rainbow, RD, McCarron, JG. Ion channels in smooth muscle: Regulation by the sarcoplasmic reticulum and mitochondria. *Cell Calcium* 42: 447–466, 2007.
2. McCarron, JG, Chalmers, S, Bradley, KN, Macmillan, D, Muir, TC. Ca^{2+} microdomains in smooth muscle. *Cell Calcium* 40: 461–493, 2006.
3. Sanders, KM. Invited review: Mechanisms of calcium handling in smooth muscles. *J Appl Physiol* 91: 1438–1449, 2001.
4. Macmillan, D, McCarron, JG. The phospholipase C inhibitor U-73122 inhibits Ca(2+) release from the intracellular sarcoplasmic reticulum Ca(2+) store by inhibiting Ca(2+) pumps in smooth muscle. *Br J Pharmacol* 160: 1295–1301, 2010.
5. McCarron, JG, Olson, ML. A single luminally continuous sarcoplasmic reticulum with apparently separate Ca^{2+} stores in smooth muscle. *J Biol Chem* 283: 7206–7218, 2008.
6. Rainbow, RD, Macmillan, D, McCarron, JG. The sarcoplasmic reticulum Ca^{2+} store arrangement in vascular smooth muscle. *Cell Calcium* 46: 313–322, 2009.
7. Berridge, MJ. Inositol trisphosphate and calcium signalling mechanisms. *Biochim Biophys Acta* 1793: 933–940, 2009.
8. Lemmon, MA, Schlessinger, J. Cell signaling by receptor tyrosine kinases. *Cell* 141: 1117–1134, 2010.
9. Burdyga, T, Wray, S. Action potential refractory period in ureter smooth muscle is set by Ca sparks and BK channels. *Nature* 436: 559–562, 2005.
10. McCarron, JG, Craig, JW, Bradley, KN, Muir, TC. Agonist-induced phasic and tonic responses in smooth muscle are mediated by $InsP_3$. *J Cell Sci* 115: 2207–2218, 2002.
11. Nelson, MT, Cheng, H, Rubart, M, Santana, LF, Bonev, AD, Knot, HJ, Lederer, WJ. Relaxation of arterial smooth muscle by calcium sparks. *Science* 270: 633–637, 1995.
12. ZhuGe, R, Tuft, RA, Fogarty, KE, Bellve, K, Fay, FS, Walsh, JV, Jr. The influence of sarcoplasmic reticulum Ca^{2+} concentration on Ca^{2+} sparks and spontaneous transient outward currents in single smooth muscle cells. *J Gen Physiol* 113: 215–228, 1999.
13. Aharon, S, Bercovier, M, Parnas, H. Parallel computation enables precise description of Ca^{2+} distribution in nerve terminals. *Bull Math Biol* 58: 1075–1097, 1996.
14. Llinas, R, Sugimori, M, Silver, RB. Microdomains of high calcium concentration in a presynaptic terminal. *Science* 256: 677–679, 1992.
15. Marsault, R, Murgia, M, Pozzan, T, Rizzuto, R. Domains of high Ca^{2+} beneath the plasma membrane of living A7r5 cells. *Embo J* 16: 1575–1581, 1997.
16. Naraghi, M, Neher, E. Linearized buffered Ca^{2+} diffusion in microdomains and its implications for calculation of $[Ca^{2+}]$ at the mouth of a calcium channel. *J Neurosci* 17: 6961–6973, 1997.
17. Schneggenburger, R, Neher, E. Intracellular calcium dependence of transmitter release rates at a fast central synapse. *Nature* 406: 889–893, 2000.
18. Kamishima, T, McCarron, JG. Ca^{2+} removal mechanisms in rat cerebral resistance size arteries. *Biophys J* 75: 1767–1773, 1998.

19. McGeown, JG, McCarron, JG, Drummond, RM, Fay, FS. Calcium-calmodulin-dependent mechanisms accelerate calcium decay in gastric myocytes from Bufo marinus. *J Physiol* 506 (Pt 1): 95–107, 1998.
20. Bao, R, Lifshitz, LM, Tuft, RA, Bellve, K, Fogarty, KE, ZhuGe, R. A close association of RyRs with highly dense clusters of Ca^{2+}-activated Cl^- channels underlies the activation of STICs by Ca^{2+} sparks in mouse airway smooth muscle. *J Gen Physiol* 132: 145–160, 2008.
21. Kargacin, GJ. Responses of Ca^{2+}-binding proteins to localized, transient changes in intracellular $[Ca^{2+}]$. *J Theor Biol* 221: 245–258, 2003.
22. Macrez, N, Mironneau, J. Local Ca^{2+} signals in cellular signalling. *Curr Mol Med* 4: 263–275, 2004.
23. Zhuge, R, Bao, R, Fogarty, KE, Lifshitz, LM. Ca^{2+} sparks act as potent regulators of excitation-contraction coupling in airway smooth muscle. *J Biol Chem* 285: 2203–2210, 2009.
24. Bai, Y, Edelmann, M, Sanderson, MJ. The contribution of inositol 1,4,5-trisphosphate and ryanodine receptors to agonist-induced Ca^{2+} signaling of airway smooth muscle cells. *Am J Physiol Lung Cell Mol Physiol* 297: L347–L361, 2009.
25. Balemba, OB, Heppner, TJ, Bonev, AD, Nelson, MT, Mawe, GM. Calcium waves in intact guinea pig gallbladder smooth muscle cells. *Am J Physiol Gastrointest Liver Physiol* 291: G717–G727, 2006.
26. Boittin, FX, Macrez, N, Halet, G, Mironneau, J. Norepinephrine-induced Ca^{2+} waves depend on $InsP_3$ and ryanodine receptor activation in vascular myocytes. *Am J Physiol* 277: C139–C151, 1999.
27. Gordienko, DV, Harhun, MI, Kustov, MV, Pucovsky, V, Bolton, TB. Sub-plasmalemmal $[Ca^{2+}]_i$ upstroke in myocytes of the guinea-pig small intestine evoked by muscarinic stimulation: IP_3R-mediated Ca^{2+} release induced by voltage-gated Ca^{2+} entry. *Cell Calcium* 43: 122–141, 2008.
28. Jaggar, JH, Nelson, MT. Differential regulation of Ca^{2+} sparks and Ca^{2+} waves by UTP in rat cerebral artery smooth muscle cells. *Am J Physiol Cell Physiol* 279: C1528–C1539, 2000.
29. McCarron, JG, Chalmers, S, MacMillan, D, Olson, ML. Agonist-evoked Ca^{2+} wave progression requires Ca^{2+} and IP_3. *J Cell Physiol* 244: 334–344, 2010.
30. McCarron, JG, MacMillan, D, Bradley, KN, Chalmers, S, Muir, TC. Origin and mechanisms of Ca^{2+} waves in smooth muscle as revealed by localized photolysis of caged inositol 1,4,5-trisphosphate. *J Biol Chem* 279: 8417–8427, 2004.
31. Bai, Y, Sanderson, MJ. Airway smooth muscle relaxation results from a reduction in the frequency of Ca^{2+} oscillations induced by a cAMP-mediated inhibition of the IP_3 receptor. *Respir Res* 7: 34, 2006.
32. Berridge, MJ, Lipp, P, Bootman, MD. The versatility and universality of calcium signalling. *Nat Rev Mol Cell Biol* 1: 11–21, 2000.
33. Bootman, MD, Berridge, MJ. The elemental principles of calcium signaling. *Cell* 83: 675–678, 1995.
34. Marchant, JS, Parker, I. Role of elementary Ca^{2+} puffs in generating repetitive Ca^{2+} oscillations. *Embo J* 20: 65–76, 2001.
35. McCarron, JG, Olson, ML, Currie, S, Wright, AJ, Anderson, KI, Girkin, JM. Elevations of intracellular calcium reflect normal voltage-dependent behavior, and not constitutive activity, of voltage-dependent calcium channels in gastrointestinal and vascular smooth muscle. *J Gen Physiol* 133: 439–457, 2009.
36. Parekh, AB. Store-operated Ca^{2+} entry: Dynamic interplay between endoplasmic reticulum, mitochondria and plasma membrane. *J Physiol* 547: 333–348, 2003.
37. Rizzuto, R, Brini, M, Murgia, M, Pozzan, T. Microdomains with high Ca^{2+} close to IP_3-sensitive channels that are sensed by neighboring mitochondria. *Science* 262: 744–747, 1993.

38. Arnaudeau, S, Kelley, WL, Walsh, JV, Jr., Demaurex, N. Mitochondria recycle Ca^{2+} to the endoplasmic reticulum and prevent the depletion of neighboring endoplasmic reticulum regions. *J Biol Chem* 276: 29430–29439, 2001.
39. Drummond, RM, Fay, FS. Mitochondria contribute to Ca^{2+} removal in smooth muscle cells. *Pflugers Arch* 431: 473–482, 1996.
40. Drummond, RM, Mix, TC, Tuft, RA, Walsh, JV, Jr., Fay, FS. Mitochondrial Ca^{2+} homeostasis during Ca^{2+} influx and Ca^{2+} release in gastric myocytes from Bufo marinus. *J Physiol* 522: 375–390, 2000.
41. Drummond, RM, Tuft, RA. Release of Ca^{2+} from the sarcoplasmic reticulum increases mitochondrial [Ca^{2+}] in rat pulmonary artery smooth muscle cells. *J Physiol* 516 (Pt 1): 139–147, 1999.
42. Olson, ML, Chalmers, S, McCarron, JG. Mitochondrial Ca^{2+} uptake increases Ca^{2+} release from inositol 1,4,5-trisphosphate receptor clusters in smooth muscle cells. *J Biol Chem* 285: 2040–2050, 2010.
43. McCarron, JG, Olson, ML, Chalmers, S. Mitochondrial regulation of cytosolic Ca^{2+} signals in smooth muscle. *Pflugers Arch* 464: 51–62, 2012.
44. McCarron, JG, Muir, TC. Mitochondrial regulation of the cytosolic Ca^{2+} concentration and the $InsP_3$-sensitive Ca^{2+} store in guinea-pig colonic smooth muscle. *J Physiol* 516: 149–161, 1999.
45. Chalmers, S, McCarron, JG. The mitochondrial membrane potential and Ca^{2+} oscillations in smooth muscle. *J Cell Sci* 121: 75–85, 2008.
46. Kamishima, T, Davies, NW, Standen, NB. Mechanisms that regulate $[Ca^{2+}]_i$ following depolarization in rat systemic arterial smooth muscle cells. *J Physiol* 522: 285–295, 2000.
47. Kamishima, T, McCarron, JG. Regulation of the cytosolic Ca^{2+} concentration by Ca^{2+} stores in single smooth muscle cells from rat cerebral arteries. *J Physiol* 501: 497–508, 1997.
48. Chalmers, S, Saunter, C, Wilson, C, Coats, P, Girkin, JM, McCarron, JG. Mitochondrial motility and vascular smooth muscle proliferation. *Arterioscler Thromb Vasc Biol* 32: 3000–3011, 2012.
49. Chalmers, S, Saunter, CD, Girkin, JM, McCarron, JG. Age decreases mitochondrial motility and increases mitochondrial size in vascular smooth muscle. *J Physiol* 594: 4283–4295, 2016.
50. Chalmers, S, Saunter, CD, Girkin, JM, McCarron, JG. Flicker-assisted localization microscopy reveals altered mitochondrial architecture in hypertension. *Nat Sci Rep* 30: 2000–2013, 2015.
51. McCarron, J, Wilson, C, Sandison, ME, Olson, ML, Girkin, JM, Saunter, C, Chalmers, S. From structure to function: Mitochondrial morphology, motion and shaping in vascular smooth muscle. *J Vasc Res* 5: 357–371, 2013.
52. Lawrie, AM, Rizzuto, R, Pozzan, T, Simpson, AW. A role for calcium influx in the regulation of mitochondrial calcium in endothelial cells. *J Biol Chem* 271: 10753, 1996.
53. Hoth, M, Button, DC, Lewis, RS. Mitochondrial control of calcium-channel gating: A mechanism for sustained signaling and transcriptional activation in T lymphocytes. *Proc Natl Acad Sci USA* 97: 10607–10612, 2000.
54. Hoth, M, Fanger, CM, Lewis, RS. Mitochondrial regulation of store-operated calcium signaling in T lymphocytes. *J Cell Biol* 137: 633–648, 1997.
55. Varadi, A, Cirulli, V, Rutter, GA. Mitochondrial localization as a determinant of capacitative Ca^{2+} entry in HeLa cells. *Cell Calcium* 36: 499, 2004.
56. Barstow, KL, Locknar, SA, Merriam, LA, Parsons, RL. The modulation of action potential generation by calcium-induced calcium release is enhanced by mitochondrial inhibitors in mudpuppy parasympathetic neurons. *Neuroscience* 124: 327, 2004.
57. Hernandez-Guijo, JM, Maneu-Flores, VE, Ruiz-Nuno, A, Villarroya, M, Garcia, AG, Gandia, L. Calcium-dependent inhibition of L, N, and P/Q Ca^{2+} channels in chromaffin cells: Role of mitochondria. *J Neurosci* 21: 2553, 2001.

58. Sanchez, JA, Garcia, MC, Sharma, VK, Young, KC, Matlib, MA, Sheu, SS. Mitochondria regulate inactivation of L-type Ca^{2+} channels in rat heart. *J Physiol* 536: 387–396, 2001.
59. Greenwood, IA, Helliwell, RM, Large, WA. Modulation of Ca^{2+}-activated Cl^- currents in rabbit portal vein smooth muscle by an inhibitor of mitochondrial Ca^{2+} uptake. *J Physiol* 505: 53–64, 1997.
60. Cheranov, SY, Jaggar, JH. Mitochondrial modulation of Ca^{2+} sparks and transient KCa currents in smooth muscle cells of rat cerebral arteries. *J Physiol* 556: 755–771, 2004.
61. Ochi, R, Dhagia, V, Lakhkar, A, Patel, D, Wolin, MS, Gupte, SA. Rotenone-stimulated superoxide release from mitochondrial complex I acutely augments L-type Ca2+ current in A7r5 aortic smooth muscle cells. *Am J Physiol Heart Circ Physiol* 310: H1118–H1128, 2016.
62. Rizzuto, R, Pinton, P, Carrington, W, Fay, FS, Fogarty, KE, Lifshitz, LM, Tuft, RA, Pozzan, T. Close contacts with the endoplasmic reticulum as determinants of mitochondrial Ca^{2+} responses. *Science* 280: 1763–1766, 1998.
63. Dai, J, Kuo, KH, Leo, JM, van Breemen, C, Lee, CH. Rearrangement of the close contact between the mitochondria and the sarcoplasmic reticulum in airway smooth muscle. *Cell Calcium* 37: 333–340, 2005.
64. Nixon, GF, Mignery, GA, Somlyo, AV. Immunogold localization of inositol 1,4,5-trisphosphate receptors and characterization of ultrastructural features of the sarcoplasmic reticulum in phasic and tonic smooth muscle. *J Muscle Res Cell Motil* 15: 682–700, 1994.
65. Csordas, G, Renken, C, Varnai, P, Walter, L, Weaver, D, Buttle, KF, Balla, T, Mannella, CA, Hajnoczky, G. Structural and functional features and significance of the physical linkage between ER and mitochondria. *J Cell Biol* 174: 915–921, 2006.
66. Boncompagni, S, Rossi, AE, Micaroni, M, Beznoussenko, GV, Polishchuk, RS, Dirksen, RT, Protasi, F. Mitochondria are linked to calcium stores in striated muscle by developmentally regulated tethering structures. *Mol Biol Cell* 20: 1058–1067, 2009.
67. Szabadkai, G, Bianchi, K, Varnai, P, De Stefani, D, Wieckowski, MR, Cavagna, D, Nagy, AI, Balla, T, Rizzuto, R. Chaperone-mediated coupling of endoplasmic reticulum and mitochondrial Ca^{2+} channels. *J Cell Biol* 175: 901–911, 2006.
68. Ma, JH, Shen, S, Wang, JJ, He, Z, Poon, A, Li J, Qu, J, Zhang, SX. Comparative proteomic analysis of the mitochondria-associated ER membrane (MAM) in a long-term type 2 diabetic rodent model. *Sci Rep* 7: 2062, 2017.
69. de Brito, OM, Scorrano, L. Mitofusin 2 tethers endoplasmic reticulum to mitochondria. *Nature* 456: 605–610, 2008.
70. Simmen, T, Aslan, JE, Blagoveshchenskaya, AD, Thomas, L, Wan, L, Xiang, Y, Feliciangeli, SF, Hung, CH, Crump, CM, Thomas, G. PACS-2 controls endoplasmic reticulum-mitochondria communication and Bid-mediated apoptosis. *Embo J* 24: 717–729, 2005.
71. Hayashi, T, Su, TP. Sigma-1 receptor chaperones at the ER-mitochondrion interface regulate Ca^{2+} signaling and cell survival. *Cell* 131: 596–610, 2007.
72. Bezprozvanny, I, Watras, J, Ehrlich, BE. Bell-shaped calcium-response curves of Ins(1,4,5)P3- and calcium-gated channels from endoplasmic reticulum of cerebellum. *Nature* 351: 751–754, 1991.
73. Iino, M. Biphasic Ca^{2+} dependence of inositol 1,4,5-trisphosphate-induced Ca^{2+} release in smooth muscle cells of the guinea pig taenia caeci. *J Gen Physiol* 95: 1103–1122, 1990.
74. Iino, M, Tsukioka, M. Feedback control of inositol trisphosphate signalling by calcium. *Mol Cell Endocrinol* 98: 141–146, 1994.

75. Oancea, E, Meyer, T. Reversible desensitization of inositol trisphosphate-induced calcium release provides a mechanism for repetitive calcium spikes. *J Biol Chem* 271: 17253–17260, 1996.
76. De Giorgi, F, Lartigue, L, Ichas, F. Electrical coupling and plasticity of the mitochondrial network. *Cell Calcium* 28: 365–370, 2000.
77. Guillery, O, Malka, F, Frachon, P, Milea, D, Rojo, M, Lombes, A. Modulation of mitochondrial morphology by bioenergetics defects in primary human fibroblasts. *Neuromuscul Disord* 18: 319–330, 2008.
78. Mitra, K, Wunder, C, Roysam, B, Lin, G, Lippincott-Schwartz, J. A hyperfused mitochondrial state achieved at G1-S regulates cyclin E buildup and entry into S phase. *Proc Natl Acad Sci USA* 106: 11960–11965, 2009.
79. Mannella, CA. Structural diversity of mitochondria: Functional implications. *Ann N Y Acad Sci* 1147: 171–179, 2008.
80. Frey, TG, Mannella, CA. The internal structure of mitochondria. *Trends Biochem Sci* 25: 319–324, 2000.
81. Betzig, E, Patterson, GH, Sougrat, R, Lindwasser, OW, Olenych, S, Bonifacino, JS, Davidson, MW, Lippincott-Schwartz, J, Hess, HF. Imaging intracellular fluorescent proteins at nanometer resolution. *Science* 313: 1642–1645, 2006.
82. Burnette, DT, Sengupta, P, Dai, Y, Lippincott-Schwartz, J, Kachar, B. Bleaching/blinking assisted localization microscopy for superresolution imaging using standard fluorescent molecules. *Proc Natl Acad Sci USA* 108: 21081–21086, 2011.
83. Cox, S, Rosten, E, Monypenny, J, Jovanovic-Talisman, T, Burnette, DT, Lippincott-Schwartz, J, Jones, GE, Heintzmann, R. Bayesian localization microscopy reveals nanoscale podosome dynamics. *Nat Methods* 9: 195–200, 2012.
84. Egner, A, Jakobs, S, Hell, SW. Fast 100-nm resolution three-dimensional microscope reveals structural plasticity of mitochondria in live yeast. *Proc Natl Acad Sci USA* 99: 3370–3375, 2002.
85. Hirvonen, LM, Wicker, K, Mandula, O, Heintzmann, R. Structured illumination microscopy of a living cell. *Eur Biophys J* 38: 807–812, 2009.
86. Huang, B, Jones, SA, Brandenburg, B, Zhuang, X. Whole-cell 3D STORM reveals interactions between cellular structures with nanometer-scale resolution. *Nat Methods* 5: 1047–1052, 2008.
87. Jans, DC, Wurm, CA, Riedel, D, Wenzel, D, Stagge, F, Deckers, M, Rehling, P, Jakobs, S. STED super-resolution microscopy reveals an array of MINOS clusters along human mitochondria. *Proc Natl Acad Sci USA* 110: 8936–8941, 2013.
88. Rosenbloom, AB, Lee, SH, To, M, Lee, A, Shin, JY, Bustamante, C. Optimized two-color super resolution imaging of Drp1 during mitochondrial fission with a slow-switching Dronpa variant. *Proc Natl Acad Sci USA* 111: 13093–13098, 2014.
89. Shao, L, Kner, P, Rego, EH, Gustafsson, MG. Super-resolution 3D microscopy of live whole cells using structured illumination. *Nat Methods* 8: 1044–1046, 2011.
90. Giorgio, V, von Stockum, S, Antoniel, M, Fabbro, A, Fogolari, F, Forte, M, Glick GD et al. Dimers of mitochondrial ATP synthase form the permeability transition pore. *Proc Natl Acad Sci USA* 110: 5887–5892, 2013.
91. Halestrap, AP, Richardson, AP. The mitochondrial permeability transition: A current perspective on its identity and role in ischaemia/reperfusion injury. *J Mol Cell Cardiol* 78: 129–141, 2015.
92. O'Reilly, CM, Fogarty, KE, Drummond, RM, Tuft, RA, Walsh, JV, Jr. Quantitative analysis of spontaneous mitochondrial depolarizations. *Biophys J* 85: 3350–3357, 2003.

93. O'Reilly, CM, Fogarty, KE, Drummond, RM, Tuft, RA, Walsh, JV, Jr. Spontaneous mitochondrial depolarizations are independent of SR Ca^{2+} release. *Am J Physiol Cell Physiol* 286: C1139–C1151, 2004.
94. Duchen, MR, Leyssens, A, Crompton, M. Transient mitochondrial depolarizations reflect focal sarcoplasmic reticular calcium release in single rat cardiomyocytes. *J Cell Biol* 142: 975–988, 1998.
95. Hajnoczky, G, Csordas, G, Madesh, M, Pacher, P. The machinery of local Ca^{2+} signalling between sarco-endoplasmic reticulum and mitochondria. *J Physiol* 529 Pt 1: 69, 2000.
96. Narayanan, D, Xi, Q, Pfeffer, LM, Jaggar, JH. Mitochondria control functional CaV1.2 expression in smooth muscle cells of cerebral arteries. *Circ Res* 107: 631–641, 2010.

10 ORAI Channels in Vascular Smooth Muscle

Maxime Guéguinou, Xuexin Zhang, Trayambak Pathak, Scott Emrich, Ryan Yoast, and Mohamed Trebak

CONTENTS

10.1 INTRODUCTION

Calcium (Ca^{2+}) is a highly versatile second messenger that controls a plethora of biological process, including cell secretion, proliferation, growth, apoptosis, immune function, neurotransmitter transmission, and muscle contraction.[1] In order to control the aforementioned physiological responses, cells have developed an elaborate molecular toolkit to carefully regulate Ca^{2+} concentrations within the cytosol and intracellular organelles.[2] Cells can directly control internal Ca^{2+} concentrations by means of Ca^{2+} permeable channels, Ca^{2+} pumps, and Ca^{2+} exchangers. Through changes in plasma membrane potential mediated by the activity of channels conducting other ions such as potassium (K^+), sodium (Na^+), and chloride (Cl^-), cells can indirectly regulate intracellular Ca^{2+} concentrations.[3–5] Ca^{2+} buffering proteins can also substantially contribute to the regulation of Ca^{2+} concentrations within the cell. For instance, calsequestrin is a major Ca^{2+} buffer in the cisternae of the sarcoplasmic reticulum (SR) of skeletal muscles, which regulate muscle contraction.[6]

Vascular smooth muscle cells (VSMCs) are one of the major cell types of the blood vessel wall that play a crucial role in maintaining blood pressure and vascular tone.[3–5] In these fully differentiated quiescent VSMCs, changes in cytosolic Ca^{2+} arise from two distinct sources: release from internal stores, especially the SR *via* the Ryanodine Receptor (RyR) and entry from the extracellular milieu through plasma membrane voltage-gated Ca^{2+} channel of the high-voltage-activated type (L-type).[7] These channels are major contributors to the global rise in cytoplasmic Ca^{2+} concentration, which leads to VSMC contraction. In certain pathological conditions, including atherosclerosis, hypertension, and restenosis, quiescent contractile VSMCs undergo a phenotypic switch into a proliferative and motile phenotype commonly referred to as *synthetic*.[8] This phenotypic modulation is characterized by dramatic changes in expression and activation of many signaling pathways, including Ca^{2+} channels, pumps, transporters, Ca^{2+}-dependent enzymes, and Ca^{2+}-handling proteins.[3] For instance, synthetic VSMCs downregulate L-type Ca^{2+} channels while upregulating receptor-activated Ca^{2+} channels, such as members of the canonical transient receptor potential channels (TRPC) and ORAI channels.[9]

Using optical and electrophysiological methods, ORAI channel activity is essentially undetected in quiescent VSMCs but becomes prominent in synthetic VSMCs.[9] To date, ORAI channels in synthetic VSMCs were shown to mediate two different Ca^{2+} influx pathways: the ubiquitous store-operated Ca^{2+} entry (SOCE) and a less characterized store-independent Ca^{2+} entry pathway.[10] In this chapter, we will briefly overview how to study the activities of these ORAI channels in synthetic VSMCs.

10.2 SOCE IN VASCULAR SMOOTH MUSCLE CELLS

The original concept that Ca^{2+} entry across the plasma membrane is linked to the state of filling of the internal reserves of the endoplasmic reticulum (ER) was proposed over three decades ago by Putney in his capacitative Ca^{2+} entry model.[11] Agonists acting on G protein-coupled receptors (GPCRs) or receptor tyrosine kinases (RTKs) lead to the activation of phospholipase C isoforms PLCβ and PLCγ, respectively.[12,13] PLC catalyzes the hydrolysis of phosphatidylinositol 4,5-bisphopshate (PIP_2) into diacylglycerol (DAG) and inositol-1,4,5-trisphosphate (IP_3).[14] The latter binds to IP_3 receptors (IP_3R) on the membrane of the ER, resulting in release of Ca^{2+} from ER stores into the cytosol and subsequent store depletion.[14] Depletion of ER stores is sensed by resident ER membrane proteins stromal interaction molecules 1 and 2 (STIM1 and STIM2), which contain N-terminal low-affinity EF hands located in the lumen of the ER. Store depletion triggers the loss of Ca^{2+} binding to the EF hands and results in multimerization and trafficking of STIM1/2 to highly specialized nanodomains where the ER is in close proximity of the plasma membrane.[15–19] This physical interaction is necessary for gating ORAI1 channels through conformational changes, causing activation of Ca^{2+} entry into the cytosol, which is then subsequently pumped into the ER through the action of the sarcoplasmic/endoplasmic reticulum Ca^{2+} -ATPase (SERCA) pump.[13,20–23] ORAI1 is the major SOCE channel. ORAI1 has two mammalian gene homologs ORAI2 and ORAI3, with ORAI3 found exclusively in mammals.[9,10,24,25]

In quiescent VSMCs, the expression of ORAI and STIM proteins is either very low or absent and pharmacological store depletion (using the SERCA pump inhibitor thapsigargin) failed to activate SOCE.[26] In contrast, ORAI1 and STIM1 are upregulated in synthetic VSMCs, which are found in diseases of vessel remodeling. These synthetic VSMCs show robust SOCE in response to thapsigargin or PLC-coupled receptor agonists such as the platelet-derived growth factor (PDGF).[26,27] STIM2, ORAI2, and ORAI3 proteins were also upregulated in synthetic VSMCs.[26–29] Preventing the upregulation of either STIM1 or ORAI1 in a rat model of carotid artery restenosis using shRNA-encoding lentiviral transduction reduced CRAC currents and inhibited VSMC proliferation, migration and neointima hyperplasia.[30] These results demonstrated a pivotal role for ORAI channels in VSMC remodeling during vascular injury.

10.3 STORE-INDEPENDENT Ca^{2+} ENTRY IN VASCULAR SMOOTH MUSCLE CELLS

Shuttleworth and colleagues identified a conductance in HEK293 cells activated by a relatively low concentration of exogenous arachidonic acid (AA) (around 8 μM), or by low concentrations of a muscarinic agonist (0.2 ~ 1 μM Carbachol) and named it Arachidonate-Regulated Ca^{2+} (ARC) channel.[31–33] The same group showed that ARC channels are heteromultimers of ORAI1 and ORAI3 proteins, are regulated by STIM1, and suggested that the N-terminus of ORAI3 is required for the sensitivity of the channel to AA.[32,34] Subsequent studies identified a conductance in primary aortic VSMC named Leukotriene-C_4-Regulated Ca^{2+} (LRC) channel.[28] LRC channels can be activated by exogenous AA or intracellular Leukotriene-C_4 (LTC_4, an AA metabolite), which is independent of the action of LTC_4 or its downstream metabolites on cysteinyl leukotriene receptors.[28,35,36] The enzymatic activity of LTC_4 synthase is required to convert the AA to LTC_4.[35] Dialysis of LTC_4 through the patch pipette in whole-cell patch recordings activated a Ca^{2+} conductance in synthetic VSMCs independent of store depletion.[28,29,35] Activation of protease-activated receptor 1 (PAR1) by the protease thrombin in VSMC activated LRC currents through AA generation and metabolism into LTC_4 by the sequential actions of PLC, DAG lipase, 5-lipooxygenase, and LTC_4 synthase.[35] Importantly, RNA interference-mediated protein knockdown showed that LRC currents depended on expression of ORAI1, ORAI3, and STIM1 proteins.[28] A detailed comparative study in HEK293 cells and synthetic VSMCs showed that ARC and LRC currents represent the same channel encoded by the same cellular populations of ORAI1, ORAI3, and STIM1.[36] Therefore, for the sake of simplicity we will use the original ARC nomenclature when discussing this conductance. ARC activity in HEK293 and VSMC could be rescued by STIM1 constructs that do not traffic to the plasma membrane, suggesting that internal ER-resident STIM1 is necessary and sufficient for ARC channel activation.[36] Additionally, a specific and constitutive interaction of the second C-terminal coiled-coil of STIM1 with ORAI3 C-terminal region is necessary for ARC channel activation by LTC_4.[35] *In vivo* restenosis studies in rats similar to those discussed earlier showed that ORAI3 is upregulated in medial and neointimal VSMCs after vascular injury.[28,29] Furthermore, *in vivo* knockdown of ORAI3 after vascular injury in rat carotids inhibited ARC currents activated by intracellular LTC_4 and reduced neointimal hyperplasia.[28] *In vivo* knockdown of LTC_4 synthase dramatically inhibited neointimal hyperplasia in injured rat carotid arteries subjected to balloon-mediated injury.[29] These *in vitro* and *in vivo* studies highlight the importance of upregulation of ARC currents in VSMC pathophysiology.

10.4 STIM/ORAI EXPRESSION AND PHARMACOLOGY

The crystal structure of drosophila ORAI has been solved at 3.35 angstrom resolution and has revealed that the channel is organized into a hexameric structure. The central pore is composed of six glutamate residues (E180 in drosophila; E106 in human) representing the channel's selectivity filter.[37] Atomic resolution structures of mammalian ORAI isoforms are currently undefined, but based on the drosophila structure and studies using concatenated oligomers of ORAI1, functional mammalian ORAI channels are likely hexamers.[38,39] Based on messenger RNA, STIM, and ORAI isoforms are widely expressed in most, if not all, mammalian tissues and cell types. While specific antibodies against STIM1 and STIM2 are available, the lack of reliable antibodies against ORAI isoforms has hampered the characterization of their protein expression in different tissues; this is especially true for ORAI2 where no specific antibodies are currently available. Based on Western blots and assessment of SOCE function, STIM1 and ORAI1 proteins are widely expressed and functional in all non-excitable cells, with notable prominence in hematopoietic cells such as lymphocytes and mast cells.[20] STIM1 and ORAI1 are also functional in skeletal muscle where SOCE mediated by these proteins plays a critical role in preventing skeletal muscle fatigue during prolonged periods of activity.[40,41] Healthy control cardiomyocytes and VSMCs do not appear to express significant levels of ORAI1 and STIM1 proteins and have negligible SOCE, although STIM1 protein expression in these cell types is more pronounced compared to that of ORAI1.[26,30,42] However, STIM1 and ORAI1 expression and SOCE are markedly increased in remodeled cardiomyocytes and VSMCs during animal models of cardiac hypertrophy and vascular injury, respectively.[26,30,42,43] A similar upregulation has also been described for ORAI3 expression and the ARC pathway in animal models of cardiac hypertrophy and vascular injury, as described earlier.[28,29,43] The low basal expression of STIM1 in healthy cardiomyocytes and VSMCs is intriguing and likely plays an ORAI-independent role in these cell types. STIM1 is increasingly implicated in regulating additional channels and pumps, including members of the broader TRP family of channels.[44]

To date, there are no selective pharmacological inhibitors of ORAI1 channels, much less antagonists that can distinguish between different ORAI isoforms. Furthermore, mammalian ORAI isoforms are likely to form heterohexamers yielding channels with distinct pharmacology and electrophysiology that would enhance the diversity of ORAI-mediated Ca^{2+} signaling. Two novel compounds termed GSK-7975A and GSK-5503A have been shown to potently inhibit ORAI1 as well as ORAI3 channels with an IC_{50} of 4 μM without affecting STIM1–STIM1 oligomerization or STIM1–ORAI1 interactions, suggesting that they might act on an allosteric site within ORAI isoform selectivity filter{Derler, 2013 #74}. Fairly selective inhibitors of ORAI channel isoforms are the trivalent lanthanides (Gd^{3+} or La^{3+}) at relatively low concentrations ranging from 1–5 μM.[26,45] Early studies pre-dating the discovery of STIM and ORAI proteins showed that SOCE in HEK293 cells is inhibited by Gd^{3+} with an IC50 of ~200 nM.[46] Several different pharmacological compounds have been used to characterize SOCE and ORAI channel function in both fluorescence Ca^{2+} imaging and patch clamp electrophysiology. While these inhibitors (when used at relatively low concentrations) can be quite useful in determining ORAI channel activation in these assays, these protocols should be complemented by the use of molecular tools aimed at either reducing or abrogating ORAI isoform expression with either RNA silencing or CRISPR/Cas9-mediated gene knockout, respectively. Data obtained from the use of these inhibitors in cell biological assays such as those assessing cell migration, proliferation, chemotaxis, or apoptosis should be interpreted with extreme caution and the use of these inhibitors *in vivo* in mice is, in our opinion, not warranted. These non-specific pharmacological inhibitors include SKF96365 and the pyrazole compound BTP2 both shown to fully inhibit SOCE when used at 10 μM.[47,48] The myosin light chain kinase (MLCK) inhibitor ML-9 also fully abrogates SOCE when used at a concentration of 50 μM.[49] Another inhibitor called 2-Aminoethyldiphenyl borate (2-APB), which was first described as an inhibitor of the IP_3 receptor when used at 100 μM,[50] can inhibit native ORAI1-mediated SOCE at lower concentrations (30 μM) but have a potentiating effect on SOCE at 5–10 μM.[51] This potentiating effect of 2-APB is due to strengthening the aggregation of STIM proteins and interaction with ORAI1, while the inhibitory effect of 30–50 μM of 2-APB and 50 μM of ML-9 are due to disruption of STIM1 aggregation and puncta formation by these compounds.[49] Additionally, the inhibition of SOCE by high concentrations of 2-APB most likely occurs through blocking the pore of the ORAI1 channel directly.[51] Interestingly, although 2-APB is clearly a non-specific drug targeting many distinct channel families, it is a useful tool to distinguish between different ORAI isoforms in imaging or electrophysiological assays. In ectopic expression systems where ORAI isoforms are individually co-expressed with STIM1 in HEK293 cells, the bimodal effect of 2-APB on ORAI1 potentiation and inhibition contrasted with enhanced activity of ORAI3 by 2-APB (even at higher concentrations; e.g., 50 μM) and partial inhibition of ORAI2 channels without potentiating effects.[51]

10.5 STIM/ORAI ELECTROPHYSIOLOGY

The SOCE current mediated by ORAI1 channels is termed the Ca^{2+} release-activated Ca^{2+} (CRAC) current.[52] CRAC currents can be elicited through dialysis of a high concentration of a Ca^{2+} chelator (10–20 mM of either EGTA or BAPTA) through the patch pipette.[52] Addition of either IP_3 (10 μM) or adenophostin A (2 μM) in the patch pipette to activate the IP_3 receptor accelerates CRAC current activation.[53] Under conditions where Ca^{2+} in the pipette is buffered to physiological cytosolic Ca^{2+} concentrations (100–150 nM) using 10 mM BAPTA, CRAC currents can be promoted through passive store depletion by addition of 2 μM thapsigargin or 500 nM Ionomycin to the bath solution. Upon store depletion, native CRAC currents in synthetic VSMCs are small (less than 0.5 pA/pF), slow developing inward rectifying, and highly Ca^{2+} selective currents that show a positive reversal potential around ~+50 mV.[26] One useful protocol aimed at enhancing the size of CRAC currents and the confidence in recordings is to use divalent-free bath (DVF) solutions.[48,54] Under DVF solutions, CRAC currents are carried by Na^+ ions generating substantially bigger currents that could be easily detected. Furthermore, CRAC currents recorded in DVF bath solutions show a peculiar characteristic in that they quickly inactivate; this inactivation of DVF currents is termed *depotentiation*, which seems to result from reduction in the number of active channels rather than a change in their open probability.[48,55]

CRAC channels are tightly regulated through complex mechanisms involving fast and slow Ca^{2+}-mediated negative feedback, a process called Ca^{2+}-dependent inactivation (CDI), whereby Ca^{2+} entering through the mouth of CRAC channels regulates its activity.[56,57] Fast and slow CDI likely occur through two distinct mechanisms; fast CDI is observed within 100 milliseconds while slow inactivation develops over tens of seconds. Typically, fast and slow CDI are studied under patch clamp protocols making use of 10 mM of the slow chelator EGTA (as opposed to the fast chelator BAPTA). As is the case of SOCE recorded by optical methods (described in the following), CRAC current measurements are blocked by all the inhibitors listed earlier at the recommended concentrations and by protein knockdown of either STIM1 or ORAI1. CRAC single-channel recordings are not readily accessible since CRAC currents have a very small single-channel conductance estimated through noise analysis (around 700 fS for Na^+ currents).[55]

ARC currents are store-independent currents and therefore recordings should be performed with pipette solutions where Ca^{2+} is buffered to a physiological 100–150 nM with either 10 mM of EGTA or BAPTA.[36] ARC currents can be elicited through either the inclusion of LTC_4 (100 nM) in the patch pipette or addition of 5–8 μM of AA to the bath solution after establishing whole-cell configuration and baseline current.[36] ARC currents can also be amplified in DVF solutions; however, these currents do not show depotentiation.[36] Quite interestingly, ARC currents are neither potentiated nor inhibited by 50 μM 2-APB, providing means for distinction between ARC and CRAC currents.[28,36]

10.6 FLUORESCENCE MEASUREMENTS OF Ca^{2+} ENTRY THROUGH ORAI CHANNELS

The most common Ca^{2+} indicator used to measure global cytosolic Ca^{2+} signals generated by the SOCE pathway is the indicator Fura2, originally developed by Tsien and collaborators.[58] Fura2 is a powerful Ca^{2+} indicator that allows for ratiometric measurements of Ca^{2+} signals that accounts for variability of dye loading (yielding different dye concentrations within the cytosol of different cells) and for differences in cell size. Fura2 has two excitation wavelengths characteristic of its Ca^{2+} unbound (380 nm) and Ca^{2+} bound (340 nm) forms. Fura2 fluorescence emission light is typically measured at around 510 nm. When cytosolic Ca^{2+} concentration increases, emitted fluorescence upon excitation at 340 nm increases while fluorescence upon excitation at 380 nm decreases. Typically, the ratio of emitted fluorescence ($R_{F340/F380}$) is proportional to cytosolic Ca^{2+} concentrations. A widespread Fura2 protocol that is used to specifically measure Ca^{2+} through SOCE after pharmacological (e.g., thapsigargin) or receptor-mediated (e.g., PDGF; 10 ng/ml in VSMCs) store depletion calls for the separation of the

Ca^{2+} release portion of the signal from the Ca^{2+} entry phase.[26] To achieve this, at the beginning of the recordings Ca^{2+}-containing bath solutions are exchanged with nominally Ca^{2+}-free bath solutions and the store depleting stimulus. This will reveal the typical Ca^{2+} transient due to Ca^{2+} release from the ER followed by return of the $R_{F340/F380}$ signal to basal levels. Nominally Ca^{2+}-free bath solutions with or without a small concentration of EGTA (~50 μM) are ideal; inclusion of high concentrations of EGTA can accelerate store depletion and generate inconsistent results. A subsequent exchange of bath solutions with stimulus- and Ca^{2+}-containing 2mM solutions will reveal a sustained plateau of the $R_{F340/F380}$ signal that corresponds to Ca^{2+} entry across the plasma membrane through SOCE. Subsequent addition of a pharmacological blocker would inhibit SOCE and return the $R_{F340/F380}$ signal back to baseline (see traces A-C in Fig. 1.[26]). For each set of experiments, we recommend performing leak controls where cells are subjected to the same protocol but without inclusion of stimulus. In some instances, investigators are interested in performing experiments to study cytosolic oscillatory Ca^{2+} signals through SOCE. These Ca^{2+} oscillations are typically observed when cells are challenged with submaximal concentrations of agonists (e.g., 5 μM of the muscarinic agonist charbacol in HEK293 cells). In this case, the use of the ratiometric dye Fura5F (a variant of Fura2) would generate more reliable Ca^{2+} oscillations due to Fura5F higher dissociation constant (400 nM for Fura5F vs. 140 nM for Fura2) and a faster Ca^{2+} off rate.

Recently, the development of genetically encoded Ca^{2+} indicators (GECI) to measure Ca^{2+} signals has made possible measurements of local Ca^{2+} signals in cells and specific tissues of whole animals by targeting of these indicators to different organelles or to different regions of the cytosol and plasma membrane.[59] One such Ca^{2+} indicator is GCaMP.[60] GCaMP consists of circularly permuted green fluorescent protein, the Ca^{2+}-binding protein calmodulin (CaM), and a CaM-interacting M13 peptide sequence from MLCK.[60] GCaMP proteins have been subsequently modified to progressively improve the range of the fluorescence signal, resulting in versions ranging from GCaMP3 through GCaMP9. When Ca^{2+} binds CaM it undergoes a conformational change, and its hinge region is able to bind the M13 peptide. In the absence of Ca^{2+}, the circularly permutated fluorescent proteins exist in a poorly fluorescent state due to a water pathway that enables protonation of the chromophore and poor absorbance at the excitation wavelengths. Ca^{2+} binding to CaM results in a structural shift that causes bright fluorescence. The use of a GCaMP6 version targeted to the plasma membrane by a CAAX motif[61] allows for measurements of Ca^{2+} entry through SOCE within the cytoplasmic nanodomain adjacent to the membrane.[54] Recently, Cahalan and colleagues fused genetically encoded GCaMP6 indicator to ORAI1 to specifically measure locally generated ORAI1-associated Ca^{2+} influx. This allowed for the optical recording of ORAI1 single-channel function, thus far not feasible with electrophysiological recordings.[62]

The activation of Ca^{2+} entry through ORAI channels requires physical interactions between STIM and ORAI proteins.[20] To study these interactions within living cells, fluorescence resonance energy transfer (FRET) and bioluminescence resonance energy transfer (BRET) techniques have proved to be of great value. FRET is a distance-dependent physical process by which energy is transferred from an excited donor molecular fluorophore (CFP or GFP) to another acceptor fluorophore (YFP or RFP, respectively). FRET provides measurements of molecular proximity (10 ~ 100 Å) and can be highly efficient if the donor and acceptor are positioned within the Förster radius, the distance at which 50% of the excitation energy of the donor is transferred to the acceptor (generally between 3 ~ 6 nm).[63] To examine the association and interaction between STIM and ORAI, the C-termini of STIM and ORAI, both of which are cytosolic and required for STIM/ORAI interactions, should be fluorescently tagged. The advantage of this method is that the investigator can follow spatiotemporal interactions between STIM and ORAI after stimulation with different agonists and their reversal after agonist washout or application of an inhibitor. The drawback of this live cell FRET system is that it can only be performed using overexpressed proteins. FRET imaging microscopy suffers from various issues such as auto-fluorescence, photo-bleaching, and spectral bleed through with contributions of donor and acceptor fluorescence emission to the FRET channel. BRET involves resonance energy transfer between a bioluminescent donor and a fluorescent accepter. Therefore, BRET avoids some of the problems inherent to FRET, such as photo bleaching and auto fluorescence.[64] BRET is more sensitive than FRET and the use of an enzyme to excite the luminescent donor precludes the need for an external excitation source.[65] One disadvantage of BRET is that the signal-to-noise ratio is affected by the overlap of the emission spectra of the Luciferase substrate and YFP. Similar to FRET, live cell BRET is limited to overexpressed proteins.

For studying interactions between endogenous ORAI and STIM proteins, the use of proximity ligation assays would allow for detection and quantification of protein–protein interactions in either fixed cells or tissue samples.[66,67] One drawback, however, is that it is not possible to study live cells or follow the kinetics of protein-protein interactions over time. Furthermore, success of proximity ligation assays relies on the existence of reliable primary antibodies against target proteins and the accessibility of the epitope in fixed cells or tissues.

Typically, primary target-specific antibodies are applied to fixed samples. It is necessary to select two primary antibodies produced in two different host species so that it can be used together with the generic Duolink species-specific secondary antibodies. The secondary antibodies contain unique DNA strands that work as a template for added oligonucleotides. If the two secondary antibodies are in close proximity (<40 nm), the oligonucleotides will be ligated by a ligase to form a circular template.[66,67] This template, still anchored to the antibody, is subsequently amplified and detected using complementary labeled oligonucleotide probes. Detection is performed either with a fluorescent label for fluorescence microscopy or with horseradish peroxidase for bright field detection.[67]

10.7 CONCLUSION

In the past decade since the discovery of STIM and ORAI proteins as critical players in receptor-regulated Ca^{2+} entry pathways, significant advances have been made regarding the physiological and pathophysiological roles of these proteins.[68] Valuable knowledge was obtained from the generation and study of tissue-specific knockout mice for individual STIM and ORAI isoforms. While STIM1 and ORAI1 knockout mice have been relatively well characterized, knowledge inferred from STIM2, ORAI2, or ORAI3 knockout mice is either scarce or absent. ORAI1-deficient patients were diagnosed several decades ago (before the discovery of STIM and ORAI proteins) for lacking SOCE in their T lymphocytes.[69] Patients deficient in functional STIM1 or ORAI1 proteins lack SOCE and CRAC currents and display a collection of symptoms, most notably a severe combined immunodeficiency (SCID) syndrome associated with autoimmunity, in addition to muscle hypotonia and ectodermal dysplasia.[70] These patients have relatively short life spans unless they receive a bone marrow transplant to correct their immune dysfunction.[71] Remarkably, global knockout mice for ORAI1 and STIM1 phenocopy human patients and unless these mice are generated on a mixed background, they die either before birth or shortly thereafter. Tissue-specific knockout mice for STIM1 and ORAI1 (and also STIM1/STIM2 double knockout) have established roles for these proteins in a number of physiological functions. Such roles include immune cell development, differentiation and activation, skeletal muscle development, and resistance to fatigue and fluid secretion by exocrine glands, most notably secretion of saliva, sweat, tears, milk, and pancreatic enzymes.[68] In the cardiovascular system, arteries from smooth muscle-specific STIM1 knockout mice showed a significant reduction in contraction in response to phenylephrine but contraction in response to the thromboxane analog U46619 while depolarization by KCl was unaffected.[72] Whether this reflects an ORAI-independent role of STIM1 in vascular contractility remains unexplored. As mentioned earlier, increased protein expression of STIM1, ORAI1, and ORAI3 and corresponding CRAC and ARC currents are crucial for development of cardiac hypertrophy, vascular remodeling, and hypertension, and preventing the upregulation of these molecules has a protective effects in these disease models.[28,30,42,43] Specifically, smooth muscle-specific STIM1 knockout mice subjected to the angiotensin II model of hypertension showed inhibited hypertension, cardiac hypertrophy and perivascular fibrosis, and endothelial dysfunction compared with littermate control mice.[73] The upregulation of STIM1 in smooth muscle and hearts of wild type mice treated with angiotensin II is necessary for increased ER stress and induces cardiovascular dysfunction; smooth muscle-specific STIM1 knockout mice show inhibited ER stress after angiotensin II treatment.[73]

In summary, we have described herein commonly used experimental approaches to study ORAI channel function and dynamics. ORAI channels have emerged as important contributors to smooth muscle remodeling under different disease conditions.[9,25,74] While the contribution of ORAI channels to VSMC contractility appears to be marginal, the role of STIM1, ORAI1, and ORAI3 in smooth muscle remodeling is well established.[9,25] The bulk of the published literature has focused heavily on ORAI1 at the expense of ORAI2 and ORAI3 and future studies are needed to better understand the contributions of these proteins to vascular function. ORAI isoforms are expected to form heteromultimeric channels with distinct modes of regulation, tissue expression and physiological function. Future structural studies will likely reveal the different oligomeric states of native channels formed by ORAI isoforms. The use of new state-of-the art methods such as CRISPR/Cas9 gene knockout/editing, cryo-electron microcopy and super-resolution optical microscopy will likely provide important information of the structure, function and regulation of ORAI channels. Ultimately, this will contribute to a better understanding of their role in smooth muscle physiology and pathophysiology.

ACKNOWLEDGMENTS

Research in this laboratory is supported by grants R01HL123364, R01HL097111, and R21AG050072 from the National Institutes of Health and grant NPRP8-110-3-021 from the Qatar National Research Fund (QNRF) to MT.

REFERENCES

1. Berridge, MJ. Calcium signalling in health and disease. *Biochem Biophys Res Commun*. 2017;485:5.
2. Clapham, DE. Calcium signaling. *Cell*. 2007;131:1047–1058.
3. House, SJ, Potier, M, Bisaillon, J, Singer, HA, and Trebak, M. The non-excitable smooth muscle: Calcium signaling and phenotypic switching during vascular disease. *Pflugers Arch*. 2008;456:769–785.
4. Longden, TA, Hill-Eubanks, DC, and Nelson, MT. Ion channel networks in the control of cerebral blood flow. *J Cereb Blood Flow Metab*. 2016;36:492–512.
5. Earley, S, and Brayden, JE. Transient receptor potential channels in the vasculature. *Physiol Rev*. 2015;95:645–690.
6. Dulhunty, AF, Wei-LaPierre, L, Casarotto, MG, and Beard, NA. Core skeletal muscle ryanodine receptor calcium release complex. *Clin Exp Pharmacol Physiol*. 2017;44:3–12.
7. Catterall, WA. Voltage-gated calcium channels. *Cold Spring Harb Perspect Biol*. 2011;3:a003947.
8. Owens, GK. Molecular control of vascular smooth muscle cell differentiation and phenotypic plasticity. *Novartis Found Symp*. 2007;283:174–191; discussion 191–193, 238–241.
9. Trebak, M. STIM/Orai signalling complexes in vascular smooth muscle. *J Physiol*. 2012;590:4201–4208.
10. Trebak, M, and Putney, JW, Jr. ORAI Calcium Channels. *Physiology* (*Bethesda*). 2017;32:332–342.
11. Putney, JW, Jr. A model for receptor-regulated calcium entry. *Cell Calcium*. 1986;7:1–12.
12. Trebak, M, Lemonnier, L, Smyth, JT, Vazquez, G, and Putney, JW, Jr. Phospholipase C-coupled receptors and activation of TRPC channels. *Handb Exp Pharmacol*. 2007:593–614.
13. Putney, JW, and Tomita, T. Phospholipase C signaling and calcium influx. *Adv Biol Regul*. 2012;52:152–164.
14. Berridge, MJ. The Inositol Trisphosphate/Calcium Signaling Pathway in Health and Disease. *Physiol Rev*. 2016;96:1261–1296.
15. Feske, S, Gwack, Y, Prakriya, M, Srikanth, S, Puppel, SH, Tanasa, B, Hogan, PG, Lewis, RS, Daly, M, and Rao, A. A mutation in Orai1 causes immune deficiency by abrogating CRAC channel function. *Nature*. 2006;441:179–185.
16. Liou, J, Kim, ML, Heo, WD, Jones, JT, Myers, JW, Ferrell, JE, Jr., and Meyer, T. STIM is a Ca^{2+} sensor essential for Ca^{2+}-store-depletion-triggered Ca^{2+} influx. *Curr Biol*. 2005;15:1235–1241.
17. Roos, J, DiGregorio, PJ, Yeromin, AV, Ohlsen, K, Lioudyno, M, Zhang, S, Safrina, O et al. STIM1, an essential and conserved component of store-operated Ca^{2+} channel function. *J Cell Biol*. 2005;169:435–445.

18. Vig, M, Peinelt, C, Beck, A, Koomoa, DL, Rabah, D, Koblan-Huberson, M, Kraft, S et al. CRACM1 is a plasma membrane protein essential for store-operated Ca^{2+} entry. *Science*. 2006;312:1220–1223.
19. Zhang, SL, Yeromin, AV, Zhang, XH, Yu, Y, Safrina, O, Penna, A, Roos, J, Stauderman, KA, and Cahalan, MD. Genome-wide RNAi screen of Ca(2+) influx identifies genes that regulate Ca(2+) release-activated Ca(2+) channel activity. *Proc Natl Acad Sci USA*. 2006;103:9357–9362.
20. Prakriya, M, and Lewis, RS. Store-Operated Calcium Channels. *Physiol Rev*. 2015;95:1383–1436.
21. Potier, M, and Trebak, M. New developments in the signaling mechanisms of the store-operated calcium entry pathway. *Pflugers Arch*. 2008;457:405–415.
22. Schindl, R, Muik, M, Fahrner, M, Derler, I, Fritsch, R, Bergsmann, J, and Romanin, C. Recent progress on STIM1 domains controlling Orai activation. *Cell Calcium*. 2009;46:227–232.
23. Frischauf, I, Fahrner, M, Jardin, I, and Romanin, C. The STIM1: Orai Interaction. *Adv Exp Med Biol*. 2016;898:25–46.
24. Motiani, RK, Stolwijk, JA, Newton, RL, Zhang, X, and Trebak, M. Emerging roles of Orai3 in pathophysiology. *Channels (Austin)*. 2013;7:392–401.
25. Spinelli, AM, and Trebak, M. Orai channel-mediated Ca^{2+} signals in vascular and airway smooth muscle. *Am J Physiol Cell Physiol*. 2016;310:C402–C413.
26. Potier, M, Gonzalez, JC, Motiani, RK, Abdullaev, IF, Bisaillon, JM, Singer, HA, and Trebak, M. Evidence for STIM1- and Orai1-dependent store-operated calcium influx through ICRAC in vascular smooth muscle cells: Role in proliferation and migration. *FASEB J*. 2009;23:2425–2437.
27. Berra-Romani, R, Mazzocco-Spezzia, A, Pulina, MV, and Golovina, VA. Ca^{2+} handling is altered when arterial myocytes progress from a contractile to a proliferative phenotype in culture. *Am J Physiol Cell Physiol*. 2008;295:C779–C790.
28. Gonzalez-Cobos, JC, Zhang, X, Zhang, W, Ruhle, B, Motiani, RK, Schindl, R, Muik, M et al. Store-independent Orai1/3 channels activated by intracrine leukotriene C4: Role in neointimal hyperplasia. *Circ Res*. 2013;112:1013–1025.
29. Zhang, W, Zhang, X, Gonzalez-Cobos, JC, Stolwijk, JA, Matrougui, K, and Trebak, M. Leukotriene-C4 synthase, a critical enzyme in the activation of store-independent Orai1/Orai3 channels, is required for neointimal hyperplasia. *J Biol Chem*. 2015;290:5015–5027.
30. Zhang, W, Halligan, KE, Zhang, X, Bisaillon, JM, Gonzalez-Cobos, JC, Motiani, RK, Hu, G et al. Orai1-mediated I (CRAC) is essential for neointima formation after vascular injury. *Circ Res*. 2011;109:534–542.
31. Mignen, O, and Shuttleworth, TJ. I(ARC), a novel arachidonate-regulated, noncapacitative Ca(2+) entry channel. *J Biol Chem*. 2000;275:9114–9119.
32. Mignen, O, Thompson, JL, and Shuttleworth, TJ. Both Orai1 and Orai3 are essential components of the arachidonate-regulated Ca^{2+}-selective (ARC) channels. *J Physiol*. 2008;586:185–195.
33. Thompson, JL, Mignen, O, and Shuttleworth, TJ. The ARC channel—An endogenous store-independent Orai channel. *Curr Top Membr*. 2013;71:125–148.
34. Thompson, J, Mignen, O, and Shuttleworth, TJ. The N-terminal domain of Orai3 determines selectivity for activation of the store-independent ARC channel by arachidonic acid. *Channels (Austin)*. 2010;4:398–410.
35. Zhang, X, Gonzalez-Cobos, JC, Schindl, R, Muik, M, Ruhle, B, Motiani, RK, Bisaillon, JM et al. Mechanisms of STIM1 activation of store-independent leukotriene C4-regulated Ca^{2+} channels. *Mol Cell Biol*. 2013;33:3715–3723.
36. Zhang, X, Zhang, W, Gonzalez-Cobos, JC, Jardin, I, Romanin, C, Matrougui, K, and Trebak, M. Complex role of STIM1 in the activation of store-independent Orai1/3 channels. *J Gen Physiol*. 2014;143:345–359.

37. Hou, X, Pedi, L, Diver, MM, and Long, SB. Crystal structure of the calcium release-activated calcium channel Orai. *Science*. 2012;338:1308–1313.
38. Cai, X, Zhou, Y, Nwokonko, RM, Loktionova, NA, Wang, X, Xin, P, Trebak, M, Wang, Y, and Gill, DL. The Orai1 store-operated calcium channel functions as a hexamer. *J Biol Chem*. 2016;291:25764–25775.
39. Yen, M, Lokteva, LA, and Lewis, RS. Functional analysis of Orai1 concatemers supports a hexameric stoichiometry for the CRAC channel. *Biophys J*. 2016;111:1897–1907.
40. Trebak, M, Zhang, W, Ruhle, B, Henkel, MM, Gonzalez-Cobos, JC, Motiani, RK, Stolwijk, JA, Newton, RL, and Zhang, X. What role for store-operated Ca(2)(+) entry in muscle? *Microcirculation*. 2013;20:330–336.
41. Wei-Lapierre, L, Carrell, EM, Boncompagni, S, Protasi, F, and Dirksen, RT. Orai1-dependent calcium entry promotes skeletal muscle growth and limits fatigue. *Nat Commun*. 2013;4:2805.
42. Hulot, JS, Fauconnier, J, Ramanujam, D, Chaanine, A, Aubart, F, Sassi, Y, Merkle, S et al. Critical role for stromal interaction molecule 1 in cardiac hypertrophy. *Circulation*. 2011;124:796–805.
43. Saliba, Y, Keck, M, Marchand, A, Atassi, F, Ouille, A, Cazorla, O, Trebak, M et al. Emergence of Orai3 activity during cardiac hypertrophy. *Cardiovasc Res*. 2015;105:248–259.
44. Choi, S, Maleth, J, Jha, A, Lee, KP, Kim, MS, So, I, Ahuja, M, and Muallem, S. The TRPCs-STIM1-Orai interaction. *Handb Exp Pharmacol*. 2014;223:1035–1054.
45. Abdullaev, IF, Bisaillon, JM, Potier, M, Gonzalez, JC, Motiani, RK, and Trebak, M. Stim1 and Orai1 mediate CRAC currents and store-operated calcium entry important for endothelial cell proliferation. *Circ Res*. 2008;103:1289–1299.
46. Trebak, M, Bird, GS, McKay, RR, and Putney, JW, Jr. Comparison of human TRPC3 channels in receptor-activated and store-operated modes. Differential sensitivity to channel blockers suggests fundamental differences in channel composition. *J Biol Chem*. 2002;277:21617–21623.
47. Stolwijk, JA, Zhang, X, Gueguinou, M, Zhang, W, Matrougui, K, Renken, C, and Trebak, M. Calcium signaling is dispensable for receptor regulation of endothelial barrier function. *J Biol Chem*. 2016;291:22894–22912.
48. Prakriya, M, and Lewis, RS. Separation and characterization of currents through store-operated CRAC channels and Mg^{2+}-inhibited cation (MIC) channels. *J Gen Physiol*. 2002;119:487–507.
49. Smyth, JT, Dehaven, WI, Bird, GS, and Putney, JW, Jr. Ca^{2+}-store-dependent and -independent reversal of Stim1 localization and function. *J Cell Sci*. 2008;121:762–772.
50. Ma, HT, Patterson, RL, van Rossum, DB, Birnbaumer, L, Mikoshiba, K, and Gill, DL. Requirement of the inositol trisphosphate receptor for activation of store-operated Ca2+ channels. *Science*. 2000;287:1647–1651.
51. DeHaven, WI, Smyth, JT, Boyles, RR, Bird, GS, and Putney, JW, Jr. Complex actions of 2-aminoethyldiphenyl borate on store-operated calcium entry. *J Biol Chem*. 2008;283:19265–19273.
52. Hoth, M, and Penner, R. Depletion of intracellular calcium stores activates a calcium current in mast cells. *Nature*. 1992;355:353–356.
53. Broad, LM, Armstrong, DL, and Putney, JW, Jr. Role of the inositol 1,4,5-trisphosphate receptor in Ca(2+) feedback inhibition of calcium release-activated calcium current (I(crac)). *J Biol Chem*. 1999;274:32881–32888.
54. Ben-Kasus Nissim, T, Zhang, X, Elazar, A, Roy, S, Stolwijk, JA, Zhou, Y, Motiani, RK et al. Mitochondria control store-operated Ca^{2+} entry through Na^{+} and redox signals. *EMBO J*. 2017;36:797–815.
55. Prakriya, M, and Lewis, RS. Regulation of CRAC channel activity by recruitment of silent channels to a high open-probability gating mode. *J Gen Physiol*. 2006;128:373–86.

56. Zweifach, A, and Lewis, RS. Slow calcium-dependent inactivation of depletion-activated calcium current. Store-dependent and -independent mechanisms. *J Biol Chem.* 1995;270:14445–14451.
57. Zweifach, A, and Lewis, RS. Rapid inactivation of depletion-activated calcium current (ICRAC) due to local calcium feedback. *J Gen Physiol.* 1995;105:209–226.
58. Grynkiewicz, G, Poenie, M, and Tsien, RY. A new generation of Ca^{2+} indicators with greatly improved fluorescence properties. *J Biol Chem.* 1985;260:3440–3450.
59. Kotlikoff, MI. Genetically encoded Ca^{2+} indicators: Using genetics and molecular design to understand complex physiology. *J Physiol.* 2007;578:55–67.
60. Nakai, J, and Ohkura, M. Probing calcium ions with biosensors. *Biotechnol Genet Eng Rev.* 2003;20:3–21.
61. Tsai, FC, Seki, A, Yang, HW, Hayer, A, Carrasco, S, Malmersjo, S, and Meyer, T. A polarized Ca^{2+}, diacylglycerol and STIM1 signalling system regulates directed cell migration. *Nat Cell Biol.* 2014;16:133–144.
62. Dynes, JL, Amcheslavsky, A, and Cahalan, MD. Genetically targeted single-channel optical recording reveals multiple Orai1 gating states and oscillations in calcium influx. *Proc Natl Acad Sci USA.* 2016;113:440–445.
63. Sun, Y, Rombola, C, Jyothikumar, V, and Periasamy, A. Forster resonance energy transfer microscopy and spectroscopy for localizing protein-protein interactions in living cells. *Cytometry A.* 2013;83:780–793.
64. Xie, Q, Soutto, M, Xu, X, Zhang, Y, and Johnson, CH. Bioluminescence resonance energy transfer (BRET) imaging in plant seedlings and mammalian cells. *Methods Mol Biol.* 2011;680:3–28.
65. Xu, Y, Piston, DW, and Johnson, CH. A bioluminescence resonance energy transfer (BRET) system: Application to interacting circadian clock proteins. *Proc Natl Acad Sci USA.* 1999;96:151–156.
66. Fredriksson, S, Gullberg, M, Jarvius, J, Olsson, C, Pietras, K, Gustafsdottir, SM, Ostman, A, and Landegren, U. Protein detection using proximity-dependent DNA ligation assays. *Nat Biotechnol.* 2002;20:473–477.
67. Soderberg, O, Gullberg, M, Jarvius, M, Ridderstrale, K, Leuchowius, KJ, Jarvius, J, Wester, K et al. Direct observation of individual endogenous protein complexes in situ by proximity ligation. *Nat Methods.* 2006;3:995–1000.
68. Putney, JW, Steinckwich-Besancon, N, Numaga-Tomita, T, Davis, FM, Desai, PN, D'Agostin, DM, Wu, S, and Bird, GS. The functions of store-operated calcium channels. *Biochim Biophys Acta.* 2017;1864:900–906.
69. Partiseti, M, Le Deist, F, Hivroz, C, Fischer, A, Korn, H, and Choquet, D. The calcium current activated by T cell receptor and store depletion in human lymphocytes is absent in a primary immunodeficiency. *J Biol Chem.* 1994;269:32327–32335.
70. Lacruz, RS, and Feske, S. Diseases caused by mutations in ORAI1 and STIM1. *Ann N Y Acad Sci.* 2015;1356:45–79.
71. Feske, S. CRAC channelopathies. *Pflugers Arch.* 2010;460:417–435.
72. Kassan, M, Zhang, W, Aissa, KA, Stolwijk, J, Trebak, M, and Matrougui, K. Differential role for stromal interacting molecule 1 in the regulation of vascular function. *Pflugers Arch.* 2015;467:1195–1202.
73. Kassan, M, Ait-Aissa, K, Radwan, E, Mali, V, Haddox, S, Gabani, M, Zhang, W et al. Essential role of smooth muscle STIM1 in hypertension and cardiovascular dysfunction. *Arterioscler Thromb Vasc Biol.* 2016;36:1900–1909.
74. Spinelli, AM, Gonzalez-Cobos, JC, Zhang, X, Motiani, RK, Rowan, S, Zhang, W, Garrett, J et al. Airway smooth muscle STIM1 and Orai1 are upregulated in asthmatic mice and mediate PDGF-activated SOCE, CRAC currents, proliferation, and migration. *Pflugers Arch.* 2012;464:481–492.

11 G Protein-Coupled Receptors in Airway Smooth Muscle Function and Obstructive Lung Disease

Tonio Pera and Raymond B. Penn

CONTENTS

11.1 INTRODUCTION

11.1.1 Why G Protein-Coupled Receptors Are Important in Airway Smooth Muscle Biology and Airway Disease

G protein-coupled receptors (GPCRs) are the largest protein family in the human genome. They translate a myriad of diverse extracellular stimuli into intracellular signals that regulate cell functions. Thus, they are arguably the most important regulators of physiology, and accordingly they are the targets of most prescribed drugs.

GPCRs are prevalent in all cell types, including airway smooth muscle (ASM). ASM represents an interesting cell type to the GPCR biologist because of the numerous different GPCRs they express, many being of physiological consequence. Almost 15 years ago we published a comprehensive list of GPCRs known to be expressed in ASM [1]; that list has since grown (Table 11.1) and its length clearly demonstrates ASM to be a cell capable of surveying and responding to a complex environment. We have posited that the control of the most important function of ASM—contraction—reflects a balance of GPCR signaling, most frequently a competition between pro-contractile Gq-coupled receptor signaling and pro-relaxant Gs-coupled receptor signaling [2,3]. Moreover, ASM growth, which contributes to pathological airway remodeling in obstructive lung disease (OLD), is similarly increased by Gq-coupled receptor signaling and inhibited by Gs-coupled receptor signaling. A third important but less-studied ASM function—its ability to regulate the immunological response during allergic lung inflammation—is also regulated by GPCRs in a poorly understood manner. Thus, the challenge to the airway biologist is to identify the various GPCRs, expressed in ASM, determine their capacity to influence ASM functions, and identify the various means by which these receptors and their signaling pathways are activated, inhibited, and regulated.

11.1.2 Current Pressing Questions Regarding G Protein-Coupled Receptors in Airway Smooth Muscle Biology

Whereas initial studies of ASM (dating back to the 1960s) focused largely on characterizing ASM mechanical and contractile properties, as noted earlier we now appreciate a more diverse nature of ASM function. Accordingly, the current study of GPCRs in ASM attempts to explain the role of a given GPCR in regulating a wide breadth of GPCR functions, under both normal and patho-physiological conditions.

TABLE 11.1
Expression and Function of GPCRs in ASM

GPCR	G Protein	References	Function
β2AR	Gs	[4–7]	RLXN, Cyt, GI
EP2	Gs	[8–10]	RLXN, Cyt, GI
EP4	Gs	[9,11–15]	?
A2b	Gs	[16]	RLXN
VIP	Gs	[17–19]	GI
Prostacyclin	Gs	[20,21]	GI
Dopamine D1	Gs	[22]	RLXN
OGR1	Gq, Gs	[23,24]	CXN, RLXN (?)
M3 mAChR	Gq	[25–30]	CXN, GP
H1	Gq	[31,32]	CXN, GP
BK B2	Gq	[33–38]	CXN
CLT1R	Gq	[39–43]	CXN, GP
ET-A/B	Gq	[44–49]	CXN, GP
ET-B	Gq	[50]	RLXN
EP1	Gq	[51,52]	CXN
NK2	Gq	[53,54]	CXN, GP
SP (NK1)	Gq	[55–57]	CXN, GP
alpha1	Gq	[57–60]	unclear
TP	Gq	[20,61,62]	CXN, GP
CaSR	Gq	[63]	CXN
A3 adenosine	Gq	[64]	unclear
PAR-1,2,3	Gq	[65,66]	GS, GP
P2 purinergic	Gq	[64]	unclear
FFAR1,4	Gq	[67]	CXN
AT1 (angiotensin II)	Gq	[68]	CXN
MAS1	Gq	[69]	RLXN
oxytocin receptor	Gq	[70]	CXN
M2 mAChR	Gi	[25,71]	unclear
CXCR1,2	Gi	[72]	CXN, MGRTN
5-HT 2c rat	Gi	[73]	CXN, GP
EDG 1-7	Gi	[74–76]	GS, Cyt
5-HTx	Gi	[77–79]	CXN, GP
A1 AdenosineR	Gi	[16,80,81]	CXN
EP3	Gi	[82–84]	GP
GABAb	Gi	[85]	unclear
Dopamine D2	Gi	[86]	unclear
BLT1,2	Gi	[87]	GS
FZD	G?	[88–90]	ECM prod., contractility
BTR	Gustducin	[91,92]	RLXN
olfactory receptors	Golf	[93]	GI
LPNH1,3	G?	[94]	CXN

Thus, from the perspective of the ASM biologist, the pressing questions with respect to GPCRs are:

1. Which ASM functions can be regulated by (which) GPCRs?
2. Which endogenous GPCR agonists matter, in both physiology and pathophysiology?
3. Which exogenous GPCR ligands have therapeutic value?

Within the last 15–20 years we have witnessed some profound developments in GPCR biology. These include discovery of numerous intracellular regulatory mechanisms (and their mediators) affecting GPCR signaling (desensitization and resensitization), the pleiotropic nature of arrestin proteins as regulators of desensitization, resensitization, and G protein-independent signaling, and advances in pharmacology (allosteric modulators, biased agonism) (reviewed in [95]).

Accordingly, from the perspective of the GPCR biologist, the pressing questions with respect to GPCRs in ASM are similar to those posed by the ASM biologist yet emphasize the aforementioned receptor-centric issues:

1. Which GPCRs are on ASM; how do they signal and what ASM function do these signals regulate?
2. How is ASM GPCR signaling regulated? Is this regulation of consequence in physiology, disease, or therapy?
3. How diverse is the signaling of a given ASM GPCR? How do the different signals affect functions; which are beneficial, and which might contribute to disease? Can the receptor be *tuned* to activate certain signaling and not others?
4. How can GPCR ligands most effectively cooperate in combination therapy?

Thus, the issues important to the airway and GPCR biologist overlap, and collectively define an exciting new era of discovery for both fields.

11.2 APPROACHES FOR STUDYING G PROTEIN-COUPLED RECEPTORS IN AIRWAY SMOOTH MUSCLE BIOLOGY AND OBSTRUCTIVE LUNG DISEASE THERAPY

11.2.1 A Spectrum of Model Systems Is Required

The (near-) ideal model system for studying a *cell* like ASM provides the means to capture *all the cell does* in its *relevant context*. This means providing the ability to simultaneously assess:

1. All relevant *in vivo* function, including contraction, growth, size change, secretion/absorption, metabolism
2. All signaling events
3. Gene regulation

in real time, in the human body, with the means to specifically regulate the expression/activity of any given GPCR or downstream molecule in a given cell at any time. Then, one only needs to account for the variability in subjects conferred by biologic and genetic variability, and environmental influences, by performing a sufficient number of experiments to overcome these drivers of variance, recognizing you must know what these factors are *a priori* in order to appropriately design and power your study. If one desires to understand the more integrative system relative to that cell (i.e., tissue or organ system), the ideal system enables a similar comprehensive analysis of all interactive factors (other cell types, tissues, and organ systems) and cooperation.

Once we achieve these abilities in science, we will likely understand all of existence, at which point God will likely pull the plug and end *her* experiment. Until that time, we must employ a scientific strategy that utilizes less-than-ideal models and interpret our studies the best we can.

So, what's the best we can do currently with the study of ASM? All experimental model systems represent a compromise. A fundamental truth in model systems of biological science is that *in vivo* systems enable insight into physiological relevance yet limit the ability to explore mechanism. Moreover, the complexity of the *in vivo* system renders control of variables difficult. Alternatively, the more reductionist a model system (e.g., tissue, cell-based, and cell-free systems), the greater the ability to control variables and explore mechanism exists, all at the cost of loss of physiological context/relevance. Accordingly, a *comprehensive* approach to the study of ASM biology in health and disease, employing multiple less-than-ideal model systems, represents an attractive strategy, with successful insight dependent on understanding the power and limitations of the systems employed [96].

Our laboratories and many others have embraced this strategy in the study of many GPCRs in ASM biology. A generic strategy for understanding the expression, signaling, regulation, and function of a given GPCR in ASM involves the following approaches outlined briefly in the following text and organized in a comprehensive manner in Table 11.2.

11.2.2 Cell-Based Models

Cell-based models using ASM cultures derived from either humans [55] or various animals [120] represent arguably the most useful model for understanding GPCR expression and signaling in ASM. In addition, recent methodological advances have enabled insightful study of most ASM functions, despite an obvious loss of physiological context more readily captured in tissue and *in vivo* models.

Refinement of culturing techniques for both human and non-human ASM cultures has produced a reliable system for studying ASM pharmacology, biology, physiology, and the roles of GPCRs in each of these disciplines. Among the advantages of ASM cultures are their ability to maintain many of the features of ASM cells *in vivo*, including maintenance of structural features/morphology, and the maintenance of expression

TABLE 11.2
Techniques for Studying GPCR Expression and Function in ASM Cells and Tissues

	Type	Approach	References	Strengths	Limitations
Cell-Based (ASM Cultures)					
GPCR expression	Protein	Radioligand binding (cell lysates or membranes)	[97–99]	Pharmacology of most radio- and *cold* ligands well established	Radioactivity work increasingly less supported; lack of specific ligands
	Protein	Antibody detection via flow, immunoblot, ICC	[100]	Can often discern cellular location	Often poorly quantitative
Signaling, morphology	cAMP	RIA, ELISA	[99]	Multiple approaches available, most are accurate and sensitive	Expensive
	Ca2+	Flexstation, scope-based imaging	[101]	Multiple approaches available, most are accurate and sensitive	
				Assay relatively inexpensive, increasing number of sensors available	Equipment expensive
	Intracellular proteins/ phospho-proteins	Immunoblotting: ECL film, LiCOR, Wes	[102–104]	Multiple approaches available, most are accurate and sensitive	Poor antibodies for many proteins, especially GPCRs
		ICC	[105,106]	Multiplexing (other proteins, nucleotides) capable in several systems	Certain equipment expensive
		Flow cytometry	[100]	Dual simultaneous holo- and phospho-protein detection capable	
		Protein chips, multiplex	[107,108]	Increasing ability of computational analysis with other omics	
		SILAC, ITRAK	[109]		
Function	Contraction	MTC, FTTM, cell shape, other	[91,110,111]	Highly controlled models; amenable to pharmacologic, molecular strategies	Expensive infrastructure, obvious lack of context
	Proliferation	Thymidine incorporation, AlamarBlue, CyQuant	[112–116]	Well established for multiple cell types including ASM	Several analyses only indirect measure

(Continued)

TABLE 11.2 (*Continued*)
Techniques for Studying GPCR Expression and Function in ASM Cells and Tissues

	Type	Approach	References	Strengths	Limitations
	Hypertrophy	Microscopy, ICC	[117–119]	Increasingly improving imaging capabilities	
	Migration	Boyden chamber	[120,121]		
		Scratch assay	[122]		
	Gene transcription	qPCR, microarray, Quantigene,	[123–128]	Increasingly improving accuracy, sensitivity, ease of use Some forms inexpensive	mRNA not protein Gene variation can confound interpretation Some forms expensive
Tissue-Based (PLCS, Airways, Strips)					
GPCR expression	Protein	Radioligand binding (tissue lysates or membranes)	[98]		
	Protein	Antibody detection via immunoblot or IHC	[85,98,101]		
	cAMP				
Signaling	Ca2+		[129]		Inherently difficult, expensive equipment
	Intracellular proteins/phospho-proteins	Immunoblotting of tissue lysates	[26,85,130]		
		IHC	[131]		
Function	Contraction	Organ bath large and miniaturized	[6,28,29,120,132,133]		
	Proliferation	Thymidine incorporation	[26]		
	Hypertrophy	Not explored in tissue culture			

(*Continued*)

TABLE 11.2 (*Continued*)
Techniques for Studying GPCR Expression and Function in ASM Cells and Tissues

	Type	Approach	References	Strengths	Limitations
In Vivo					
GPCR expression	Protein	Radioligand binding *in vivo*/PET	[134]		
Signaling	Phospho-protein induction	Feasible if performing analysis of harvested tissue			Doable, but difficult; requires rapid tissue excision and sample stabilization
Function	Contraction	Multiple invasive measures of airway resistance	[98,135,136]		Representative of human lung function
		penH	[137]		More reflective of upper airway/nasal resistance
	Proliferation	Analysis of cell number, proliferation markers in harvested tissue	[113]		
	Hypertrophy	Morphology analysis of cell number, proliferation markers in harvested tissue	[117,119,138]		
	Gene transcription, translation	Laser microdissection human ASM followed by RNASeq	[139]	Feasible if performing analysis of harvested tissue	Analysis of harvested tissue may fail to capture effects imposed *in vivo*, acute effects may be lost with processing

of many GPCRs and their downstream pathway signaling elements that retain the ability to regulate ASM functions [55]. Importantly, ASM cultures can take advantage of the majority of experimental approaches for studying GPCR expression, regulation, and signaling used in more artificial systems (e.g., HEK293 or COS cells) expressing either endogenous or (transfected, heterologously expressed) GPCRs. These experiment approaches include radioligand binding, most proteomic and transcriptomic approaches, and second/intracellular messenger (most notably cAMP, calcium) analyses. Recent advances in imaging technologies further allow for visualization of subcellular GPCR-mediated signaling events [140]. And although the application of molecular biology approaches (including transfection/infection to express or knockdown proteins) has lagged beyond their use in more artificial cell systems, such approaches are now routinely used in many ASM/airway biology labs, as are the use of genetically modified mice and ASM cells cultured from them [120,141–163]. Finally, the development and refinement of several new cell-based approaches for measuring ASM cell contraction [156] has provided the opportunity to closely link cellular biological/biochemical events with function, in the same (matched) system.

The limitations of ASM cell cultures as a model system are several, however. By pure virtue of being cultured, these cells are being selected for proliferative ability, and thus a switch towards the proliferative phenotype (and away from a contractile phenotype) evolves with time in culture, although this can be stalled/reverse under conditions of serum deprivation [157]. The expression of certain GPCRs is also lost as cells expand; for example, the expression of both the m3 muscarinic acetylcholine receptor and the cysteinyl leukotriene type 1 receptor wanes significantly by the third or fourth passage of human ASM cultures [98,120,157]. Moreover, the rate of cell proliferation in human ASM also tends to wane significantly by the eighth or later passage of the culture (Penn, unpublished observations). Interestingly, infection of human ASM cultures with hTERT (human telomerase reverse transcriptase) appears to significantly extend the useful lifespan of ASM cultures [112], and in some instances, enables retention or expression of (otherwise downregulated) GPCRs such as the m3 muscarinic acetylcholine receptor [98].

Additional limitations include the lack of physiological context: surrounding cells, matrix, architecture, and neurohumoral influences present in tissue/*in vivo*. Use of ASM cultures derived from animals have been useful, and those from mice take advantage of mouse genetic approaches to manipulate proteins of interest. Yet species differences exist on many levels, including species-specific expression of several GPCRs of importance in human ASM biology. For example, murine ASM does not express (or lacks any functional response to agonists of) either H1 histamine receptor [158] or CysLT1R [120], while the function of EP1, EP2, EP3, and EP4 receptor subtypes with respect to regulation of contraction vary significantly among species [11,159]. From a practical standpoint, murine ASM cultures are difficult to maintain beyond four passages (Penn and Deshpande, unpublished observations).

11.2.3 Tissue-Based Models

Tissue-based models, for decades restricted to isolated airway preparations or ASM tissue strips, have recently been complemented by precision-cut lung slices (PCLS)

[160] and provide an extremely useful system for measuring ASM contraction and its regulation. The advantage of tissue-based models is the relevant physiological context, in which ASM contractile regulation occurs in an architectural and cellular context that more closely replicates the *in vivo* condition than ASM cells/cultures do. Indeed, for decades most of the ASM pharmacology that drove the development and ultimate clinical use of asthma/COPD drugs was studied using isolated airways ASM strips (reviewed in [135]).

ASM strips or isolated airways from animals of multiple species have been used for decades, although with the increased availability of human donor lung deemed not suitable for transplant, viable airways from human lung (typically 4–5 generation bronchi) can be isolated and studied in organ bath/myograph systems. These, as well as airways from animals, retain responsiveness to pharmacological stimuli, and neural activation of ASM contraction can be mimicked by electric field stimulation [161,162].

More recently, an alternative tissue model has been developed in which thin (~250 micron) slices of lung termed *precision-cut lung slices* (PCLS) are generated and, after a brief period of recovery in culture, enable analysis of regulation of ASM contraction by imaging [160]. Imaging approaches in the PCLS have gained increasing acceptance since 1992 [163] and can be used for analysis of both human and animal airways. Like isolated airways, PCLS have the advantage of examining ASM regulation in a relevant multi-cellular tissue context [160] and like isolated airways, can be denuded of epithelia to explore epithelia-ASM interactions. Moreover, unlike ASM strips or isolated airways, PCLS are amenable to cryopreservation as well as higher throughput analyses, such as analysis of frozen-thawed slices in a 12-well [164] or even a 96-well plate format [165].

ASM tissue models have limited utility for studying anything other than regulation of contraction, although with recent application of pharmacological or molecular interventions in tissue culture (see the following), the capacity to gain mechanistic insight has expanded. Additional limitations of ASM tissue models involve the complication of variable viability of isolated tissue (less problematic in PCLS); thus, when studying GPCR agonist-induced contraction, it is important to assess non-GPCR-mediated contraction (e.g., KCl-induced) to control for viability, and variability in fundamental contractile properties independent of pharmacomechanical coupling. Moreover, mechanical or other forces acting on airway epithelia within the tissue can contribute to variability in the contractile response, as can the presence of embedded mast cells, which can release both histamine and LTC/D4 that influence basal or stimulated ASM tone [166]. A specific limitation of isolated murine airways is that any airway other than the trachea is difficult to isolate and analyze, due to both the small size and the fact that the mouse has an airway *tree* much different from that of the human, with few large conducting airways.

11.2.4 *In Vivo* Models

In vivo models provide the best models for testing the effects of GPCR ligands on airway resistance and, if human, the ultimate model for assessing utility of a GPCR ligand as lung disease therapeutic. In addition, the ability of GPCRs to regulate ASM hyperplasia or hypertrophy, caused by either allergic inflammation [167] or chronic ASM contraction independent of inflammation [168], is readily assessed in *in vivo* models using mouse, rat (reviewed in [169,170]), or guinea pig [135]. Assessing any feature of airway remodeling in humans, however, is extremely problematic, due to the several limitations including the slow-developing nature of airway remodeling, limited access to tissue, and insufficient imaging techniques [171]. In addition, it is difficult to gain insight into ASM GPCR function without complementary insight from more reductionist (tissue- or cell-based) models for further assessment of ASM function and signaling. This is because multiple cell types, either resident lung cells or cells that infiltrate the lung (inflammatory cells) can be the targets of either inhaled or systemic GPCR ligands.

Although animal *in vivo* models, and particularly murine models, enable some mechanistic insight into the actions and regulation of GPCRs by virtue of their ability to be genetically manipulated, species-specific differences remain a significant limitation. The species-specific limitations noted earlier for ASM cells and tissue similarly apply to *in vivo* models, and the specific limitations of mouse models with respect GPCRs, ASM, and multiple other regulators/cells have been discussed extensively elsewhere [2,135].

11.2.5 Experimental Strategies and Tools Applied to the Study of G Protein-Coupled Receptors Signaling and Function in Airway Smooth Muscle Model Systems

For each of these aforementioned model systems (cell, tissue, *in vivo*) a large number of experimental approaches/tools can be applied, the most common being pharmacological, molecular, and genetic. Most strategies involve expressing, inhibiting, or knocking down/out various proteins to gain insight to regulatory control of signaling and functions. Not surprisingly, the key signaling elements linked to the three major ASM functions have received the most attention to date. GPCR-mediated regulation of contraction/contractile signaling has focused primarily in understanding transducers of pharmaco-mechanical coupling, including all elements upstream of MLC20 phosphorylation (regulators of intracellular calcium, calcium sensitization, and calcium-activated targets), although recently a growing appreciation of the role of dynamic cytoskeleton regulation in ASM contraction has brought attention the growing number of regulatory proteins and mechanisms affecting the actin cytoskeleton and its connections to the crossbridge cycling machinery and the extracellular environment [172]. Studies of GPCR regulation of ASM proliferation, of demonstrating positive (again, predominately Gq-coupled [102,173]) or negative (primarily Gs-coupled [82]), cooperativity with (mitogenic) receptor tyrosine kinases, have focused on regulation of canonical MAPK pathways, although recently other mechanisms have been demonstrated, for example, for bitter tastant receptors [174,175]. Lastly, GPCR regulation of gene products that function in a paracrine manner, serving either an immunomodulatory or airway remodeling function, encompasses analysis and manipulation of a host of molecules whose characterization is served by increasing powerful *omic* approaches [123] and strategies to manipulate gene expression at multiple levels ranging from DNA sequences to translated protein.

11.2.5.1 Pharmacological Strategies/Tools

Pharmacological ligands, be they agonists (partial, full, or inverse) or antagonists (competitive, noncompetitive), are employed to identify GPCRs expressed in a given cell and to elucidate their signaling. For several GPCRs, including the β_2AR, very specific ligands exist with minimal ability to interact with other receptors. Such specific ligands are obviously useful experimentally and are often necessary for therapeutic drugs to achieve an acceptable therapeutic ratio. Unfortunately, for many GPCRs and GCPR subfamilies, the lack of specific ligands significantly hampers the study of these receptors. This is particularly true of those GPCRs outside of the class A rhodopsin family, and the expression, let alone function, of these (Class B-F) receptors in ASM are poorly understood. Yet even among the Class A family, many subfamilies of GPCRs, particularly purinergic, chemokine, neuropeptide, LPA/S1P, eicosanoid, and chemosensory receptors, are poorly characterized in ASM due the lack of useful ligands.

The other prevalent pharmacological tool in GPCR biology is the enzyme inhibitor, potentially useful in implicating various enzymes, most often kinases, in GPCR-mediated signaling or feedback regulation. In some instances in the study of ASM, kinase inhibitors (when adequately specific) have been useful in defining GPCR signaling and its role in various ASM functions. For example, subtype-specific PKC inhibitors have helped discern important roles for classical PKC isoforms in regulating calcium sensitization mechanisms by both GPCRs and calcium, and in the feedback regulation (desensitization) of the CysLT1R [101,120] in ASM. However, the majority of pharmacological kinase inhibitors lack sufficient specificity to confidently implicate their intended target in signaling/functional processes in most cell, tissue, and *in vivo* experiments. The reasons for such a lack of specificity are many (reviewed in [176]), but most frequently because they target the ATP-binding site of kinases, which tends to be similar among kinases with a given family. Given the limited utility of most pharmacological enzyme inhibitors, a more comprehensive approach (i.e., use of additional [molecular or genetic] strategies) is often required in order to adequately delineate signaling events, and their functional consequences, in most model systems.

11.2.5.2 Molecular Strategies/Tools

Early studies of GPCR biology benefited greatly from the molecular biology strategies employed in artificial (transformed or immortalized) cell systems such as human embryonic kidney (HEK)293, COS, and Chinese hamster ovary (CHO) cells, enabling expression of proteins as either GPCRs, or mediators or inhibitors of GPCR signaling/feedback regulation. However, transfection/infection techniques allowing heterologous protein/peptide expression initially worked very poorly in most primary cells, including ASM. Transfection efficiency (i.e., the percentage of cells expressing the transfected construct) was not only very low (less than 10%), but expression was transient with primary ASM cells (and their limited lifespan) not amenable to selection strategies enabling stable expression. Early application of transfection strategies in ASM were limited to reporter constructs [177], whose readout was not highly dependent on transfection efficiency. However, as transfection/infection strategies and reagents improved, transfection efficiency improved to the point that meaningful experiments could be conducted. Our group was among the first to successfully transfect primary ASM cells in analyses of GPCR function. Building on early studies demonstrating the feasibility of adenovirus-mediated infection of cultured rat tracheal myocytes [178], we used adenovirus particles to augment DEAE–dextran-mediated transfection and achieved ~50% transfection efficiency when co-transfecting pCDNA3-GRK2 with a GFP-expressing construct. We subsequently sorted GFP+ cells to achieve a homogenous population of cells overexpressing GRK2, which exhibited enhanced β_2AR desensitization [99]. Using (solely) a lipid-based transfection reagent (Fugene), we were later able to achieve even higher transfection efficiency to more readily obtain a near 100% population of GFP—or GFP-chimera—expressing cells ([179] and Figure 11.1). Since this time, we have established the ability to infect constructs, using retro-viral or lentiviral infection of ASM cultures, with >90% efficiency, and with antibiotic selection a 100% homogenous population of cells is often achieved in a week [110,140]. Moreover, we have also achieved high efficiency with transient construct expression with lentivirus in ASM tissue, with associated functional consequences [110].

The ability to knockdown proteins via transfection/infection with si/sh RNA constructs, existing for almost a decade now, has significantly advanced the study of ASM biology in all primary cell types, including ASM. Transient transfection of siRNA has been employed to knockdown various GPCRs and regulatory molecules (e.g., GRKs [154] and arrestins [97,98]) in human ASM cells to delineate GPCR signaling, its regulation, and role in regulating ASM functions (see the following). Stable expression of shRNA to achieve stable knockdown in ASM cells or tissue, however, has been somewhat difficult to achieve to date. Our experience is that although shRNA-mediated stable knockdown can be achieved, the level of knockdown necessary to cause a functional effect is typically insufficient. For example, although we have been able to achieve siRNA-mediated knockdown of arrestin proteins in ASM cells to the level of 70%–80%, which results in inhibition of β_2AR desensitization and functional consequences, attempts to achieve stable knockdown using lentiviral-mediated shRNA constructs targeting arrestins results in a maximal 50% knockdown, which does not significantly affect β_2AR signaling or function. Ultimate success with stable knockdown in both ASM cells and tissue is probably only a matter of time, given the tremendous rate of technical advance this approach has exhibited in the last few years.

11.2.5.3 Genetic Strategies/Tools

The use of transgenic and knockout (and occasionally, knock-in) mice is prevalent in the study of GPCR function. Although also prevalent in the study of allergic lung disease/asthma, most mouse genetics strategies for these studies focus on understanding the role of inflammatory molecules (e.g., using overexpression [180] or knockout [181] of various cytokines) in the inflammation process, with ASM being only indirectly affected. Our group has employed various genetic approaches in mice in attempts to understand both

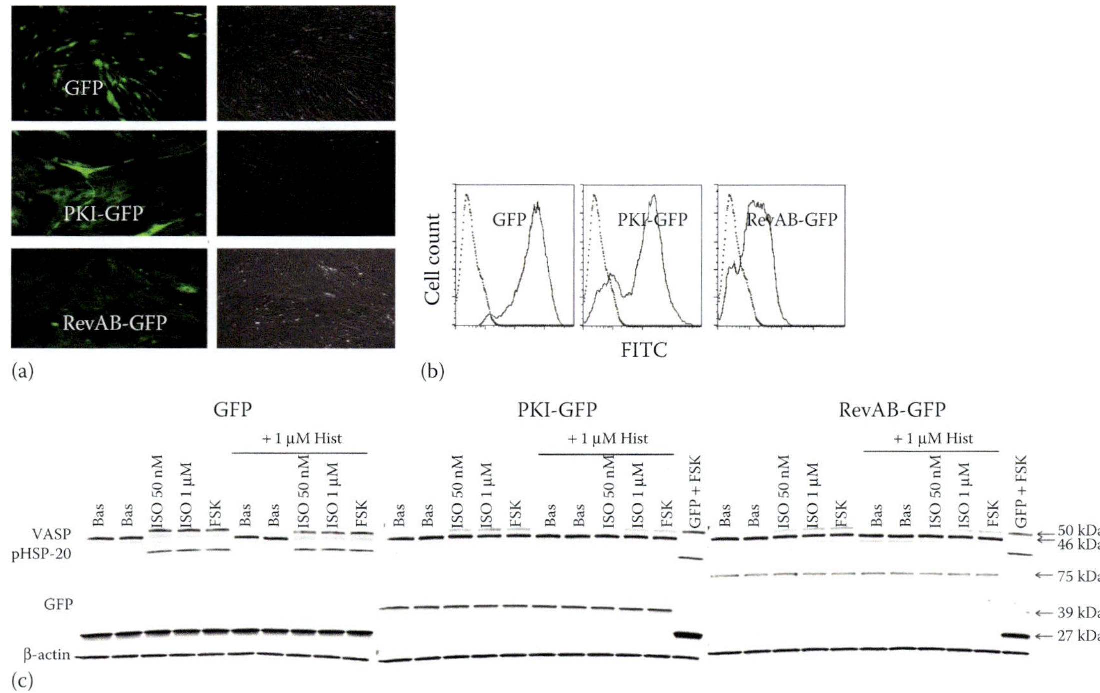

FIGURE 11.1 Establishing PKA as the predominant effector mediating β_2AR, Gs-coupled GPCR regulation of ASM functions. (a) Images depicting retroviral expression of PKA inhibitory constructs PKI-GFP and RevAB-GFP in human ASM cultures. (b) Flow cytometry analysis (FITC) of GFP/GFP chimera expression of cells in (a). (c) PKI-GFP, RevAB-GFP expression in human ASM cells inhibits the phosphorylation of the PKA substrates vasodilator specific phosphoprotein (VASP) and HSP20, by beta-agonist and forskolin (FSK) in human ASM. (*Continued*)

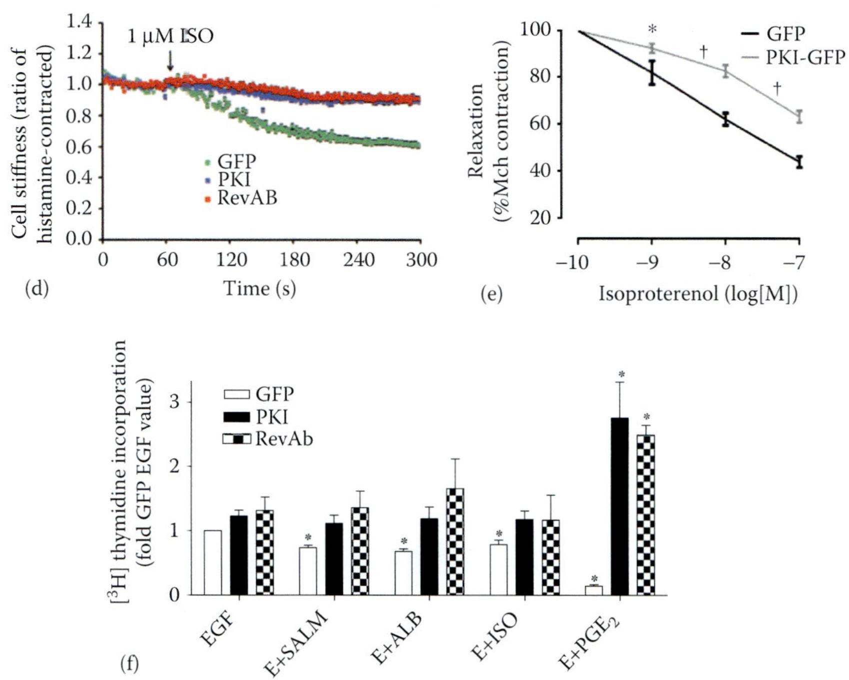

FIGURE 11.1 (Continued) Establishing PKA as the predominant effector mediating β_2AR, Gs-coupled GPCR regulation of ASM functions. (d–f) PKI-GFP, RevAB-GFP expression reverses the inhibitory effect of beta-agonist on human ASM contraction in cells ([d], assess via Magnetic Twisting Cytometry [111]) and tissue (e), and on mitogen-induced DNA synthesis (f) in human ASM cells. (Data from [a and b] Guo, M. et al., *Biochemistry*, 44, 13771–13782, 2005; [c and d] Morgan, S.J. et al., *J Biol. Chem.*, 289, 23065–23074, 2014; Yan, H. et al., *FASEB J.*, 25, 389–397, 2011.)

the function and regulation of various GPCRs in ASM, and their contribution to the asthma phenotype. Because mice express little or no CysLT1R, we employed a mouse engineered to express human CysLT1R [182], to examine the desensitization mechanisms of CysLT1R, and demonstrated an agonist-specific receptor desensitization dependent on protein kinase C [120]. In other studies, we have employed mice lacking the beta-arrestin1 or beta-arrestin2 genes for *in vivo*, tissue, and cell-based studies of GPCR regulation of ASM function. These studies demonstrated the differential ability of arrestin isoforms to regulate different GPCRs [97,98,120], and the physiological relevance of arrestin-mediated β_2AR desensitization in ASM [97,98].

Others have also taken advantage of mice expressing CRE under a smooth muscle specific promoter (SMP8, alpha-SMA) to generate smooth muscle-specific expression of various genes [154,182,183].

The obvious advantage of employing genetically modified mice is in the ability to gain mechanistic insight into a gene in an *in vivo* model, in a specific and powerful manner that other strategies (e.g., pharmacological) cannot provide. Moreover, ASM tissue and cells can be harvested and cultured from these mice, providing a nice complement to studies of human ASM tissue and cells for which limited manipulative interventions exist. The obvious drawback with genetically engineered mice is that these are *mice*, not humans. Not only can GPCR expression and function vary significantly between mouse and human in a cell-specific manner, so can the expression and function of molecules, and complex regulatory phenomena (e.g., inflammation) that regulate GPCRs and ASM. For example, murine ASM does express (or at least does not respond to agonists of) the H1 histamine receptor, a GPCR that is extremely valuable in the study of human ASM contraction given that histamine remains highly expressed and an effective contractile agent in ASM cultures that undergo extended passage. In addition, the principal EP receptor on mouse ASM that mediates relaxation of ASM by PGE_2 is the EP2 subtype; in human (and rat) ASM, the EP4 receptor is predominant [11]. Thus, these differences highlight the need to use a wide range of models, and sometimes multiple species, to understand relevant GPCR function in ASM.

Lastly, an emerging tool for knocking down/out and even knocking in endogenously expressed proteins in primary cells is the *clustered regularly interspaced short palindromic repeats* (CRISPR)/*CRISPR-associated system 9* (Cas9) system [184]. The system works by genomic editing, thus could be discussed in the following section on *Genetic strategies/tools*. To date, we know of no studies of ASM employing this technique, but several are likely on the horizon as this strategy becomes mastered in more labs.

11.3 CASE STUDIES HIGHLIGHTING APPROACHES FOR STUDYING G PROTEIN-COUPLED RECEPTORS IN AIRWAY SMOOTH MUSCLE: BETA-2-ADRENOCEPTOR (β_2AR) SIGNALING, FUNCTION, AND REGULATION IN AIRWAY SMOOTH MUSCLE

Numerous studies from our lab and others have employed the various model systems and strategies for investigating GPCRs in ASM discussed earlier, with studies of the β_2AR being the most comprehensive and insightful.

Prior to the 1990s, studies of ASM reflected the clinical need to identify potent and efficacious bronchodilators for use in OLD treatment or prophylaxis (and in many ways, this holds true for today's research as well). Many studies, using isolated muscle strips or airways from animals in *ex vivo* analyses of contraction, identified numerous GPCR ligands capable of inducing or inhibiting ASM tension. Agonists of the beta-adrenoceptors, used clinically to manage asthma for over 100 years [95], were shown to be effective relaxant agents of contracted ASM, and for over 30 years now the pursuit of a better beta-agonists has dominated ASM and asthma/COPD pharmacology. *Better* has been defined as: (1) more potent, reflecting efficacious relaxation at low doses; (2) more selective (for the β_2-AR subtype), given ASM relaxation to beta-agonists is mediated overwhelmingly by the beta-2 subtype, and systemic beta-1-adrenoceptor agonism contributes to unwanted clinical side effects including tachycardia and tremor [185,186]; (3) safer, a condition not easily addressed in anything other than human clinical trials; and (4) longer acting, based on the universally-accepted premise that greater convenience (including less frequent dosing) of drug administration improves adherence, and therefore, better clinical outcomes.

The aforementioned studies constitute an extremely large body of research into beta-agonist/βAR pharmacology, yet numerous questions remain, several spurred by the evolution of the field of GPCR biology, and by several findings specific to the beta-adrenoceptors. Studies from our lab have focused on several questions important to both GPCR and asthma biology/pharmacology, presented in the following.

11.3.1 How Exactly Do Beta-Agonists/β_2ARs Work to Regulate Airway Smooth Muscle Functions?

Well before the existence of the β_2AR was discovered, numerous agents (reviewed in [95]) now known to activate the β_2AR were shown to relieve shortness of breath associated with asthma. Studies dating back to the 1970s, typically using isolated guinea pig airways or muscle strips to assess regulation of ASM contractile state, characterized the potency and efficacy of numerous agents in relaxing ASM. When the β_1AR [187] and β_2AR [188] (and later, the β_3AR [189]) were cloned, and subtype-specific radioligands for the β_1AR and β_2AR subtypes were established (reviewed in [190,191]), it became possible to accurately identify and quantify β_1AR and β_2AR expression in cells and tissues, which in turn facilitated development of beta-adrenoceptor subtype-specific ligands and their capacity to signal and regulate function in cells, tissues, and organ systems.

We now know the β_2AR subtype is the predominant beta-adrenoceptor subtype expressed in ASM. In cultures of human ASM cells derived from human trachea, radioligand binding assays estimate ~90% of beta-adrenoceptors expressed are β_2AR [99]. In ASM tissue from murine tracheae, unpublished studies from our lab suggest that beta-adrenoceptor subtype expression is split approximately 50:50 between the β_1AR and β_2AR. However, signaling events by β_1AR and β_2AR in cultures of murine ASM derived from murine trachea appear to be similar (unpublished data), and studies from the Walker lab and others find airways distal to the murine trachea preferentially express the β_2AR subtype, and that beta-agonist effects of murine airway resistance are β_2AR-dependent [192].

It has long been assumed that the prorelaxant signaling mediated by beta-agonist occurred via the β_2AR/Gs/adenylyl cyclase (AC)/cAMP →protein kinase A (PKA) pathway. For decades this notion was accepted in the field of airway biology, despite being based completely on associative data, with no direct evidence supporting it. This lack of direct evidence existed due to the lack of reliable experimental approaches for inhibiting intracellular PKA (reviewed in [176]), and in the difficulty in characterizing intracellular PKA activity and confirming its inhibition. In 2011, Zieba et al. [193] challenged the long-held dogma of PKA mediating the relaxant effect of beta-agonist by demonstrating that small molecule activators of the cAMP effector *Epac* could relax ASM from various species. Our lab subsequently attempted to clarify the relative contributions of Epac and PKA in the relaxant effect of beta-agonist on ASM (Figure 11.1). To overcome limitations of the (previously failed) strategies for inhibiting intracellular PKA, we generated plasmids encoding two different peptide proteins capable of PKA inhibition. One was the PKA inhibitor peptide (PKI), the other the mutant regulatory PKA subunit RevAB (which does not bind cAMP and thus cannot release the catalytic PKA subunit). To readily assess expression of these constructs in ASM, they were created as GFP chimeras. To enable high-efficiency expression in ASM, we cloned these constructs into both retroviral and lentiviral vectors and generated retro/lenti—virus for infection of ASM cells and tissue. To assess PKA activation (and its inhibition), we analyzed the phosphorylation state of multiple known intracellular PKA substrates in lysates of stimulated ASM cells. Using this in multiple studies, we established the ability of the expressed PKA-inhibitory constructs to (1) inhibit human ASM cell and murine ASM tissue relaxation [110] by beta-agonist using Magnetic Twisting Cytometry (MTC) and myograph analyses, respectively; and (2) also demonstrate that the anti-mitogenic effect [82,179] of activated Gs-coupled GPCRs (β_2AR, EP2/4 receptors) is predominantly PKA dependent. Moreover, direct biochemical assays failed to demonstrate Epac activation by multiple GPCR agonists and other inducers of intracellular cAMP accumulation in ASM. Thus, at both the cell and tissue level, using refined molecular approaches we have been able to establish the predominate role of PKA as the mediator of inhibition of both ASM contraction and relaxation. Currently, we are employing genetic strategies enabling inducible, ASM-specific expression of PKI-GFP to explore this question in an *in vivo* model [194].

11.3.2 Does β_2AR Desensitization in the Airway Occur, and Does It Matter?

It has been estimated that up to 55% of all asthmatics are inadequately managed [195], raising the question of why beta-agonists (along with inhaled corticosteroids, the primary asthma therapeutic) are inadequate treatment or prophylaxis for some many patients. For asthmatics whose disease is treated with beta-agonist for an extended period of time, a loss of efficacy for inhaled beta-agonists has been documented in the form of *loss of bronchoprotective effect* [196]. Loss of bronchoprotective effect is defined as a reduced ability of beta-agonist, when inhaled prior to a methacholine challenge of the lung, to inhibit the increase in airway resistance the methacholine inhalation induces.

Understandably, β_2AR desensitization has been frequently speculated as a contributing mechanism to the observed loss of beta-agonist efficacy with chronic beta-agonist use. For many GPCRs, their desensitization has been shown to occur in studies using various cell-free and cell-based systems. And given the numerous incidences of demonstrated loss of *in vivo* efficacy of agonists targeting various GPCRs (most extensively studied and appreciated for opioid receptors, and of course, rhodopsin), it is a small leap of faith to ascribe receptor desensitization to diminished *in vivo* effectiveness of a drug.

The discovery of regulatory proteins that mediate GPCR desensitization created the opportunity to manipulate desensitization and assess whether GPCR desensitization *per se* translated in altered GPCR-mediated functions. Studies by Lefkowitz and colleagues pioneered the discovery and characterization GPCR kinases (*GRKs*) and *arrestin* proteins. GRKs specifically phosphorylate certain agonist-occupied GPCRs, and along with arrestins promote GPCR uncoupling from G protein, diminish signaling throughput, and (often) promote receptor internalization and degradation (reviewed in [197]). We have employed various tools, originally applied to cell-free system or artificial cellular expression systems (HEK293, COS cells, etc.), to establish the functional relevance of β_2AR desensitization in primary ASM cells in culture, tissue, and *in vivo* (Figure 11.2). First, we expressed a kinase-dead mutant [198], or an (inhibitory) C-terminal domains [154,155], of GRK2/3 in human ASM cultures to gain insight into which ASM GPCRs are regulated by GRKs. Interestingly, we found that β_2AR signaling, and to a lesser extent A2b adenosine receptor signaling, could be augmented by inhibiting GRK2/3, whereas signaling by PGE_2 (EP2, EP4 receptors) was not. Similar results were obtained with GRK2/3 or arrestin3 knockdown in human ASM cultures, and in ASM cultures derived from ASM tissues from mice lacking the arrestin3 gene [97,98,155] (of note, GRK2 knockout is embryonic lethal). Thus, at least on the level of signaling (cAMP accumulation), β_2AR signaling was constrained by canonical GRK2/3- and arrestin-mediated desensitization.

To link these observed effects on β_2AR signaling to functional desensitization, we employed genetic strategies in mice. Smooth muscle-specific expression of the (GRK-inhibiting) GRK2 C-terminus enabled greater beta-agonist-mediated relaxation of contracted murine airways *ex vivo* [154], and of methacholine-induced airway resistance *in vivo* (Penn and Deshpande, manuscript in preparation). The same *ex vivo* and *in vivo* effects were observed using arrestin3- (but not arrestin2-) knockout mice [97,98]. Current studies in our lab are looking to extend these observations to human ASM airways (isolated 3rd–4th generation airway rings, and PCLS) infected to express either shRNA or GRK and arrestin inhibitory constructs. Collectively, these studies strongly suggest that β_2AR desensitization does indeed occur in ASM and reduces the beta-agonist ability to relax ASM and bronchodilate. Whether this phenomenon can ultimately be targeted in human asthmatics to improve the therapeutic efficacy is unclear. However, current efforts to develop small-molecule GRK or arrestin

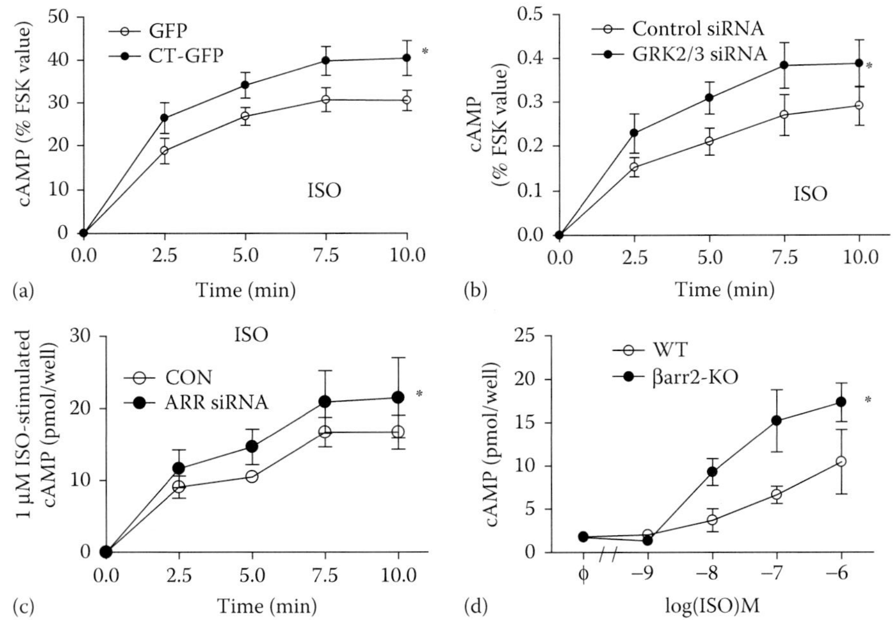

FIGURE 11.2 β_2AR desensitization occurs in ASM, and it matters. Inhibition, knockdown, or knockout of regulatory proteins that mediate β_2AR desensitization cause an increase in the signaling and functional effects of the β_2AR in ASM. (a,b). Inhibition of GRK2/3 via either retroviral expression of inhibitory construct GRKct (a) or via GRK2/3 knockdown increases beta-agonist-mediated signaling effects on signaling. (c,d). *(Continued)*

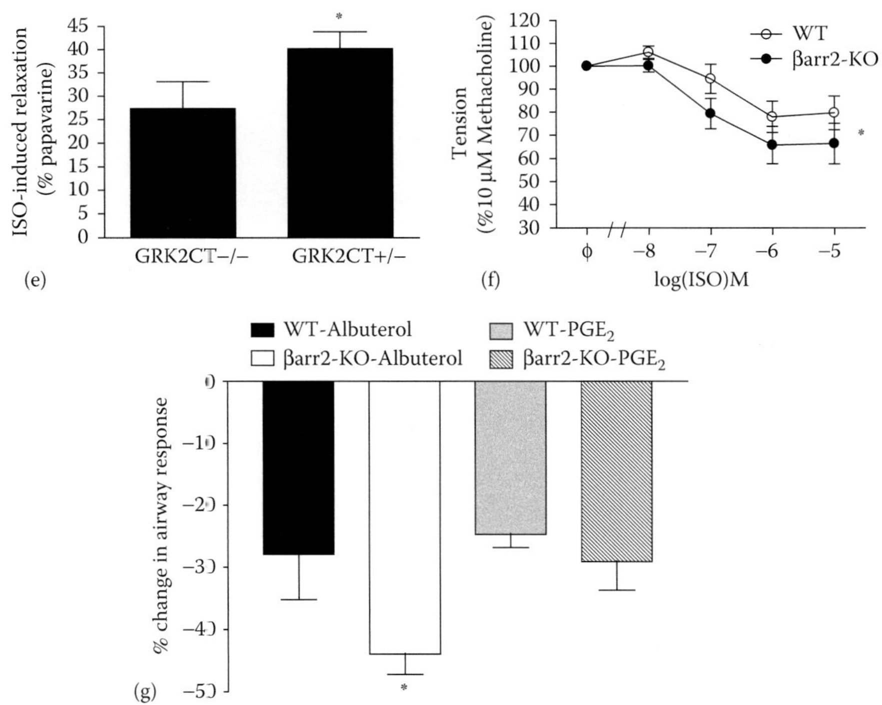

FIGURE 11.2 (Continued) Similarly, beta-arrestin2 knockdown in human ASM cultures (c) or knockout (d) in murine ASM cultures increases beta-agonist-mediated signaling. (e,f). GRKct expression (e) or beta-arrestin2 knockout (f) in murine ASM causes increased relaxation of contracted airways *ex vivo*. (g) Beta-arrestin2 knockout increases beta-agonist—but not PGE_2—mediated inhibition of *in vivo* airway resistance. (Data from [a] Kong, K.C. et al., *Biochemistry*, 47, 9279–9288, 2008; [b, e, g] Deshpande, D.A. et al., *FASEB J.*, 22, 2134–2141, 2008; [c, d, f] Deshpande, D.A. et al., *FASEB J.*, 28, 956–965, 2014.)

inhibitors [199–201] or biased ligands or peptides [202,203] enabling β_2ARs to avoid GRK/arrestin regulation, offer some hope that β_2AR desensitization can be targeted in humans.

11.3.3 A Story in Development: Is β_2AR Signaling in Airway Smooth Muscle and Asthma More Complicated than the β_2AR/Gs/AC/cAMP/PKA Canonical Pathway? What Does Non-Canonical Signaling Do (Good or Bad)?

Although the majority of our and others' work in β_2AR signaling in ASM has focused on the canonical β_2AR/Gs/AC/cAMP →PKA (and to a limited extent, →Epac) pathway, as noted above the field of GPCR biology has determined that GPCRs can be fairly promiscuous molecules. For some time now, we have appreciated the ability of numerous GPCRs to couple to more than one G protein [204]. With respect to the β_2AR, coupling to Gi has been reported in artificial systems as well as in some primary cell types [205–207]. In 1997, Daaka et al. identified a potential mechanism promoting β_2AR-Gi coupling, observing that PKA phosphorylation of the β_2AR caused a specificity switch of the receptor favoring Gi coupling [207]. The extent to which β_2AR-Gi coupling in ASM is of consequence is poorly understood; we have failed to see a significant effect of pertussis-toxin (inhibits Gi) treatment of ASM on β_2AR signaling, although studies from the Liggett lab suggest β_2AR-Gi could contribute to β_2AR desensitization [208].

More compelling is the concept that non-canonical β_2AR signaling can occur as G protein-*independent* signaling, particularly that dependent on arrestin proteins. There is now a wealth of literature demonstrating that arrestins, originally identified as regulators of GPCR desensitization/resensitization and internalization, also function as scaffolds that initiate signaling that is independent of G proteins. Wisler et al. [209] originally found that in HEK-293 cells expressing the β_2AR, the βAR ligand carvedilol had inverse efficacy for stimulating G(s)-dependent adenylyl cyclase yet recruited beta-arrestin to the β_2AR, caused β_2AR internalization, and stimulated ERK1/2 phosphorylation. Subsequent studies, using more artificial systems, demonstrated the capacity of various GPCRs to activate multiple pathways, some G protein-dependent and others G protein-independent/arrestin-dependent [210]. Moreover, the emerging field of biased ligand pharmacology exploits this knowledge in the search for new (or identifying existing) orthosteric ligands or allosteric modulators that *skew* or *bias* signaling towards a specific pathway. Not surprisingly, the study of GPCRs in any cell type, tissue, organ system or disease is presently dominated by the search to: (1) understand the relevance of the multiple, previously unappreciated non-canonical signaling events mediated by a GPCR; and (2) identify existing, or develop new, biased ligands that can function as therapeutics by selectively activating the beneficial pathways while avoiding the pathogenic pathways promoted by a GPCR.

Unfortunately, the study of qualitative signaling and biased ligand pharmacology to date has employed primarily cell-free and artificial cell systems in the attempt to identify signaling diversity as well as ligands or modulators capable of biasing signaling. Studies in primary cell types and more integrative models have been limited. To date, we have not been able to identify any physiologically relevant signaling by the β_2AR in ASM other than canonical β_2AR/Gs/AC/cAMP →PKA signaling. There is, however, compelling evidence that ASM is indirectly affected by non-canonical β_2AR signaling in the airway, and that developing biased ligands of the β_2AR, and of other GPCRs will have clinical value for asthma. Although arrestin-dependent β_2AR signaling and its functional consequences in ASM remain poorly understood, the beta-arrestin2 KO mouse exhibits resistance to allergen-induced lung pathology including both airway inflammation and airway hyperresponsiveness [211], a phenotype that is unlikely to result solely from any attenuation of desensitization of the β_2AR or any other GPCR. Indeed, a number of studies by the Bond lab have attributed a permissive effect of β_2AR agonism to the allergen-induced asthma phenotype in mice, and their most recent data suggest that non-canonical, arrestin-dependent β_2AR signaling may be the culprit. Exogenous beta-agonist causes an increase in the airway mucus production caused by allergic lung inflammation, but delivery of the β_2AR antagonist nadolol inhibits the induction of mucus caused by allergen challenge ([212] and Figure 11.3). Mice lacking either the ADRB2 (β_2AR) gene [213] or systemic epinephrine [214] are protected from developing the allergen-induced asthma phenotype. Moreover, when mice lacking systemic epinephrine are treated with unbiased beta-agonists, or carvedilol,

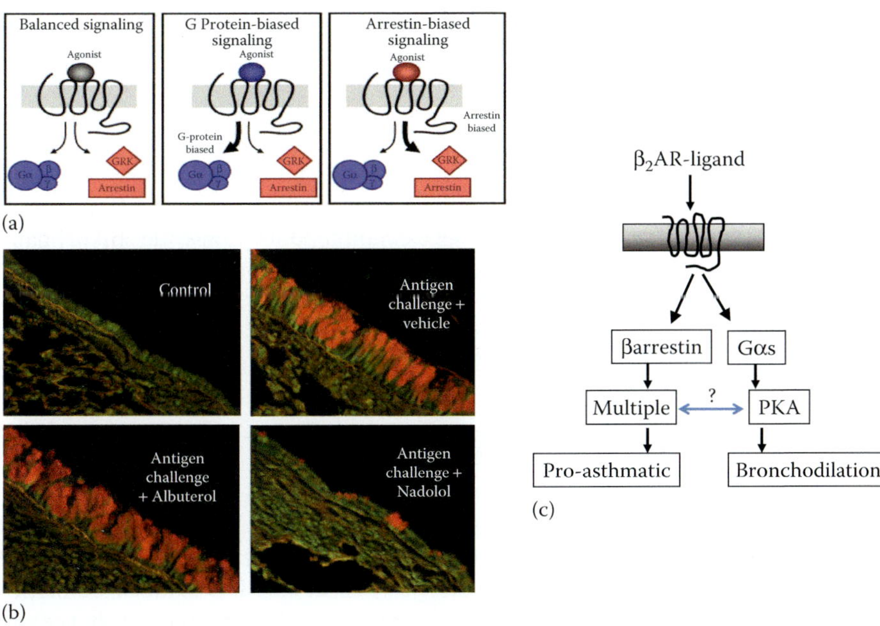

FIGURE 11.3 Qualitative signaling and biased agonism of the β_2AR in asthma pathobiology. (a) Concept figure biased agonism. (b) Exogenous beta-agonist (albuterol) increases allergen-induced mucus production in the airway *in vivo*, whereas the unbiased antagonist nadolol inhibits it. (c) Potential role of biased agonism in promoting airway physiology, disease. (Data from [b] Nguyen, L.P. et al., *Am. J. Respir. Cell Mol. Biol.*, 38, 256–262, 2008.)

which is an antagonist of canonical β_2AR/Gs/AC/cAMP →PKA signaling, yet an agonist of β_2AR-arrestin signaling (i.e., an arrestin-biased ligand), the phenotype is restored [214]. In addition, agents such as phosphodiesterase inhibitors that augment cAMP-dependent signaling independent of GPCRs, mitigate the disease phenotype [215]. Thus, the overall effect of canonical β_2AR/Gs/AC/cAMP →PKA appears good/beneficial/therapeutic, whereas β_2AR-arrestin-dependent signaling appears deleterious/pathological. The specific role of β_2AR-arrestin-dependent signaling in *ASM* in these effects remains unclear. Alternatively, airway epithelium appears to be a strong candidate in effecting the asthma phenotype via β_2AR-arrestin-dependent signaling, as inhibition of β_2AR signaling with nadolol, which blocks both Gs- and arrestin-mediated signaling, eliminates mucus production in airway epithelia, whereas treatment with carvedilol, which antagonizes β_2AR-Gs signaling but promotes β_2AR-arrestin signaling, enables mucous production.

Other GPCRs implicated in the pathogenesis of asthma also appear capable of diverse/qualitative signaling that may contribute differentially to disease or therapy. Nichols et al. demonstrated that in a murine model of allergic inflammation, the effects of protease-activated receptor-2 (PAR-2) agonism could be separated into arrestin-dependent pathogenic (e.g., pro-inflammatory) effects, whereas beneficial/therapeutic effects (e.g., bronchorelaxation) were arrestin-independent (G-protein-dependent) [216]. Recently, we demonstrated the proton-sensing GPCR *OGR1/GPR68* to be capable of activating multiple G protein-dependent (including both bronchoconstricting Gq and bronchorelaxing Gs) pathways in ASM in response to modest changes in intracellular pH [23] and have discovered certain benzodiazepines can function as OGR1 allosteric modulators [217], some exhibiting bias [218]. Finally, exciting findings from the Tobin lab have identified qualitative signaling by the M3 to regulate several physiological processes [219,220], including ASM contraction [221]. Using mice expressing a mutated M3 receptor that displays impaired ligand-induced receptor phosphorylation and arrestin recruitment—but retains its ability to undergo Gq coupling and mediate calcium mobilization—the authors demonstrated that M3 receptor phosphorylation is required for ASM contraction [221]. These studies suggest that M3-arrestin recruitment may mediate activation of Rho-pathways, which contribute to ASM contraction *ex vivo* and *in vivo*. These findings point to the potential therapeutic utility of a yet-to-be discovered/developed biased M3 ligand for the myriad of diseases involving smooth muscle dysfunction.

The study of qualitative GPCR signaling and the signaling and functional properties of biased ligands in ASM will require several of the tools and strategies employed in the aforementioned studies of ASM GPCR signaling and regulation, under both physiological and disease conditions. Full characterization of GPCR signaling events will require more than the standard assays of second messenger (calcium, cAMP) accumulation, particularly when attempting to measure subtler cryptic signaling that may be of functional significance. To this end, reporter assays assessing luciferase activity induced by stimulation of a transfected luciferase reporter activated by various transcription factors (e.g., CRE, AP-1) downstream of various signaling pathways, can provide evidence of signaling pathways in circumstances in which analysis of upstream signaling elements may be difficult. Also, comprehensive or targeted analyses of phosphoproteins that extend beyond that afforded by standard immunoblotting is an approach our lab is currently pursuing. Alternatives to conventional western blotting such as the WesSimple system enables sensitive analysis of extremely (as little as 0.5 µg of lysate) small protein amounts, thus being well suited for small tissue samples (such as those obtained with biopsy) or cellular fractions [103]. Additionally, strategies for assessing compartmentalized signaling in a cell, which may be obscured when assessing changes in global cellular protein, include analysis of cell fractions and new imaging techniques that enable visualization of localized signaling elements such as cAMP and calcium [222–226], and proteins/phosphoproteins as well [227]. However, real-time imaging of the spatial patterns of cAMP signals and protein phosphorylation typically requires FRET-based probes, which have inherently low signal-to-noise ratios [228,229]. Both lifetime and spectral imaging approaches have been successfully implemented to increase the signal-to-noise ratio of FRET measurements. Recent implementations of spectral imaging approaches have allowed automated tracking of cAMP signals in moving cells, as well as measurement of cAMP-sustained signals in three spatial dimensions [230–232].

With these tools in hand, the ability to characterize diverse signaling events and their location can enable novel associations of signals and function, and also identify targets to manipulate as a therapeutic means.

11.3.3.1 Where Are We Headed?

As discussed previously, the current study of GPCRs in ASM is driven by both advances in technology and experimental approaches, and the growing appreciation that GPCRs can possess diverse signaling capabilities that can be *tuned* by biased pharmacology. Within the last 20 years, model systems for studying ASM have emerged and evolved, and an even bigger boon to the field has been the application of many new tools and strategies, in particular those that enable molecular genetic manipulation of protein expression. These tools will undoubtedly be further refined, and together with the seemingly explosive growth of imaging capabilities in both cells and tissues, our understanding of the GPCRs that regulate airway functions will continue to advance rapidly and almost certainly translate into more efficacious and safe drugs for the management of OLDs.

ACKNOWLEDGMENTS

Studies in the Penn and Pera labs are supported by NIH grants HL58506, HL136209, HL114471, and AI110007 (RBP) and HL140064 (TP).

REFERENCES

1. Billington, C. K., and R. B. Penn. 2003. Signaling and regulation of G protein-coupled receptors in airway smooth muscle. *Respir Res* 4: 2–2.
2. Deshpande, D. A., and R. B. Penn. 2006. Targeting G protein-coupled receptor signaling in asthma. *Cell Signal* 18: 2105–2120.
3. Pera, T., and R. B. Penn. 2016. Bronchoprotection and bronchorelaxation in asthma: New targets, and new ways to target the old ones. *Pharmacol Ther* 164: 82–96.

4. Kume, H., I. P. Hall, R. J. Washabau, K. Tagaki, and M. I. Kotlikoff. 1994. β-adrenergic agonists regulate KCa channels in airway smooth muscle by cAMP-dependent and -independent mechanisms. *J Clin Invest* 93: 371–379.
5. Goldie, R. G., D. Spina, P. J. Henry, K. M. Lulich, and J. W. Paterson. 1986. In vitro responsiveness of human asthmatic bronchus to carbachol, histamine, beta-adrenoceptor agonists and theophylline. *Br J Clin Pharmacol* 22: 669–676.
6. Zaagsma, J., P. J. van der Heijden, M. W. van der Schaar, and C. M. Bank. 1984. Differentiation of functional adrenoceptors in human and guinea pig airways. *Eur J Respir Dis* 135: 16–33.
7. Hall, I. P., S. Widdop, P. Townsend, and K. Daykin. 1992. Control of cyclic AMP levels in primary cultures of human tracheal smooth muscle cells. *Br J Pharmacol* 107: 422–428.
8. Sheller, J. R., D. Mitchell, B. Meyrick, J. Oates, and R. Breyer. 2000. EP(2) receptor mediates bronchodilation by PGE(2) in mice. *J Appl Physiol* 88: 2214–2218.
9. Penn, R. B., R. M. Pascual, Y.-M. Kim, S. J. Mundell, V. P. Krymskaya, R. A. Panettieri Jr, and J. L. Benovic. 2001. Arrestin specificity for G protein-coupled receptors in human airway smooth muscle. *J Biol Chem* 276: 32648–32656.
10. Fortner, C. N., R. M. Breyer, and R. J. Paul. 2001. EP2 receptors mediate airway relaxation to substance P, ATP, and PGE2. *Am J Physiol Lung Cell Mol Physiol* 281: 469–474.
11. Buckley, J., M. A. Birrell, S. A. Maher, A. T. Nials, D. L. Clarke, and M. G. Belvisi. 2011. EP(4) receptor as a new target for bronchodilator therapy. *Thorax* 66: 1029–1035.
12. Clarke, D. L., M. G. Belvisi, E. Hardaker, R. Newton, and M. A. Giembycz. 2005. E-Ring 8-isoprostanes are agonists at EP2- and EP4-prostanoid receptors on human airway smooth muscle cells and regulate the release of colony-stimulating factors by activating cAMP-dependent protein kinase. *Mol Pharmacol* 67: 383–393.
13. Clarke, D. L., M. G. Belvisi, S. J. Smith, E. Hardaker, M. H. Yacoub, K. K. Meja, R. Newton, D. M. Slater, and M. A. Giembycz. 2005. Prostanoid receptor expression by human airway smooth muscle cells and regulation of the secretion of granulocyte colony-stimulating factor. *Am J Physiol Lung Cell Mol Physiol* 288: L238–L250.
14. Benyahia, C., I. Gomez, L. Kanyinda, K. Boukais, C. Danel, G. Leséche, D. Longrois, and X. Norel. 2012. PGE2 receptor (EP4) agonists: Potent dilators of human bronchi and future asthma therapy? *Pulm Pharmacol Ther* 25: 115–118.
15. Aso, H., S. Ito, A. Mori, N. Suganuma, M. Morioka, N. Takahara, M. Kondo, and Y. Hasegawa. 2013. Differential regulation of airway smooth muscle cell migration by E-prostanoid receptor subtypes. *Am J Respir Cell Mol Biol* 48: 322–329.
16. Mundell, S. J., M. E. Olah, R. A. Panettieri, J. L. Benovic, and R. B. Penn. 2000. Regulation of G protein-coupled receptor-adenylyl cyclase responsiveness in human airway smooth muscle by exogenous and endogenous adenosine. *Am J Respir Cell Mol Biol* 24: 155–163.
17. Carstairs, J. R., and P. J. Barnes. 1986. Visualization of vasoactive intestinal peptide receptors in human and guinea pig lung. *J Pharmacol Exp Ther* 239: 249–255.
18. Lazarus, S. C., C. B. Basbaum, P. J. Barnes, and W. M. Gold. 1986. cAMP immunocytochemistry provides evidence for functional VIP receptors in trachea. *Am J Physiol* 251: 115–119.
19. Maruno, K., A. Absood, and S. I. Said. 1995. VIP inhibits basal and histamine-stimulated proliferation of human smooth muscle cells. *Am J Physiol* 268: L1047–L1051.
20. Belvisi, M. G., M. Saunders, M. Yacoub, and J. A. Mitchell. 1998. Expression of cyclo-oxygenase-2 in human airway smooth muscle is associated with profound reductions in cell growth. *Br J Pharmacol* 125: 1102–1108.

21. Pascual, R. M., C. K. Billington, I. P. Hall, R. A. Panettieri, J. E. Fish, S. P. Peters, and R. B. Penn. 2000. Comparison of chronic cytokine versus PGE2 pretreatment effects on G protein-coupled receptor (GPCR) signaling in human airway smooth muscle (HASM). *Am J Respir Criti Care Med* 161: A696.
22. Mizuta, K., Y. Zhang, D. Xu, F. Mizuta, F. D'Ovidio, E. Masaki, and C. W. Emala. 2013. The dopamine D(1) receptor is expressed and facilitates relaxation in airway smooth muscle. *Respir Res* 14: 89.
23. Saxena, H., D. Deshpande, B. Tiegs, H. Yan, R. Battafarano, W. Burrows, G. Damera et al. 2012. The GPCR OGR1 (GPR68) mediates diverse signalling and contraction of airway smooth muscle in response to small reductions in extracellular pH. *Br J Pharmacol* 166: 981–990.
24. Ichimonji, I., H. Tomura, C. Mogi, K. Sato, H. Aoki, T. Hisada, K. Dobashi, T. Ishizuka, M. Mori, and F. Okajima. 2010. Extracellular acidification stimulates IL-6 production and Ca(2+) mobilization through proton-sensing OGR1 receptors in human airway smooth muscle cells. *Am J Physiol Lung Cell Mol Physiol* 299: L567–L577.
25. Mak, J. C., J. N. Baraniuk, and P. J. Barnes. 1992. Localization of muscarinic receptor subtype mRNAs in human lung. *Am J Respir Cell Mol Biol* 7: 344–348.
26. Gosens, R., M. M. Bromhaar, A. Tonkes, D. Schaafsma, J. Zaagsma, S. A. Nelemans, and H. Meurs. 2004. Muscarinic M(3) receptor-dependent regulation of airway smooth muscle contractile phenotype. *Br J Pharmacol* 141: 943–950.
27. Gosens, R., S. A. Nelemans, M. M. Grootte Bromhaar, S. McKay, J. Zaagsma, and H. Meurs. 2003. Muscarinic M3-receptors mediate cholinergic synergism of mitogenesis in airway smooth muscle. *Am J Respir Cell Mol Biol* 28: 257–262.
28. Roffel, A. F., C. R. Elzinga, and J. Zaagsma. 1990. Muscarinic M3 receptors mediate contraction of human central and peripheral airway smooth muscle. *Pulm Pharmacol* 3: 47–51.
29. Roffel, A. F., H. Meurs, C. R. Elzinga, and J. Zaagsma. 1990. Characterization of the muscarinic receptor subtype involved in phosphoinositide metabolism in bovine tracheal smooth muscle. *Br J Pharmacol* 99: 293–296.
30. Watson, N., P. J. Barnes, and J. Maclagan. 1992. Actions of methoctramine, a muscarinic M2 receptor antagonist, on muscarinic and nicotinic cholinoceptors in guinea-pig airways in vivo and in vitro. *Br J Pharmacol* 105: 107–112.
31. Daykin, K., S. Widdop, and I. P. Hall. 1993. Control of histamine induced inositol phospholipid hydrolysis in cultured human tracheal smooth muscle cells. *Eur J Pharmacol* 246: 135–140.
32. Grandordy, B. M., and P. J. Barnes. 1987. Airway smooth muscle and disease workshop: Phosphoinositide turnover. *Am Rev Respir Dis* 136: 17–20.
33. Farmer, S. G., J. E. Ensor, and R. M. Burch. 1991. Evidence that cultured airway smooth muscle cells contain bradykinin B2 and B3 receptors. *Am J Respir Cell Mol Biol* 4: 273–277.
34. Marsh, K. A., and S. J. Hill. 1992. Bradykinin B2 receptor-mediated phosphoinositide hydrolysis in bovine cultured tracheal smooth muscle cells. *Br J Pharmacol* 107: 443–447.
35. Mak, J. C., and P. J. Barnes. 1991. Autoradiographic visualization of bradykinin receptors in human and guinea pig lung. *Eur J Pharmacol* 194: 37–43.
36. Pang, L., and A. J. Knox. 1997. PGE2 release by bradykinin in human airway smooth muscle cells: Involvement of cyclooxygenase-2 induction. *Am J Physiol* 273: 1132–1140.
37. Pyne, S., and N. J. Pyne. 1994. Bradykinin-stimulated phosphatidate and 1,2-diacylglycerol accumulation in guinea-pig airway smooth muscle: Evidence for regulation 'down-stream' of phospholipases. *Cell Signal* 6: 269–277.

38. Tomasic, M., J. P. Boyle, J. F. Worley, and M. I. Kotlikoff. 1992. Contractile agonists activate voltage-dependent calcium channels in airway smooth muscle cells. *Am J Physiol* 263: 106–113.
39. Panettieri, R. A., E. M. Tan, V. Ciocca, M. A. Luttmann, T. B. Leonard, and D. W. Hay. 1998. Effects of LTD4 on human airway smooth muscle cell proliferation, matrix expression, and contraction in vitro: Differential sensitivity to cysteinyl leukotriene receptor antagonists. *Am J Respir Cell Mol Biol* 19: 453–461.
40. Jonsson, E. W. 1998. Functional characterisation of receptors for cysteinyl leukotrienes in smooth muscle. *Acta Physiol Scand Suppl* 641: 1–55.
41. Lynch, K. R., G. P. O'Neill, Q. Liu, D. S. Im, N. Sawyer, K. M. Metters, N. Coulombe et al. 1999. Characterization of the human cysteinyl leukotriene CysLT1 receptor. *Nature* 399: 789–793.
42. Sarau, H. M., R. S. Ames, J. Chambers, C. Ellis, N. Elshourbagy, J. J. Foley, D. B. Schmidt et al. 1999. Identification, molecular cloning, expression, and characterization of a cysteinyl leukotriene receptor. *Mol Pharmacol* 56: 657–663.
43. Figueroa, D. J., R. M. Breyer, S. K. Defoe, S. Kargman, B. L. Daugherty, K. Waldburger, Q. Liu et al. 2001. Expression of the cysteinyl leukotriene 1 receptor in normal human lung and peripheral blood leukocytes. *Am J Respir Criti Care Med* 163: 226–233.
44. D'Agostino, B., L. Gallelli, M. Falciani, P. Di Pierro, F. Rossi, and A. Filippelli. 1999. Endothelin-1 induced bronchial hyperresponsiveness in the rabbit: An ET(A) receptor-mediated phenomenon. *Naunyn-Schmiedeberg's Arch Pharmacol* 360: 665–669.
45. Fehr, J. J., C. A. Hirshman, and C. W. Emala. 2000. Cellular signaling by the potent bronchoconstrictor endothelin-1 in airway smooth muscle. *Crit Care Med* 28: 1884–1888.
46. Goldie, R. G., P. J. Henry, P. G. Knott, G. J. Self, M. A. Luttmann, and D. W. Hay. 1995. Endothelin-1 receptor density, distribution, and function in human isolated asthmatic airways. *Am J Respir Criti Care Med* 152: 1653–1658.
47. Hay, D. W., M. A. Luttmann, R. M. Muccitelli, and R. G. Goldie. 1999. Endothelin receptors and calcium translocation pathways in human airways. *Naunyn-Schmiedeberg's Arch Pharmacol* 359: 404–410.
48. Kizawa, Y., N. Ohuchi, K. Saito, T. Kusama, and H. Murakami. 2001. Effects of endothelin-1 and nitric oxide on proliferation of cultured guinea pig bronchial smooth muscle cells. *Comp Biochem Physiol C Toxicol Pharmacol* 128: 495–501.
49. Carratu, P., M. Scuri, J. L. Styblo, A. Wanner, and M. K. Glassberg. 1997. ET-1 induces mitogenesis in ovine airway smooth muscle cells via ETA and ETB receptors. *Am J Physiol* 272: 1021–1024.
50. El-Mowafy, A. M., and D. F. Biggs. 2001. ETB receptor activates adenylyl cyclase via a c-PLA2-dependent mechanism: A novel counterregulatory mechanism of ET-induced contraction in airway smooth muscle. *Biochem Biophys Res Commun* 286: 388–393.
51. Ndukwu, I. M., S. R. White, A. R. Leff, and R. W. Mitchell. 1997. EP1 receptor blockade attenuates both spontaneous tone and PGE2-elicited contraction in guinea pig trachealis. *Am J Physiol Lung Cell Mol Physiol* 273: L626–L633.
52. McGraw, D. W., K. A. Mihlbachler, M. R. Schwarb, F. F. Rahman, K. M. Small, K. F. Almoosa, and S. B. Liggett. 2006. Airway smooth muscle prostaglandin-EP1 receptors directly modulate beta2-adrenergic receptors within a unique heterodimeric complex. *J Clin Invest* 116: 1400–1409.
53. Grandordy, B. M., N. Frossard, K. J. Rhoden, and P. J. Barnes. 1988. Tachykinin-induced phosphoinositide breakdown in airway smooth muscle and epithelium: Relationship to contraction. *Mol Pharmacol* 33: 515–519.
54. Mak, J. C., M. Astolfi, X. L. Zhang, S. Evangelista, S. Manzini, and P. J. Barnes. 1996. Autoradiographic mapping of pulmonary NK1 and NK2 tachykinin receptors and changes after repeated antigen challenge in guinea pigs. *Peptides* 17: 1389–1395.

55. Panettieri, R. A., R. K. Murray, L. R. DePalo, P. A. Yadvish, and M. I. Kotlikoff. 1989. A human smooth muscle cell line that retains physiological responsiveness. *Am J Physiol* 256: C329–C335.
56. Noveral, J. P., and M. M. Grunstein. 1995. Tachykinin regulation of airway smooth muscle cell proliferation. *Am J Physiol* 269: 339–343.
57. Goldie, R. G. 1990. Receptors in asthmatic airways. *Am Rev Respir Dis* 141: 151–156.
58. Noveral, J. P., and M. M. Grunstein. 1994. Adrenergic receptor-mediated regulation of cultures rabbit airway smooth muscle cell regulation. *Am J Physiol* 267: L291–L299.
59. Barnes, P. J., and C. B. Basbaum. 1983. Mapping of adrenergic receptors in the trachea by autoradiography. *Exp Lung Res* 5: 183–192.
60. Barnes, P. J., C. B. Basbaum, and J. A. Nadel. 1983. Autoradiographic localization of autonomic receptors in airway smooth muscle. Marked differences between large and small airways. *Am Rev Respir Dis* 127: 758–762.
61. Noveral, J. P., and M. M. Grunstein. 1992. Role and mechanism of thromboxane-induced proliferation of cultured airway smooth muscle cells. *Am J Physiol* 263: 555–561.
62. Armour, C. L., P. R. Johnson, M. L. Alfredson, and J. L. Black. 1989. Characterization of contractile prostanoid receptors on human airway smooth muscle. *Eur J Pharmacol* 165: 215–222.
63. Yarova, P. L., A. L. Stewart, V. Sathish, R. D. Britt, M. A. Thompson, P. Lowe, M. Freeman et al. 2015. Calcium-sensing receptor antagonists abrogate airway hyper-responsiveness and inflammation in allergic asthma. *Sci Transl Med* 7: 284ra260.
64. Michoud, M. C., B. Tolloczko, and J. G. Martin. 1997. Effects of purine nucleotides and nucleoside on cytosolic calcium levels in rat tracheal smooth muscle cells. *Am J Respir Cell Mol Biol* 16: 199–205.
65. Berger, P., D. W. Perng, H. Thabrew, S. J. Compton, J. A. Cairns, A. R. McEuen, R. Marthan, J. M. Tunon De Lara, and A. F. Walls. 2001. Tryptase and agonists of PAR-2 induce the proliferation of human airway smooth muscle cells. *J Appl Physiol* 91: 1372–1379.
66. Hauck, R. W., C. Schulz, A. Schomig, R. K. Hoffman, and R. A. Panettieri Jr. 1999. alpha-Thrombin stimulates contraction of human bronchial rings by activation of protease-activated receptors. *Am J Physiol* 277: 22–29.
67. Mizuta, K., Y. Zhang, F. Mizuta, H. Hoshijima, T. Shiga, E. Masaki, and C. W. Emala. 2015. Novel identification of the free fatty acid receptor FFAR1 that promotes contraction in airway smooth muscle. *Am J Physiol Lung Cell Mol Physiol* 309: L970–L982.
68. Gosens, R., D. Schaafsma, M. M. Grootte Bromhaar, B. Vrugt, J. Zaagsma, H. Meurs, and S. A. Nelemans. 2004. Growth factor-induced contraction of human bronchial smooth muscle is Rho-kinase-dependent. *Eur J Pharmacol* 494: 73–76.
69. Li, N., R. Cai, Y. Niu, B. Shen, J. Xu, and Y. Cheng. 2012. Inhibition of angiotensin II-induced contraction of human airway smooth muscle cells by angiotensin-(1-7) via downregulation of the RhoA/ROCK2 signaling pathway. *Int J Mol Med* 30: 811–818.
70. Amrani, Y., F. Syed, C. Huang, K. Li, V. Liu, D. Jain, S. Keslacy et al. 2010. Expression and activation of the oxytocin receptor in airway smooth muscle cells: Regulation by TNFα and IL-13. *Respir Res* 11: 104–104.
71. Widdop, S., K. Daykin, and I. P. Hall. 1993. Expression of muscarinic M2 receptors in cultured human airway smooth muscle cells. *Am J Respir Cell Mol Biol* 9: 541–546.
72. Govindaraju, V., M. C. Michoud, M. Al-Chalabi, P. Ferraro, W. S. Powell, and J. G. Martin. 2006. Interleukin-8: Novel roles in human airway smooth muscle cell contraction and migration. *Am J Physiol Cell Physiol* 291: C957–C965.
73. Tolloczko, B., Y. L. Jia, and J. G. Martin. 1995. Serotonin-evoked calcium transients in airway smooth muscle cells. *Am J Physiol* 269: 234–240.

74. Cerutis, D. R., M. Nogami, J. L. Anderson, J. D. Churchill, D. J. Romberger, S. I. Rennard, and M. L. Toews. 1997. Lysophosphatidic acid and EGF stimulate mitogenesis in human airway smooth muscle cells. *Am J Physiol* 273: L10–L15.
75. Nogami, M., S. M. Whittle, D. J. Romberger, S. I. Rennard, and M. Toews. 1995. Lysophosphatidic acid regulation of cyclic AMP accumulation in cultured human airway smooth muscle cells. *Mol Pharmacol* 48: 766–773.
76. Toews, M. L., E. E. Ustinova, and H. D. Schultz. 1997. Lysophosphatidic acid enhances contractility of isolated airway smooth muscle. *J Appl Physiol* 83: 1216–1222.
77. Yang, C. M., Y. L. Yo, J. T. Hsieh, and R. Ong. 1994. 5-Hydroxytryptamine receptor-mediated phosphoinositide hydrolysis in canine cultured tracheal smooth muscle cells. *Br J Pharmacol* 111: 777–786.
78. Tao, F. C., B. Tolloczko, C. A. Mitchell, W. S. Powell, and J. G. Martin. 2000. Inositol (1,4,5)trisphosphate metabolism and enhanced calcium mobilization in airway smooth muscle of hyperresponsive rats. *Am J Respir Cell Mol Biol* 23: 514–520.
79. Zacour, M. E., and J. G. Martin. 1996. Enhanced growth response of airway smooth muscle in inbred rats with airway hyperresponsiveness. *Am J Respir Cell Mol Biol* 15: 590–599.
80. Abebe, W., and S. J. Mustafa. 1998. A1 adenosine receptor-mediated Ins(1,4,5)P3 generation in allergic rabbit airway smooth muscle. *Am J Physiol* 275: 990–997.
81. Nyce, J. W., and W. J. Metzger. 1997. DNA antisense therapy for asthma in an animal model. *Nature* 385: 721–725.
82. Yan, H., D. A. Deshpande, A. M. Misior, M. C. Miles, H. Saxena, E. C. Riemer, R. M. Pascual, R. A. Panettieri, and R. B. Penn. 2011. Anti-mitogenic effects of beta-agonists and PGE2 on airway smooth muscle are PKA dependent. *FASEB J* 25: 389–397.
83. Burgess, J. K., Q. Ge, S. Boustany, J. L. Black, and P. R. A. Johnson. 2004. Increased sensitivity of asthmatic airway smooth muscle cells to prostaglandin E2 might be mediated by increased numbers of E-prostanoid receptors. *J Allergy Clin Immunol* 113: 876–881.
84. Mori, A., S. Ito, M. Morioka, H. Aso, M. Kondo, M. Sokabe, and Y. Hasegawa. 2011. Effects of specific prostanoid EP receptor agonists on cell proliferation and intracellular Ca^{2+} concentrations in human airway smooth muscle cells. *Eur J Pharmacol* 659: 72–78.
85. Osawa, Y., D. Xu, D. Sternberg, J. R. Sonett, J. D'Armiento, R. A. Panettieri, and C. W. Emala. 2006. Functional expression of the GABAB receptor in human airway smooth muscle. *Am J Physiol Lung Cel Mol Physiol* 291: L923–L931.
86. Mizuta, K., Y. Zhang, D. Xu, E. Masaki, R. A. Panettieri, and C. W. Emala. 2012. The dopamine D(2) receptor is expressed and sensitizes adenylyl cyclase activity in airway smooth muscle. *Am J Physiol Lung Cell Mol Physiol* 302: L316–L324.
87. Watanabe, S., A. Yamasaki, K. Hashimoto, Y. Shigeoka, H. Chikumi, Y. Hasegawa, T. Sumikawa et al. 2009. Expression of functional leukotriene B(4) receptors on human airway smooth muscle cells. *J Allergy Clin Immunol* 124: 59–65, e51–53.
88. Kumawat, K., M. H. Menzen, I. S. T. Bos, H. A. Baarsma, P. Borger, M. Roth, N. Tamm et al. 2013. Noncanonical WNT-5A signaling regulates TGF-β-induced extracellular matrix production by airway smooth muscle cells. *FASEB J* 27: 1631–1643.
89. Kumawat, K., T. Koopmans, M. H. Menzen, A. Prins, M. Smit, A. J. Halayko, and R. Gosens. 2016. Cooperative signaling by TGF-β1 and WNT-11 drives sm-α-actin expression in smooth muscle via Rho kinase-actin-MRTF-A signaling. *Am J Physiol Lung Cell Mol Physiol* 311: L529–L537.
90. Koopmans, T., K. Kumawat, A. J. Halayko, and R. Gosens. 2016. Regulation of actin dynamics by WNT-5A: Implications for human airway smooth muscle contraction. *Sci Rep* 6: 30676.

91. Deshpande, D. A., W. C. Wang, E. L. McIlmoyle, K. S. Robinett, R. M. Schillinger, S. S. An, J. S. Sham, and S. B. Liggett. 2010. Bitter taste receptors on airway smooth muscle bronchodilate by localized calcium signaling and reverse obstruction. *Nat Med* 16: 1299–1304.
92. Robinett, K. S., D. A. Deshpande, M. M. Malone, and S. B. Liggett. 2011. Agonist-promoted homologous desensitization of human airway smooth muscle bitter taste receptors. *Am J Respir Cell Mol Biol* 45: 1069–1074.
93. Aisenberg, W. H., J. Huang, W. Zhu, P. Rajkumar, R. Cruz, L. Santhanam, N. Natarajan et al. 2016. Defining an olfactory receptor function in airway smooth muscle cells. *Sci Rep* 6: 38231.
94. Faiz, A., C. Donovan, M. A. Nieuwenhuis, M. van den Berge, D. S. Postma, S. Yao, C. Y. Park et al. 2017. Latrophilin receptors: Novel bronchodilator targets in asthma. *Thorax* 72: 74–82.
95. Walker, J. K., R. B. Penn, N. A. Hanania, B. F. Dickey, and R. A. Bond. 2011. New perspectives regarding beta(2)-adrenoceptor ligands in the treatment of asthma. *Br J Pharmacol* 163: 18–28.
96. Penn, R. B., V. E. Ortega, and E. R. Bleecker. 2007. A roadmap to functional genomics. *Physiol Genomics* 30: 82–88.
97. Deshpande, D. A., B. S. Theriot, R. B. Penn, and J. K. Walker. 2008. β-Arrestins specifically constrain β2-adrenergic receptor signaling and function in airway smooth muscle. *FASEB J* 22: 2134–2141.
98. Pera, T., A. Hegde, D. A. Deshpande, S. J. Morgan, B. C. Tiegs, B. S. Theriot, Y. H. Choi, J. K. Walker, and R. B. Penn. 2015. Specificity of arrestin subtypes in regulating airway smooth muscle G protein-coupled receptor signaling and function. *FASEB J* 29: 4227–4235.
99. Penn, R. B., R. A. Panettieri Jr, and J. L. Benovic. 1998. Mechanisms of acute desensitization of the β_2AR-adenylyl cyclase pathway in human airway smooth muscle. *Am J Respir Cell Mol Biol* 19: 338–348.
100. Amrani, Y., P. E. Moore, R. Hoffman, S. A. Shore, and R. A. Panettieri Jr. 2001. Interferon-gamma modulates cysteinyl leukotriene receptor-1 expression and function in human airway myocytes. *Am J Respir Criti Care Med* 164: 2098–2101.
101. Naik, S., C. K. Billington, R. M. Pascual, D. A. Deshpande, F. P. Stefano, T. A. Kohout, D. M. Eckman, J. L. Benovic, and R. B. Penn. 2005. Regulation of cysteinyl leukotriene type 1 receptor internalization and signaling. *J Biol Chem* 280: 8722–8732.
102. Billington, C. K., K. C. Kong, R. Bhattacharyya, P. B. Wedegaertner, R. A. Panettieri, T. O. Chan, and R. B. Penn. 2005. Cooperative regulation of p70S6 kinase by receptor tyrosine kinases and G protein-coupled receptors augments airway smooth muscle growth. *Biochemistry* 44: 14595–14605.
103. Harris, V. M. 2015. Protein detection by simple western™ analysis. In *Western Blotting: Methods and Protocols*, B. T. Kurien and R. H. Scofield (Eds.). New York: Springer, pp. 465–468.
104. Deshpande, D. A., B. Tiegs, M. Ippolito, H. Schorsch, T. Murphy, and R. Penn. 2016. Desmethylbenzodiazepines induce OGR1 signaling in airway smooth muscle cells. In *A29. Inflammation and Mechanisms of Airway Smooth Muscle Contraction*, pp. A1269–A1269.
105. Bogard, A. S., P. Adris, and R. S. Ostrom. 2012. Adenylyl cyclase 2 selectively couples to E prostanoid type 2 receptors, whereas adenylyl cyclase 3 is not receptor-regulated in airway smooth muscle. *J Pharmacol Exp Ther* 342: 586–595.
106. Tang, D. D., C. E. Turner, and S. J. Gunst. 2003. Expression of non-phosphorylatable paxillin mutants in canine tracheal smooth muscle inhibits tension development. *J Physiol* 553: 21–35.

107. Gosens, R., D. Rieks, H. Meurs, D. K. Ninaber, K. F. Rabe, J. Nanninga, S. Kolahian, A. J. Halayko, P. S. Hiemstra, and S. Zuyderduyn. 2009 Muscarinic M3 receptor stimulation increases cigarette smoke-induced IL-8 secretion by human airway smooth muscle cells. *Eur Respir J* 34: 1436–1443.
108. Eidelman, O., N. Marozkina, M. Srivastava, W. Huang, W. Guilford, R. Panettieri, B. Gaston, and H. Pollard. 2009. S-nitrosylation targets in human airway smooth muscle cells: Simultaneous measurement of protein expression and cysteine modification on a large scale antibody microarray platform. *Am J Respir Criti Care Med* 1: A2059.
109. Vivieros, J., S. Sidoli, G. Cao, B. A. Garcia, A. Reynold, and J. Panettieri. 2016. Elevated histone H3K27 methylation mediates intrinsic hypercontractility in human airway smooth muscle cells from fatal asthma subjects. In *A77. Airway Smooth Muscle: Tuning, Tinkering, and Treating*, pp. A2462–A2462.
110. Morgan, S. J., D. A. Deshpande, B. C. Tiegs, A. M. Misior, H. Yan, A. V. Hershfeld, T. C. Rich, R. A. Panettieri, S. S. An, and R. B. Penn. 2014. β-Agonist-mediated relaxation of airway smooth muscle is protein kinase A-dependent. *J Biol Chem* 289: 23065–23074.
111. An, S. S., R. E. Laudadio, J. Lai, R. A. Rogers, and J. J. Fredberg. 2002. Stiffness changes in cultured airway smooth muscle cells. *Am J Physiol Cell Physiol* 283: C792–C801.
112. Gosens, R., G. L. Stelmack, G. Dueck, K. D. McNeill, A. Yamasaki, W. T. Gerthoffer, H. Unruh, A. S. Gounni, J. Zaagsma, and A. J. Halayko. 2006. Role of caveolin-1 in p42/p44 MAP kinase activation and proliferation of human airway smooth muscle. *Am J Physiol Lung Cell Mol Physiol* 291: L523–L534.
113. Hassan, M., T. Jo, P. A. Risse, B. Tolloczko, C. Lemière, R. Olivenstein, Q. Hamid, and J. G. Martin. 2010. Airway smooth muscle remodeling is a dynamic process in severe long-standing asthma. *J Allergy Clin Immunol* 125: 1037–1045.e1033.
114. Sharma, P., A. Panebra, T. Pera, B. C. Tiegs, A. Hershfeld, L. C. Kenyon, and D. A. Deshpande. 2016. Antimitogenic effect of bitter taste receptor agonists on airway smooth muscle cells. *Am J Physiol Lung Cell Mol Physiol* 310: L365–L376.
115. Orsini, M. J., V. P. Krymskaya, A. J. Eszterhas, J. L. Benovic, R. A. Panettieri, and R. B. Penn. 1999. MAPK superfamily activation in human airway smooth muscle: Prolonged p42/p44 activation required for mitogenesis. *Am J Physiol Lung Cell Mol Physiol* 277: 479–488.
116. Gosens, R., D. Schaafsma, H. Meurs, J. Zaagsma, and S. A. Nelemans. 2004. Role of Rho-kinase in maintaining airway smooth muscle contractile phenotype. *Eur J Pharmacol* 483: 71–78.
117. Gosens, R., G. L. Stelmack, S. T. Bos, G. Dueck, M. M. Mutawe, D. Schaafsma, H. Unruh et al. 2011. Caveolin-1 is required for contractile phenotype expression by airway smooth muscle cells. *J Cell Mol Med* 15: 2430–2442.
118. Sasse, D. S. K., D. V. Kadiyala, T. Danhorn Jr., T. L. Phang, and D. A. N. Gerber. 2017. Glucocorticoid receptor ChIP-seq identifies PLCD1 as a KLF15 target that represses airway smooth muscle hypertrophy. *Am J Respir Cell Mol Biol* 57: 226–237.
119. Ma, L., M. Brown, P. Kogut, K. Serban, X. Li, J. McConville, B. Chen et al. 2011. Akt activation induces hypertrophy without contractile phenotypic maturation in airway smooth muscle. *Am J Physiol Lung Cell Mol Physiol* 300: L701–L709.
120. Deshpande, D. A., R. M. Pascual, S. Wang, D. M. Eckman, E. C. Riemer, C. D. Funk, and R. B. Penn. 2007. PKC-dependent regulation of the receptor locus dominates functional consequences of cysteinyl leukotriene type 1 receptor activation. *FASEB J* 21: 2335–2342.
121. Goncharova, E. A., D. A. Goncharov, H. Zhao, R. B. Penn, V. P. Krymskaya, and R. A. Panettieri Jr. 2012. beta2-adrenergic receptor agonists modulate human airway smooth muscle cell migration via vasodilator-stimulated phosphoprotein. *Am J Respir Cell Mol Biol* 46: 48–54.

122. Goncharova, E. A., D. A. Goncharov, and V. P. Krymskaya. 2007. Assays for in vitro monitoring of human airway smooth muscle (ASM) and human pulmonary arterial vascular smooth muscle (VSM) cell migration. *Nat Protoc* 1: 2933–2939.
123. Misior, A. M., D. A. Deshpande, M. J. Loza, R. M. Pascual, J. D. Hipp, and R. B. Penn. 2009. Glucocorticoid- and protein kinase A-dependent transcriptome regulation in airway smooth muscle. *Am J Respir Cell Mol Biol* 41: 24–39.
124. Yick, C. Y., A. H. Zwinderman, P. W. Kunst, K. Grünberg, T. Mauad, K. Fluiter, E. H. Bel, R. Lutter, F. Baas, and P. J. Sterk. 2013. Glucocorticoid-induced changes in gene expression of airway smooth muscle in patients with asthma. *Am J Respir Criti Care Med* 187: 1076–1084.
125. Dragon, S., S. J. Hirst, T. H. Lee, and A. S. Gounni. 2014. IL-17A mediates a selective gene expression profile in asthmatic human airway smooth muscle cells. *Am J Respir Cell Mol Biol* 50: 1053–1063.
126. Masuno, K., S. M. Haldar, D. Jeyaraj, C. M. Mailloux, X. Huang, R. A. Panettieri, M. K. Jain, and A. N. Gerber. 2011. Expression profiling identifies Klf15 as a glucocorticoid target that regulates airway hyperresponsiveness. *Am J Respir Cell Mol Biol* 45: 642–649.
127. Morten, B. C., R. J. Scott, and K. A. Avery-Kiejda. 2016. Comparison of the QuantiGene 2.0 assay and real-time RT-PCR in the detection of p53 isoform mRNA expression in formalin-fixed paraffin-embedded tissues—A preliminary study. *PLoS One* 11: e0165930.
128. Warrior, U., Y. Fan, C. A. David, J. A. Wilkins, E. M. McKeegan, J. L. Kofron, and D. J. Burns. 2000. Application of QuantiGene nucleic acid quantification technology for high throughput screening. *J Biomol Screen* 5: 343–352.
129. Bergner, A., and M. J. Sanderson. 2002. Acetylcholine-induced calcium signaling and contraction of airway smooth muscle cells in lung slices. *J Gen Physiol* 119: 187–198.
130. Gosens, R., I. S. Bos, J. Zaagsma, and H. Meurs. 2005. Protective effects of tiotropium bromide in the progression of airway smooth muscle remodeling. *Am J Respir Criti Care Med* 171: 1096–1102.
131. Xie, Y., H. Jiang, H. Nguyen, S. Jia, A. Berro, R. A. Panettieri Jr, D. W. Wolff, P. W. Abel, T. B. Casale, and Y. Tu. 2012. Regulator of G protein signaling 2 is a key modulator of airway hyperresponsiveness. *J Allergy Clin Immunol* 130: 968–976. e963.
132. Meurs, H., A. F. Roffel, J. B. Postema, A. Timmermans, C. R. Elzinga, H. F. Kauffman, and J. Zaagsma. 1988. Evidence for a direct relationship between phosphoinositide metabolism and airway smooth muscle contraction induced by muscarinic agonists. *Eur J Pharmacol* 156: 271–274.
133. Roffel, A. F., C. R. Elzinga, and J. Zaagsma. 1993. Cholinergic contraction of the guinea pig lung strip is mediated by muscarinic M2-like receptors. *Eur J Pharmacol* 250: 267–279.
134. van Waarde, A., B. Maas, P. Doze, R. H. Slart, H. W. Frijlink, W. Vaalburg, and P. H. Elsinga. 2005. Positron emission tomography studies of human airways using an inhaled β-adrenoceptor antagonist, S-11 C-CGP 12388. *Chest* 128: 3020–3027.
135. Meurs, H., R. E. Santing, R. Remie, T. W. van der Mark, F. J. Westerhof, A. B. Zuidof, I. S. T. Bos, and J. Zaagsma. 2006. A guinea pig model of acute and chronic asthma using permanently instrumented and unrestrained animals. *Nat Protoc* 1: 840–847.
136. Martin, T. R., N. P. Gerard, S. J. Galli, and J. M. Drazen. 1988. Pulmonary responses to bronchoconstrictor agonists in the mouse. *J Appl Physiol* 64: 2318–2323.
137. Hamelmann, E., J. Schwarze, K. Takeda, A. Oshiba, G. L. Larsen, C. G. Irvin, and E. W. Gelfand. 1997. Noninvasive measurement of airway responsiveness in allergic mice using barometric plethysmography. *Am J Respir Criti Care Med* 156: 766–775.

138. Plant, P. J., M. L. North, A. Ward, M. Ward, N. Khanna, J. Correa, J. A. Scott, and J. Batt. 2012. Hypertrophic airway smooth muscle mass correlates with increased airway responsiveness in a Murine model of asthma. *Am J Respir Cell Mol Biol* 46: 532–540.
139. Yick, C. Y., A. H. Zwinderman, P. W. Kunst, K. Grünberg, T. Mauad, S. Chowdhury, E. H. Bel, F. Baas, R. Lutter, and P. J. Sterk. 2014. Gene expression profiling of laser microdissected airway smooth muscle tissue in asthma and atopy. *Allergy* 69: 1233–1240.
140. Horvat, S. J., D. A. Deshpande, H. Yan, R. A. Panettieri, J. Codina, T. D. DuBose Jr, W. Xin, T. C. Rich, and R. B. Penn. 2012. A-kinase anchoring proteins regulate compartmentalized cAMP signaling in airway smooth muscle. *FASEB J* 26: 3670–3679.
141. Li, J., R. Wang, O. J. Gannon, A. C. Rezey, S. Jiang, B. D. Gerlach, G. Liao, and D. D. Tang. 2016. Polo-like kinase 1 regulates vimentin phosphorylation at Ser-56 and contraction in smooth muscle. *J Biol Chem* 291: 23693–23703.
142. Chachi, L., Abbasian, M., A. Gavrila, A. Alzahrani, O. Tliba, P. Bradding, A. J. Wardlaw, C. Brightling, and Y. Amrani. 2017. Protein phosphatase 5 mediates corticosteroid insensitivity in airway smooth muscle in patients with severe asthma. *Allergy* 72: 126–136.
143. Koziol-White, C. J., E. J. Yoo, G. Cao, J. Zhang, E. Papanikolaou, I. Pushkarsky, A. Andrews. 2016. Inhibition of PI3K promotes dilation of human small airways in a rho kinase-dependent manner. *Br J Pharmacol* 173: 2726–2738.
144. Schuliga, M., S. Langenbach, Y. C. Xia, C. Qin, J. S. L. Mok, T. Harris, G. A. Mackay, R. L. Medcalf, and A. G. Stewart. 2013. Plasminogen-stimulated inflammatory cytokine production by airway smooth muscle cells is regulated by annexin A2. *Am J Respir Cell Mol Biol* 49: 751–758.
145. Chen, C., M. Kudo, F. Rutaganira, H. Takano, C. Lee, A. Atakilit, K. S. Robinett et al. 2012. Integrin $\alpha(9)\beta(1)$ in airway smooth muscle suppresses exaggerated airway narrowing. *J Clin Invest* 122: 2916–2927.
146. Zhu, M., L. Flynt, S. Ghosh, M. Mellema, A. Banerjee, E. Williams, R. A. Panettieri, and S. A. Shore. 2011. Anti-inflammatory effects of thiazolidinediones in human airway smooth muscle cells. *Am J Respir Cell Mol Biol* 45: 111–119.
147. Jansen, S. R., A. M. Van Ziel, H. A. Baarsma, and R. Gosens. 2010. β-Catenin regulates airway smooth muscle contraction. *Am J Physiol Lung Cell Mol Physiol* 299: L204–L214.
148. Billington, C. K., I. J. Le Jeune, K. W. Young, and I. P. Hall. 2007. A major functional role for phosphodiesterase 4D5 in human airway smooth muscle cells. *Am J Respir Cell Mol Biol* 38: 1–7.
149. Yocum, G. T., G. Gallos, Y. Zhang, R. Jahan, M. R. Stephen, Z. Varagic, R. Puthenkalam, M. Ernst, J. M. Cook, and C. W. Emala. 2015. Targeting the GABAA receptor $\alpha 4$ subunit in airway smooth muscle to alleviate bronchoconstriction. *Am J Respir Cell Mol Biol* 54: 546–553.
150. Birrell, M. A., S. J. Bonvini, M. A. Wortley, J. Buckley, L. Yew-Booth, S. A. Maher, N. Dale, E. D. Dubuis, and M. G. Belvisi. 2015. The role of adenylyl cyclase isoform 6 in beta-adrenoceptor signalling in murine airways. *Br J Pharmacol* 172: 131–141.
151. Aravamudan, B., S. K. VanOosten, L. W. Meuchel, P. Vohra, M. Thompson, G. C. Sieck, Y. S. Prakash, and C. M. Pabelick. 2012. Caveolin-1 knockout mice exhibit airway hyperreactivity. *Am J Physiol Lung Cell Mol Physiol* 303: L669–L681.
152. Holden, N. S., M. J. Bell, C. F. Rider, E. M. King, D. D. Gaunt, R. Leigh, M. Johnson et al. 2011. $\beta(2)$-Adrenoceptor agonist-induced RGS2 expression is a genomic mechanism of bronchoprotection that is enhanced by glucocorticoids. *Proc Natl Acad Sci USA* 108: 19713–19718.

153. Allen, C., J. Hartney, T. Coffman, R. B. Penn, J. Wess, and B. H. Koller. 2005. Thromboxane A2 induces airway constriction through an M3 muscarinic acetylcholine receptor dependent mechanism. *Am J Physiol Lung Cell Mol Physiol* 21: 21.
154. Deshpande, D. A., H. Yan, K. C. Kong, B. C. Tiegs, S. J. Morgan, T. Pera, R. A. Panettieri, A. D. Eckhart, and R. B. Penn. 2014. Exploiting functional domains of GRK2/3 to alter the competitive balance of pro- and anticontractile signaling in airway smooth muscle. *FASEB J* 28: 956–965.
155. Kong, K. C., U. Gandhi, T. J. Martin, C. B. Anz, H. Yan, A. M. Misior, R. M. Pascual, D. A. Deshpande, and R. B. Penn. 2008. Endogenous Gs-coupled receptors in smooth muscle exhibit differential susceptibility to GRK2/3-mediated desensitization. *Biochemistry* 47: 9279–9288.
156. Wright, D., P. Sharma, M. H. Ryu, P. A. Rissé, M. Ngo, H. Maarsingh, C. Koziol-White, A. Jha, A. J. Halayko, and A. R. West. 2013. Models to study airway smooth muscle contraction in vivo, ex vivo and in vitro: Implications in understanding asthma. *Pulm Pharmacol Ther* 26: 24–36.
157. Halayko, A. J., B. Camoretti-Mercado, S. M. Forsythe, J. E. Vieira, R. W. Mitchell, M. E. Wylam, M. B. Hershenson, and J. Solway. 1999. Divergent differentiation paths in airway smooth muscle culture: Induction of functionally contractile myocytes. *Am J Physiol* 276: 197–206.
158. Held, H.-D., C. Martin, and S. Uhlig. 1999. Characterization of airway and vascular responses in murine lungs. *Br J Pharmacol* 126: 1191–1199.
159. Birrell, M. A., S. A. Maher, B. Dekkak, V. Jones, S. Wong, P. Brook, and M. G. Belvisi. 2015. Anti-inflammatory effects of PGE2 in the lung: Role of the EP4 receptor subtype. *Thorax* 70: 740–747.
160. Sanderson, M. J. 2011. Exploring lung physiology in health and disease with lung slices. *Pulm Pharmacol Ther* 24: 452–465.
161. Foster, R. W. 1964. A note on the electrically transmurally stimulated isolated trachea of the guinea-pig. *J Pharm Pharmacol* 16: 125–128.
162. Schlepütz, M., S. Uhlig, and C. Martin. 2011. Electric field stimulation of precision-cut lung slices. *J Appl Physiol* 110: 545–554.
163. Stefaniak, M. S. 1992. Biochemical and histological characterization of agar-filled precision-cut rat lung slices in dynamic organ culture as an in vitro tool. *In Vitro Toxicol* 5: 7.
164. Rosner, S. R., S. Ram-Mohan, J. R. Paez-Cortez, T. L. Lavoie, M. L. Dowell, L. Yuan, X. Ai et al. 2014. Airway contractility in the precision-cut lung slice after cryopreservation. *Am J Respir Cell Mol Biol* 50: 876–881.
165. Watson, C. Y., F. Damiani, S. Ram-Mohan, S. Rodrigues, P. de Moura Queiroz, T. C. Donaghey, J. H. Rosenblum Lichtenstein, J. D. Brain, R. Krishnan, and R. M. Molina. 2016. Screening for chemical toxicity using cryopreserved precision cut lung slices. *Toxicol Sci* 150: 225–233.
166. Cooper, P. R., J. Zhang, G. Damera, T. Hoshi, D. A. Zopf, and R. A. Panettieri Jr. 2011. C-027 Inhibits IgE-mediated passive sensitization bronchoconstriction and acts as a histamine and serotonin antagonist in human airways. *Allergy Asthma Proc* 32: 359–365.
167. Sharma, P., R. Yi, A. P. Nayak, N. Wang, F. Tang, M. J. Knight, S. Pan, B. Oliver, and D. A. Deshpande. 2017. Bitter taste receptor agonists mitigate features of allergic asthma in mice. *Sci Rep* 7: 46166.
168. Grainge, C. L., L. C. K. Lau, J. A. Ward, V. Dulay, G. Lahiff, S. Wilson, S. Holgate, D. E. Davies, and P. H. Howarth. 2011. Effect of bronchoconstriction on airway remodeling in asthma. *N Engl J Med* 364: 2006–2015.
169. Zosky, G. R., and P. D. Sly. 2007. Animal models of asthma. *Clin Exp Allergy* 37: 973–988.
170. Nials, A. T., and S. Uddin. 2008. Mouse models of allergic asthma: Acute and chronic allergen challenge. *Dis Models Mech* 1: 213–220.

171. Prakash, Y. S., A. J. Halayko, R. Gosens, R. A. Panettieri Jr, B. Camoretti-Mercado, and R. B. Penn. 2017. An official American thoracic society research statement: Current challenges facing research and therapeutic advances in airway remodeling. *Am J Respir Criti Care Med* 195: e4–e19.
172. Tang, D. D. 2015. Critical role of actin-associated proteins in smooth muscle contraction, cell proliferation, airway hyperresponsiveness and airway remodeling. *Respir Res* 16: 134.
173. Kong, K. C., C. K. Billington, U. Gandhi, R. A. Panettieri Jr., and R. B. Penn. 2006. Cooperative mitogenic signaling by G protein-coupled receptors and growth factors is dependent on G(q/11). *FASEB J* 20: 1558–1560.
174. Tan, X., and M. J. Sanderson. 2014. Bitter tasting compounds dilate airways by inhibiting airway smooth muscle calcium oscillations and calcium sensitivity. *Br J Pharmacol* 171: 646–662.
175. Zhang, C.-H., L. M. Lifshitz, K. F. Uy, M. Ikebe, K. E. Fogarty, and R. ZhuGe. 2013. The cellular and molecular basis of bitter tastant-induced bronchodilation. *PLoS Biol* 11: e1001501.
176. Penn, R. B., J. L. Parent, A. N. Pronin, R. A. Panettieri Jr., and J. L. Benovic. 1999. Pharmacological inhibition of protein kinases in intact cells: Antagonism of beta adrenergic receptor ligand binding by H-89 reveals limitations of usefulness. *J Pharmacol Exp Ther* 288: 428–437.
177. Karpova, A. Y., M. K. Abe, J. Li, P. Liu, J. M. Rhee, W. L. Kuo, and M. B. Hershenson. 1997. MEK1 is required for PDGF-induced ERK activation and DNA synthesis in tracheal myocytes. *Am J Physiol* 16: L558–L565.
178. Yu, M. F., J. I. Ewaskiewicz, S. Adda, K. Bailey, V. Harris, D. Sosnoski, M. Tomasic, J. Wilson, and M. I. Kotlikoff. 1996. Gene transfer by adenovirus in smooth muscle cells. *Respir Physiol* 105: 155–162.
179. Guo, M., R. M. Pascual, S. Wang, M. F. Fontana, C. A. Valancius, R. A. Panettieri Jr, S. L. Tilley, and R. B. Penn. 2005. Cytokines regulate beta-2-adrenergic receptor responsiveness in airway smooth muscle via multiple PKA- and EP2 receptor-dependent mechanisms. *Biochemistry* 44: 13771–13782.
180. Zhu, Z., R. J. Homer, Z. Wang, Q. Chen, G. P. Geba, J. Wang, Y. Zhang, and J. A. Elias. 1999. Pulmonary expression of interleukin-13 causes inflammation, mucus hypersecretion, subepithelial fibrosis, physiologic abnormalities, and eotaxin production. *J Clin Invest* 103: 779–788.
181. Pauwels, R. A., G. J. Brusselle, and J. C. Kips. 1997. Cytokine manipulation in animal models of asthma. *Am J Respir Criti Care Med* 156: S78–S81.
182. Yang, G., A. Haczku, H. Chen, V. Martin, H. Galczenski, Y. Tomer, C. R. Van Besien, J. F. Evans, R. A. Panettieri, and C. D. Funk. 2004. Transgenic smooth muscle expression of the human CysLT1 receptor induces enhanced responsiveness of murine airways to leukotriene D4. *Am J Physiol. Lung Cell Mol Physiol* 286: 992–1001.
183. McGraw, D. W., S. L. Forbes, L. A. Kramer, D. P. Witte, C. N. Fortner, R. J. Paul, and S. B. Liggett. 1999. Transgenic overexpression of beta(2)-adrenergic receptors in airway smooth muscle alters myocyte function and ablates bronchial hyperreactivity. *J Biol Chem* 274: 32241–32247.
184. Mali, P., K. M. Esvelt, and G. M. Church. 2013. Cas9 as a versatile tool for engineering biology. *Nat Meth* 10: 957–963.
185. Lulich, K. M., R. G. Goldie, G. Ryan, and J. W. Paterson. 1986. Adverse reactions to beta 2-agonist bronchodilators. *Med Toxicol* 1: 286–299.
186. Ortega, V. E., and S. P. Peters. 2010. Beta-2 adrenergic agonists: Focus on safety and benefits versus risks. *Curr Opin Pharmacol* 10: 246–253.
187. Frielle, T., S. Collins, K. W. Daniel, M. G. Caron, R. J. Lefkowitz, and B. K. Kobilka. 1987. Cloning of the cDNA for the human beta 1-adrenergic receptor. *Proc Natl Acad Sci USA* 84: 7920–7924.

188. Dixon, R. A. F., B. K. Kobilka, D. J. Strader, J. L. Benovic, H. G. Dohlman, T. Frielle, M. A. Bolanowski et al. 1986. Cloning of the gene and cDNA for mammalian beta-adrenergic receptor and homology with rhodopsin. *Nature* 321: 75–79.
189. Emorine, L. J., S. Marullo, M. M. Briend-Sutren, G. Patey, K. Tate, C. Delavier-Klutchko, and D. Strosberg. 1989. Molecular characterization of the human beta 3-adrenergic receptor. *Science* 245: 1118–1121.
190. Lefkowitz, R. J. 1979. Direct binding studies of adrenergic receptors: Biochemical, physiologic, and clinical implications. *Ann Intern Med* 91: 450–458.
191. Dunigan, C. D., P. K. Curran, and P. H. Fishman. 2000. Detection of beta-adrenergic receptors by radioligand binding. *Meth Mol Biol* 126: 329–343.
192. Lin, R., S. Degan, B. S. Theriot, B. M. Fischer, R. T. Strachan, J. Liang, R. A. Pierce et al. 2012. Chronic treatment in vivo with beta-adrenoceptor agonists induces dysfunction of airway beta(2) -adrenoceptors and exacerbates lung inflammation in mice. *Br J Pharmacol* 165: 2365–2377.
193. Zieba, B. J., M. V. Artamonov, L. Jin, K. Momotani, R. Ho, A. S. Franke, R. L. Neppl et al. 2011. The cAMP-responsive Rapl guanine nucleotide exchange factor, Epac, induces smooth muscle relaxation by down-regulation of RhoA activity. *J Biol Chem* 286: 16681–16692.
194. Michael, J. V., A. P. Nayak, R. Yi, T. O. Chan, and R. B. Penn. 2017. A unique floxed PKI mouse reveals protein kinase a dependence of bronchorelaxing agents. *Am J Respir Criti Care Med* 195: A3159.
195. Peters, S. P., C. A. Jones, T. Haselkorn, D. R. Mink, D. J. Valacer, and S. T. Weiss. 2007. Real-world evaluation of asthma control and treatment (REACT): Findings from a national web-based survey. *J Allergy Clin Immunol* 119: 1454–1461.
196. Yates, D. H., S. A. Kharitonov, and P. J. Barnes. 1996. An inhaled glucocorticoid does not prevent tolerance to the bronchoprotective effect of a long-acting inhaled beta 2-agonist. *Am J Respir Criti Care Med* 154: 1603–1607.
197. Premont, R. T., and R. R. Gainetdinov. 2007. Physiological roles of G protein-coupled receptor kinases and arrestins. *Annu Rev Physiol* 69: 511–534.
198. Kong, G., R. Penn, and J. L. Benovic. 1994. A beta-adrenergic receptor kinase dominant negative mutant attenuates desensitization of the beta2-adrenergic receptor. *J Biol Chem* 269: 13084–13087.
199. Waldschmidt, H. V., K. T. Homan, O. Cruz-Rodríguez, M. C. Cato, J. Waninger-Saroni, K. M. Larimore, A. Cannavo et al. 2016. Structure-based design, synthesis, and biological evaluation of highly selective and potent G protein coupled receptor kinase 2 inhibitors. *J Med Chem* 59: 3793–3807.
200. Homan, K. T., K. M. Larimore, J. M. Elkins, M. Szklarz, S. Knapp, and J. J. G. Tesmer. 2015. Identification and structure–function analysis of subfamily selective G protein-coupled receptor kinase inhibitors. *ACS Chem Biol* 10: 310–319.
201. Dogra, S., C. Sona, A. Kumar, and P. N. Yadav. 2016. Chapter 12–Tango assay for ligand-induced GPCR–β-arrestin2 interaction: Application in drug discovery. In *Methods in Cell Biology*, K. S. Arun (Ed.). Cambridge, MA: Academic Press, pp. 233–254.
202. Carr, R., Y. Du, J. Quoyer, R. A. Panettieri, J. M. Janz, M. Bouvier, B. K. Kobilka, and J. L. Benovic. 2014. Development and characterization of pepducins as G(s)-biased allosteric agonists. *J Biol Chem* 289: 35668–35684.
203. Carr, R., C. Koziol-White, J. Zhang, H. Lam, S. S. An, G. G. Tall, R. A. Panettieri, and J. L. Benovic. 2016. Interdicting G(q) activation in airway disease by receptor-dependent and receptor-independent mechanisms. *Mol Pharmacol* 89: 94–104.
204. Kukkonen, J. P., J. Näsman, and K. E. O. Åkerman. 2001. Modelling of promiscuous receptor–Gi/Gs-protein coupling and effector response. *Trends Pharmacol Sci* 22: 616–622.

205. McGraw, D. W., J. M. Elwing, K. M. Fogel, W. C. Wang, C. B. Glinka, K. A. Mihlbachler, M. E. Rothenberg, and S. B. Liggett. 2007. Crosstalk between Gi and Gq/Gs pathways in airway smooth muscle regulates bronchial contractility and relaxation. *J Clin Invest* 117: 1391–1398.
206. Heubach, J. F., U. Ravens, and A. J. Kaumann. 2004. Epinephrine activates both Gs and Gi pathways, but norepinephrine activates only the Gs pathway through human β2-adrenoceptors overexpressed in mouse heart. *Mol Pharmacol* 65: 1313–1322.
207. Daaka, Y., L. M. Luttrell, and R. J. Lefkowitz. 1997. Switching of the coupling of the beta2-adrenergic receptor to different G proteins by protein kinase A. *Nature* 390: 88–91.
208. Tepe, N. M., and S. B. Liggett. 2000. Functional receptor coupling to Gi is a mechanism of agonist-promoted desensitization of the beta2-adrenergic receptor. *J Recept Signal Transduct Res* 20: 75–85.
209. Wisler, J. W., S. M. DeWire, E. J. Whalen, J. D. Violin, M. T. Drake, S. Ahn, S. K. Shenoy, and R. J. Lefkowitz. 2007. A unique mechanism of beta-blocker action: Carvedilol stimulates beta-arrestin signaling. *Proc Natl Acad Sci USA* 104: 16657–16662.
210. Luttrell, L. M., S. Maudsley, and L. M. Bohn. 2015. Fulfilling the promise of "Biased" G protein–Coupled receptor agonism. *Mol Pharmacol* 88: 579–588.
211. Chen, M., A. Hegde, Y. H. Choi, B. S. Theriot, R. T. Premont, W. Chen, and J. K. L. Walker. 2015. Genetic deletion of βarrestin-2 and the mitigation of established airway hyperresponsiveness in a murine asthma model. *Am J Respir Cell Mol Biol* 53: 346–354.
212. Nguyen, L. P., O. Omoluabi, S. Parra, J. M. Frieske, C. Clement, Z. Ammar-Aouchiche, S. B. Ho et al. 2008. Chronic exposure to beta-blockers attenuates inflammation and mucin content in a murine asthma model. *Am J Respir Cell Mol Biol* 38: 256–262.
213. Nguyen, L. P., R. Lin, S. Parra, O. Omoluabi, N. A. Hanania, M. J. Tuvim, B. J. Knoll, B. F. Dickey, and R. A. Bond. 2009. Beta2-adrenoceptor signaling is required for the development of an asthma phenotype in a murine model. *Proc Natl Acad Sci USA* 106: 2435–2440.
214. Thanawala, V. J., G. S. Forkuo, N. Al-Sawalha, Z. Azzegagh, L. P. Nguyen, J. L. Eriksen, M. J. Tuvim, T. W. Lowder, B. F. Dickey, B. J. Knoll, J. K. Walker, and R. A. Bond. 2013. beta2-Adrenoceptor agonists are required for development of the asthma phenotype in a murine model. *Am J Respir Cell Mol Biol* 48: 220–229.
215. Forkuo, G. S., H. Kim, V. J. Thanawala, N. Al-Sawalha, D. Valdez, R. Joshi, S. Parra et al. 2016. Phosphodiesterase 4 inhibitors attenuate the asthma phenotype produced by beta2-adrenoceptor agonists in phenylethanolamine N-methyltransferase-knockout mice. *Am J Respir Cell Mol Biol* 55: 234–242.
216. Nichols, H. L., M. Saffeddine, B. S. Theriot, A. Hegde, D. Polley, T. El-Mays, H. Vliagoftis et al. 2012. beta-Arrestin-2 mediates the proinflammatory effects of proteinase-activated receptor-2 in the airway. *Proc Natl Acad Sci USA* 109: 16660–16665.
217. Huang, X.-P., J. Karpiak, W. K. Kroeze, H. Zhu, X. Chen, S. S. Moy, K. A. Saddoris et al. 2015. Allosteric ligands for the pharmacologically dark receptors GPR68 and GPR65. *Nature* 527: 477–483.
218. Pera, T., D. A. Deshpande, M. Ippolito, B. Wang, A. Gavrila, J. V. Michael, A. P. Nayak et al. 2018. Biased signaling of the proton-sensing receptor OGR1 by benzodiazepines. *FASEB J 32* (2): 862–874. doi: 10.1096/fj.201700555R. Epub 2018 Jan 4. PMID: 29042451.
219. Poulin, B., A. Butcher, P. McWilliams, J.-M. Bourgognon, R. Pawlak, K. C. Kong, A. Bottrill et al. 2010. The M(3)-muscarinic receptor regulates learning and memory in a receptor phosphorylation/arrestin-dependent manner. *Proc Natl Acad Sci USA* 107: 9440–9445.

220. Kong, K. C., A. J. Butcher, P. McWilliams, D. Jones, J. Wess, F. F. Hamdan, T. Werry, E. M. Rosethorne et al. 2010. M(3)-muscarinic receptor promotes insulin release via receptor phosphorylation/arrestin-dependent activation of protein kinase D1. *Proc Natl Acad Sci USA* 107: 21181–21186.
221. Bradley, S. J., C. H. Wiegman, M. M. Iglesias, K. C. Kong, A. J. Butcher, B. Plouffe et al. 2016. Mapping physiological G protein-coupled receptor signaling pathways reveals a role for receptor phosphorylation in airway contraction. *Proc Natl Acad Sci USA* 113: 4524–4529.
222. Willoughby, D., and D. M. Cooper. 2006. Ca^{2+} stimulation of adenylyl cyclase generates dynamic oscillations in cyclic AMP. *J Cell Sci* 119: 828–836.
223. Harbeck, M. C., O. Chepurny, V. O. Nikolaev, M. J. Lohse, G. G. Holz, and M. W. Roe. 2006. Simultaneous optical measurements of cytosolic Ca^{2+} and cAMP in single cells. *Science STKE* 353: l6.
224. Mongillo, M., T. McSorley, S. Evellin, A. Sood, V. Lissandron, A. Terrin, E. Huston et al. 2004. Fluorescence resonance energy transfer-based analysis of cAMP dynamics in live neonatal rat cardiac myocytes reveals distinct functions of compartmentalized phosphodiesterases. *Circ Res* 95: 67–75.
225. Ponsioen, B., J. Zhao, J. Riedl, F. Zwartkruis, G. van der Krogt, M. Zaccolo, W. H. Moolenaar, J. L. Bos, and K. Jalink. 2004. Detecting cAMP-induced Epac activation by fluorescence resonance energy transfer: Epac as a novel cAMP indicator. *EMBO Rep* 5: 1176–1180.
226. Giepmans, B. N., S. R. Adams, M. H. Ellisman, and R. Y. Tsien. 2006. The fluorescent toolbox for assessing protein location and function. *Science* 312: 217–224.
227. Zhang, J., Y. Ma, S. S. Taylor, and R. Y. Tsien. 2001. Genetically encoded reporters of protein kinase A activity reveal impact of substrate tethering. *Proc Natl Acad Sci USA* 98: 14997–15002.
228. Leavesley, S. J., and T. C. Rich. 2016. Overcoming limitations of FRET measurements. *Cytometry A* 89: 325–327.
229. Rich, T. C., K. J. Webb, and S. J. Leavesley. 2014. Perspectives on cyclic nucleotide microdomains and signaling specificity: Can we decipher the information content contained within cyclic nucleotide signals? *J Gen Physiol* 143: 17–27. PMCID:PMC3874573.
230. Leavesley, S. J., A. L. Britain, L. K. Cichon, V. Nikolaev, and T. C. Rich. 2013. Assessing FRET using spectral techniques. *Cytometry A* 83: 898–912. PCMID:PMC4374658.
231. Leavesley, S. J., A. Nakhmani, Y. Gao, and T. C. Rich. 2015. Automated image analysis of FRET signals for subcellular cAMP quantification. *Meth Mol Biol* 1294: 59–70.
232. Rich, T. C., N. Annamdevula, K. Trinh, A. L. Britain, S. A. Mayes, J. R. Griswold, J. Deal, C. Hoffman, S. West, and S. J. Leavesley 2017. 5D imaging approaches reveal the formation of distinct intracellular cAMP spatial gradients. *Proc. SPIE 10070, Three-Dimensional and Multidimensional Microscopy: Image Acquisition and Processing XXIV*, 100700R. doi: 10.1117/12.2253164; https://doi.org/10.1117/12.2253164.

12 Brain-Derived Neurotrophic Factor in Airway Smooth Muscle

Contributions to Asthma

Michael Thompson, Rodney Britt Jr, Anne Roesler, Katelyn Cummings, Christina M. Pabelick, and Y. S. Prakash

CONTENTS

12.1 INTRODUCTION

Bronchial airways are key elements of respiratory structure and function throughout life. Airway diseases such as asthma and chronic obstructive pulmonary disease (COPD) arise from intrinsic factors such as fetal and neonatal developmental abnormalities, and importantly from lifelong exposures to allergens, inflammatory mediators, and environmental insults such as pollutants and tobacco smoke. Understanding the mechanisms that regulate normal airway growth and maintenance in the context of factors contributing to airway disease pathogenesis becomes important for developing targeted therapeutics for asthma and COPD. In this regard, heterogeneity of cell types within the bronchial airway (epithelium, airway smooth muscle [ASM], interstitial fibroblasts, immune cells, nerves) highlights the complex and interactive pathways that likely play a role. This makes it imperative to determine whether common mechanisms that cross different airway diseases can be identified, allowing for a wide range of novel therapeutic approaches to be developed.

In the context of shared molecular etiologies for lung diseases, an emerging aspect is growth factors. Here, neurotrophins may be a particularly relevant topic. Significant and longstanding information on neurotrophins derives from their expression and function in the brain, where they play key roles in neurotransmission, synaptic plasticity, neural growth and regeneration, and the important roles they play in diseases such as depression and Alzheimer's [1–9] . In recent years, there has been increasing interest in understanding neurotrophins in other organ systems including the lung [10–20]. Here, it is becoming clear that expression and functional heterogeneity of neurotrophins are key to understanding specific neurotrophins in airway structure and function. Particular interest lies in brain-derived neurotrophic factor (BDNF) [12,18,19] although the cell sources versus targets of BDNF, and importantly its physiological or pathophysiological significance, are still being explored. This review discusses the role of neurotrophins in the airway and the pathophysiology of asthma and COPD, with a specific focus on BDNF, which extensive studies have revealed to have a pivotal role in airway diseases.

12.2 BASICS OF BRAIN-DERIVED NEUROTROPHIC FACTOR EXPRESSION AND SIGNALING

The first neurotrophin, nerve growth factor (NGF), was recognized over 60 years ago by Rita Levi-Montalcini and Viktor Hamburger in the context of nerve regrowth in embryonic limb buds of the chick [21–26]. Since that time, the neurotrophin family has classically been considered to consist of NGF, BDNF, neurotrophin-3 (NT3), and neurotrophin-4 (NT4) [25–27]. Generally, neurotrophins are considered a nutritive, target-derived factor produced by other neurons or muscles to promote neuronal growth and survival, plasticity, and other neuronal functions [3,4,7–9,22,27–29]. Thus, understanding the upstream and downstream aspects of BDNF expression and signaling, information largely derived from studies in the nervous system, may provide insight into how things work in the airway. However, it is also important to understand that there may be cell- and context-specific heterogeneity. In this regard, relevant to the airway, neurotrophins such as BDNF may act as more than just a growth/nutritive factor, with their expression and signaling being connected to regulatory pathways such as other neurotrophins, steroids (including glucocorticoids), and importantly inflammation [6,30–37]. In the present review, we provide a brief background of BDNF production and signaling in the broad context, followed by discussion on the current state of knowledge regarding how BDNF may be expressed and function within the airway.

BDNF Production: The BDNF gene has multiple promoters allowing for multiple mRNA transcripts [38–40], resulting in complex regulation of BDNF generation that cannot be succinctly summarized here. Most relevant to potential roles in the airway, agonists (or electrical stimulation) that lead to increased intracellular Ca^{2+} ($[Ca^{2+}]_i$) can induce BDNF transcription [40] since there are Ca^{2+}-responsive elements in regulatory regions of BDNF exons: cAMP response element binding protein (CREB), protein kinase A (PKA), calmodulin-dependent protein kinases, NFκB and NFAT. These pathways are relevant since they are themselves activated by $[Ca^{2+}]_i$ (e.g., PKA, CREB) and furthermore are known to regulate several processes in the airway, particularly in inflammation.

The BDNF protein is synthesized as pre-pro-BDNF (~27 kDa) [41–45] that is intracellularly cleaved to pro-BDNF, which can either be secreted into the extracellular space in this pro-form (e.g., via vesicles) or further cleaved within vesicles to mature BDNF and then released. Extracellular pro-BDNF is cleaved by factors such as plasmin and importantly matrix metalloproteinases (MMPs) into mature BDNF (Figure 12.1). Thus, multiple levels of regulation via mRNA and protein can contribute to the end availability of BDNF for subsequent signaling.

BDNF Signaling: Both pro- and mature BDNF can signal through two distinct, major receptor types: the high-affinity tropomyosin related kinase (Trk) receptor (~140 KDa), with BDNF (and NT4) specifically binding to TrkB [46–48] and the low-affinity *pan-neurotrophin* receptor p75NTR (Figure 12.1). Other neurotrophins bind to different Trks: NGF to TrkA, and NT3 to TrkC [22,24,25].

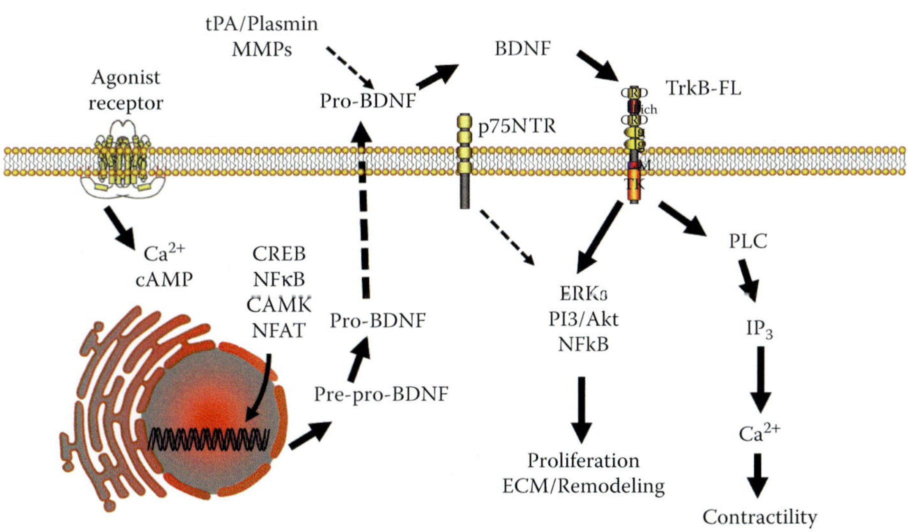

FIGURE 12.1 Production and effects of BDNF. Factors such as Ca^{2+} and a number of signaling intermediates can stimulate production of pre-pro-BDNF, which is subsequently cleaved to pro-BDNF that is secreted. Pro-BDNF can be extracellularly cleaved to mature BDNF that then acts on high-affinity TrkB-FL receptors or low-affinity p75NTR receptors. In ASM, TrkB-FL appears to be of more functional importance. TrkB activation leads to increased $[Ca^{2+}]_i$ via the IP3 pathway as well as increased ASM proliferation and ECM production via different signaling intermediates. Thus, ASM-derived BDNF has the potential for autocrine effects on smooth muscle structure and function.

Cells that respond to BDNF express full-length versions of the transmembrane receptor; in the case of BDNF, TrkB-FL contains an essential intracellular tyrosine kinase domain [49,50]. With BDNF binding, TrkB-FL undergoes autophosphorylation, and substrate binding of Shc, GRB2, ATP, and PLCγ. Signaling occurs through three tyrosine kinase mediated pathways [50–54]: PLCγ1/PKC (resulting in IP_3 and Ca^{2+} release from intracellular stores), MAPK-ERK pathway, and PI3K/Akt, the latter two being important for cell survival and proliferation. Separately, a truncated TrkB-T1 isoform lacking the tyrosine kinase domain can act as a dominant negative receptor, or can sequester BDNF and thus limit signaling [55–58].

Both pro-BDNF and mature BDNF also bind to the low-affinity, pan-neurotrophin p75NTR receptor that is a member of the TNF receptor superfamily. P75NTR can modulate a variety of intracellular pathways including PI3/Akt, NFκB, MAPKs, JNK, RhoA, PKA, and HIF, all of which are relevant to airway contractility, proliferation, growth, and responsiveness to factors such as inflammation [59–66]. BDNF binding to p75NTR can also enhance TrkB receptor signaling through MAPK or Akt, and lead to enhanced cell survival through NFκB, overall potentiating BDNF effects. Thus, p75NTR effects are diverse and context-specific, and can often lead to opposing effects for BDNF effects.

In summary, BDNF signaling occurs through both pro- and mature ligand forms, via two major receptors, the high affinity TrkB-FL (with a suppressive role for TrkB-T1) and the low-affinity p75NTR. Accordingly, in any specific cell type, depending on the relative levels of pro- versus mature BDNF, and expression of TrkBs versus p75NTR, the end effect of BDNF can vary. Downstream to receptor binding, a number of downstream signaling cascades lead to genomic effects via transcription factors such as NFκB, AP-1, and CREB, which are also highly relevant to airway inflammation [59,60,65,66]. BDNF can also stimulate production of cyclic nucleotides such as cAMP and cGMP. Such genomic effects are consistent with the canonical role of BDNF (and other neurotrophins) as growth and nutritive factors. However, in addition to genomic effects (again relevant to the airway), there is now increasing recognition that BDNF has rapid, likely *non-genomic* effects in seconds to minutes utilizing TrkB receptors [23,67–69]. Here, BDNF can modulate plasma membrane receptors, Ca^{2+} influx channels, and voltage-gated Na^{+} channels, at least in neurons [23,67–69].

12.3 CONCEPT OF "LOCAL" BRAIN-DERIVED NEUROTROPHIC FACTOR IN THE CONTEXT OF THE AIRWAY

Consistent with the idea of neurotrophins serving to promote neuronal structure and function, innervated organs such as skin, skeletal muscles, epithelia, and smooth muscles should and do express neurotrophins including BDNF [70,71]. However, locally-produced BDNF can result in autocrine/paracrine effects on cell types other than nerves within an organ that express relevant receptors. Indeed, there is now increasing data that Trk receptors are widely distributed in non-neuronal tissues [11,12,19,72] making such tissues the targets of either circulating or locally produced BDNF. Indeed, BDNF and Trks, as well as p75NTR have now been localized to different lung compartments including airway innervation, resident immune cells, bronchial and alveolar epithelium, smooth muscle, fibroblasts, and endothelium [19,73,74]. The relevance of BDNF to airways in particular lies in the recent recognition that circulating BDNF levels as well as local receptor expression are increased in asthma, and clinical evidence for increased BDNF in sputum and bronchoalveolar lavages following smoking or cigarette smoke exposure, viral infections, allergens, and other insults [16,75–78] While these studies are necessarily associative, they highlight the concept of *local* BDNF as a source as well as a target. Thus, it becomes important to understand the cellular sources of BDNF within the airways, and changes thereof with disease, and the cell types that are influenced by such locally-produced BDNF.

12.4 AIRWAY SMOOTH MUSCLE AS SOURCE AND TARGET OF BRAIN-DERIVED NEUROTROPHIC FACTOR

Considering the fact that within airways, the ASM represents a major fraction of total area as well as mass, BDNF produced by ASM could reach physiologically-relevant levels. Conversely, if such ASM-derived BDNF (or that produced by surrounding cells) acts on ASM itself (i.e., autocrine/paracrine effects), then such local BDNF becomes of physiological significance. Indeed, there is now increasing evidence that ASM is a significant source of neurotrophins including BDNF. Human lungs show constitutive expression of BDNF within the smooth muscle layer per se [73], as do isolated human ASM cells. BDNF can be localized to ASM bundles of larger airways as well as in small airways that determine bronchial tone. While such constitutive ligand expression can result in effects on other cell types, it turns out TrkB is also expressed in human ASM cells and tissues [79–83], as is p75NTR [79]. These data highlight the idea that ASM is both a source and target of BDNF in the airway.

ASM Production of BDNF: Interestingly, as in neurons [46,84], BDNF secretion occurs in human ASM when $[Ca^{2+}]_i$ is elevated [83]. In this regard, mechanisms that regulate $[Ca^{2+}]_i$ in ASM also play a role in modulating BDNF secretion. For example, canonical transient receptor potential channels (TRPC), well-known to be important in $[Ca^{2+}]_i$ regulation of ASM [83–86], are also involved in exocytosis of other intracellular proteins [87], and TRPC3 appears to be important for vesicular BDNF secretion by human ASM, as shown using GFP-tagged BDNF and real-time fluorescence imaging [83]. Furthermore, a number of Ca^{2+} regulatory mechanisms as well as those involved in asthma and COPD appear to be involved in regulation of BDNF production, thus highlighting the relevance of ASM-derived BDNF to airway disease. For example, we recently showed that mechanisms such as PLCγ, TRPC6, Rho kinase, PI3K, ERK, and PKC regulate BDNF production and secretion in human ASM [88]. At the genomic levels, transcription factors such as CREB, NFKB, and NFAT appear to be involved [88]. Thus, it is possible that during bronchoconstriction, agonist stimulation and resultant $[Ca^{2+}]_i$ increases can enhance BDNF secretion in the airway, and factors such as inflammation can result in increased BDNF (see the following). The question then becomes what autocrine/paracrine effects BDNF has on ASM (or other surrounding cells).

BDNF Effects on ASM: BDNF is now shown to have both rapid, non-genomic as well as prolonged, genomic effects on ASM. Exogenous BDNF rapidly enhances $[Ca^{2+}]_i$ in human ASM cells and potentiates $[Ca^{2+}]_i$ responses to bronchoconstrictor agonist [79]: effects enhanced in the presence of cytokines [80]. Such rapid BDNF effects involve increases in plasma membrane Ca^{2+} influx [79], particularly via store-operated Ca^{2+} entry that involves TRPC3, the Ca^{2+} regulatory protein, stromal interaction molecule 1 (STIM1), and the ion channel, Orai1 [82]. As expected from the involvement of PLCγ, BDNF enhances Ca^{2+} release via IP_3 receptor channels. Such non-genomic BDNF effects are mediated exclusively by TrkB, and do not involve p75NTR [80]. Prolonged exposure to BDNF results in increased expression of $[Ca^{2+}]_i$ regulatory proteins [82] which should also lead to increased $[Ca^{2+}]_i$ responses to bronchoconstrictor. Furthermore, consistent with activation of pathways such as MAPK, PI3K/Akt, and NFκB, BDNF also increases human ASM proliferation [89].

ASM BDNF and Inflammation: Several factors relevant to asthma or COPD enhance ASM BDNF and receptor levels [80,81,83,90–92], making BDNF a mediator as well as a potential modulator of insults such as inflammation or environmental exposures in airway disease. Pro-inflammatory cytokines such as IL-1β [82], TNFα, and IL-13 [89], but not IL-4, increase BDNF as well as TrkB-FL in human ASM cells, as do airway irritants such as Substance P [81] and cigarette smoke [92]. Conversely, BDNF appears to be important in the effects of TNFα or IL-13 on ASM Ca^{2+} responses to bronchoconstrictor agonist [80] as well as ASM proliferation [82], as suggested by suppression of cytokine-induced proliferation by the BDNF chelator TrkB-Fc (a chimeric protein)[89]. Again, genomic BDNF effects appear to largely involve TrkB.

In addition to effects on $[Ca^{2+}]_i$ mechanisms and proliferation, BDNF influences airway remodeling that occurs in diseases such as asthma and COPD [60,93–95] by enhancing extracellular matrix (ECM) production [96]. The link between ECM and BDNF is symbiotic in that MMPs, particularly the gelatinases MMP-2 and MMP-9, extracellularly cleave pro-BDNF [97–99], while MMPs that are key to breakdown and recomposition of ECM proteins (collagens and fibronectin) are produced by ASM and thus modulated by BDNF. In human ASM cells, BDNF increases MMP-9 but not MMP-2 secretion [100], and does so via enhanced MMP-9 mRNA, protein, and extracellular levels [96] Furthermore, cytokines such as TNFα enhance MMP-9 expression and activity via BDNF [96]. Overall, via non-genomic and genomic effects, ASM-derived BDNF can influence a range of ASM functions relevant to normal structure and function, as well as to hyperreactivity and remodeling in diseases such as asthma and COPD (Figure 12.2).

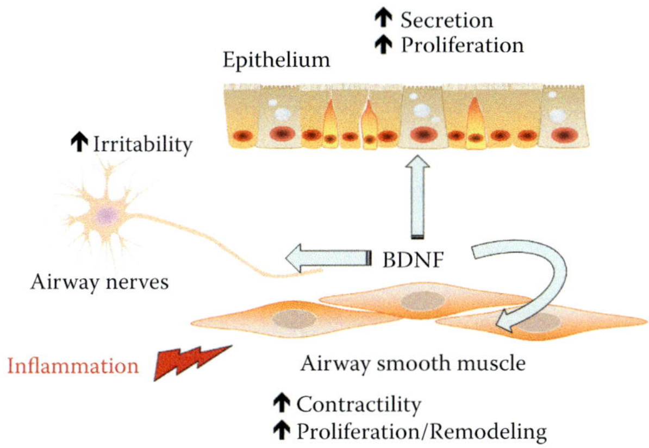

FIGURE 12.2 ASM-derived BDNF can be increased by insults such as inflammation, for example in the context of asthma, leading to autocrine effects, as well as paracrine effects on epithelial proliferation and secretion, and airway irritability by cholinergic innervation. Thus, targeting ASM BDNF may be beneficial in asthma.

12.5 BRAIN-DERIVED NEUROTROPHIC FACTOR EFFECTS ON OTHER AIRWAY CELL TYPES

ASM-derived BDNF has the potential for paracrine effects on a number of cell types. Certainly, airway nerves are important, given extensive innervation of the bronchi [101–106]. The relevance of the different neural pathways regulating airway tone lies in airway hyperresponsiveness and irritability in asthma resulting from increased cholinergic output of the parasympathetic nervous system [107–110]. Importantly, airway nerve fibers show immunoreactivity for TrkB and p75NTR, and can potentially respond to neuron- or ASM-derived BDNF. However, there is currently no data on whether and how ASM-derived BDNF influences airway neuronal function.

The airway epithelium serves as a barrier to allergens and infectious agents, but also as a modulator of how the airway responds to insults, and a source for bronchoconstrictive and bronchodilatory factors [111–115]. Accordingly, BDNF function in this important cell type is relevant to basic airway function and its modulation in disease. However, there are very few studies on neurotrophins in airway epithelium in general, and those few are mostly focused on NGF [73,116,117]. Immunocytochemistry shows that airway epithelium constitutively expresses NGF, BDNF, and NT3. Increased expression of epithelial BDNF has been reported in mouse models of allergic airway hyperresponsiveness [12,116]. While these data show epithelial production of BDNF, its function is less clear. As with ASM, BDNF may also have rapid (i.e., non-genomic) effects on the airway epithelium. For example, in isolated human bronchial epithelial cells, we found that BDNF, acting via TrkB, can elevate $[Ca^{2+}]_i$, phosphorylate eNOS and induce nitric oxide production within minutes [118]. Furthermore, in ongoing studies using human bronchial epithelial cells, we have found that BDNF enhances expression of arginase 2 via a TrkB and

NFκB-mediated mechanism (RD Britt Jr., MA Thompson, and YS Prakash, unpublished observations). The significant relevance of this novel data lies in the fact that arginase can limit availability of L-arginine for production of NO, and furthermore enhance production of proline, a precursor to collagen formation. Thus, these limited data at least suggest that BDNF (regardless of source) can influence epithelial structure and function (Figure 12.2). Certainly, more studies are needed to determine the contribution of BDNF signaling and of its receptors in normal epithelium and role in airway diseases.

12.6 BRAIN-DERIVED NEUROTROPHIC FACTOR AND ASTHMA

Asthma is a chronic condition involving inflammation-driven changes to bronchial airway structure and function, represented by hyperresponsiveness to bronchoconstrictors, impaired bronchodilation, and a largely irreversible remodeling of the airway leading to obstruction. With evidence for BDNF expression and signaling in airway innervation, ASM, and perhaps epithelium (as earlier), there is significant interest in the potential role for targeting BDNF in asthma [15,16,80,100,107,119,120–123]. In this regard, there is some evidence for an association between BDNF and asthma, particularly in children [78,124–126]. Furthermore, in humans as well as in mouse models of asthma, allergen challenge increases BDNF levels in bronchoalveolar lavage fluid [90,127]. Certainly, such increases may be a result of enhanced BDNF secretion by a number of resident airway and immune cells. Increased BDNF may then promote airway irritability via the nerves, modulate epithelium-derived bronchodilator responses, or increase ASM $[Ca^{2+}]_i$ and contractility, especially in the presence of inflammatory cytokines [80]. Finally, BDNF may modulate airway remodeling via proliferation, migration, and secretion of inflammatory mediators and modulators such as MMPs [95,96,112]. Indeed, BDNF does increase MMP9 and to a lesser extent MMP2 [96] and enhances smooth muscle migration [100]. We have shown increased human ASM cell proliferation by BDNF in the presence of cytokines such as TNFα and IL-13 that are relevant to asthma pathophysiology [89]. In a recent study, we have also observed that BDNF expression is higher in ASM of asthmatic patients [96]. Based on these diverse data, it is very likely that BDNF plays a role in asthma pathophysiology. However, more work is needed to identify the potential mechanisms it activates. Nonetheless, given that BDNF can be targeted via TrkB-Fc, it may represent a potential therapeutic avenue in asthma.

12.7 BRAIN-DERIVED NEUROTROPHIC FACTOR AND ENVIRONMENTAL EXPOSURES

A range of environmental exposures such as pollutants are risk factors for development of allergic diseases. Active smoking as well as exposure to secondhand smoke are well-known causes of COPD and can furthermore potentiate effects of other allergens and insults. While much of the focus has been on inflammation, oxidant stress, and other recognized factors in diseases such as COPD, the role of BDNF expression and/or signaling in this context has been examined only to a limited extent, particularly in the lung. Smoking elevates BDNF levels the tears of patients with atopic keratoconjuctivitis [128]. In rats, nicotine increases expression of p75NTR, although surprisingly not Trk receptors [129]. In a recent study using human ASM, we found that cigarette smoke increases both TrkB and p75NTR expression, and that BDNF potentiates cigarette smoke enhancement of ASM $[Ca^{2+}]_i$ responses and thus contractility [92]. Thus, the possibility exists that BDNF can enhance airway inflammation and reactivity following environmental exposures. Here, as with asthma, BDNF (regardless of source) may influence neuronal or immune effects in the airway as well as influence airway remodeling, airway irritability, and hyperresponsiveness. Again, as with asthma, targeting of BDNF may be an attractive therapeutic avenue.

12.8 BRAIN-DERIVED NEUROTROPHIC FACTOR AS A THERAPEUTIC TARGET

Targeting BDNF and/or its receptors represents a novel and attractive area in neuroscience and may be quite relevant to the lung. However, the pleiotropic effects of BDNF on different airway components, while attractive in terms of broad-based therapy, also make it a challenge to target specific cell types, especially if only certain cell types (e.g., ASM) are found to be relevant to disease pathophysiology, and furthermore if these cell types are not as easy to access. A prime example is targeting ASM via inhalational routes without adversely affecting or disrupting the epithelium. Simple nebulization or intratracheal administration of BDNF has been performed in animal studies [12,130], but the extent of lung penetration is not clear, nor are the local versus systemic levels, degradation versus elimination. Novel drug delivery technologies such as nanoparticles, cell-specific markers to facilitate preferential uptake, and viral or other vectors are appealing future approaches. Small molecule activators (or inhibitors) of Trks versus p75NTR have been developed as potential options [54,131–134] but have not been applied to the lung. Molecules such as TrkB-Fc that bind extracellular BDNF and terminate its action could be helpful, but again may be most useful in the airway lumen but less useful if specific cell types are to be targeted.

Regardless of the approaches used to target BNDF or its receptors, the complexity of signaling may be an issue. For example, depending on the relative expression of TrkB-FL versus TrkB-T1 versus p75NTR, small molecules that act on TrkB may have differential effects on cell survival versus cell death. Here, small, synthetic peptides that can selectively activate specific aspects of TrkB or p75NTR and thus mimic BNDF [132] may be appealing, as long as limitations including stability, bioavailability, and barrier penetration can be overcome. Ligands for TrkB have also been developed [132], in particular the specific agonists 7,8-dihydroxyflavone [135] or deoxygedunin [136], but remain to be tested in the lung.

While these exciting developments in approaches to modulating BDNF/TrkB are very interesting, their applicability to airway diseases is less clear for the simple reason that in the airways, BDNF appears to be pro-inflammatory, pro-contractility, and pro-remodeling; therefore, inhibition of BDNF expression and signaling is needed. However, in the context of neurological diseases where loss of trophic influences is the problem much of the focus on BDNF-based therapies is geared towards enhancing rather than suppressing BDNF/TrkB effects. Accordingly, there is much need for identifying novel approaches to inhibiting BDNF/TrkB in the context of airway diseases.

12.9 SUMMARY AND CONCLUSIONS

While signaling via BDNF in different airway components has now been recognized (Figure 12.2), a complete, integrated physiological model for BDNF function in the airway needs to be developed. Here, a number of questions can be put forth: (1) What are the normal, major sources of BDNF in the airway? Here, ASM appears to be particularly important; (2) What are the primary target cell types for BDNF? As discussed in this review, it appears that epithelium, ASM, and nerves are all potential targets. What is less clear is whether BDNF acts on each of these cells under normal circumstances; that is, if BDNF is widely expressed in the airway, from an evolutionary standpoint, what is its intended function: modulation of neurogenic airway tone versus airway structure (remodeling) versus immune function? (3) If BDNF is important in neuronal development, is there a role in lung development as well? (4) Under disease conditions, is altered BDNF expression and function secondary or due to initiating insults such as inflammation or infection, or is BDNF itself a trigger of disease? Ongoing research integrating *in vitro* and *in vivo* approaches should help address many of these issues. Given substantial homology of BDNF, TrkB, and p75NTR across species, corroboration of animal experiments with human samples—especially patients with specific, well-defined airway diseases—will be particularly helpful.

REFERENCES

1. Allen, S. J., J. J. Watson, D. K. Shoemark, N. U. Barua, and N. K. Patel. 2013. GDNF, NGF and BDNF as therapeutic options for neurodegeneration. *Pharmacol Ther* 138: 155–175.
2. Dooley, D., P. Vidal, and S. Hendrix. 2013. Immunopharmacological intervention for successful neural stem cell therapy: New perspectives in CNS neurogenesis and repair. *Pharmacol Ther* 141: 21–31.
3. He, Y. Y., X. Y. Zhang, W. H. Yung, J. N. Zhu, and J. J. Wang. 2013. Role of BDNF in central motor structures and motor diseases. *Mol Neurobiol* 48: 783–793.
4. Leal, G., D. Comprido, and C. B. Duarte. 2013. BDNF-induced local protein synthesis and synaptic plasticity. *Neuropharmacology* 76: 639–656.
5. Nowacka, M., and E. Obuchowicz. 2013. BDNF and VEGF in the pathogenesis of stress-induced affective diseases: An insight from experimental studies. *Pharmacol Rep* 65: 535–546.
6. Numakawa, T., N. Adachi, M. Richards, S. Chiba, and H. Kunugi. 2013. Brain-derived neurotrophic factor and glucocorticoids: Reciprocal influence on the central nervous system. *Neuroscience* 239: 157–172.
7. Park, H., and M. M. Poo. 2013. Neurotrophin regulation of neural circuit development and function. *Nat Rev Neurosci* 14: 7–23.
8. Suliman, S., S. M. Hemmings, and S. Seedat. 2013. Brain-Derived Neurotrophic Factor (BDNF) protein levels in anxiety disorders: Systematic review and meta-regression analysis. *Front Integr Neurosci* 7: 55.

9. Zagrebelsky, M., and M. Korte. 2013. Form follows function: BDNF and its involvement in sculpting the function and structure of synapses. *Neuropharmacology* 76: 628–638.
10. Braun, A., M. Lommatzsch, and H. Renz. 2000. The role of neurotrophins in allergic bronchial asthma. *Clin Exp Allergy* 30: 178–86.
11. Hoyle, G. W. 2003. Neurotrophins and lung disease. *Cytokine Growth Factor Rev* 14: 551–558.
12. Lommatzsch, M., A. Braun, and H. Renz. 2003. Neurotrophins in allergic airway dysfunction: What the mouse model is teaching us. *Ann N Y Acad Sci* 992: 241–249.
13. Piedimonte, G. 2003. Contribution of neuroimmune mechanisms to airway inflammation and remodeling during and after respiratory syncytial virus infection. *Pediatr Infect Dis J* 22: S66–S74; discussion S74–S75.
14. Jacoby, D. B. 2004. Pathophysiology of airway viral infections. *Pulm Pharmacol Ther* 17: 333–336.
15. Renz, H., S. Kerzel, and W. A. Nockher. 2004. The role of neurotrophins in bronchial asthma: Contribution of the pan-neurotrophin receptor p75. *Prog Brain Res* 146: 325–333.
16. Rochlitzer, S., C. Nassenstein, and A. Braun. 2006. The contribution of neurotrophins to the pathogenesis of allergic asthma. *Biochem Soc Trans* 34: 594–599.
17. Taylor-Clark, T., and B. J. Undem. 2006. Transduction mechanisms in airway sensory nerves. *J Appl Physiol* 101: 950–959.
18. Yao, Q., S. I. Zaidi, M. A. Haxhiu, and R. J. Martin. 2006. Neonatal lung and airway injury: A role for neurotrophins. *Semin Perinatol* 30: 156–162.
19. Prakash, Y., M. A. Thompson, L. Meuchel, C. M. Pabelick, C. B. Mantilla, S. Zaidi, and R. J. Martin. 2010. Neurotrophins in lung health and disease. *Expert Rev Respir Med* 4: 395–411.
20. Aven, L., and X. Ai. 2013. Mechanisms of respiratory innervation during embryonic development. *Organogenesis* 9: 194–198.
21. Levi-Montalcini, R. 1998. The saga of the nerve growth factor. *Neuroreport* 9: R71–R83.
22. Teng, K. K., and B. L. Hempstead. 2004. Neurotrophins and their receptors: Signaling trios in complex biological systems. *Cell Mol Life Sci* 61: 35–48.
23. Blum, R., and A. Konnerth. 2005. Neurotrophin-mediated rapid signaling in the central nervous system: Mechanisms and functions. *Physiology* (*Bethesda*) 20: 70–78.
24. Lu, B., P. T. Pang, and N. H. Woo. 2005. The yin and yang of neurotrophin action. *Nat Rev Neurosci* 6: 603–614.
25. Reichardt, L. F. 2006. Neurotrophin-regulated signalling pathways. *Philos Trans R Soc Lond B Biol Sci* 361: 1545–1564.
26. Hennigan, A., R. M. O'Callaghan, and A. M. Kelly. 2007. Neurotrophins and their receptors: Roles in plasticity, neurodegeneration and neuroprotection. *Biochem Soc Trans* 35: 424–427.
27. Chao, M. V., R. Rajagopal, and F. S. Lee. 2006. Neurotrophin signalling in health and disease. *Clin Sci (London)* 110: 167–173.
28. Kalb, R. 2005. The protean actions of neurotrophins and their receptors on the life and death of neurons. *Trends Neurosci* 28: 5–11.
29. Lykissas, M. G., A. K. Batistatou, K. A. Charalabopoulos, and A. E. Beris. 2007. The role of neurotrophins in axonal growth, guidance, and regeneration. *Curr Neurovasc Res* 4: 143–151.
30. Miranda, R. C., F. Sohrabji, and D. Toran-Allerand. 1994. Interactions of estrogen with the neurotrophins and their receptors during neural development. *Horm Behav* 28: 367–375.

31. Simpkins, J. W., P. S. Green, K. E. Gridley, M. Singh, N. C. de Fiebre, and G. Rajakumar. 1997. Role of estrogen replacement therapy in memory enhancement and the prevention of neuronal loss associated with Alzheimer's disease. *Am J Med* 103: 19S–25S.
32. Schaaf, M. J., E. R. De Kloet, and E. Vreugdenhil. 2000. Corticosterone effects on BDNF expression in the hippocampus. Implications for memory formation. *Stress* 3: 201–208.
33. Kapfhammer, J. P. 2004. Cellular and molecular control of dendritic growth and development of cerebellar Purkinje cells. *Prog Histochem Cytochem* 39: 131–182.
34. Tabakman, R., S. Lecht, S. Sephanova, H. Arien-Zakay, and P. Lazarovici. 2004. Interactions between the cells of the immune and nervous system: Neurotrophins as neuroprotection mediators in CNS injury. *Prog Brain Res* 146: 387–401.
35. Nagatsu, T., and M. Sawada. 2005. Inflammatory process in Parkinson's disease: Role for cytokines. *Curr Pharm Des* 11: 999–1016.
36. Babayan, A. H., and E. A. Kramar. 2013. Rapid effects of oestrogen on synaptic plasticity: Interactions with actin and its signaling proteins. *J Neuroendocrinol* 25: 1163–1172.
37. Pluchino, N., M. Russo, A. N. Santoro, P. Litta, V. Cela, and A. R. Genazzani. 2013. Steroid hormones and BDNF. *Neuroscience* 239: 271–279.
38. Aid, T., A. Kazantseva, M. Piirsoo, K. Palm, and T. Timmusk. 2007. Mouse and rat BDNF gene structure and expression revisited. *J Neurosci Res* 85: 525–535.
39. Boulle, F., D. L. van den Hove, S. B. Jakob, B. P. Rutten, M. Hamon, J. van Os, K. P. Lesch, L. Lanfumey, H. W. Steinbusch, and G. Kenis. 2012. Epigenetic regulation of the BDNF gene: Implications for psychiatric disorders. *Mol Psychiatry* 17: 584–596.
40. Zheng, F., X. Zhou, C. Moon, and H. Wang. 2012. Regulation of brain-derived neurotrophic factor expression in neurons. *Int J Physiol Pathophysiol Pharmacol* 4: 188–200.
41. McDonald, N. Q., and M. V. Chao. 1995. Structural determinants of neurotrophin action. *J Biol Chem* 270: 19669–19672.
42. Robinson, R. C., C. Radziejewski, D. I. Stuart, and E. Y. Jones. 1995. Structure of the brain-derived neurotrophic factor/neurotrophin 3 heterodimer. *Biochemistry* 34: 4139–4146.
43. Butte, M. J., P. K. Hwang, W. C. Mobley, and R. J. Fletterick. 1998. Crystal structure of neurotrophin-3 homodimer shows distinct regions are used to bind its receptors. *Biochemistry* 37: 16846–16852.
44. Lessmann, V., K. Gottmann, and M. Malcangio. 2003. Neurotrophin secretion: Current facts and future prospects. *Prog Neurobiol* 69: 341–374.
45. Lessmann, V., and T. Brigadski. 2009. Mechanisms, locations, and kinetics of synaptic BDNF secretion: An update. *Neurosci Res* 65: 11–22.
46. Barker, P. A. 2004. p75NTR is positively promiscuous: Novel partners and new insights. *Neuron* 42: 529–533.
47. Blochl, A., and R. Blochl. 2007. A cell-biological model of p75NTR signaling. *J Neurochem* 102: 289–305.
48. Chen, Y., J. Zeng, L. Cen, Y. Chen, X. Wang, G. Yao, W. Wang, W. Qi, and K. Kong. 2009. Multiple roles of the p75 neurotrophin receptor in the nervous system. *J Int Med Res* 37: 281–288.
49. Barbacid, M. 1994. The Trk family of neurotrophin receptors. *J Neurobiol* 25: 1386–1403.
50. Ohira, K., and M. Hayashi. 2009. A new aspect of the TrkB signaling pathway in neural plasticity. *Curr Neuropharmacol* 7: 276–285.
51. Barbacid, M. 1995. Structural and functional properties of the TRK family of neurotrophin receptors. *Ann N Y Acad Sci* 766: 442–458.
52. Chao, M. V., and B. L. Hempstead. 1995. p75 and Trk: A two-receptor system. *Trends Neurosci* 18: 321–326.

53. Pitts, E. V., S. Potluri, D. M. Hess, and R. J. Balice-Gordon. 2006. Neurotrophin and Trk-mediated signaling in the neuromuscular system. *Int Anesthesiol Clin* 44: 21–76.
54. Thiele, C. J., Z. Li, and A. E. McKee. 2009. On Trk—The TrkB signal transduction pathway is an increasingly important target in cancer biology. *Clin Cancer Res* 15: 5962–5967.
55. Li, Y. X., Y. Xu, D. Ju, H. A. Lester, N. Davidson, and E. M. Schuman. 1998. Expression of a dominant negative TrkB receptor, T1, reveals a requirement for presynaptic signaling in BDNF-induced synaptic potentiation in cultured hippocampal neurons. *Proc Natl Acad Sci USA* 95: 10884–10889.
56. Rose, C. R., R. Blum, B. Pichler, A. Lepier, K. W. Kafitz, and A. Konnerth. 2003. Truncated TrkB-T1 mediates neurotrophin-evoked calcium signalling in glia cells. *Nature* 426: 74–78.
57. Cheng, A., T. Coksaygan, H. Tang, R. Khatri, R. J. Balice-Gordon, M. S. Rao, and M. P. Mattson. 2007. Truncated tyrosine kinase B brain-derived neurotrophic factor receptor directs cortical neural stem cells to a glial cell fate by a novel signaling mechanism. *J Neurochem* 100: 1515–1530.
58. Fenner, B. M. 2012. Truncated TrkB: Beyond a dominant negative receptor. *Cytokine Growth Factor Rev* 23: 15–24.
59. Halayko, A. J., and J. Solway. 2001. Molecular mechanisms of phenotypic plasticity in smooth muscle cells. *J Appl Physiol* 90: 358–368.
60. Halayko, A. J., and Y. Amrani. 2003. Mechanisms of inflammation-mediated airway smooth muscle plasticity and airways remodeling in asthma. *Respir Physiol Neurobiol* 137: 209–222.
61. Chiba, Y., and M. Misawa. 2004. The role of RhoA-mediated Ca^{2+} sensitization of bronchial smooth muscle contraction in airway hyperresponsiveness. *J Smooth Muscle Res* 40: 155–167.
62. Panettieri, R. A. Jr. 2004. Effects of corticosteroids on structural cells in asthma and chronic obstructive pulmonary disease. *Proc Am Thorac Soc* 1: 231–234.
63. Hirota, S., P. B. Helli, A. Catalli, A. Chew, and L. J. Janssen. 2005. Airway smooth muscle excitation-contraction coupling and airway hyperresponsiveness. *Can J Physiol Pharmacol* 83: 725–732.
64. James, A. 2005. Airway remodeling in asthma. *Curr Opin Pulm Med* 11: 1–6.
65. Gosens, R., J. Zaagsma, H. Meurs, and A. J. Halayko. 2006. Muscarinic receptor signaling in the pathophysiology of asthma and COPD. *Respir Res* 7: 73.
66. Damera, G., O. Tliba, and R. A. Panettieri, Jr. 2009. Airway smooth muscle as an immunomodulatory cell. *Pulm Pharmacol Ther* 22: 353–359.
67. Kovalchuk, Y., K. Holthoff, and A. Konnerth. 2004. Neurotrophin action on a rapid timescale. *Curr Opin Neurobiol* 14: 558–563.
68. Rose, C. R., R. Blum, K. W. Kafitz, Y. Kovalchuk, and A. Konnerth. 2004. From modulator to mediator: Rapid effects of BDNF on ion channels. *Bioessays* 26: 1185–1194.
69. Carvalho, A. L., M. V. Caldeira, S. D. Santos, and C. B. Duarte. 2008. Role of the brain-derived neurotrophic factor at glutamatergic synapses. *Br J Pharmacol* 153 Suppl 1: S310–S324.
70. Sariola, H. 2001. The neurotrophic factors in non-neuronal tissues. *Cell Mol Life Sci* 58: 1061–1066.
71. Nockher, W. A., and H. Renz. 2005. Neurotrophins in clinical diagnostics: pathophysiology and laboratory investigation. *Clin Chim Acta* 352: 49–74.
72. Nassenstein, C., S. Kerzel, and A. Braun. 2004. Neurotrophins and neurotrophin receptors in allergic asthma. *Prog Brain Res* 146: 347–367.
73. Ricci, A., L. Felici, S. Mariotta, F. Mannino, G. Schmid, C. Terzano, G. Cardillo, F. Amenta, and E. Bronzetti. 2004. Neurotrophin and neurotrophin receptor protein expression in the human lung. *Am J Respir Cell Mol Biol* 30: 12–19.

74. Ricci, A., P. Graziano, E. Bronzetti, C. Saltini, S. Sciacchitano, E. Cherubini, E. Renzoni, R. M. Du Bois, J. C. Grutters, and S. Mariotta. 2007. Increased pulmonary neurotrophin protein expression in idiopathic interstitial pneumonias. *Sarcoidosis Vasc Diffuse Lung Dis* 24: 13–23.
75. Ricci, A., S. Mariotta, C. Saltini, C. Falasca, M. R. Giovagnoli, F. Mannino, P. Graziano, S. Sciacchitano, and F. Amenta. 2005. Neurotrophin system activation in bronchoalveolar lavage fluid immune cells in pulmonary sarcoidosis. *Sarcoidosis Vasc Diffuse Lung Dis* 22: 186–194.
76. Tortorolo, L., A. Langer, G. Polidori, G. Vento, B. Stampachiacchere, L. Aloe, and G. Piedimonte. 2005. Neurotrophin overexpression in lower airways of infants with respiratory syncytial virus infection. *Am J Respir Crit Care Med* 172: 233–237.
77. Dagnell, C., J. Grunewald, M. Kramar, H. Haugom-Olsen, G. P. Elmberger, A. Eklund, and C. Olgart Hoglund. 2010. Neurotrophins and neurotrophin receptors in pulmonary sarcoidosis-granulomas as a source of expression. *Respir Res* 11: 156.
78. Andiappan, A. K., P. N. Parate, R. Anantharaman, B. K. Suri, Y. Wang de, and F. T. Chew. 2011. Genetic variation in BDNF is associated with allergic asthma and allergic rhinitis in an ethnic Chinese population in Singapore. *Cytokine* 56: 218–223.
79. Prakash, Y. S., A. Iyanoye, B. Ay, C. B. Mantilla, and C. M. Pabelick. 2006. Neurotrophin effects on intracellular Ca^{2+} and force in airway smooth muscle. *Am J Physiol Lung Cell Mol Physiol* 291: L447–L456.
80. Prakash, Y. S., M. A. Thompson, and C. M. Pabelick. 2009. Brain-derived neurotrophic factor in TNF-alpha modulation of Ca^{2+} in human airway smooth muscle. *Am J Respir Cell Mol Biol* 41: 603–611.
81. Meuchel, L. W., A. Stewart, D. F. Smelter, A. J. Abcejo, M. A. Thompson, S. I. Zaidi, R. J. Martin, and Y. S. Prakash. 2011. Neurokinin-neurotrophin interactions in airway smooth muscle. *Am J Physiol Lung Cell Mol Physiol* 301: L91–L98.
82. Abcejo, A. J., V. Sathish, D. F. Smelter, B. Aravamudan, M. A. Thompson, W. R. Hartman, C. M. Pabelick, and Y. S. Prakash. 2012. Brain-derived neurotrophic factor enhances calcium regulatory mechanisms in human airway smooth muscle. *PLoS One* 7: e44343.
83. Vohra, P. K., M. A. Thompson, V. Sathish, A. Kiel, C. Jerde, C. M. Pabelick, B. B. Singh, and Y. S. Prakash. 2013. TRPC3 regulates release of brain-derived neurotrophic factor from human airway smooth muscle. *Biochim Biophys Acta* 1833: 2953–2960.
84. Nakahashi, T., H. Fujimura, C. A. Altar, J. Li, J. Kambayashi, N. N. Tandon, and B. Sun. 2000. Vascular endothelial cells synthesize and secrete brain-derived neurotrophic factor. *FEBS Lett* 470: 113–117.
85. White, T. A., A. Xue, E. N. Chini, M. Thompson, G. C. Sieck, and M. E. Wylam. 2006. Role of TRPC3 in tumor necrosis factor-alpha enhanced calcium influx in human airway myocytes. *Am J Respir Cell Mol Biol* 35: 243–251.
86. Amrani, Y. 2007. TNF-alpha and calcium signaling in airway smooth muscle cells: A never-ending story with promising therapeutic relevance. *Am J Respir Cell Mol Biol* 36: 387–388.
87. Bollimuntha, S., S. Selvaraj, and B. B. Singh. 2011. Emerging roles of canonical TRP channels in neuronal function. *Adv Exp Med Biol* 704: 573–593.
88. Aravamudan, B., M. A. Thompson, C. M. Pabelick, and Y. S. Prakash. 2016. Mechanisms of BDNF regulation in asthmatic airway smooth muscle. *Am J Physiol Lung Cell Mol Physiol* 311: L270–L279.
89. Aravamudan, B., M. Thompson, C. Pabelick, and Y. S. Prakash. 2012. Brain-derived neurotrophic factor induces proliferation of human airway smooth muscle cells. *J Cell Mol Med* 16: 812–823.

90. Braun, A., M. Lommatzsch, G. R. Lewin, J. C. Virchow, and H. Renz. 1999. Neurotrophins: A link between airway inflammation and airway smooth muscle contractility in asthma? *Int Arch Allergy Immunol* 118: 163–165.
91. Kemi, C., J. Grunewald, A. Eklund, and C. O. Hoglund. 2006. Differential regulation of neurotrophin expression in human bronchial smooth muscle cells. *Respir Res* 7: 18.
92. Sathish, V., S. K. Vanoosten, B. S. Miller, B. Aravamudan, M. A. Thompson, C. M. Pabelick, R. Vassallo, and Y. S. Prakash. 2013. Brain-derived neurotrophic factor in cigarette smoke-induced airway hyperreactivity. *Am J Respir Cell Mol Biol* 48: 431–438.
93. Holgate, S. T. 2002. Airway inflammation and remodeling in asthma: Current concepts. *Mol Biotechnol* 22: 179–89.
94. Joubert, P., and Q. Hamid. 2005. Role of airway smooth muscle in airway remodeling. *J Allergy Clin Immunol* 116: 713–716.
95. Lagente, V., and E. Boichot. 2009. Role of matrix metalloproteinases in the inflammatory process of respiratory diseases. *J Mol Cell Cardiol* 48: 440–444.
96. Bai, T. R. 2010. Evidence for airway remodeling in chronic asthma. *Curr Opin Allergy Clin Immunol* 10: 82–86.
97. Lee, R., P. Kermani, K. K. Teng, and B. L. Hempstead. 2001. Regulation of cell survival by secreted proneurotrophins. *Science* 294: 1945–1948.
98. Ethell, I. M., and D. W. Ethell. 2007. Matrix metalloproteinases in brain development and remodeling: Synaptic functions and targets. *J Neurosci Res* 85: 2813–2823.
99. Soleman, S., M. A. Filippov, A. Dityatev, and J. W. Fawcett. 2013. Targeting the neural extracellular matrix in neurological disorders. *Neuroscience* 253: 194–213.
100. Dagnell, C., C. Kemi, J. Klominek, P. Eriksson, C. M. Skold, A. Eklund, J. Grunewald, and C. Olgart Hoglund. 2007. Effects of neurotrophins on human bronchial smooth muscle cell migration and matrix metalloproteinase-9 secretion. *Transl Res* 150: 303–310.
101. Racke, K., and S. Matthiesen. 2004. The airway cholinergic system: Physiology and pharmacology. *Pulm Pharmacol Ther* 17: 181–198.
102. Canning, B. J. 2006. Reflex regulation of airway smooth muscle tone. *J Appl Physiol* 101: 971–985.
103. Undem, B. J., and C. Nassenstein. 2009. Airway nerves and dyspnea associated with inflammatory airway disease. *Respir Physiol Neurobiol* 167: 36–44.
104. Verhein, K. C., A. D. Fryer, and D. B. Jacoby. 2009. Neural control of airway inflammation. *Curr Allergy Asthma Rep* 9: 484–490.
105. Kc, P., and R. J. Martin. 2010. Role of central neurotransmission and chemoreception on airway control. *Respir Physiol Neurobiol* 173: 213–222.
106. Undem, B. J., and C. Potenzieri. 2012. Autonomic neural control of intrathoracic airways. *Compr Physiol* 2: 1241–1267.
107. Nockher, W. A., and H. Renz. 2006. Neurotrophins and asthma: Novel insight into neuroimmune interaction. *J Allergy Clin Immunol* 117: 67–71.
108. Butler, C. A., and L. G. Heaney. 2007. Neurogenic inflammation and asthma. *Inflamm Allergy Drug Targets* 6: 127–132.
109. Pisi, G., D. Olivieri, and A. Chetta. 2009. The airway neurogenic inflammation: Clinical and pharmacological implications. *Inflamm Allergy Drug Targets* 8: 176–181.
110. Nassini, R., S. Materazzi, G. De Siena, F. De Cesaris, and P. Geppetti. 2010. Transient receptor potential channels as novel drug targets in respiratory diseases. *Curr Opin Investig Drugs* 11: 535–542.
111. Crystal, R. G., S. H. Randell, J. F. Engelhardt, J. Voynow, and M. E. Sunday. 2008. Airway epithelial cells: Current concepts and challenges. *Proc Am Thorac Soc* 5: 772–777.

112. Dekkers, B. G., H. Maarsingh, H. Meurs, and R. Gosens. 2009. Airway structural components drive airway smooth muscle remodeling in asthma. *Proc Am Thorac Soc* 6: 683–692.
113. Gras, D., P. Chanez, I. Vachier, A. Petit, and A. Bourdin. 2013. Bronchial epithelium as a target for innovative treatments in asthma. *Pharmacol Ther* 140: 290–305.
114. Hirota, N., and J. G. Martin. 2013. Mechanisms of airway remodeling. *Chest* 144: 1026–1032.
115. Holgate, S. T. 2013. Mechanisms of asthma and implications for its prevention and treatment: A personal journey. *Allergy Asthma Immunol Res* 5: 343–347.
116. Hahn, C., A. P. Islamian, H. Renz, and W. A. Nockher. 2006. Airway epithelial cells produce neurotrophins and promote the survival of eosinophils during allergic airway inflammation. *J Allergy Clin Immunol* 117: 787–794.
117. Othumpangat, S., L. F. Gibson, L. Samsell, and G. Piedimonte. 2009. NGF is an essential survival factor for bronchial epithelial cells during respiratory syncytial virus infection. *PLoS One* 4: e6444.
118. Meuchel, L. M., E. A. Townsend, M. A. Thompson, C. M. Pabelick, and Y. S. Prakash. 2010. Effect of neurotrophins on NO generation in airway epithelial cells. In *International Conference of the American Thoracic Society*, New Orleans, LA.
119. Frossard, N., V. Freund, and C. Advenier. 2004. Nerve growth factor and its receptors in asthma and inflammation. *Eur J Pharmacol* 500: 453–465.
120. Nassenstein, C., D. Dawbarn, K. Pollock, S. J. Allen, V. J. Erpenbeck, E. Spies, N. Krug, and A. Braun. 2006. Pulmonary distribution, regulation, and functional role of Trk receptors in a murine model of asthma. *J Allergy Clin Immunol* 118: 597–605.
121. Freund-Michel, V., and N. Frossard. 2008. The nerve growth factor and its receptors in airway inflammatory diseases. *Pharmacol Ther* 117: 52–76.
122. Abram, M., M. Wegmann, V. Fokuhl, S. Sonar, E. O. Luger, S. Kerzel, A. Radbruch, H. Renz, and M. Zemlin. 2009. Nerve growth factor and neurotrophin-3 mediate survival of pulmonary plasma cells during the allergic airway inflammation. *J Immunol* 182: 4705–4712.
123. Bennedich Kahn, L., L. E. Gustafsson, and C. Olgart Hoglund. 2008. Brain-derived neurotrophic factor enhances histamine-induced airway responses and changes levels of exhaled nitric oxide in guinea pigs in vivo. *Eur J Pharmacol* 595: 78–83.
124. Szczepankiewicz, A., M. J. Rose-Zerilli, S. J. Barton, S. T. Holgate, and J. W. Holloway. 2009. Association analysis of brain-derived neurotrophic factor gene polymorphisms in asthmatic families. *Int Arch Allergy Immunol* 149: 343–349.
125. Koskela, H. O., M. K. Purokivi, and J. Romppanen. 2010. Neurotrophins in chronic cough: Association with asthma but not with cough severity. *Clin Respir J* 4: 45–50.
126. Szczepankiewicz, A., A. Breborowicz, P. Sobkowiak, and A. Popiel. 2010. Association of BDNF gene polymorphism with asthma in Polish children. *World Allergy Organ J* 3: 235–238.
127. Virchow, J. C., P. Julius, M. Lommatzsch, W. Luttmann, H. Renz, and A. Braun. 1998. Neurotrophins are increased in bronchoalveolar lavage fluid after segmental allergen provocation. *Am J Respir Crit Care Med* 158: 2002–2005.
128. Kimata, H. 2004. Passive smoking elevates neurotrophin levels in tears. *Hum Exp Toxicol* 23: 215–217.
129. Urrego, F., M. Scuri, A. Auais, L. Mohtasham, and G. Piedimonte. 2009. Combined effects of chronic nicotine and acute virus exposure on neurotrophin expression in rat lung. *Pediatr Pulmonol* 44: 1075–1084.
130. Braun, A., M. Lommatzsch, U. Neuhaus-Steinmetz, D. Quarcoo, T. Glaab, G. P. McGregor, A. Fischer, and H. Renz. 2004. Brain-derived neurotrophic factor (BDNF) contributes to neuronal dysfunction in a model of allergic airway inflammation. *Br J Pharmacol* 141: 431–440.

131. Longo, F. M., and S. M. Massa. 2008. Small molecule modulation of p75 neurotrophin receptor functions. *CNS Neurol Disord Drug Targets* 7: 63–70.
132. Longo, F. M., and S. M. Massa. 2013. Small-molecule modulation of neurotrophin receptors: A strategy for the treatment of neurological disease. *Nat Rev Drug Discov* 12: 507–525.
133. Skaper, S. D. 2008. The biology of neurotrophins, signalling pathways, and functional peptide mimetics of neurotrophins and their receptors. *CNS Neurol Disord Drug Targets* 7: 46–62.
134. Webster, N. J., and M. C. Pirrung. 2008. Small molecule activators of the Trk receptors for neuroprotection. *BMC Neurosci* 9 Suppl 2: S1.
135. Jang, S. W., X. Liu, M. Yepes, K. R. Shepherd, G. W. Miller, Y. Liu, W. D. Wilson et al. 2010b. A selective TrkB agonist with potent neurotrophic activities by 7,8-dihydroxyflavone. *Proc Natl Acad Sci USA* 107: 2687–2692.
136. Jang, S. W., X. Liu, C. B. Chan, S. A. France, I. Sayeed, W. Tang, X. Lin et al. 2010a. Deoxygedunin, a natural product with potent neurotrophic activity in mice. *PLoS One* 5: e11528.

13 Antibody-Based Approaches for Protein Analysis of Smooth Muscles

Brian A. Perrino and Yeming Xie

CONTENTS

13.1 INTRODUCTION

Protein quantitation and detection methods typically rely on antibody-based or mass spectrometry-based technologies. These technologies provide good information regarding relative protein levels as well as protein modifications in response to stimuli, but methods for precise and accurate quantitation of absolute protein levels in complex biological samples face many challenges [1,2]. Quantification standards are difficult to establish in ELISA and antibody array assays due to endogenous signal interference, and poor assay reproducibility is not uncommon with conventional SDS-PAGE western approaches. Though mass spectrometry provides a high-precision analysis platform, it may have low accuracy due to the prevalence of interference from other peptides and small molecules in the sample matrix. Mass spectrometry methods also require complicated sample preparation procedures, and instrumentation that is typically only available from dedicated core laboratories or facilities. SDS-PAGE and western blotting remains the most widely used approach to detect and analyze proteins and protein phosphorylation in cell or tissue extracts, due to its long history of use, relatively inexpensive equipment and reagents, ease of data acquisition utilizing imaging techniques, and the vast number of antibodies now available for protein detection [3]. However, more detailed analyses of relative protein levels as well as protein modifications from smooth muscle tissues can be achieved by utilizing complementary antibody-based approaches.

13.2 SDS-PAGE AND WESTERN BLOTTING

Western blotting approaches have diverse applications for investigating protein expression and abundance, kinase and phosphatase activities, cellular localization, protein–protein interactions, and post-translational modifications (PTMs) [4,5]. Western blotting, also referred to as immunoblotting (IB), is the separation of native or denatured proteins by vertical gel electrophoresis, and then transferring these separated proteins to a protein binding membrane by horizontal electrophoresis [6]. The target protein is subsequently detected with a primary antibody specific to the target protein and a secondary antibody specific for the host species in which the primary antibody was raised. With the advent of the chemiluminescent substrates luminol and peroxide for horseradish peroxidase (HRP), secondary antibodies are typically conjugated to HRP and the protein band detected by the visible light (428 nm) produced by the reaction. Chemiluminescent western blotting (also called enhanced chemiluminescence; ECL) is replacing radioactive probes and less-sensitive colorimetric assays. Traditionally, the emitted light was captured and documented on X-rays films. However, a major drawback to film is that it is only semi-quantitative [7]. Film has a narrow linear range and is easily saturated by moderate-to-high signal intensities [7]. In addition, using film for western blots is expensive and inconvenient. Multiple exposures must be done by directly placing the X-ray film against the plastic-wrapped blot to obtain an image in the linear range. The film then has to be developed, typically by using a common developer in a dark room. Along with the cost of purchasing the X-ray film and developing chemicals, there is now also the cost for the hazardous waste disposal of the developing chemicals. Thus, highly sensitive CCD camera systems with greater dynamic ranges and no chemical wastes are replacing X-ray film as the method of choice for imaging western blot data, due to the ease and convenience of direct generation of a digital image, and lower operating costs.

Detecting proteins and monitoring PTMs by SDS-PAGE and western blotting involves multiple complex steps. Important quality control and attention to detail are required at every step of the western blot procedure because each step involves multiple variables, which if overlooked, will adversely affect the outcome. Every step in the western blot process must be understood and carefully considered to minimize the time required to determine the optimal conditions and allow for the greatest accuracy and interpretation of data. Smooth muscles offer considerable challenges for western blot analysis of proteins in that along with the thousands of different proteins and enzymes present in any animal cell, the cytoplasm of smooth muscle additionally contains large amounts of specialized myofilament proteins (e.g., actin, myosin, α-actinin, etc.). Thus, some degree of subcellular fractionation of smooth muscle lysates is usually necessary to enrich for the protein(s) of interest by the removal of unwanted proteins [8,9]. Other specialized differential centrifugation techniques utilizing density gradients (e.g., sucrose, ficoll, percoll) are required for more specific subcellular membrane fractionation [10,11].

As an example of careful consideration of the importance of sample preparation, determining the phosphorylation status of myosin light chain (MLC), MYPT1, or CPI-17 by SDS-PAGE and western blotting of smooth muscle lysates requires specific sample processing procedures. To preserve the phosphorylation status of these proteins the smooth muscle samples must be submerged into cold acetone/TCA/DTT solutions immediately after the experimental treatments, followed by snap freezing in liquid nitrogen for storage at −80°C [12–15]. If these steps are omitted, the phosphorylation status of these proteins as indicated by the subsequent western blot analysis will likely not reflect the original phosphorylation status of the protein in the live smooth muscle tissue [13]. Typically, the phosphorylation levels are underestimated, due to uneven and unpredictable dephosphorylation of these proteins during tissue lysis and processing. The culprit seems to be MLC phosphatase, which appears to revive its activity, even in the presence of lysis buffer phosphatase inhibitors, as the frozen tissue thaws in the lysis buffer at 4°C [13]. Although the particular order of treatment and temperature is still debated, it is clear from a number of studies of different smooth muscles from different laboratories that an acetone/TCA/DTT treatment step is necessary prior to lysis to completely inactivate harmful phosphatase activity towards these phosphorylated proteins.

Since the development of the western blot almost 40 years ago, many protocol modifications and technologies have emerged for increasing the sensitivity, speed, and applications of the method. Newer detection technologies and more rapid analyses will continue to be offered but will still require the same important experimental and sample processing considerations to be taken into account when conducting western blot analysis. For example, emerging technologies will still require researchers to optimize lysis conditions, amount of protein loaded, antisera concentrations, blocking buffers, membrane type and porosity, membrane background fluorescence, washes, incubation times, the ECL substrate, and exposure times. The ultimate goal will always be the production of the most accurate and reproducible experimental data possible.

13.2.1 Phos-Tag SDS-PAGE Analysis of Protein Phosphorylation

Phos-tag SDS-PAGE is a fairly recent example of a modification of SDS-PAGE and western blotting. Phos-tag SDS-PAGE involves phosphate affinity electrophoresis, in which the Phos-tag molecules are incorporated into a general reducing polyacrylamide resolving gel to chelate the phosphate (PO_4^-) group of phosphorylated Ser, Thr, or Tyr, and retard the mobility of the phosphorylated protein through the resolving gel [16]. This permits the separation of the phosphorylated and nonphosphorylated forms of a protein, and the subsequent detection of changes in the protein's phosphorylation status by western blotting, using only an antibody against the protein of interest. The preparation of protein samples, the overall procedure, and the reagents used for Phos-tag SDS-PAGE are almost identical to those for conventional SDS-PAGE, except that the acrylamide-pendant Phos-tag, which binds to the phosphate groups, is copolymerized (along with micromolar Mn^{2+} or Zn^{2+}) with the resolving gel in Phos-tag SDS-PAGE [16]. Another factor that must be considered with the Phos-tag SDS-PAGE western blot approach is that the efficiency of the electrotransfer of target proteins from the Phos-tag resolving gel onto the blotting membrane is reduced, due to the tendency of the phosphorylated proteins to remain in the gel because of the strong interaction between Phos-tag and phosphorylated proteins [17]. However, this problem can be overcome by simply incubating the Phos-tag gel in transfer buffer, which contains 1 mM EDTA, to chelate the metal ion coordinated to the Phos-tag molecule [17]. In conventional SDS-PAGE, the mobility of a given protein in the resolving gel is often the same whether it is non-phosphorylated or phosphorylated, and they are therefore not efficiently separated. However, because the phosphorylated protein has a slower mobility in the Phos-tag gel, the requirement for an antibody against the phospho-site is eliminated. In the western blot, the antibody will bind to both the phosphorylated and non-phosphorylated forms of a protein equally, resulting in a doublet (if the protein is only singly phosphorylated), and the signal intensities from the phosphorylated and non-phosphorylated bands can be directly compared, avoiding discrepancies that can arise when two antibodies with unknown affinities are used to detect phosphorylated and non-phosphorylated protein bands. Unlike urea-glycerol SDS-PAGE, which specifically separates phosphorylated myosin from non-phosphorylated myosin, Phos-tag electrophoresis permits the separation of any phosphorylated protein from its non-phosphorylated counterpart and can be applied in a range of various protein kinase/phosphatase profiling studies for quantitative analysis. This technique also allows for the quantitative analyses of multiple phosphorylation statuses of individual cellular proteins. Thus, if a protein molecule has multiple phosphorylation sites and if it exists in multiple phosphorylation states, the resulting differences in the electrophoretic mobilities of the various phosphorylated forms of the protein in a lysate sample result in the formation of several bands that can be individually detected. As the number of phosphorylated sites on a protein increases, the slower the mobility of the protein in the Phos-tag resolving gel [18,19]. For example, the Phos-tag SDS approach has been used to analyze the complex pattern of the multiple phosphorylation sites on the myosin-targeting subunit 1 of MLC phosphatase (MYPT1) in rat caudal artery smooth muscles [20].

Because the various phosphorylated forms of a target protein can be detected as multiple migration bands on the Phos-tag gel by gel staining or IB procedures, Phos-tag SDS-PAGE permits detailed quantitative analysis of the phosphorylated protein of interest. However, to subsequently identify the sites of phosphorylation of multiple up-shifted phosphorylated species, it is necessary to utilize downstream analyses, such as IB using site-specific anti-phosphoprotein antibodies (if the sites are known) or MS analysis after in-gel digestion (if the sites are unknown). If used in conjunction with site-specific anti-phosphoprotein antibodies, the Phos-tag acrylamide gel-based electrophoretic separation procedure can be used to map the sites on each of the phosphoprotein species that are formed during cell signaling [21].

13.3 MICROFLUIDIC SDS-PAGE AND IMMUNODETECTION OF IMMOBILIZED PROTEINS

In 2014, ProteinSimple introduced Wes, an innovative, automated platform based on microfluidic capillary electrophoresis to separate, immobilize, and immunodetect proteins from cell or tissue lysates [22–24]. Conceptually, detection of protein targets with the Wes system is the same as conventional western analysis. Sample proteins are separated by size, immobilized to a solid surface, and detected using target specific primary antibodies, and a chemiluminescent signal is generated by HRP-conjugated secondary antibodies.

A key aspect of the Wes system is the use of capillary electrophoresis for protein separation. Protein separation, immobilization, and immunodetection are all accomplished in the capillary tube. Thus, each microcapillary tube is equivalent to an independent experiment, making it possible to perform up to 24 independent experiments in a single three-hour run. This means that 24 different protein targets can be detected in one sample in a single run or controls and experimental treatments probing one protein can be run simultaneously in a single run. The rapid time to results is a great advantage over conventional western blotting. There is also the added power that these analyses are quantifiable and can be directly compared to each other. The use of microcapillary electrophoresis also greatly reduces the sample volumes required and results in very consistent separations without effects from neighboring samples, edge effects or smiling. Another innovation of the Wes system is the immobilization of the separated proteins directly to the capillary prior to immunodetection. In conventional western blotting the sample proteins have to be electrophoretically transferred from the gel and immobilized onto a nitrocellulose or polyvinylidene difluoride (PVDF) membrane by non-covalent interactions (i.e., hydrophobic and electrostatic) [25]. The Wes system utilizes UV light to catalyze the covalent crosslinking of the proteins to the capillary itself, directly immobilizing each protein in place at the point of their migration. This feature completely eliminates the inefficiency associated with the conventional process of electrophoretic transfer and non-covalent immobilization [26]. Immunodetection is achieved by sequential application of the primary antibodies, secondary antibodies, and wash buffers to the immobilized proteins within the capillary tubes, and target protein detection accomplished by application of the chemiluminescence substrate (luminol-peroxide) to the capillary and direct capture of the light emitted during the enzyme catalyzed reaction.

Technically, carrying out Wes analyses of proteins in a tissue lysate is fairly straightforward and actually easier than the conventional SDS-PAGE and western blot workflow. The investigator decides which proteins are to be analyzed, and then follows the Wes plate loading pattern with the tissue lysate, the proper primary antibodies, the correct HRP-conjugated secondary antibodies, and the luminol-peroxide substrate. The plate and the capillary cartridges are placed into the Wes instrument, and the results acquired 3 h later. Where scientific expertise is required is in the sample preparation techniques, designing the composition of the tissue lysis buffers and the method of tissue lysis, and the titration of lysate concentrations and antibody dilutions for optimization of the output signals. Careful consideration of these parameters before the Wes analysis can result in rapid determination of a set

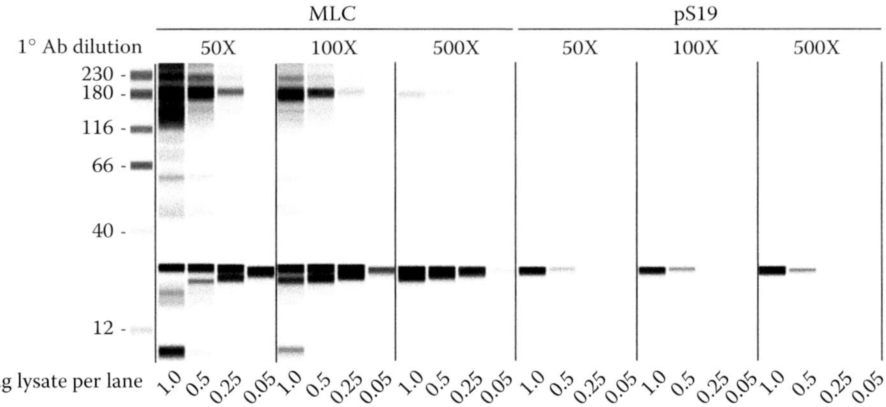

FIGURE 13.1 Detection of MLC and pS19 in human gastric fundus lysates with the Wes system. Three dilutions of the antibodies (Ab) against MLC and pS19 were tested against four different amounts of human gastric fundus lysates, using one cartridge containing 25 microcapillaries.

of *standard conditions* for protein immunodetection, due to the ability to utilize 24 individual separation and detection capillaries in each Wes analysis.

For example, using the Wes system, we were able to determine the optimal antibody dilutions and lysate concentrations to detect MLC (MYL9) and MYL9 phosphorylated at Ser19 (MYL9 pS19) from human gastric fundus smooth muscle samples in one three-hour run (Figure 13.1). Our strategy utilized four different concentrations of lysate probed with three different dilutions of each antibody, to determine that 250 ng of lysate and a 500-fold dilution of antibody was sufficient for detection of MYL9, and 500 ng of lysate and a 500-fold dilution of antibody was sufficient for detection of pS19. In addition, we are utilizing the Wes system to monitor changes in the phosphorylation of MYPT1 (myosin phosphatase targeting subunit 1; PPP1R12A) and CPI-17 (PKC-potentiated inhibitory protein of PP1; PPP1R14A), two key proteins involved in the Ca^{2+} sensitization response of smooth muscles to contractile agonists [27].

13.4 CO-IMMUNOPRECIPITATION AND *IN SITU* PROXIMITY LIGATION ANALYSIS

In many cases it is no longer enough to determine whether a particular protein is present or not within a particular cell type; it is more meaningful to determine how protein expression impacts other proteins by examining its interactions with other proteins. The long-established method of co-immunoprecipitation (co-IP) presents multiple advantages such as low-cost, ease-of-use, compatibility with downstream analytical methods, and relatively high specificity. The approaches employed in the investigation of protein–protein interactions have advantages as well as disadvantages and, therefore, it is important to analyze protein–protein interactions with more than one approach, for both confirmation and validation, in order to avoid false-positive results [28]. Co-IP has traditionally been one of the strongest and relatively convenient methods to identify physical interactions between two or more proteins *in vivo* [29]. The principle of co-IP is based upon *fishing* a protein complex from cell lysates or homogenates with an antibody directed against one of the interacting partners and subsequent isolation of the IP complex using immobilized protein A or protein G, typically onto small, inert, uniform-sized agarose or Sepharose beads [29]. With the appropriate buffer ionic and detergent components for maximizing specific protein–protein interactions, while washing away *non-specific* interactions, specific protein complexes can be *pulled out* of cell or tissue lysates by the bead-antibody complex with a few low-speed spins in a microcentrifuge. For downstream analyses, the proteins are eluted by boiling the beads in an SDS-PAGE sample buffer or incubating the beads in a low-pH buffer to release the IP complex from the beads. The major disadvantage to these approaches is that the IP antibodies are also released, resulting in contamination of the sample with the IP antibodies. Using IP and IB antibodies from different host species helps to avoid interference from the IP antibody during subsequent IB detection. Alternatively, and becoming increasingly popular, is the covalent crosslinking of antibodies either to the protein A/G moiety, or to the beads themselves using chemical cross linkers such as the amine reactive N-hydroxy-succinimide esters (NHS ester) [30]. More recently, chemical crosslinking to magnetic beads (Dynabeads) with surface chemistries designed for direct crosslinking of antibodies to the beads has emerged as a more convenient option. Following elution by boiling or low pH, only the target proteins are released, resulting in a cleaner final preparation [31]. In addition, due to the smaller, uniform size of magnetic Dynabeads, and the use of a magnet to immobilize the bead-protein complex, non-specific interactions are greatly minimized [31].

Co-IP approaches have been employed to detect voltage-dependent K^+ (Kv) channel subunits in renal and cerebral arteries [32,33], aberrant D1 and D3 dopamine receptor transregulation in mesenteric arteries in hypertension [34], activation of soluble guanylyl cyclase by Hsp70 in pulmonary arteries [35], inhibition of Smad7 by serum response factor in airway smooth muscle cells (SMCs) [36], the binding of FKBP12 (FK506 binding protein 1a) and calcineurin with the ryanodine receptor 2 (RyR2) in colonic SMCs [37], palladin binding to myocardin-related transcription factors during SMC differentiation [38], and the interaction of STIM1 with TRPC1 during intracellular Ca^{2+} store depletion in vascular SMCs (VSMCs) [39]. These representative examples of studies employing co-IP strategies in different SMC systems highlight the versatility of the technique.

However, investigating proteins, and their interactions and PTMs directly in cells and tissues has advantages over co-IP approaches, and can offer meaningful insights into cellular processes and their alterations in disease. Protein functions that are regulated by PTMs and by specific binding interactions with other partners often result in the transient assembly of molecular switches whose activities depend upon the interacting protein components [40,41]. Monitoring the formation, presence, and localization of protein complexes and PTMs can provide important spatial and temporal information about the cell types in which the proteins are located and in what subcellular or organelle compartment these interactions occur [42]. Several methods for visualizing protein interactions *in situ* (e.g., FRET), involving genetic constructs, have been developed and used successfully in assays of live cells maintained in tissue culture [42].

While genetically modified cells or mice are powerful means for investigating protein function, such approaches are impossible to apply to human tissues, or when genetic modifications are not possible. In contrast, methods based on specific protein recognition by antibodies can be applied to almost any sample and used to detect protein interactions, PTMs, or the subcellular co-localization of proteins. Using well-established immunohistochemistry techniques, the *in situ* proximity ligation assay (*is*PLA) is a powerful method for localizing individual proteins, detecting neighboring proteins, and visualizing PTMs in tissue sections and cell lines *in vitro* [43]. Protein targets can be readily detected and localized *in situ* with single molecule resolution and objectively quantified in fixed cells and tissues. Transient interactions are revealed *in situ* and specific cell populations can be identified by their different patterns of protein co-expression [44]. Importantly, the results from conventional co-IP and co-localization techniques can be independently confirmed by *is*PLA. *is*PLA makes use of pairs of primary antibodies from different host species to recognize a protein–protein complex, a target protein PTM, or a target protein of interest. Species-specific secondary antibodies with a short complementary DNA oligonucleotide attached, called PLA probes, bind to the primary antibodies. When the PLA probes are in close proximity, the DNA strands interact through the subsequent addition of two other circle-forming DNA oligonucleotides. After joining of the two added oligonucleotides by enzymatic ligation, the DNA circles are copied via rolling circle amplification. In the amplification reaction, long DNA strands containing several hundred complements of the DNA circle are formed. The output signals are generated by fluorescently-labeled complementary oligonucleotide probes and visualized by microscopy as brightly fluorescent spots. In general, proteins that are within ~40 nm of each other can be detected by *is*PLA [44,45] The pairwise antibody binding in *is*PLA generates highly selective detection, since individual PLA probes fail to give rise to detection signals, while the requirement for recognition by pairs of antibodies results in the formation of a brightly fluorescent spot that can be imaged by microscopy.

Different protein–protein interactions within several intracellular signaling pathways of different smooth muscles have been examined using the *is*PLA approach. For example, *is*PLA has been utilized to analyze the co-localization of Orai1 and TRPC1 with $Ca_V1.2$ channels [46], the interaction of Orai1 with BKCa channels [47], the endocytic recycling of focal adhesion proteins [48], the vesicular ATP transporter and myosin Va [49], and the interactions of paxillin with Pak at adhesion junctions [50].

We are utilizing the *is*PLA approach to examine the dynamic phosphorylation events and protein–protein interactions that occur during the contractile responses of gastric fundus smooth muscles. To complement the SDS-PAGE and western blot approach, we are utilizing *is*PLA to detect and quantify increases in MYL9 phosphorylation gastric fundus smooth muscle in response to different modes of cholinergic stimulation. As shown in Figure 13.2, we are able to monitor changes in the phosphorylation of MYPT1, CPI-17, and MLC (MYL9) in

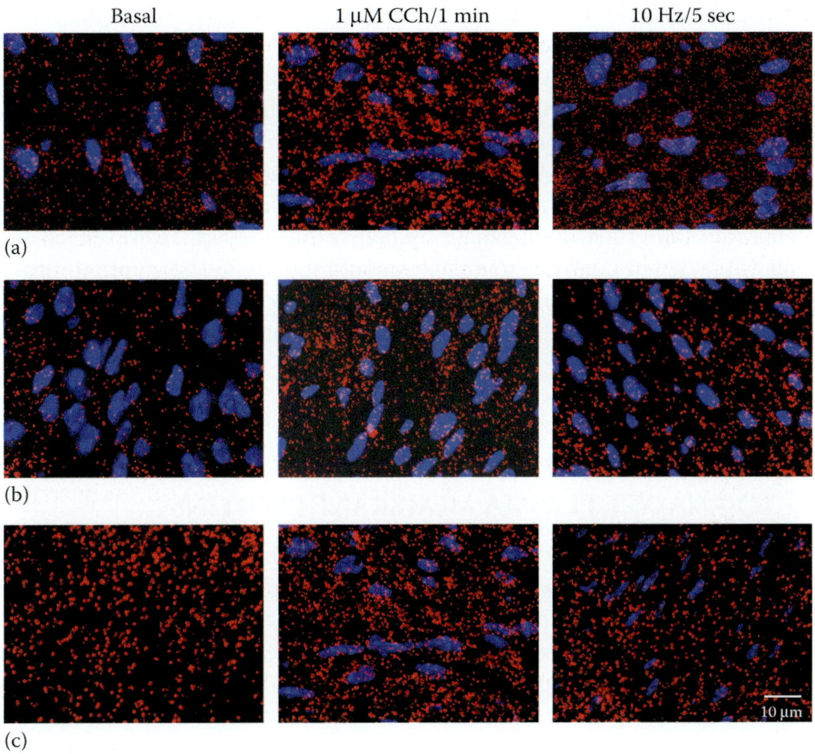

FIGURE 13.2 *is*PLA detection of (a) MYPT1-T853, (b) CPI-17-T38, and (c) MLC-S19 phosphorylation in fixed sections of murine gastric fundus smooth muscles. Images were obtained from unstimulated muscles (Basal), or muscle strips stimulated with 1 μM CCh for 1 min (in the presence of 0.3 μM tetrodotoxin), muscle strips stimulated by EFS at 10 Hz for 5 s (10 Hz/5 s), in the presence of 100 μM L-NNA and 1 μM MRS2500.

response to bath-applied carbachol (CCh) or by neurally released (acetylcholine) ACh in response to electrical field stimulation (EFS), using fixed sections of gastric fundus smooth muscles.

Bath-applied CCh increased the phosphorylation of MYPT1 at T853, CPI-17 at T38, and MLC at S19, as indicated by the increased number of red fluorescent spots compared to unstimulated muscles (Basal). In contrast, EFS of neurally released ACh only increased CPI-17 phosphorylation at T38. Thus, it is likely after quantitation of the *is*PLA signals that the *is*PLA approach will verify our previous findings using conventional SDS-PAGE and western blotting to monitor MYPT1, CPI-17, and MLC phosphorylation in smooth muscles [27].

In summary, *is*PLA is a versatile technique for detecting and measuring multiple protein interactions or PTMs in smooth muscle tissues or individual cells. Although it cannot be used with live cell imaging, *is*PLA approaches can expand studies of smooth muscle physiology into the analysis of the localization and roles of specific protein complexes and PTMs within discrete subcellular locations during processes such as development, proliferation, differentiation, contraction, and pathological responses. In spite of these strong points, there are weaknesses that should be considered. As always, any method that relies on antibodies can present specificity issues since it is not always possible to acquire antibodies of high specificity and avidity. In addition, with co-IP there is some degree of concern for false-positives arising from the release of proteins normally confined to discrete cellular compartments into the crude homogenate by cell or tissue lysis. However, co-IP and proximity assays will continue to enable smooth muscle investigators to provide a more coherent view of protein modifications and interactions underlying cellular responses compared with methods that mainly measure expression levels of individual proteins.

13.5 SUBCELLULAR FRACTIONATION APPROACHES TO ANALYZE PLASMA MEMBRANE PROTEINS

The plasma membrane (PM) only represents about 5%–10% of the total membrane surface area of the cell. However, the integral transmembrane proteins expressed in the PM of SMCs are essential for the functional and physiological requirements of smooth muscles [51]. About 25% of the genes in the human genome encode PM proteins [52]. In addition, approximately 60% of FDA-approved drugs currently on the market target PM proteins [53]. A detailed understanding of the expression patterns of the ion channels, receptors, mechanotransducers, lipid-protein domains, and transporters expressed in the PM of SMCs is essential to a more complete understanding of SMC homeostasis, and to better identify additional targets for therapeutic treatments of human diseases. The endoplasmic reticulum and Golgi membranes comprise almost 80% of the cellular membrane surface area. Thus, PM proteins comprise a small fraction of total cellular membrane proteins, and cell fractionation techniques are usually required to enrich the final sample in PM proteins for expression and phosphorylation analyses. Typically, low-speed centrifugations (500 × g to 1000 × g) will move cell nuclei into the pellet. Spinning the post-nuclear supernatant in a microcentrifuge (16,000 × g to 20,000 × g) will remove most mitochondria from the supernatant, leaving a *cytoplasmic fraction* consisting of soluble proteins, and microsomes originating from the PM and organelle membranes. Usually, this degree of separation is sufficient for most western blot-based approaches for protein and phospho-protein analysis [8,9]. A final purification of the PM fraction can be accomplished by density gradient centrifugation in sucrose gradients or other density media. Further specialized differential centrifugation techniques utilizing density gradients (e.g., sucrose, ficoll, percoll) are required for more specific-subcellular membrane fractionation [10,11]. Subcellular fractionation techniques can typically yield samples containing up to of 40%–50% PM proteins [54,55]. Because the PM fraction is a small percentage of the total cellular volume, these approaches usually require large amounts of starting material, and therefore it is almost limited to studies on cultured cells and tissues that are available in large amounts.

13.6 CELL SURFACE BIOTINYLATION OF PLASMA MEMBRANE PROTEINS

For samples that are not available in large amounts, biotinylation of surface proteins for PM enrichment can be an effective strategy [53,56,57]. This approach involves chemical coupling of amine- or thiol-reactive biotinylation reagents to proteins located on the cell surface. After solubilization of the membranes and the protein digestion, biotinylated peptides are affinity purified on resins with immobilized avidin. Similar to the PM coating strategy, the surface biotinylation approach is more suitable for analysis of cells in suspension rather than for tissues. Care must be taken when biotinylating cells enzymatically dispersed from tissues, as possible contaminations by cytoplasmic and nuclear proteins originating from broken cells can seriously affect the results of the analysis.

A chemical reagent to be used for protein biotinylation experiments is typically comprised of the following building blocks: (1) The biotin moiety, for the interaction of biotinylated proteins with avidin/streptavidin (SA)-based affinity matrix reagents. Using biotin in conjunction with SA yields the highest binding affinity and allows the use of strong detergents during the purification of relatively insoluble proteins. (2) A spacer of sufficient length to allow protein capture by immobilized SA. The spacer is usually cleavable by chemical or physical agents to facilitate protein release after capture. For example, sulfo-NHS-SS-biotin is a water-soluble biotinylation reagent that reacts with primary amines, does not permeate membranes so only surface proteins are labeled, and contains a disulfide bond in the spacer arm to elute the biotinylated proteins from the SA affinity matrix using a reducing agent such as DTT [58]. (3) The reactive moiety for the covalent binding of biotin to the protein(s). The most common reactive groups are reactive esters, like the NHS group, which undergoes a nucleophilic substitution reaction in the presence of primary amines (e.g., typically exposed lysine residues), or reactive imides, such as the maleimido group, which reacts with free thiol groups (e.g., exposed cysteine residues).

Following enrichment of PM proteins by the biotin SA interaction, PM proteins can by analyzed using shotgun mass spectrometry approaches, or target PM proteins of interest analyzed by SDS-PAGE western blot approaches [59]. Several PM-dependent processes and PM proteins of SMCs have been examined using cell surface biotinylation, such as receptor and ion channel distribution, endocytosis, and recycling in VSMCs [60–68], ceramide-induced increase in PM ABCA1 (ATP-binding cassette subfamily member 1), and ceramide-induced decrease in PM HERG (K^+ voltage-gated channel subfamily member 2; KCNH2) channel [69,70], localization of cathepsin S with integrin $\alpha v\beta 3$ at the PM of migrating VSMC [71], and the function of $Ca_V1.2$ $\alpha 2$-$\delta 1$ subunits [72].

13.7 TRANSGENE APPROACHES FOR OBTAINING HOMOGENOUS SMOOTH MUSCLE CELL POPULATIONS FROM SMOOTH MUSCLES

Smooth muscles are complex tissues containing a variety of cells in addition to muscle cells, which provide important regulatory controls in both the normal and pathophysiological responses of smooth muscles [73,74]. Thus, it is inherently difficult to deduce the roles of specific cell types in tissues as complex as smooth muscles. Functional studies using muscle strips and expression studies on whole muscle extracts do not distinguish which cells express specific genes or are responsible for specific functions. Having fluorescent reporters expressed in different cell types makes it possible to isolate each cellular component for genomic, proteomic, and functional analyses. The generation of transgenic mice with smooth muscle-specific expression of fluorescent reporter proteins has greatly facilitated the study of smooth muscle physiology *in vivo* [75,76]. Xin et al., were the first to utilize the smooth muscle myosin heavy chain (smMHC) promoter to direct the expression of a bi-cistronic transgene consisting of Cre recombinase and enhanced green fluorescent protein (eGFP) coding sequences [76]. Mice expressing this transgene have been shown display fluorescence that originates from SMCs in all smooth muscles examined [76].

Since then, mouse strains with eGFP expression driven by the c-kit or Ano1 promoter, the PDGFRα promoter, and smMHC promoter have now been extensively characterized and used to investigate the functional roles of interstitial cells of Cajal (ICC), PDGFRα^+ cells, and SMCs, respectively, in several different smooth muscle types [77–80]. Enzymatic dissociation of smooth muscles from cell-specific reporter mice yield viable, fluorescent cells that can be sorted by FACS and collected for smooth muscle-specific functional and gene expression studies [77,81,82]. With the advent of highly sensitive single-cell western and proteomic analytical approaches, it may finally be possible to use fluorescently sorted SMCs for protein expression analyses and link the gene expression data with SMC-specific protein expression to obtain a more complete picture of how smooth muscle function is generated.

ACKNOWLEDGMENTS

This work is supported by NIH grants 14GHSU1397, GM110767, and a Takeda Innovation Center Grant.

REFERENCES

1. Ellington, A. A., I. J. Kullo, K. R. Bailey, and G. G. Klee. 2010. Antibody-based protein multiplex platforms: Technical and operational challenges. *Clin Chem* 56: 186–193.
2. Gillette, M. A., and S. A. Carr. 2013. Quantitative analysis of peptides and proteins in biomedicine by targeted mass spectrometry. *Nat Meth* 10: 28–34.
3. Gorr, T. A., and J. Vogel. 2015. Western blotting revisited: Critical perusal of underappreciated technical issues. *Proteomics Clin Appl* 9: 396–405.
4. Bass, J. J., D. J. Wilkinson, D. Rankin, B. E. Phillips, N. J. Szewczyk, K. Smith, and P. J. Atherton. 2016. An overview of technical considerations for Western blotting applications to physiological research. *Scand J Med Sci Sports* 27: 4–25.
5. Weiel, J., N. Ahn, R. Seger, and E. G. Krebs. 1990. Communication between protein tyrosine and protein serine/threonine phosphorylation. *Adv Second Messenger Phosphoprotein Res* 24: 182–195.
6. Towbin, H., T. Staehelin, and J. Gordon. 1979. Electrophoretic transfer of proteins from polyacrylamide gels to nitrocellulose sheets: Procedure and some applications. *Proc Natl Acad Sci USA* 76: 4350–4354.

7. Khoury, M. K., I. Parker, and D. W. Aswad. 2010. Acquisition of chemiluminescent signals from immunoblots with a digital SLR camera. *Anal Biochem* 397: 129–131.
8. Michelsen, U., and J. von Hagen. 2009. Chapter 19—Isolation of subcellular organelles and structures. In: Richard, RB, Murray, PD (Eds.). *Meth Enzymol* 463: 305–328.
9. Yates, J. R. 3rd, A. Gilchrist, K. E. Howell, and J. J. Bergeron. 2005. Proteomics of organelles and large cellular structures. *Nat Rev Mol Cell Biol* 6: 702–714.
10. Araùjo, M., L. A. Hube, and T. Stasyk. 2008. Isolation of endocitic organelles by density gradient centrifugation. *Methods Mol Biol* 424: 317–331.
11. Yan, W., D. Hwang, and R. Aebersold. 2008. Quantitative proteomic analysis to profile dynamic changes in the spatial distribution of cellular proteins. *Methods Mol Biol* 389–401.
12. Bhetwal, B. P., C. L. An, S. A. Fisher, and B. A. Perrino. 2011. Regulation of basal LC20 phosphorylation by MYPT1 and CPI-17 in murine gastric antrum, gastric fundus, and proximal colon smooth muscles. *Neurogastroenterol Motil* 23: e425–e436.
13. Johnson, R. P., A. F. El-Yazbi, K. Takeya, E. J. Walsh, M. P. Walsh, and W. C. Cole. 2009. Ca^{2+} sensitization via phosphorylation of myosin phosphatase targeting subunit at threonine-855 by Rho kinase contributes to the arterial myogenic response. *J Physiol* 587: 2537–2553.
14. Tsai, M. H., K. E. Kamm, and J. T. Stull. 2012. Signalling to contractile proteins by muscarinic and purinergic pathways in neurally stimulated bladder smooth muscle. *J Physiol* 590: 5107–5121.
15. Wang, Y., X. R. Zheng, N. Riddick, M. Bryden, W. Baur, X. Zhang, and H. K. Surks. 2009. ROCK isoform regulation of myosin phosphatase and contractility in vascular smooth muscle cells. *Circ Res* 104: 531–540.
16. Kinoshita, E., E. Kinoshita-Kikuta, K. Takiyama, and T. Koike. 2006. Phosphate-binding tag, a new tool to visualize phosphorylated proteins. *Mol Cell Proteomics* 5: 749–757.
17. Kinoshita-Kikuta, E., E. Kinoshita, A. Matsuda, and T. Koike. 2014. Tips on improving the efficiency of electrotransfer of target proteins from Phos-tag SDS-PAGE gel. *Proteomics* 14: 2437–2442.
18. Hosokawa, T., T. Saito, A. Asada, K. Fukunaga, and S. Hisanaga. 2010. Quantitative measurement of in vivo phosphorylation states of Cdk5 activator p35 by Phos-tag SDS-PAGE. *Mol Cell Proteomics* 9: 1133–1143.
19. Kimura, T., H. Hatsuta, M. Masuda-Suzukake, M. Hosokawa, K. Ishiguro, H. Akiyama, S. Murayama, M. Hasegawa, and S. Hisanaga. 2016. The abundance of nonphosphorylated Tau in mouse and human tauopathy brains revealed by the use of Phos-Tag method. *Am J Pathol* 186: 398–409.
20. Sutherland, C., J. A. MacDonald, and M. P. Walsh. 2016. Analysis of phosphorylation of the myosin-targeting subunit of myosin light chain phosphatase by Phos-tag SDS-PAGE. *Am J Physiol-Cell Physiol* 310: C681–C691.
21. Kinoshita, E., E. Kinoshita-Kikuta, Y. Kubota, M. Takekawa, and T. Koike. 2016. A Phos-tag SDS-PAGE method that effectively uses phosphoproteomic data for profiling the phosphorylation dynamics of MEK1. *Proteomics* 16: 1825–1836.
22. Chen, J. Q., J. H. Lee, M. A. Herrmann, K.-S. Park, M. R. Heldman, P. K. Goldsmith, Y. Wang, and G. Giaccone. 2013. Capillary isoelectric-focusing immunoassays to study dynamic oncoprotein phosphorylation and drug response to targeted therapies in non-small cell lung cancer. *Mol Cancer Ther* 12: 2601–2613.
23. Sabnis, H., H. L. Bradley, S. T. Bunting, T. Cooper, and K. Bunting. 2014. Capillary nano-immunoassay for Akt 1/2/3 and 4EBP1 phosphorylation in acute myeloid leukemia. *J Transl Med* 12: 166.

24. Xu, D., S. Mane, and Z. Sosic. 2015. Characterization of a biopharmaceutical protein and evaluation of its purification process using automated capillary Western blot. *Electrophoresis* 36: 363–370.
25. Goldman, A., and D. W. Speicher. 2015. Electroblotting from polyacrylamide gels. *Curr Protoc Protein Sci* 82: 10.7.1–16.
26. Janes, K. A. 2015. An analysis of critical factors for quantitative immunoblotting. *Sci Signal* 8: rs2.
27. Bhetwal, B. P., K. M. Sanders, C. An, D. M. Trappanese, R. S. Moreland, and B. A. Perrino. 2013. Ca^{2+} sensitization pathways accessed by cholinergic neurotransmission in the murine gastric fundus. *J Physiol* 591(Pt 12): 2971–2986.
28. Ngounou Wetie, A. G., I. Sokolowska, A. G. Woods, U. Roy, K. Deinhardt, and C. C. Darie. 2014. Protein–protein interactions: Switch from classical methods to proteomics and bioinformatics-based approaches. *Cell Mol Life Sci* 71: 205–228.
29. Ren, L., D. Emery, B. Kaboord, E. Chang, and M. W. Qoroonflen. 2003. Improved immunomatrix methods to detect protein:protein interactions. *J Biochem Biophys Meth* 57: 143–157.
30. Qoronfleh, M. W., L. Ren, D. Emery, M. Perr, and B. Kaboord. 2003. Use of immunomatrix methods to improve protein-protein interaction detection. *J Biomed Biotechnol* 2003: 291–298.
31. Sousa, M. M. L., K. W. Steen, L. Hagen, and G. Slupphaug. 2011. Antibody cross-linking and target elution protocols used for immunoprecipitation significantly modulate signal-to-noise ratio in downstream 2D-PAGE analysis. *Proteome Sci* 9: 45–45.
32. Albarwani, S., L. T. Nemetz, J. A. Madden, A. A. Tobin, S. K. England, P. F. Pratt, and N. J. Rusch. 2003. Voltage-gated K(+) channels in rat small cerebral arteries: Molecular identity of the functional channels. *J Physiol* 551: 751–763.
33. Fergus, D. J., J. R. Martens, and S. K. England. 2003. Kv channel subunits that contribute to voltage-gated K+ current in renal vascular smooth muscle. *Pflügers Archiv* 445: 697–704.
34. Zeng, C., D. Wang, Z. Yang, Z. Wang, L. D. Asico, C. S. Wilcox, G. M. Eisner, W. J. Welch, R. A. Felder, and P. A. Jose. 2004. Aberrant D1 and D3 dopamine receptor transregulation in hypertension. *Hypertension* 43: 654–660.
35. Balashova, N., F.-J. Chang, M. Lamothe, Q. Sun, and A. Beuve. 2005. Characterization of a novel type of endogenous activator of soluble guanylyl cyclase. *J Biol Chem* 280: 2186–2196.
36. Camoretti-Mercado, B., D. J. Fernandes, S. Dewundara, J. Churchill, L. Ma, P. C. Kogut, J. F. McConville, M. S. Parmacek, and J. Solway. 2006. Inhibition of transforming growth factor β-enhanced serum response factor-dependent transcription by SMAD7. *J Biol Chem* 281: 20383–20392.
37. MacMillan, D., S. Currie, and J. G. McCarron. 2008. FK506-binding protein (FKBP12) regulates ryanodine receptor-evoked Ca^{2+} release in colonic but not aortic smooth muscle. *Cell Calcium* 43: 539–549.
38. Jin, L., Q. Gan, B. J. Zieba, S. M. Goicoechea, G. K. Owens, C. A. Otey, and A. V. Somlyo. 2010. The actin associated protein palladin is important for the early smooth muscle cell differentiation. *PLoS One* 5: e12823.
39. Shi, J., F. Miralles, L. Birnbaumer, W. A. Large, and A. P. Albert. 2017. Store-operated interactions between plasmalemmal STIM1 and TRPC1 proteins stimulate PLCβ1 to induce TRPC1 channel activation in vascular smooth muscle cells. *J Physiol* 595: 1039–1058.
40. Bradshaw, J. M. 2010. The Src, Syk, and Tec family kinases: Distinct types of molecular switches. *Cell Signal* 22: 1175–1184.
41. Vázquez-Prado, J., P. Casas-González, and J. A. Garcia-Sainz. 2003. G protein-coupled receptor cross-talk: Pivotal roles of protein phosphorylation and protein–protein interactions. *Cell Signal* 15: 549–557.

42. Pacchiana, R., M. Abbate, U. Armato, I. D. Prà, and A. Chiarini. 2014. Combining immunofluorescence with in situ proximity ligation assay: A novel imaging approach to monitor protein–protein inter-actions in relation to subcellular localization. *Histochem Cell Biol* 142: 593–600.
43. Koos, B., L. Andersson, C.-M. Clausson, K. Grannas, A. Klaesson, and G. Cane. 2014. Analysis of protein interactions in situ by proximity ligation assays. *Curr Top Microbiol Immunol* 377: 111–126.
44. Söderberg, O., K.-J. Leuchowius, M. Gullberg, M. Jarvius, I. Weibrecht, L. G. Larsson, and U. Landegren. 2008. Characterizing proteins and their interactions in cells and tissues using the in situ proximity ligation assay. *Methods* 45: 227–232.
45. Gauthier, T., A. Claude-Taupin, R. Delage-Mourroux, M. Boyer-Guittaut, and E. Hervouet. 2015. Proximity ligation in situ assay is a powerful tool to monitor specific ATG protein interactions following autophagy induction. *PLoS One* 10: e0128701.
46. Ávila-Medina, J., E. Calderón-Sánchez, P. González-Rodríguez, F. Monje-Quiroga, J. A. Rosado, A. Castellano, A. Ordonez, and T. Smani. 2016. Orai1 and TRPC1 proteins co-localize with CaV1.2 channels to form a signal complex in vascular smooth muscle cells. *J Biol Chem* 291: 21148–21159.
47. Chen, M., J. Li, F. Jiang, J. Fu, X. Xia, J. Du, M. Hu, J. Huang, and B. Shen. 2016. Orai1 forms a signal complex with BKCa channel in mesenteric artery smooth muscle cells. *Physiol Rep* 4: e12682.
48. Poythress, R. H., C. Gallant, S. Vetterkind, and K. G. Morgan. 2013. Vasoconstrictor-induced endocytic recycling regulates focal adhesion protein localization and function in vascular smooth muscle. *Am J Physiol Cell Physiol* 305: C215–C227.
49. Chaudhury, A., X.-D. He, and R. K. Goyal. 2012. Role of myosin Va in purinergic vesicular neurotransmission in the gut. *Am J Physiol Gastrointest Liver Physiol* 302: G598–G607.
50. Zhang, W., Y. Huang, and S. J. Gunst. 2016. p21-Activated kinase (Pak) regulates airway smooth muscle contraction by regulating paxillin complexes that mediate actin polymerization. *J Physiol* 594: 4879–4900.
51. Heldin, C. H., and A. Moustakas. 2016. Signaling receptors for TGF-β family members. *Cold Spring Harb Perspect Biol* 8: 1–34.
52. Lander, E. S., L. M. Linton, B. Birren, C. Nusbaum, M. C. Zody, J. Baldwin, K. Devon et al. 2001. Initial sequencing and analysis of the human genome. *Nature* 409: 860–921.
53. Zhao, Y., W. Zhang, Y. Kho, and Y. Zhao. 2004. Proteomic analysis of integral plasma membrane proteins. *Anal Chem* 76: 1817–1823.
54. Lund, R., R. Leth-Larsen, O. N. Jensen, and H. J. Ditzel. 2009. Efficient isolation and quantitative proteomic analysis of cancer cell plasma membrane proteins for identification of metastasis-associated cell surface markers. *J Proteome Res* 8: 3078–3090.
55. Zhang, L., J. Xie, X. Wang, X. Liu, X. Tang, R. Cao, W. Hu, S. Nie, C. Fan, and S. Liang. 2005. Proteomic analysis of mouse liver plasma membrane: Use of differential extraction to enrich hydrophobic membrane proteins. *Proteomics* 5: 4510–4524.
56. Nunomura, K., K. Nagano, C. Itagaki, M. Taoka, N. Okamura, Y. Yamauchi, S. Sugano, N. Takahashi, T. Izumi, and T. Isobe. 2005. Cell surface labeling and mass spectrometry reveal diversity of cell surface markers and signaling molecules expressed in undifferentiated mouse embryonic stem cells. *Mol Cell Proteomics* 4: 1968–1976.
57. Sostaric, E., A. S. Georgiou, C. H. Wong, P. F. Watson, W. V. Holt, and A. Fazeli. 2006. Global profiling of surface plasma membrane proteome of oviductal epithelial cells. *J Proteome Res* 5: 3029–3037.
58. Elia, G. 2008. Biotinylation reagents for the study of cell surface proteins. *Proteomics* 8: 4012–4024.

59. Sidibe, A., X. Yin, E. Tarelli, Q. Xiao, A. Zampetaki, Q. Xu, and M. Mayr. 2007. Integrated membrane protein analysis of mature and embryonic stem cell-derived smooth muscle cells using a novel combination of CyDye/biotin Labeling. *Mol Cell Proteomics* 6: 1788–1797.
60. Adebiyi, A., G. Zhao, D. Narayanan, C. M. Thomas, J. P. Bannister, and J. H. Jaggar. 2010. Isoform-selective physical coupling of TRPC3 channels to IP3 receptors in smooth muscle cells regulates arterial contractility. *Circ Res* 106: 1603–1612.
61. Bannister, J. P., S. Bulley, M. D. Leo, M. W. Kidd, and J. H. Jaggar. 2016. Rab25 influences functional Cav1.2 channel surface expression in arterial smooth muscle cells. *Am J Physiol Cell Physiol* 310: C885–C893.
62. Crnich, R., G. C. Amberg, M. D. Leo, A. L. Gonzales, M. M. Tamkun, J. H. Jaggar, and S. Earley. 2010. Vasoconstriction resulting from dynamic membrane trafficking of TRPM4 in vascular smooth muscle cells. *Am J Physiol Cell Physiol* 299: C682–C694.
63. Evanson, K. W., J. P. Bannister, M. D. Leo, and J. H. Jaggar. 2014. LRRC26 is a functional BK channel auxiliary γ subunit in arterial smooth muscle cells. *Circ Res* 115: 423–431.
64. Jiao, J., V. Garg, B. Yang, T. S. Elton, and K. Hu. 2008. Protein kinase C-ε induces caveolin-dependent internalization of vascular adenosine 5′-triphosphate-sensitive K^+ channels. *Hypertension* 52: 499–506.
65. Leo, M. D., S. Bulley, J. P. Bannister, K. P. Kuruvilla, D. Narayanan, and J. H. Jaggar. 2015. Angiotensin II stimulates internalization and degradation of arterial myocyte plasma membrane BK channels to induce vasoconstriction. *Am J Physiol Cell Physiol* 309: C392–C402.
66. Nourian, Z., M. Li, M. D. Leo, J. H. Jaggar, A. P. Braun, and M. A. Hill. 2014. Large conductance Ca^{2+}-activated K+ channel (BKCa) α-subunit splice variants in resistance arteries from rat cerebral and skeletal muscle vasculature. *PLoS One* 9: e98863.
67. Thomas-Gatewood, C., Z. P. Neeb, S. Bulley, A. Adebiyi, J. P. Bannister, M. D. Leo, and J. H. Jaggar. 2011. TMEM16A channels generate Ca^{2+}-activated Cl^- currents in cerebral artery smooth muscle cells. *Am J Physiol Heart and Circulat Physiol* 301: H1819–H1827.
68. Weaver, A. M., M. McCabe, I. Kim, M. M. Allietta, and S. L. Gonias. 1996. Epidermal growth factor and platelet-derived growth factor-BB induce a stable increase in the activity of low density lipoprotein receptor-related protein in vascular smooth muscle cells by altering receptor distribution and recycling. *J Biol Chem* 271: 24894–24900.
69. Chapman, H., C. Ramström, L. Korhonen, M. Laine, K. T. Wann, D. Lindholm, M. Pasternack, and K. Törnquist. 2005. Downregulation of the HERG (KCNH2) K+ channel by ceramide: Evidence for ubiquitin-mediated lysosomal degradation. *J Cell Sci* 118: 5325–5334.
70. Witting, S. R., J. N. Maiorano, and W. S. Davidson. 2003. Ceramide enhances cholesterol efflux to apolipoprotein A-I by increasing the cell surface presence of ATP-binding cassette transporter A1. *J Biol Chem* 278: 40121–40127.
71. Cheng, X. W., M. Kuzuya, K. Nakamura, Q. Di, Z. Liu, T. Sasaki, S. Kanda et al. 2006. Localization of cysteine protease, cathepsin S, to the surface of vascular smooth muscle cells by association with integrin $\alpha v\beta 3$. *Am J Pathol* 168: 685–694.
72. Bannister, J. P., A. Adebiyi, G. Zhao, D. Narayanan, C. M. Thomas, J. Y. Feng, and J. H. Jaggar. 2009. Smooth muscle cell $\alpha 2\delta$-1 subunits are essential for vasoregulation by CaV1.2 channels. *Circ Res* 105: 948–955.
73. Sanders, K. M., S. D. Koh, S. Ro, and S. M. Ward. 2012. Regulation of gastrointestinal motility-insights from smooth muscle biology. *Nat Rev Gastroenterol Hepatol* 9: 633–645.
74. Sanders, K. M., S. M. Ward, and S. D. Koh. 2014. Interstitial cells: Regulators of smooth muscle function. *Physiol Rev* 94: 859–907.

75. Wamhoff, B. R., S. Sinha, and G. K. Owens. 2007. Conditional mouse models to study developmental and pathophysiological gene function in muscle. *Handb Exp Pharmacol* 178: 441–468.
76. Xin, H., K. Deng, M. Rishniw, G. Ji, and M. I. Kotlikoff. 2002. Smooth muscle expression of Cre recombinase and eGFP in transgenic mice. *Physiol Genomics* 10: 211–215.
77. Cobine, C. A., A. G. Sotherton, L. E. Peri, K. M. Sanders, S. M. Ward, and K. D. Keef. 2014. Nitrergic neuromuscular transmission in the mouse internal anal sphincter is accomplished by multiple pathways and postjunctional effector cells. *Am J Physiol Gastrointest Liver Physiol* 307: G1057–G1072.
78. Harhun, M. I., V. Pucovský, O. V. Povstyan, D. V. Gordienko, and T. B. Bolton. 2005. Interstitial cells in the vasculature. *J Cell Mol Med* 9: 232–243.
79. Herring, B. P., A. M. Hoggatt, C. Burlak, and S. Offermanns. 2014. Previously differentiated medial vascular smooth muscle cells contribute to neointima formation following vascular injury. *Vascul Cell* 6: 21.
80. Huang, F., J. R. Rock, B. D. Harfe, T. Cheng, X. Huang, Y. N. Jan, and L. Y. Jan. 2009. Studies on expression and function of the TMEM16A calcium-activated chloride channel. *Proc Natl Acad Sci USA* 106: 21413–21418.
81. Lee, H., B. H. Koh, L. E. Peri, K. M. Sanders, and S. D. Koh. 2013. Functional expression of SK channels in murine detrusor PDGFRα+ cells. *J Physiol* 591: 503–513.
82. Lee, H., B. H. Koh, L. E. Peri, K. M. Sanders, and S. D. Koh. 2014. Purinergic inhibitory regulation of murine detrusor muscles mediated by PDGFRα+ interstitial cells. *J Physiol* 592: 1283–1293.

14 Role of Ion Channels in Urinary Bladder Smooth Muscle Function

Georgi V. Petkov

CONTENTS

14.1 INTRODUCTION

The physiological functions of the urinary bladder—storage and voiding of urine—are facilitated by the contraction and relaxation of the urinary bladder smooth muscle (UBSM), which comprises the bladder wall. UBSM, also known as detrusor smooth muscle, exhibits spontaneous action potentials that initiate Ca^{2+} transients and spontaneous phasic contractions [1–4]. The physiological significance of these spontaneous, non-voiding smooth muscle phasic contractions of the bladder has remained a subject of controversy for many years. Overactivity of the detrusor is a common clinical problem associated with increased phasic contractions and intravesical bladder pressure that often results in overactive bladder (OAB) and related lower urinary tract symptoms (LUTS) [5]. Many forms of LUTS are directly linked to UBSM dysfunction. OAB, which is among the top-ten most prevalent chronic health conditions, affects approximately 17% of the Western world population and increases with age [6]. A recent study suggests that the spontaneous UBSM contractions that are linked to small transient increases in intravesical bladder pressure activate afferent nerves, thus communicating the sensation of bladder fullness to the central nervous system [7]. This key physiological mechanism of UBSM spontaneous phasic contraction-induced bursts of afferent nerve activity appears to play an important role in conveying the sensation of bladder fullness and regulating micturition frequency.

UBSM ion channels located in the cell membrane play an important role in determining the functional properties and are recognized potential novel pharmacological or genetic targets for treatment of OAB/LUTS. In fact, UBSM expresses several families of ion channels that maintain the resting membrane potential and/or the shape of spontaneous action potentials. New evidence suggests that bladder interstitial cells, such as platelet-derived growth factor receptor-α ($PDGFR\alpha^+$) cells, may also play a role in determining the pattern of UBSM spontaneous electrical activity, although their precise physiological role is just now emerging [8–11]. Since many species differences exist and because humans are the target species of interest for therapeutic intervention, my research group, in collaboration with clinicians, has focused on studying UBSM ion channel activity directly on clinically–characterized human UBSM samples and correlating ion channel activity and function with the patients' clinical phenotype. Studies on human UBSM are critical to validate animal models reflecting human bladder function, and since human UBSM is very difficult to obtain in sufficient quantities necessary to carry out quantitative investigations, such studies have a great deal of merit. This chapter discusses what is currently known about the functional and regulatory roles of UBSM ion channel families located in the cell membrane, with emphasis on recent advancements in the field.

14.2 VOLTAGE-GATED Ca^{2+} CHANNELS

L-type voltage-gated Ca^{2+} channels have critically important physiological roles in UBSM [12–16]. Ca^{2+} entry through dihydropyridine-sensitive L-type voltage-gated Ca^{2+} channels is responsible for the initial depolarization phase of the action potential and leads to an increase in the global intracellular Ca^{2+} concentration that activates UBSM phasic contractions [2,17–20]. Blocking the L-type voltage-gated Ca^{2+} channels with dihydropyridines or other channel inhibitors completely eliminates the spontaneous action potentials and related phasic contractions [2,17,18,21,22]. Although UBSM action potentials resemble the shape of neuronal action potential, UBSM lacks tetrodotoxin-sensitive voltage-gated Na^+ channels, which are known to determine the upstroke of the action potential in neurons and heart muscle [23].

T-type voltage-gated Ca^{2+} channels, which normally activate at more negative potentials, are also expressed in UBSM but their functional role is less clear [24,25]. Most likely, they add to UBSM spontaneous electrical activity by contributing to the slight transient depolarizations, known as pre-potentials.

14.3 K^+ CHANNELS

The central role played by K^+ channels in modulating UBSM function derives from their functionally antagonistic relationship with the L-type voltage-gated Ca^{2+} channels that deliver the extracellular Ca^{2+} influx necessary to activate contraction.

As key regulators of the UBSM membrane excitability, the K^+ channels provide fine-tune control of the opening and closing of the L-type voltage-gated Ca^{2+} channels, Ca^{2+} entry into the UBSM cells, and therefore UBSM contractility. In general, inhibition of the UBSM K^+ channels leads to an increased membrane excitability and contractility, whereas their activation hyperpolarizes the membrane and decreases the contractility. Two recent comprehensive reviews provide detailed information on the subject [4,26].

There are four major families of K^+ channels in UBSM: voltage-gated K^+ (K_V) channels, Ca^{2+}-activated K^+ (K_{Ca}) channels, inward-rectifying ATP-sensitive K^+ (K_{ir}, K_{ATP}) channels, and two-pore domain K^+ (K_{2P}) channels [4,27–34].

14.3.1 Voltage-Gated K^+ (K_V) Channels

The functional role of K_V channels in UBSM is a hot topic under increasing investigation [22,27–29,33,35]. In UBSM cells, K_V channels, which is the most diverse K^+ channel family, work to repolarize the membrane potential to end the action potential, and also have a role in setting the resting membrane potential [2,4,22,29,35,36]. Expression of $K_V1.3$ and $K_V1.6$ channels has been shown in human UBSM [37], and low expression of $K_V1.2$ and $K_V1.5$ has been reported in rat UBSM [38]. Expression of $K_V2.1$ and $K_V2.2$ channels has been demonstrated in rat UBSM; $K_V2.1$ seems to be the predominant channel subtype [28,38,39]. Decreased expression of mRNA for the $K_V2.1$ channel in a rat model with UBSM hyper-reflexia induced by spinal cord injury has been suggested, but the study lacked quantitative evaluation [39]. In mouse UBSM, the $K_V2.1$ subunit associates with electrically silent subunits ($K_V5.1$, $K_V6.1$, $K_V6.2$ or $K_V6.3$), forming $K_V2.1$/X heterotetramers [36]. Similar to rat UBSM [28], human UBSM cells express $K_V2.1$ and $K_V2.2$ channels along with the electrically silent $K_V9.1$ subunit [29]. Guinea pig UBSM expresses $K_V2.1$ (but not $K_V2.2$) and the electrically silent subunits $K_V6.2$, $K_V6.3$, $K_V8.2$, and $K_V9.1$–$K_V9.3$, indicating marked species differences [22]. Human, rat, and guinea pig UBSM does not express the $K_V4.2$ channel, which mediates the fast A-type K_V current [22,28,29]. A functional role for $K_V11.1$ channel has been also suggested in guinea pig UBSM [40].

Among all K_V channel families, the K_V7 family ($K_V7.1$–$K_V7.5$), also known by the gene names as KCNQ channels, appears to have the most profound role in UBSM function. In recent years, my research group has pioneered the systematic investigation of the functional and regulatory roles of the K_V7 channels in UBSM [27,33,41–45] and this line of investigations has been followed by other groups too [46–52]. K_V7 channel pore-forming α-subunits, which can form homo- or hetero-multimers, can further associate with small, single transmembrane regulatory subunits encoded by the KCNE genes [53,54]. The diversity of K_V7 channel heteromultimeric combinations, such as $K_V7.2$/$K_V7.3$, $K_V7.3$/$K_V7.5$, or $K_V7.4$/$K_V7.5$, expands their functional and biophysical properties [4,23,33]. This diversity of K_V7 channel expression and function underlies their attractiveness as potential novel therapeutic targets for OAB [4,54].

K_V7 channel pharmacological modulators include activators retigabine ($K_V7.2$–$K_V7.5$) and L-364,373 ($K_V7.1$), as well as the K_V7 channel inhibitors XE991 and linopirdine, both of which effectively block all K_V7 channels [27,33]. Additional novel K_V7 channel subtype selective openers, such as ICA-069673 and ML213, have been recently shown to dramatically reduce UBSM contractility [33,41,45,49,51]. Retigabine (Ezogabine or Trobalt), the first-in-class K_V7 channel opener marketed for epilepsy, increased micturition volume, and voiding intervals and, when given intravesically, this compound decreased capsaicin-induced detrusor overactivity [55]. In isolated UBSM strips, retigabine reduced spontaneous, pharmacologically-induced, and nerve-evoked UBSM contractions, effects that were reversed by K_V7 channel inhibitor XE991 [27,42,47–49,51]. The reported occurrence of retigabine–induced urinary retention compared with the placebo in clinical trials for epilepsy [56] further supports the concept that K_V7 openers might be very useful for the treatment of OAB. Thus, what was initially perceived as side effects from retigabine may actually lead to novel therapeutics for OAB.

A recent multi-level study, using the novel $K_V7.2/K_V7.3$ channel activator ICA-069673, provides strong evidence to suggest a critical role for the $K_V7.2$- and $K_V7.3$-containing channels in UBSM function at both cellular and tissue levels. Pharmacological activation of $K_V7.2/K_V7.3$ channels with ICA-069673 hyperpolarized the membrane potential and inhibited spontaneous action potentials in freshly-isolated guinea-pig UBSM cells, which blocked Ca^{2+} influx through L-type voltage-gated Ca^{2+} channels to reduce global intracellular Ca^{2+} concentrations, causing relaxation of UBSM isolated strips. Furthermore, several studies from different groups suggest that $K_V7.4$- and $K_V7.5$-containing channels have the highest expression levels in UBSM and that they most likely mediate most of the functional outcomes in UBSM [27,41,44,49,51].

The structural diversity of the K_V channels contributes to the variety of their pharmacological properties and makes them potentially attractive targets for pharmacological treatment in bladder diseases [4,22,27–29,33,57,58]. Future investigations in this area may lead to the development of selective pharmacological or genetic therapies for OAB [54,58,59].

14.3.2 Ca^{2+}-Activated K^+ Channels

In UBSM, the K_{Ca} channel family is represented by three major subtypes of channels, which are classified based on their single-channel conductance: large conductance voltage- and Ca^{2+}-activated K^+ (BK) channels; small conductance Ca^{2+}-activated K^+ channels (SK1–SK3); and intermediate conductance Ca^{2+}-activated K^+ (IK) channels [4]. SK and IK channels are Ca^{2+}-activated but voltage-insensitive.

14.3.2.1 BK Channels

The BK channels are the most prominent and physiologically relevant K^+ channels that regulate UBSM function in health and disease [18,35,60–70]. Besides its high single-channel conductance (~200 pS), this importance also arises from the fact that the BK channel is the only K^+ channel that is activated by both intracellular Ca^{2+} and membrane depolarization, so it is uniquely suited to serve as a Ca^{2+}/voltage signal integrator in the modulation of UBSM cell excitability [4,23,64,71,72]. The physiological role of BK channels is to provide a negative-feedback mechanism to limit the amplitude and duration of UBSM action potentials and related phasic contractions [17,18,61,64]. The BK channel has very high expression levels in UBSM [64,66,73,74]. In recent years, my group in collaboration with clinicians, have provided a methodological identification and characterization of the BK channel functional and regulatory roles in native human UBSM [64,72,75–80].

While the BK channel pore-forming α subunit is encoded by a single gene, a number of splice variants possessing differential regulatory mechanisms have been reported [64,65,81]. The β1-subunit is the primary auxiliary BK channel subunit in UBSM [66], but several studies have also identified the presence of β4-subunits [64,73,82] while confirming the absence of BK channel β2- and β3-subunits in UBSM cells [64,73]. The native human UBSM BK channels likely consist of diverse (α/α, α/β1 and α/β4) BK channel complexes [72]. Targeted deletion of the BK channel pore-forming α-subunit or regulatory β1-subunit results in increased UBSM excitability and contractility in mouse models [60,65,66]. Data from global and smooth muscle-specific BK channel α-subunit knockout (KO) mice indicate that deletion of the BK channel α-subunit dramatically increased spontaneous and nerve-evoked contractions, and these mice have elevated urination frequency [60,65,69,70,83]. While BK channel β1-subunit increases the apparent Ca^{2+} and voltage sensitivity of UBSM BK channels [66], the function of the more recently discovered β4-subunit in UBSM is less clear [64,73,82], although it might have a role in UBSM pathophysiology [76,82].

BK channels are blocked with high-affinity by iberiotoxin and charybdotoxin [84–86]. Moreover, all known BK channel splice variants are effectively inhibited by paxilline [87,88], thus making paxilline ideal for UBSM BK channel functional studies. BK channels are also inhibited non-selectively by tetraethylammonium, but with low potency [23,60,63]. Pharmacological blockade of BK channels increases the amplitude and duration of spontaneous action potentials, indicating that the repolarization phase of the UBSM action potential is mediated by BK channel activity [2,17,18,35]. Although some studies failed to report UBSM resting membrane potential depolarization upon BK channel pharmacological inhibition [2,17,35], it is now well documented that blocking BK channels depolarizes the resting membrane potential in both isolated UBSM cells and intact tissues [18,64,77,78,80,89–91]. Moreover, genetic deletion of the BK channel α-subunit causes sustained membrane depolarization in UBSM [60,69]. In UBSM isolated from various other species and humans, pharmacological inhibition of BK channels increases the amplitude, duration, and force of the spontaneous phasic contractions, as well as UBSM tone [21,40,60,61,64,66,69,76–78,80,86,89,90,92,93]. In contrast, iberiotoxin had no effect on phasic and tonic contractions of UBSM strips isolated from mice that lacked the BK channel α-subunit [60]. Pharmacological inhibition or genetic deletion of BK channels enhances the nerve-evoked contractions in UBSM strips isolated from different species [64,69,70,75,83,94–97]. On the other hand, BK channel activators increase BK channel open probability and whole-cell BK currents, causing membrane hyperpolarization and relaxation of UBSM [40,68,77,93,98–107].

Ca^{2+} is a critical BK channel regulator [23,26,71]. In UBSM, there are two major Ca^{2+} sources for BK channel activation: (1) Ca^{2+} entry through L-type voltage-gated Ca^{2+} channels and (2) Ca^{2+} release from the ryanodine receptors (RyRs) of the sarcoplasmic reticulum (SR) [62–64,67,89]. BK channels are under the local control of the so-called *Ca^{2+} sparks* caused by spontaneous highly localized and transient Ca^{2+} releases from the RyRs, which primary role is to trigger transient BK currents (TBKCs), originally known as spontaneous transient outward currents (STOCs), without affecting the global intracellular Ca^{2+} concentration [62,67,74,91]. The physiological roles of Ca^{2+} sparks and related TBKCs are well documented in UBSM of various species [60,62,63,67,74,80,89,91,108], including human UBSM [64,76,79]. At the level of the membrane potential, the TBKCs manifest as spontaneous transient hyperpolarizations (STHs) that contribute to the overall UBSM hyperpolarization [64,79,80,91,108,109]. In UBSM cells, likewise to the TBKCs, STHs are completely inhibited by BK channel blockers such as iberiotoxin or paxilline [64,78,80,90,91]. There is a tight correlation between the TBKC frequency and STH frequency in UBSM cells [80].

Bladder function is regulated by parasympathetic nerve fibers releasing acetylcholine, the primary excitatory neurotransmitter in the bladder, which activates muscarinic (M) receptors in UBSM cells, initiating physiological phasic contractions that facilitate voiding [110,111]. In rat UBSM cells, activation of M3 receptors with carbachol leads to initial appearance of large transient outward BK currents followed by sustained inhibition of the spontaneous TBKCs, STHs, and depolarization of UBSM resting membrane potential [112]. These large carbachol-induced transient outward BK currents are attributed to the inositol 1,4,5-trisphosphate receptor (IP_3R)-induced SR Ca^{2+} release as they are completely blocked by the IP_3 receptor inhibitor xestospongin-C [112]. Depletion of SR Ca^{2+} upon activation of M3 receptors reduces Ca^{2+} spark activity, which leads to inhibition of the spontaneous TBKCs and depolarization of UBSM cell membrane potential causing an increase in L-type voltage-gated Ca^{2+} channel activity and UBSM contractility. This mechanism of muscarinic inhibition of BK channels has now been confirmed to operate in human UBSM as well [113]. Collectively, these studies have revealed that M3 receptors are functionally coupled to the BK channels in rat and human UBSM [112,113]. Furthermore, the M3 receptor intracellular signaling involves activation of protein kinase C (PKC), which can also regulate BK channels in UBSM [114,115]. Specifically, PKC inhibits BK channel activity indirectly via a Ca^{2+}-dependent mechanism involving the attenuation of the Ca^{2+} release through RyRs while increasing the global intracellular Ca^{2+} necessary to activate UBSM contraction [114].

In UBSM, stimulation of β-adrenergic receptors (β-ARs) increases intracellular cAMP concentration, which activates protein kinase-A (PKA) that, in turn, phosphorylates specific proteins resulting in decreased UBSM excitability and contractility [3,60,110,116]. BK channel activity increases upon pharmacological activation of β-ARs to promote relaxation of UBSM [60,67,88,117]. PKA stimulation has been shown to activate Ca^{2+} sparks in guinea pig UBSM [67,91]. The critical role of BK channels in β-AR/PKA-mediated UBSM relaxation is demonstrated by the fact that BK channel genetic deletion in transgenic mouse models leads to a compensatory adaptive upregulation of the β-AR/PKA pathway [60,69].

UBSM expresses all three known β-AR subtypes (β1–β3) with profound species differences in the β-AR subtype expression patterns [110]. β3-ARs appear to be the most physiologically relevant in human UBSM [118]. In rat UBSM, β3-ARs and

BK channels are functionally coupled to promote UBSM relaxation by a complex Ca^{2+}-dependent mechanism [89]. In rat and human UBSM, β3-AR agonists effectively inhibit both the spontaneous and nerve-evoked UBSM contractions via a BK channel mechanism [75,89,94,119]. In contrast, although β3-ARs are expressed at the mRNA level in guinea pig UBSM, they serve a negligible role, if any, in UBSM contractility without affecting cell excitability [120]. Collectively, selective β3-AR agonists appear to be very effective in controlling UBSM function [75,89,94,121]. Intriguingly, the selective β3-AR agonist, mirabegron, has been recently approved to treat OAB [121–123].

Phosphodiesterases, which hydrolyze cAMP, are constitutively active in UBSM cells, and therefore their pharmacological inhibition can lead to a rapid increase in cellular cAMP, which in turn utilizes the PKA signaling pathway, described earlier, in activating the BK channel activity in a RyR-dependent manner [78,80,90,91]. Phosphodiesterase-1 and phosphodiesterase-4 appear to have particular functional importance in UBSM. Phosphodiesterase-1 inhibition suppresses human UBSM excitability and contractility via activation of RyRs and an increase in TBKC frequency [79,80]. Pharmacological inhibition of phosphodiesterase-4 reduces UBSM contractility by increasing the frequency of Ca^{2+} sparks and the functionally-coupled TBKCs, while simultaneously decreasing the global intracellular Ca^{2+} concentration [91].

In addition to the major BK channel regulatory mechanisms described earlier, UBSM BK channels are also regulated by various substances, hormones, and endogenous factors. Sex hormones such as estrogens and testosterone mediate UBSM relaxation [124] by directly activating the BK channels [16,125–127]. Ethanol increases BK channel activity by differential mechanisms involving dependence on intracellular Ca^{2+} and the basal level of TBKC frequency in UBSM cells [128]. Hydrogen sulfide increases UBSM contractility via direct inhibition of UBSM BK channels [129]. Moreover, prostaglandin-E2 increases UBSM excitability and contractility by selectively inhibiting TBKCs [130].

The importance of BK channels in UBSM excitability and contractility suggests that BK channel defects, alterations or mutations, may cause certain forms of LUTS, including detrusor overactivity or underactivity. Decreased BK channel expression has been reported in patients with benign prostatic hyperplasia-induced bladder outlet obstruction and associated detrusor overactivity, as well as in rabbits with bladder obstruction-induced detrusor overactivity [98]. Genetic ablation of the BK channel pore-forming α-subunit or regulatory β1-subunit causes increased UBSM contractility [60,65,66,69,70]. Overexpression of the BK channel α-subunit using gene transfer techniques can eliminate detrusor overactivity caused by bladder outlet obstruction in rats [131], an observation consistent with increased UBSM contractility in KO mice lacking BK channel subunits [60,65,66,69,70]. Genetic deletion of the BK channel β1-subunit significantly decreases single BK channel open probability, causing detrusor overactivity [66]. BK channel α-subunit mRNA expression, whole cell BK currents, and TBKCs are all decreased in UBSM from patients with neurogenic detrusor overactivity compared to control patients [76]. These molecular and cellular findings are consistent with functional studies showing significantly increased UBSM spontaneous phasic contractions and a lack of effect of iberiotoxin on spontaneous contractility in UBSM strips isolated from patients with neurogenic detrusor overactivity [76].

UBSM strips isolated from patients with neurogenic detrusor overactivity consistently did not respond to iberiotoxin or the BK channel opener NS1619, indicating BK channel dysfunction in these patients [93]. Collectively, these data suggest that neurogenic detrusor overactivity is associated with decreased UBSM BK channel expression and function, thus leading to increased UBSM excitability and contractility.

BK channel pharmacological activation with selective openers has long been explored as a pharmacological approach for LUTS treatment [21,40,68,77,97,99, 100,102,105,106,132,133]. The BK channel opener, NS1619, significantly increases the whole cell BK current, causes membrane hyperpolarization, decreases the level of global intracellular Ca^{2+}, and inhibits myogenic and nerve-evoked contractions in native human UBSM [77]. In excised membrane patches from human UBSM cells, NS1619 drastically enhances the single BK channel open probability by about 10-fold with the primary effect occurring on the channel gating [72]. Interestingly, NS1619 inhibits the UBSM action potentials and at higher concentrations also causes hyperpolarization [52]. *In vivo* administration of the BK channel activator, NS8, results in increased storage capacity of the bladder and reverses bladder hyperactivity in rats [97]. Intravenous administration of NS8 in rats also causes a reduction in micturition frequency and an increase in micturition volume [134]. BK channel openers, at lower doses, may affect bladder function without vascular effects, and indeed, there are ongoing efforts to develop more potent and selective BK channel openers [103,104,132,133,135–137]. On the other hand, targeting BK channels with selective inhibitors to stimulate UBSM contractility as a potential treatment for detrusor underactivity is a new intriguing area of investigation.

UBSM BK channel gene therapy represents another potential clinical application for treatment of LUTS [138]. Thus, the BK channel is a promising novel opportunity for therapeutic intervention in human UBSM for the treatment of OAB associated with detrusor overactivity.

14.3.2.2 IK Channels

IK channels have been shown to be expressed at the mRNA level in mouse UBSM [139]. We have also found mRNA expression of IK channels in guinea pig UBSM using single-cell RT-PCR, and further confirmed IK channel protein expression with western blot [31]. IK channel protein, however, was not found in rat and human UBSM with immunostaining [30,140]. Studies using charybdotoxin suggest that IK channels may have a functional role in mouse UBSM [35,139]; however, this requires further evidence, as charybdotoxin inhibits both IK and BK channels [86]. Although the IK channel might be expressed at the mRNA and protein levels, patch-clamp and functional studies did not confirm a role for IK channels in guinea pig, rat, and human UBSM excitability and contractility under normal physiological conditions [30,31,34,140]. It is possible that IK channels may become active under pathophysiological conditions but this requires further investigations.

14.3.2.3 SK Channels

The SK channels, regulated by Ca^{2+} and calmodulin [141], contribute to the UBSM action potential after-hyperpolarization phase [2,142,143]. The low SK channel density in UBSM cells [63], along with their Ca^{2+} sensitivity, makes these channels unique sensors for changes in global intracellular Ca^{2+} induced by Ca^{2+} entry via L-type voltage-gated Ca^{2+} channels during the action potential. Apamin inhibits all of the three known SK channel subtypes (SK1–SK3) and increases UBSM contractility [21,30,31,34,40,61,63,95,140,143–145]. Studies in SK2 KO mice suggest that SK2 channels are essential contributors to UBSM function [145]. Genetic overexpression of the SK3 channel elevated SK currents, which increased bladder capacity and decreased the frequency of UBSM phasic contractions, whereas suppression of SK3 channel expression increased the frequency of UBSM phasic contractions *in vitro* and nonvoiding contractions *in vivo* [144].

Pharmacological activation of SK channels with selective channel openers might be a novel approach to restoring normal UBSM function as shown *in vivo* and *in vitro* [30,146–148]. SK channel activation with SKA-31, the most potent and selective SK/IK channels activator, can reduce the excitability and contractility of guinea pig and human UBSM [31,34]. The SK/IK channel openers NS309 and SKA-31 dramatically increased the duration of the after-hyperpolarization and then abolished action potential firing in an apamin-sensitive manner [52]. NS309 decreases rat UBSM cell excitability and contractility by targeting the SK3 channel, the main SK channel subtype in UBSM [30]. Collectively, these studies also established that SK channels, but not IK channels, mediate the inhibitory effects on UBSM excitability and contractility [30,31,34,140]. While obviously the SK channels have fundamental roles in regulating UBSM function, knowledge about their regulatory mechanisms in human UBSM remains limited [2,34,21,149]. The role of SK channels in the whole bladder is even more complex given that non-UBSM cells within the bladder wall such as PDGFRα^+ interstitial cells also express SK channels [8–11,150,151].

14.3.3 Inward-Rectifying ATP-Sensitive K^+ Channels

In UBSM, the K_{ir} channel family is represented by the ATP-sensitive K^+ channels (K_{ATP}), also known as $K_{ir}6$ channels [152,153]. K_{ATP} channels are unique in that they link the cell metabolic state (intracellular ATP concentration) with the membrane potential and can decrease or increase the cell excitability accordingly. However, K_{ATP} channels are neither Ca^{2+} nor voltage regulated, which disconnect them from the two most important physiological cellular regulators. Although the concept of increasing K_{ATP} channel opening pharmacologically for the treatment of OAB was initially proposed in the 1980s [154], the actual discovery of the K_{ATP} channels in UBSM using the patch-clamp technique was achieved later on [155]. K_{ATP} channels have been also identified at a single-channel level in pig and human UBSM cells [152,153]. The K_{ATP} channel inhibitor, glibenclamide [32,156,157], does not affect UBSM excitability or contractility in the absence of K_{ATP} channel stimulation, questioning whether or not K_{ATP} channels contribute to UBSM resting membrane potential under normal physiological conditions [40,143,153]. On the contrary, however; pharmacological activation of K_{ATP} channels causes membrane hyperpolarization, which closes the L-type voltage-gated Ca^{2+} channels, reducing UBSM contractility [3,32,59,143,152,153,156–159].

Intriguingly, the K_{ATP} channel density is much higher in UBSM compared to other tissues [32,152,157]; thus, one way to achieve bladder-selective effects could be to apply lower doses of K_{ATP} channel openers that would not affect other tissues with lower K_{ATP} channel density. Indeed, our studies have shown that only ~1% of the K_{ATP} channels need to be activated in order to inhibit UBSM action potentials and related phasic contractions [32].

14.3.4 Two-Pore Domain (K_{2P}) Channels

The K_{2P} channels, also known as *leak K^+ channels*, have reported functional roles in UBSM under normal and pathological conditions [160–162]. They participate in stabilizing the UBSM cell resting membrane potential [162]. Mouse UBSM expresses functional K_{2P}, stretch-dependent K^+ channels of the TREK-1 subtype [161] that have also been identified in cultured human UBSM cells [163]. BL-1249, a TREK-1 channel opener, produces a membrane hyperpolarization in cultured human UBSM cells, inhibits the 30 mM KCl-induced UBSM contractions, and decreases the frequency of isovolumic contractions *in vivo*, without affecting blood pressure in rats [163]. These channels, therefore, may have a physiological role in suppressing UBSM excitability and contractility in response to stretch during bladder filling.

14.4 TRANSIENT RECEPTOR POTENTIAL CHANNELS

Transient receptor potential (TRP) channels are non-selective cation channels. As a large superfamily of 28 members, these channels are classified into seven related subfamilies: the TRPC (canonical), TRPV (vanilloid), TRPM (melastatin), TRPP (polycystin), TRPML (mucolipin), TRPN (no mechanopotential, NOMP), and TRPA (ankyrin) [164]. They respond to physical and chemical stimuli such as temperature, pH, stretch, and light, as well as intracellular stimuli such as Ca^{2+}; they constitute a fundamental mechanism by which cells respond to changes in the environment [164]. TRP channels have been reported in bladder urothelium and nerves [165–168] but their role in UBSM is just now emerging [168]. Initial microarray data from my laboratory indicated that in human UBSM TRPM4 channel expression was much higher than other TRP channels, which focused our attention toward this particular TRP channel. The TRPM4 channel is unique because it is a Ca^{2+}-activated cation channel highly selective for monovalent cations, such as Na^+ and K^+, but impermeable to anions and divalent cations including Ca^{2+} [169,170]. TRPM4 channels are voltage-dependent, have single-channel conductance ~25 pS, and exhibit high Ca^{2+} dependency [170–172]. Recently, in collaboration with Dr. Scott Earley from University of Nevada at Reno, we published the first seminal reports outlining the expression and function of TRPM4 channels in rodent UBSM [173–175]. In collaboration with clinicians, we have further extended these studies to native human UBSM using clinically-characterized specimens [176,177]. We found that the inhibition of TRPM4 channels with the novel TRPM4 channel inhibitor 9-phenanthrol [178] significantly reduced UBSM excitability. TRPM4 channel pharmacological inhibition with 9-phenanthrol significantly hyperpolarized the UBSM cell membrane potential. These data clearly demonstrated that TRPM4 is a key modulator of human UBSM under physiological conditions [177]. We have proven the novel hypothesis that TRPM4 channel activity facilitates depolarization in UBSM, and thus enhances human UBSM contractility. This line of inquiry was critical for development of novel OAB treatment strategies as prior studies have targeted hyperpolarizing currents (e.g., K^+ channels) but the depolarizing component in UBSM cells was not understood. Moreover, since initiation of bladder voiding contractions is triggered upon parasympathetic activation and the release of acetylcholine, which activates M3 receptors in UBSM cells, we performed functional studies to examine the role of TRPM4 channels when M3 receptors are activated with carbachol. The data suggest that TRPM4 channels can modulate UBSM function during the physiological process of micturition. Our comparative studies further revealed important species-related differences and suggest that TRPM4 channels have a more prominent physiological role in the regulation of UBSM function in humans in comparison to rodents [173–175,177].

As the TRPM4 channels are activated by Ca^{2+}, IP_3R-mediated Ca^{2+} release from the SR represents Ca^{2+} source for TRPM4 channel activation. We further investigated the molecular and functional interactions of the IP_3Rs and TRPM4 channels in human UBSM [179,180]. With *in situ* proximity ligation assay, we demonstrated co-localization of the TRPM4 channels and IP_3Rs in human UBSM cells [179,180]. As the TRPM4 channels and IP_3Rs must be located within close apposition to functionally interact, these findings support the concept of a potential Ca^{2+}-mediated TRPM4-IP_3R regulatory mechanism. In freshly-isolated human UBSM cells, blocking the IP_3Rs with the selective IP_3R inhibitor xestospongin-C significantly decreased TRPM4 channel-mediated transient inward cation currents (TICCs) [179,180]. The data reveal that TRPM4 channels and IP_3Rs are spatially and functionally coupled in human UBSM [179,180].

These systematic studies provide compelling evidence that TRPM4 channels are major players in UBSM function, and this channel is a very likely candidate as a therapeutic target for OAB [173–177,179,180]. The challenge of TRPM4 channels being expressed in other tissues is not unique to therapeutic targets affecting lower urinary tract function, and fortunately this obstacle has been pharmacologically overcome in a number of disease states. Considering the relatively higher TRPM4 channel mRNA expression in UBSM compared to vasculature [173], as well as their important functional roles in UBSM, pharmacological manipulation of TRPM4 channels in OAB patients may provide effective treatment with minimal adverse collateral cardiovascular effects. The potential clinical application of TRPM4 channel modulation for OAB therapy should be further validated in clinical trials.

While the TRPM4 channel's functional role in UBSM is now well established, other TRP channels may also be involved in the regulation of UBSM function. For example, it has been reported that GSK1016790A, a selective TRPV4 channel opener, increased UBSM contractions and that administration of this compound results in detrusor overactivity in wild type but not in TRPV4 channel KO mice [181]. Another study proposes that in UBSM cells, TRPV4 channels are functionally coupled with BK channels and provide Ca^{2+} influx to activate BK channels, thus limiting UBSM contractility [182]. However, novel evidence suggests that TRPV4 channels are expressed not in UBSM cells but in interstitial PDGFRα$^+$ cells and are the main source of Ca^{2+} influx to activate SK3 channels during bladder filling [150,151].

14.5 Cl^- CHANNELS

The expression and function of Cl^- channels, including Ca^{2+}–activated Cl^- channels, in UBSM are largely unexplored. The Cl^- channels are often assumed to be passive, because Cl^- is not usually at its equilibrium potential at rest, but it is possible that these channels are functional in UBSM as potentially they can modulate the membrane potential [23]. Protein and mRNA expression of Ano1/TMEM16A Cl^- channels has been reported in interstitial and UBSM cells of juvenile rat urinary bladder supported by UBSM contractility studies showing reduction of spontaneous phasic contractions by the non-selective channel inhibitors, niflumic acid, and 5-Nitro-2-(3-phenylpropylamino)benzoic acid (NPPB) [183]. NPPB also inhibits Ca^{2+} oscillations in guinea pig UBSM cells, which further suggests a possible role for these Cl^- channels in bladder function [25]. Another study confirms Ano1/TMEM16A expression in mouse bladder interstitial cells but argues against their expression in UBSM cells [184]. It is suggested that Ca^{2+}–activated Cl^- channel expression increases under pathophysiological conditions of rat bladder outlet obstruction associated with detrusor overactivity [185]. Thus, it is evident that the literature is controversial and the expression and function of Cl^- channels in UBSM require further investigations.

14.6 CONCLUSIONS

This chapter reviewed the complex interplay among the UBSM ion channels and their role in shaping the electrical and mechanical activity of UBSM in health and disease. OAB and LUTS are pressing medical issues lacking effective therapies, and therefore, there is a need to identify novel therapeutic treatments directly targeting UBSM. A key step for the development of a more effective OAB therapy is a better understanding of the ion channels that control UBSM excitability and contractility under normal and pathophysiological conditions. Thus, it is necessary to obtain further insights on the role of ion channels in UBSM function directly from human UBSM using multidisciplinary experimental approaches, allowing for a correlation of the data from molecular, cellular, and tissue levels with the patients' clinical phenotype.

ACKNOWLEDGMENTS

This study was supported by a grant from the National Institutes of Health R01 DK106964 to Georgi V. Petkov.

REFERENCES

1. Brading, A. F. 2006. Spontaneous activity of lower urinary tract smooth muscles: Correlation between ion channels and tissue function. *J Physiol* 570: 13–22.
2. Hashitani, H., and A. F. Brading. 2003. Electrical properties of detrusor smooth muscles from the pig and human urinary bladder. *Br J Pharmacol* 140: 146–158.

3. Hashitani, H., A. F. Brading, and H. Suzuki. 2004. Correlation between spontaneous electrical, calcium and mechanical activity in detrusor smooth muscle of the guinea-pig bladder. *Br J Pharmacol* 141: 183–193.
4. Petkov, G. V. 2012. Role of potassium ion channels in detrusor smooth muscle function and dysfunction. *Nat Rev Urol* 9: 30–40.
5. Wallace, K. M., and M. J. Drake. 2015. Overactive bladder. *F1000Res* 4: F1000 Faculty Rev-1406. doi:10.12688/f1000research.7131.1.
6. Coyne, K. S., C. C. Sexton, J. A. Bell, C. L. Thompson, R. Dmochowski, T. Bavendam, C. I. Chen, and J. Q. Clemens. 2013. The prevalence of lower urinary tract symptoms (LUTS) and overactive bladder (OAB) by racial/ethnic group and age: Results from OAB-POLL. *Neurourology and Urodynamics* 32: 230–237.
7. Heppner, T. J., N. R. Tykocki, D. Hill-Eubanks, and M. T. Nelson. 2016. Transient contractions of urinary bladder smooth muscle are drivers of afferent nerve activity during filling. *J Gen Physiol* 147: 323–335.
8. Koh, B. H., R. Roy, M. A. Hollywood, K. D. Thornbury, N. G. McHale, G. P. Sergeant, W. J. Hatton, S. M. Ward, K. M. Sanders, and S. D. Koh. 2012. Platelet-derived growth factor receptor-alpha cells in mouse urinary bladder: A new class of interstitial cells. *J Cell Mol Med* 16: 691–700.
9. Lee, H., B. H. Koh, L. E. Peri, K. M. Sanders, and S. D. Koh. 2013. Functional expression of SK channels in murine detrusor $PDGFR^+$ cells. *J Physiol* 591: 503–513.
10. Lee, H., B. H. Koh, L. E. Peri, K. M. Sanders, and S. D. Koh. 2014. Purinergic inhibitory regulation of murine detrusor muscles mediated by $PDGFRalpha^+$ interstitial cells. *J Physiol* 592: 1283–1293.
11. Lee, H., B. H. Koh, E. Yamasaki, N. E. George, K. M. Sanders, and S. D. Koh. 2015. UTP activates small-conductance Ca^{2+}-activated K^+ channels in murine detrusor $PDGFRalpha^+$ cells. *Am J Physiol Renal Physiol* 309: F569–F574.
12. Nakayama, S., and A. F. Brading. 1993. Evidence for multiple open states of the Ca^{2+} channels in smooth muscle cells isolated from the guinea-pig detrusor. *J Physiol* 471: 87–105.
13. Nakayama, S., and A. F. 1993. Brading. Inactivation of the voltage-dependent Ca^{2+} channel current in smooth muscle cells isolated from the guinea-pig detrusor. *J Physiol* 471: 107–127.
14. Nakayama, S., and A. F. Brading. 1996. Long Ca^{2+} channel opening induced by large depolarization and Bay K 8644 in smooth muscle cells isolated from guinea-pig detrusor. *Br J Pharmacol* 119: 716–720.
15. Nakayama, S., and A. F. Brading. 1995. Possible contribution of long open state to noninactivating Ca^{2+} current in detrusor cells. *Am J Physiol* 269: C48–C54.
16. Provence, A., K. L. Hristov, S. P. Parajuli, and G. V. Petkov. 2015. Regulation of guinea pig detrusor smooth muscle excitability by 17beta-estradiol: The role of the large conductance voltage- and Ca^{2+}-activated K^+ channels. *PLoS One* 10: e0141950.
17. Hashitani, H., and A. F. 2003. Brading. Ionic basis for the regulation of spontaneous excitation in detrusor smooth muscle cells of the guinea-pig urinary bladder. *Br J Pharmacol* 140: 159–169.
18. Heppner, T. J., A. D. Bonev, and M. T. 1997. Nelson. Ca^{2+}-activated K^+ channels regulate action potential repolarization in urinary bladder smooth muscle. *Am J Physiol* 273: C110–C117.
19. Klockner, U., and G. Isenberg. 1985. Action potentials and net membrane currents of isolated smooth muscle cells (urinary bladder of the guinea-pig). *Pflugers Arch* 405: 329–339.
20. Sui, G. P., C. Wu, and C. H. Fry. 2001. The electrophysiological properties of cultured and freshly isolated detrusor smooth muscle cells. *J Urol* 165: 627–632.
21. Darblade, B., D. Behr-Roussel, S. Oger, J. P. Hieble, T. Lebret, D. Gorny, G. Benoit, L. Alexandre, and F. Giuliano. 2006. Effects of potassium channel modulators on human detrusor smooth muscle myogenic phasic contractile activity: Potential therapeutic targets for overactive bladder. *Urology* 68: 442–448.

22. Hristov, K. L., M. Chen, R. P. Soder, S. P. Parajuli, Q. Cheng, W. F. Kellett, and G. V. Petkov. 2012. $K_V2.1$ and electrically silent K_V channel subunits control excitability and contractility of guinea pig detrusor smooth muscle. *Am J Physiol Cell Physiol* 302: C360–C372.
23. Petkov, G. V. 2009. *Pharmacology: Principles and Practice*. Chapter 16: Ion channels. Hacker, M, Messer, W, and Bachmann, K (Eds.), pp. 387–427. Amsterdam, the Netherlands: Academic Press.
24. Fry, C. H., and R. I. Jabr. 2014. T-type Ca^{2+} channels and the urinary and male genital tracts. *Pflugers Arch* 466: 781–789.
25. Martin-Cano, F. E., P. J. Gomez-Pinilla, M. J. Pozo, and P. J. Camello. 2009. Spontaneous calcium oscillations in urinary bladder smooth muscle cells. *J Physiol Pharmacol* 60: 93–99.
26. Petkov, G. V. 2014. Central role of the BK channel in urinary bladder smooth muscle physiology and pathophysiology. *Am J Physiol Regul Integr Comp Physiol* 307: R571–R584.
27. Afeli, S. A., J. Malysz, and G. V. Petkov. 2013. Molecular expression and pharmacological evidence for a functional role of Kv7 channel subtypes in Guinea pig urinary bladder smooth muscle. *PLoS One* 8: e75875.
28. Chen, M., W. F. Kellett, and G. V. Petkov. 2010. Voltage-gated K^+ channels sensitive to stromatoxin-1 regulate myogenic and neurogenic contractions of rat urinary bladder smooth muscle. *Am J Physiol Regul Integr Comp Physiol* 299: R177–R184.
29. Hristov, K. L., M. Chen, S. A. Afeli, Q. Cheng, E. S. Rovner, and G. V. Petkov. 2012. Expression and function of K_V2-containing channels in human urinary bladder smooth muscle. *Am J Physiol Cell Physiol* 302: C1599–C1608.
30. Parajuli, S. P., K. L. Hristov, R. P. Soder, W. F. Kellett, and G. V. Petkov. 2013. NS309 decreases rat detrusor smooth muscle membrane potential and phasic contractions by activating SK3 channels. *Br J Pharmacol* 168: 1611–1625.
31. Parajuli, S. P., R. P. Soder, K. L. Hristov, and G. V. Petkov. 2012. Pharmacological activation of small conductance calcium-activated potassium channels with naphtho[1,2-d]thiazol-2-ylamine decreases guinea pig detrusor smooth muscle excitability and contractility. *J Pharmacol Exp Ther* 340: 114–123.
32. Petkov, G. V., T. J. Heppner, A. D. Bonev, G. M. Herrera, and M. T. Nelson. 2001. Low levels of K_{ATP} channel activation decrease excitability and contractility of urinary bladder. *Am J Physiol Regul Integr Comp Physiol* 280: R1427–R1433.
33. Provence, A., J. Malysz, and G. V. Petkov. 2015. The novel $K_V7.2/K_V7.3$ channel opener ICA-069673 reveals subtype-specific functional roles in guinea pig detrusor smooth muscle excitability and contractility. *J Pharmacol Exp Ther* 354: 290–301.
34. Soder, R. P., S. P. Parajuli, K. L. Hristov, E. S. Rovner, and G. V. Petkov. 2013. SK channel-selective opening by SKA-31 induces hyperpolarization and decreases contractility in human urinary bladder smooth muscle. *Am J Physiol Regul Integr Comp Physiol* 304: R155–R163.
35. Hayase, M., H. Hashitani, K. Kohri, and H. Suzuki. 2009. Role of K^+ channels in regulating spontaneous activity in detrusor smooth muscle in situ in the mouse bladder. *J Urol* 181: 2355–2365.
36. Thorneloe, K. S., and M. T. Nelson. 2003. Properties and molecular basis of the mouse urinary bladder voltage-gated K^+ current. *J Physiol* 549: 65–74.
37. Davies, A. M., T. J. Batchelor, I. Eardley, and D. J. Beech. 2002. Potassium channel K_V alpha1 subunit expression and function in human detrusor muscle. *J Urol* 167: 1881–1886.
38. Ohya, S., M. Tanaka, M. Watanabe, and Y. Maizumi. 2000. Diverse expression of delayed rectifier K^+ channel subtype transcripts in several types of smooth muscles of the rat. *J Smooth Muscle Res* 36: 101–115.

39. Gan, X. G., R. H. An, Y. F. Bai, and D. B. Zong. 2008. Expressions of voltage-gated K^+ channel 2.1 and 2.2 in rat bladder with detrusor hyperreflexia. *Chin Med J (Engl)* 121: 1574–1577.
40. Imai, T., T. Okamoto, Y. Yamamoto, H. Tanaka, K. Koike, K. Shigenobu, and Y. Tanaka. 2001. Effects of different types of K^+ channel modulators on the spontaneous myogenic contraction of guinea-pig urinary bladder smooth muscle. *Acta Physiol Scand* 173: 323–333.
41. Provence, A., D. Angoli, and G. V. Petkov. 2017. The novel and selective K_V7 channel activator, ML213, reveals key physiological roles for $K_V7.4$ and $K_V7.5$ channel subtypes in guinea pig detrusor smooth muscle function. *FASEB J* 31: 690–692.
42. Provence, A., D. Angoli, E. Rovner, and G. V. Petkov. 2017. K_V7 channel pharmacological modulation in human detrusor: A promising two-way street for the potential treatment of overactive and underactive bladder. *J Urol* 197: e1353.
43. Provence, A., K. Hristov, and G. V. Petkov. 2016. Selective pharmacological activation of individual KCNQ channel subtypes in detrusor smooth muscle: Promising novel approach for overactive bladder treatment. *J Urol* 195: e412–e413.
44. Provence, A., K. L. Hristov, S. P. Parajuli, E. S. Rovner, and G. V. Petkov. 2015. Voltage-gated KCNQ channels in human detrusor smooth muscle contractility: A novel target for the pharmacological treatment of overactive bladder. *J Urol* 193: e188.
45. Provence, A., K. L. Hristov, and G. V. Petkov. 2016. Pharmacological opening of K_V7 channels by the novel activator ML-213: Role in guinea pig urinary bladder smooth muscle function. *FASEB J* 30: 1013.1011.
46. Anderson, U. A., C. Carson, L. Johnston, S. Joshi, A. M. Gurney, and K. D. McCloskey. 2013. Functional expression of KCNQ (Kv7) channels in guinea pig bladder smooth muscle and their contribution to spontaneous activity. *Br J Pharmacol* 169: 1290–1304.
47. Bientinesi, R., C. Mancuso, M. Martire, P. F. Bassi, E. Sacco, and D. Curro. 2017. K_V7 channels in the human detrusor: Channel modulator effects and gene and protein expression. *Naunyn Schmiedebergs Arch Pharmacol* 390: 127–137.
48. Rode, F., J. Svalo, M. Sheykhzade, and L. C. Ronn. 2010. Functional effects of the KCNQ modulators retigabine and XE991 in the rat urinary bladder. *Eur J Pharmacol* 638: 121–127.
49. Svalo, J., M. Bille, T. N. Parameswaran, M. Sheykhzade, J. Nordling, and P. Bouchelouche. 2013. Bladder contractility is modulated by Kv7 channels in pig detrusor. *Eur J Pharmacol* 715: 312–320.
50. Svalo, J., H. H. Hansen, L. C. Ronn, M. Sheykhzade, G. Munro, and F. Rode. 2012. Kv7 positive modulators reduce detrusor overactivity and increase bladder capacity in rats. *Basic Clin Pharmacol Toxicol* 110: 145–153.
51. Svalo, J., M. Sheykhzade, J. Nordling, C. Matras, and P. Bouchelouche. 2015. Functional and molecular evidence for Kv7 channel subtypes in human detrusor from patients with and without bladder outflow obstruction. *PLoS One* 10: e0117350.
52. Takagi, H., and H. Hashitani. 2016. Effects of K^+ channel openers on spontaneous action potentials in detrusor smooth muscle of the guinea-pig urinary bladder. *Eur J Pharmacol* 789: 179–186.
53. Gamper, N., and M. S. Shapiro. 2015. *Handbook of Ion Channels*. Chapter 20: KCNQ channels. Zheng J and Trudeau, MC (Eds.), pp. 275–306. Boca Raton, FL: Taylor & Francis Group.
54. Haick, J. M., and K. L. Byron. 2016. Novel treatment strategies for smooth muscle disorders: Targeting Kv7 potassium channels. *Pharmacol Ther* 165: 14–25.
55. Streng, T., T. Christoph, and K. E. Andersson. 2004. Urodynamic effects of the K^+ channel (KCNQ) opener retigabine in freely moving, conscious rats. *J Urol* 172: 2054–2058.

56. Brickel, N., P. Gandhi, K. VanLandingham, J. Hammond, and S. DeRossett. 2012. The urinary safety profile and secondary renal effects of retigabine (ezogabine): A first-in-class antiepileptic drug that targets KCNQ (Kv7) potassium channels. *Epilepsia* 53: 606–612.
57. Andersson, K. E., and A. J. Wein. 2004. Pharmacology of the lower urinary tract: Basis for current and future treatments of urinary incontinence. *Pharmacol Rev* 56: 581–631.
58. Wulff, H., N. A. Castle, and L. A. Pardo. 2009. Voltage-gated potassium channels as therapeutic targets. *Nat Rev Drug Discov* 8: 982–1001.
59. Gopalakrishnan, M., and C. C. Shieh. 2004. Potassium channel subtypes as molecular targets for overactive bladder and other urological disorders. *Expert Opin Ther Targets* 8: 437–458.
60. Brown, S. M., L. M. Bentcheva-Petkova, L. Liu, K. L. Hristov, M. Chen, W. F. Kellett, A. L. Meredith, R. W. Aldrich, M. T. Nelson, and G. V. Petkov. 2008. Beta-adrenergic relaxation of mouse urinary bladder smooth muscle in the absence of large-conductance Ca^{2+}-activated K^+ channel. *Am J Physiol Renal Physiol* 295: F1149–F1157.
61. Herrera, G. M., T. J. Heppner, and M. T. Nelson. 2000. Regulation of urinary bladder smooth muscle contractions by ryanodine receptors and BK and SK channels. *Am J Physiol Regul Integr Comp Physiol* 279: R60–R68.
62. Herrera, G. M., T. J. Heppner, and M. T. Nelson. 2001. Voltage dependence of the coupling of Ca^{2+} sparks to BK_{Ca} channels in urinary bladder smooth muscle. *Am J Physiol Cell Physiol* 280: C481–C490.
63. Herrera, G. M., and M. T. Nelson. 2002. Differential regulation of SK and BK channels by Ca^{2+} signals from Ca^{2+} channels and ryanodine receptors in guinea-pig urinary bladder myocytes. *J Physiol* 541: 483–492.
64. Hristov, K. L., M. Chen, W. F. Kellett, E. S. Rovner, and G. V. Petkov. 2011. Large conductance voltage- and Ca^{2+}-activated K^+ channels regulate human detrusor smooth muscle function. *Am J Physiol Cell Physiol* 301: C903–C912.
65. Meredith, A. L., K. S. Thorneloe, M. E. Werner, M. T. Nelson, and R. W. Aldrich. 2004. Overactive bladder and incontinence in the absence of the BK large conductance Ca^{2+}-activated K^+ channel. *J Biol Chem* 279: 36746–36752.
66. Petkov, G. V., A. D. Bonev, T. J. Heppner, R. Brenner, R. W. Aldrich, and M. T. Nelson. 2001. Beta1-subunit of the Ca^{2+}-activated K^+ channel regulates contractile activity of mouse urinary bladder smooth muscle. *J Physiol* 537: 443–452.
67. Petkov, G. V., and M. T. Nelson. 2005. Differential regulation of Ca^{2+}-activated K^+ channels by beta-adrenoceptors in guinea pig urinary bladder smooth muscle. *Am J Physiol Cell Physiol* 288: C1255–C1263.
68. Soder, R. P., and G. V. Petkov. 2011. Large conductance Ca^{2+}-activated K^+ channel activation with NS1619 decreases myogenic and neurogenic contractions of rat detrusor smooth muscle. *Eur J Pharmacol* 670: 252–259.
69. Sprossmann, F., P. Pankert, U. Sausbier, A. Wirth, X. B. Zhou, J. Madlung, H. Zhao et al. 2009. Inducible knockout mutagenesis reveals compensatory mechanisms elicited by constitutive BK channel deficiency in overactive murine bladder. *FEBS J* 276: 1680–1697.
70. Thorneloe, K. S., A. L. Meredith, A. M. Knorn, R. W. Aldrich, and M. T. Nelson. 2005. Urodynamic properties and neurotransmitter dependence of urinary bladder contractility in the BK channel deletion model of overactive bladder. *Am J Physiol Renal Physiol* 289: F604–F610.
71. Contreras, G. F., K. Castillo, N. Enrique, W. Carrasquel-Ursulaez, J. P. Castillo, V. Milesi, A. Neely et al. 2013. A BK (Slo1) channel journey from molecule to physiology. *Channels (Austin)* 7: 442–458.
72. Malysz, J., E. S. Rovner, and G. V. Petkov. 2013. Single-channel biophysical and pharmacological characterizations of native human large-conductance calcum-activated potassium channels in freshly isolated detrusor smooth muscle cells. *Pflugers Arch* 465: 965–975.

73. Chen, M., and G. V. Petkov. 2009. Identification of large conductance calcium activated potassium channel accessory beta4 subunit in rat and mouse bladder smooth muscle. *J Urol* 182: 374–381.
74. Ohi, Y., H. Yamamura, N. Nagano, S. Ohya, K. Muraki, M. Watanabe, and Y. Imaizumi. 2001. Local Ca^{2+} transients and distribution of BK channels and ryanodine receptors in smooth muscle cells of guinea-pig vas deferens and urinary bladder. *J Physiol* 534: 313–326.
75. Afeli, S. A., E. S. Rovner, and G. V. Petkov. 2013. BRL37344, a beta3-adrenergic receptor agonist, decreases nerve-evoked contractions in human detrusor smooth muscle isolated strips: Role of BK channels. *Urology* 82: 744, e741–e747.
76. Hristov, K. L., S. A. Afeli, S. P. Parajuli, Q. Cheng, E. S. Rovner, and G. V. Petkov. 2013. Neurogenic detrusor overactivity is associated with decreased expression and function of the large conductance voltage- and Ca^{2+}-activated K^+ channels. *PLoS One* 8: e68052.
77. Hristov, K. L., S. P. Parajuli, R. P. Soder, Q. Cheng, E. S. Rovner, and G. V. Petkov. 2012. Suppression of human detrusor smooth muscle excitability and contractility via pharmacological activation of large conductance Ca^{2+}-activated K^+ channels. *Am J Physiol Cell Physiol* 302: C1632–C1641.
78. Xin, W., Q. Cheng, R. P. Soder, E. S. Rovner, and G. V. Petkov. 2012. Constitutively active phosphodiesterase activity regulates urinary bladder smooth muscle function: Critical role of $K_{Ca}1.1$ channel. *Am J Physiol Renal Physiol* 303: F1300–F1306.
79. Xin, W., N. Li, V. S. Fernandes, B. Chen, E. S. Rovner, and G. V. Petkov. 2016. BK channel regulation by phosphodiesterase type 1: A novel signaling pathway controlling human detrusor smooth muscle function. *Am J Physiol Renal Physiol* 310: F994–F999.
80. Xin, W., R. P. Soder, Q. Cheng, E. S. Rovner, and G. V. Petkov. 2012. Selective inhibition of phosphodiesterase 1 relaxes urinary bladder smooth muscle: Role for ryanodine receptor-mediated BK channel activation. *Am J Physiol Cell Physiol* 303: C1079–C1089.
81. Chen, L., L. J. Tian, S. H. F. MacDonald, H. McClafferty, M. S. L. Hammond, J. M. Huibant, P. Ruth, H. G. Knaus, and M. J. Shipston. 2005. Functionally diverse complement of large conductance calcium- and voltage-activated potassium channel (BK) alpha-subunits generated from a single site of splicing. *J Biol Chem* 280: 33599–33609.
82. Kita, M., T. Yunoki, K. Takimoto, M. Miyazato, K. Kita, W. C. de Groat, H. Kakizaki, and N. Yoshimura. 2010. Effects of bladder outlet obstruction on properties of Ca^{2+}-activated K^+ channels in rat bladder. *Am J Physiol Regul Integr Comp Physiol* 298: R1310–R1319.
83. Werner, M. E., A. M. Knorn, A. L. Meredith, R. W. Aldrich, and M. T. Nelson. 2007. Frequency encoding of cholinergic- and purinergic-mediated signaling to mouse urinary bladder smooth muscle: Modulation by BK channels. *Am J Physiol Regul Integr Comp Physiol* 292: R616–R624.
84. Galvez, A., G. Gimenez-Gallego, J. P. Reuben, L. Roy-Contancin, P. Feigenbaum, G. J. Kaczorowski, and M. L. Garcia. 1990. Purification and characterization of a unique, potent, peptidyl probe for the high conductance calcium-activated potassium channel from venom of the scorpion Buthus tamulus. *J Biol Chem* 265: 11083–11090.
85. Miller, C., E. Moczydlowski, R. Latorre, and M. Phillips. 1985. Charybdotoxin, a protein inhibitor of single Ca^{2+}-activated K^+ channels from mammalian skeletal muscle. *Nature* 313: 316–318.
86. Suarez-Kurtz, G., M. L. Garcia, and G. J. Kaczorowski. 1991. Effects of charybdotoxin and iberiotoxin on the spontaneous motility and tonus of different guinea pig smooth muscle tissues. *J Pharmacol Exp Ther* 259: 439–443.
87. Knaus, H. G., O. B. McManus, S. H. Lee, W. A. Schmalhofer, M. Garcia-Calvo, L. M. Helms, M. Sanchez, K. Giangiacomo, J. P. Reuben, A. B. Smith et al. 1994. Tremorgenic indole alkaloids potently inhibit smooth muscle high-conductance calcium-activated potassium channels. *Biochemistry* 33: 5819–5828.

88. Saleem, F., I. C. M. Rowe, and M. J. Shipston. 2009. Characterization of BK channel splice variants using membrane potential dyes. *Br J Pharmacol* 156: 143–152.
89. Hristov, K. L., X. Cui, S. M. Brown, L. Liu, W. F. Kellett, and G. V. Petkov. 2008. Stimulation of beta3-adrenoceptors relaxes rat urinary bladder smooth muscle via activation of the large-conductance Ca^{2+}-activated K^+ channels. *Am J Physiol Cell Physiol* 295: C1344–C1353.
90. Xin, W., Q. Cheng, R. P. Soder, and G. V. Petkov. 2012. Inhibition of phosphodiesterases relaxes detrusor smooth muscle via activation of the large-conductance voltage- and Ca^{2+}-activated K^+ channel. *Am J Physiol Cell Physiol* 302: C1361–C1370.
91. Xin, W., N. Li, Q. Cheng, and G. V. Petkov. 2014. BK channel-mediated relaxation of urinary bladder smooth muscle: A novel paradigm for phosphodiesterase type 4 regulation of bladder function. *J Pharmacol Exp Ther* 349: 56–65.
92. DeFarias, F. P., M. F. Carvalho, S. H. Lee, G. J. Kaczorowski, and G. SuarezKurtz. 1996. Effects of the K^+ channel blockers paspalitrem-C and paxilline on mammalian smooth muscle. *Eur J Pharmacol* 314: 123–128.
93. Oger, S., D. Behr-Roussel, D. Gorny, J. Bernabe, E. Comperat, E. Chartier-Kastler, P. Denys, and F. Giuliano. 2011. Effects of potassium channel modulators on myogenic spontaneous phasic contractile activity in human detrusor from neurogenic patients. *BJU Int* 108: 604–611.
94. Afeli, S. A., and G. V. Petkov. 2013. Functional BK channels facilitate the beta3-adrenoceptor agonist-mediated relaxation of nerve-evoked contractions in rat urinary bladder smooth muscle isolated strips. *Eur J Pharmacol* 711: 50–56.
95. Herrera, G. M., B. Etherton, B. Nausch, and M. T. Nelson. 2005. Negative feedback regulation of nerve-mediated contractions by K_{Ca} channels in mouse urinary bladder smooth muscle. *Am J Physiol Regul Integr Comp Physiol* 289: R402–R409.
96. Kellett, W. F., and G. V. Petkov. 2010. Role of Ca^{2+}-activated K^+ channel in the neurogenic contractions induced by electrical field stimulation in detrusor smooth muscle isolated from rats and guinea pigs. *Biophysl J* 98: 125a–126a.
97. La Fuente, J. M., A. Fernandez, P. Cuevas, R. Gonzalez-Corrochano, M. X. Chen, and J. Angulo. 2014. Stimulation of large-conductance calcium-activated potassium channels inhibits neurogenic contraction of human bladder from patients with urinary symptoms and reverses acetic acid-induced bladder hyperactivity in rats. *Eur J Pharmacol* 735: 68–76.
98. Chang, S., C. M. Gomes, J. A. Hypolite, J. Marx, J. Alanzi, S. A. Zderic, B. Malkowicz, A. J. Wein, and S. Chacko. 2010. Detrusor overactivity is associated with downregulation of large-conductance calcium- and voltage-activated potassium channel protein. *Am J Physiol Renal Physiol* 298: F1416–F1423.
99. Hu, S., and H. S. Kim. 1996. On the mechanism of the differential effects of NS004 and NS1608 in smooth muscle cells from guinea pig bladder. *Eur J Pharmacol* 318: C461–C468.
100. Layne, J. J., B. Nausch, S. P. Olesen, and M. T. Nelson. 2010. BK channel activation by NS11021 decreases excitability and contractility of urinary bladder smooth muscle. *Am J Physiol Regul Integr Comp Physiol* 298: R378–R384.
101. Malysz, J., S. A. Buckner, A. V. Daza, I. Milicic, A. Perez-Medrano, and M. Gopalakrishnan. 2004. Functional characterization of large conductance calcium-activated K^+ channel openers in bladder and vascular smooth muscle. *Naunyn Schmiedebergs Arch Pharmacol* 369: 481–489.
102. Mora, T. C., and G. Suarez-Kurtz. 2005. Effects of NS1608, a BK_{Ca} channel agonist, on the contractility of guinea-pig urinary bladder in vitro. *Br J Pharmacol* 144: 636–641.
103. Roy, S., R. J. Large, A. M. Akande, A. Kshatri, T. I. Webb, C. Domene, G. P. Sergeant, N. G. McHale, K. D. Thornbury, and M. A. Hollywood. 2004. Development of GoSlo-SR-5-69, a potent activator of large conductance Ca^{2+}-activated K^+ (BK) channels. *Eur J Med Chem* 75: 426–437.

104. Roy, S., A. Morayo Akande, R. J. Large, T. I. Webb, C. Camarasu, G. P. Sergeant, N. G. McHale, K. D. Thornbury, and M. A. Hollywood. 2012. Structure-activity relationships of a novel group of large-conductance Ca^{2+}-activated K^+ (BK) channel modulators: The GoSlo-SR family. *ChemMedChem* 7: 1763–1769.
105. Sheldon, J. H., N. W. Norton, and T. M. Argentieri. 1997. Inhibition of guinea pig detrusor contraction by NS-1619 is associated with activation of BK_{Ca} and inhibition of calcium currents. *J Pharmacol Exp Ther* 283: 1193–1200.
106. Siemer, C., M. Bushfield, D. Newgreen, and S. Grissmer. 2000. Effects of NS1608 on MaxiK channels in smooth muscle cells from urinary bladder. *J Membr Biol* 173: 57–66.
107. Trivedi, S., L. Potterlee, J. H. Li, G. D. Yasay, K. Russell, J. R. Empfield, D. A. Trainor, and S. T. Kau. 1995. Calcium-dependent K-channels in guinea-pig and human urinary-bladder. *Biochem Biophys Res Commun* 213: 404–409.
108. Xin, W., N. Li, Q. Cheng, V. S. Fernandes, and G. V. Petkov. 2014. Constitutive PKA activity is essential for maintaining the excitability and contractility in guinea pig urinary bladder smooth muscle: Role of the BK channel. *Am J Physiol Cell Physiol* 307: C1142–C1150.
109. Xin, W., N. Li, V. S. Fernandes, and G. V. Petkov. 2016. Constitutively active PKA regulates neuronal acetylcholine release and contractility of guinea pig urinary bladder smooth muscle. *Am J Physiol Renal Physiol* 310: F1377–F1384.
110. Andersson, K. E., and A. Arner. 2004. Urinary bladder contraction and relaxation: Physiology and pathophysiology. *Physiol Rev* 84: 935–986.
111. Nausch, B., T. J. Heppner, and M. T. Nelson. 2010. Nerve-released acetylcholine contracts urinary bladder smooth muscle by inducing action potentials independently of IP3-mediated calcium release. *Am J Physiol Regul Integr Comp Physiol* 299: R878–R888.
112. Parajuli, S. P., and G. V. Petkov. 2013. Activation of muscarinic M3 receptors inhibits large-conductance voltage- and Ca^{2+}-activated K^+ channels in rat urinary bladder smooth muscle cells. *Am J Physiol Cell Physiol* 305: C207–C214.
113. Parajuli, S. P., K. L. Hristov, Q. Cheng, J. Malysz, E. S. Rovner, and G. V. Petkov. 2015. Functional link between muscarinic receptors and large-conductance Ca^{2+}-activated K^+ channels in freshly isolated human detrusor smooth muscle cells. *Pflugers Arch* 467: 665–675.
114. Hristov, K. L., A. C. Smith, S. P. Parajuli, J. Malysz, and G. V. Petkov. 2014. Large-conductance voltage- and Ca^{2+}-activated K^+ channel regulation by protein kinase C in guinea pig urinary bladder smooth muscle. *Am J Physiol Cell Physiol* 306: C460–C470.
115. Hypolite, J. A., Q. Lei, S. Chang, S. A. Zderic, S. Butler, A. J. Wein, A. P. Malykhina, and S. Chacko. 2013. Spontaneous and evoked contractions are regulated by PKC-mediated signaling in detrusor smooth muscle: Involvement of BK channels. *Am J Physiol Renal Physiol* 304: F451–F462.
116. Nakahira, Y., H. Hashitani, H. Fukuta, S. Sasaki, K. Kohri, and H. Suzuki. 2001. Effects of isoproterenol on spontaneous excitations in detrusor smooth muscle cells of the guinea pig. *J Urol* 166: 335–340.
117. Kobayashi, H., S. Adachi-Akahane, and T. Nagao. 2000. Involvement of BK_{Ca} channels in the relaxation of detrusor muscle via beta-adrenoceptors. *Eur J Pharmacol* 404: 231–238.
118. Yamaguchi, O., and C. R. Chapple. 2007. Beta3-adrenoceptors in urinary bladder. *Neurourol Urodyn* 26: 752–756.
119. Takemoto, J., H. Masumiya, K. Nunoki, T. Sato, H. Nakagawa, Y. Ikeda, Y. Arai, and T. Yanagisawa. 2008. Potentiation of potassium currents by beta-adrenoceptor agonists in human urinary bladder smooth muscle cells: A possible electrical mechanism of relaxation. *Pharmacology* 81: 251–258.

120. Afeli, S. A., K. L. Hristov, and G. V. Petkov. 2012. Do beta3-adrenergic receptors play a role in guinea pig detrusor smooth muscle excitability and contractility? *Am J Physiol Renal Physiol* 302: F251–F263.
121. Andersson, K. E. 2013. New developments in the management of overactive bladder: Focus on mirabegron and onabotulinumtoxinA. *Ther Clin Risk Manag* 9: 161–170.
122. Chapple, C. R., L. Cardozo, V. W. Nitti, E. Siddiqui, and M. C. Michel. 2014. Mirabegron in overactive bladder: A review of efficacy, safety, and tolerability. *Neurourol Urodyn* 33: 17–30.
123. Cui, Y. S., H. T. Zong, C. C. Yang, H. L. Yan, and Y. Zhang. 2014. The efficacy and safety of mirabegron in treating OAB: A systematic review and meta-analysis of phase III trials. *Int Urol Nephrol* 46: 275–284.
124. Yasay, G. D., S. T. Kau, and J. H. Li. 1995. Mechanoinhibitory effect of estradiol in guinea pig urinary bladder smooth muscles. *Pharmacology* 51: 273–280.
125. Hanna-Mitchell, A. T., D. Robinson, L. Cardozo, K. Everaert, and G. V. Petkov. 2016. Do we need to know more about the effects of hormones on lower urinary tract dysfunction? ICI-RS 2014. *Neurourol Urodyn* 35: 299–303.
126. Hristov, K., S. Parajuli, A. Provence, E. Rovner, and G. V. Petkov. 2016. 17β-Estradiol direct activation of large conductance voltage- and $Ca2^+$-activated K^+ channels: Novel regulatory mechanism in human detrusor smooth muscle. *J Urol* 195: e378.
127. Hristov, K. L., S. P. Parajuli, A. Provence, and G. V. Petkov. 2016. Testosterone decreases urinary bladder smooth muscle excitability via novel signaling mechanism involving direct activation of the BK channels. *Am J Physiol Renal Physiol* 311: F1253–F1259.
128. Malysz, J., S. A. Afeli, A. Provence, and G. V. Petkov. 2014. Ethanol-mediated relaxation of guinea pig urinary bladder smooth muscle: Involvement of BK and L-type Ca^{2+} channels. *Am J Physiol Cell Physiol* 306: C45–C58.
129. Fernandes, V. S., W. Xin, and G. V. Petkov. 2015. Novel mechanism of hydrogen sulfide-induced guinea pig urinary bladder smooth muscle contraction: Role of BK channels and cholinergic neurotransmission. *Am J Physiol Cell Physiol* 309: C107–C116.
130. Parajuli, S. P., A. Provence, and G. V. Petkov. 2014. Prostaglandin E2 excitatory effects on guinea pig urinary bladder smooth muscle: A novel regulatory mechanism mediated by large-conductance voltage- and Ca^{2+}-activated K^+ channels. *Eur J Pharmacol* 738: 179–185.
131. Christ, G. J., N. S. Day, M. Day, C. Santizo, W. Zhao, T. Sclafani, J. Zinman et al. 2001. Bladder injection of "naked" *hSlo*/pcDNA3 ameliorates detrusor hyperactivity in obstructed rats *in vivo*. *Am J Physiol Regul Integr Comp Physiol* 281: R1699–R1709.
132. Butera, J. A., D. J. Jenkins, J. R. Lennox, J. H. Sheldon, N. W. Norton, D. Warga, and T. M. Argentieri. 2005. Synthesis and bladder smooth muscle relaxing properties of substituted 3-amino-4-aryl-(and aralkyl-)cyclobut-3-ene-1,2-diones. *Bioorg Med Chem Lett* 15: 2495–2501.
133. Turner, S. C., W. A. Carroll, T. K. White, M. Gopalakrishnan, M. J. Coghlan, C. C. Shieh, X. F. Zhang et al. 2003. The discovery of a new class of large-conductance Ca^{2+}-activated K^+ channel opener targeted for overactive bladder: Synthesis and structure-activity relationships of 2-amino-4-azaindoles. *Bioorg Med Chem Lett* 13: 2003–2007.
134. Tanaka, M., Y. Sasaki, Y. Kimura, T. Fukui, K. Hamada, and Y. Ukai. 2003. A novel pyrrole derivative, NS-8, suppresses the rat micturition reflex by inhibiting afferent pelvic nerve activity. *BJU International* 92: 1031–1036.
135. Kshatri, A. S., Q. Li, J. Yan, R. J. Large, G. P. Sergeant, N. G. McHale, K. D. Thornbury, and M. A. 2017. Hollywood. Differential efficacy of GoSlo-SR compounds on BKalpha and BKalphagamma1-4 channels. *Channels (Austin)* 11: 66–78.
136. Large, R. J., A. Kshatri, T. I. Webb, S. Roy, A. Akande, E. Bradley, G. P. Sergeant, K. D. Thornbury, N. G. McHale, and M. A. Hollywood. 2015. Effects of the novel BK (K_{Ca}1.1) channel opener GoSlo-SR-5-130 are dependent on the presence of BKbeta subunits. *Br J Pharmacol* 172: 2544–2556.

137. Nardi, A., and S. P. Olesen. 2008. BK channel modulators: A comprehensive overview. *Curr Med Chem* 15: 1126–1146.
138. Christ, G. J. 2011. Potential applications of gene therapy/transfer to the treatment of lower urinary tract diseases/disorders. *Handb Exp Pharmacol* 255–265.
139. Ohya, S., S. Kimura, M. Kitsukawa, K. Muraki, M. Watanabe, and Y. Imaizumi. 2000. SK4 encodes intermediate conductance Ca^{2+}-activated K^+ channels in mouse urinary bladder smooth muscle cells. *Jpn J Pharmacol* 84: 97–100.
140. Afeli, S. A., E. S. Rovner, and G. V. Petkov. 2012. SK but not IK channels regulate human detrusor smooth muscle spontaneous and nerve-evoked contractions. *Am J Physiol Renal Physiol* 303: F559–F568.
141. Maylie, J., C. T. Bond, P. S. Herson, W. S. Lee, and J. P. 2004. Adelman. Small conductance Ca^{2+}-activated K^+ channels and calmodulin. *J Physiol* 554: 255–261.
142. Creed, K. E., S. Ishikawa, and Y. Ito. 1983. Electrical and mechanical activity recorded from rabbit urinary bladder in response to nerve stimulation. *J Physiol* 338: 149–164.
143. Fujii, K., C. D. Foster, A. F. Brading, and A. B. Parekh. 1990. Potassium channel blockers and the effects of cromakalim on the smooth muscle of the guinea-pig bladder. *Br J Pharmacol* 99: 779–785.
144. Herrera, G. M., M. J. Pozo, P. Zvara, G. V. Petkov, C. T. Bond, J. P. Adelman, and M. T. Nelson. 2003. Urinary bladder instability induced by selective suppression of the murine small conductance calcium-activated potassium (SK3) channel. *J Physiol* 551: 893–903.
145. Thorneloe, K. S., A. M. Knorn, P. E. Doetsch, E. S. Lashinger, A. X. Liu, C. T. Bond, J. P. Adelman, and M. T. Nelson. 2008. Small-conductance, Ca^{2+}-activated K^+ channel 2 is the key functional component of SK channels in mouse urinary bladder. *Am J Physiol Regul Integr Comp Physiol* 294: R1737–R1743.
146. Hougaard, C., M. O. Fraser, C. Chien, A. Bookout, M. Katofiasc, B. S. Jensen, F. Rode et al. 2009. A positive modulator of $K_{Ca}2$ and $K_{Ca}3$ channels, 4,5-dichloro-1,3-diethyl-1,3-dihydro-benzoimidazol-2-one (NS4591), inhibits bladder afferent firing in vitro and bladder overactivity in vivo. *J Pharmacol Exp Ther* 328: 28–39.
147. Nielsen, J. S., F. Rode, M. Rahbek, K. E. Andersson, L. C. Ronn, K. Bouchelouche, J. Nordling, and P. Bouchelouche. 2011. Effect of the SK/IK channel modulator 4,5-dichloro-1,3-diethyl-1,3-dihydro-benzoimidazol-2-one (NS4591) on contractile force in rat, pig and human detrusor smooth muscle. *BJU Int* 108: 771–777.
148. Pandita, R. K., L. C. Ronn, B. S. Jensen, and K. E. Andersson. 2006. Urodynamic effects of intravesical administration of the new small/intermediate conductance calcium activated potassium channel activator NS309 in freely moving, conscious rats. *J Urol* 176: 1220–1224.
149. Chen, M. X., S. A. Gorman, B. Benson, K. Singh, J. P. Hieble, M. C. Michel, S. N. Tate, and D. J. Trezise. 2004. Small and intermediate conductance Ca^{2+}-activated K^+ channels confer distinctive patterns of distribution in human tissues and differential cellular localisation in the colon and corpus cavernosum. *Naunyn Schmiedebergs Arch Pharmacol* 369: 602–615.
150. Lee, H., B. Koh, R. Corrigan, L. Peri, B. Perrino, T. Chai, K. Sanders, and S. D. Don Koh. 2016. Interaction of TRPV4 and SK3 channels in detrusor PDGFRα^+ cells controls bladder filling. *J Urol* 195: e352.
151. Lee, H., B. H. Koh, L. E. Peri, R. D. Corrigan, H. T. Lee, N. E. George, B. P. Bhetwal et al. 2017. Premature contractions of the bladder are suppressed by interactions between TRPV4 and SK3 channels in murine detrusor PDGFRalpha+ cells. *Sci Rep* 7: 12245.
152. Kajioka, S., S. Nakayama, H. Asano, N. Seki, S. Naito, and A. F. Brading. 2008. Levcromakalim and MgGDP activate small conductance ATP-sensitive K+ channels of K+ channel pore 6.1/sulfonylurea receptor 2A in pig detrusor smooth muscle cells: Uncoupling of cAMP signal pathways. *J Pharmacol Exp Ther* 327: 114–123.

153. Kajioka, S., N. Shahab, H. Asano, H. Morita, M. Sugihara, F. Takahashi-Yanaga, T. Yoshihara, S. Nakayama, N. Seki, and S. Naito. 2011. Diphosphate regulation of adenosine triphosphate sensitive potassium channel in human bladder smooth muscle cells. *J Urol* 186: 736–744.
154. Andersson, K. E., P. O. Andersson, M. Fovaeus, H. Hedlund, A. Malmgren, and C. Sjogren. 1988. Effects of pinacidil on bladder muscle. *Drugs* 36 Suppl 7: 41–49.
155. Bonev, A. D., and M. T. Nelson. 1993. ATP-sensitive potassium channels in smooth muscle cells from guinea pig urinary bladder. *Am J Physiol* 264: C1190–C1200.
156. Davis-Taber, R., E. J. Molinari, R. J. Altenbach, K. L. Whiteaker, C. C. Shieh, G. Rotert, S. A. 2003. Buckner et al. [125I]A-312110, a novel high-affinity 1,4-dihydropyridine ATP-sensitive K^+ channel opener: Characterization and pharmacology of binding. *Mol Pharmacol* 64: 143–153.
157. Shieh, C. C., J. Feng, S. A. Buckner, J. D. Brioni, M. J. Coghlan, J. P. Sullivan, and M. Gopalakrishnan. 2001. Functional implication of spare ATP-sensitive K^+ channels in bladder smooth muscle cells. *J Pharmacol Exp Ther* 296: 669–675.
158. Foster, C. D., K. Fujii, J. Kingdon, and A. F. Brading. 1989. The effect of cromakalim on the smooth muscle of the guinea-pig urinary bladder. *Br J Pharmacol* 97: 281–291.
159. Heppner, T. J., A. Bonev, J. H. Li, S. T. Kau, and M. T. Nelson. 1996. Zeneca ZD6169 activates ATP-sensitive K^+ channels in the urinary bladder of the guinea pig. *Pharmacology* 53: 170–179.
160. Baker, S. A., W. J. Hatton, J. Han, G. W. Hennig, F. C. Britton, and S. D. Koh. 2010. Role of TREK-1 potassium channel in bladder overactivity after partial bladder outlet obstruction in mouse. *J Urol* 183: 793–800.
161. Baker, S. A., G. W. Hennig, J. Han, F. C. Britton, T. K. Smith, and S. D. Koh. 2008. Methionine and its derivatives increase bladder excitability by inhibiting stretch-dependent K(+) channels. *Br J Pharmacol* 153: 1259–1271.
162. Beckett, E. A., I. Han, S. A. Baker, J. Han, F. C. Britton, and S. D. Koh. 2008. Functional and molecular identification of pH-sensitive K^+ channels in murine urinary bladder smooth muscle. *BJU Int* 102: 113–124.
163. Tertyshnikova, S., R. J. Knox, M. J. Plym, G. Thalody, C. Griffin, T. Neelands, D. G. Harden et al. 2005. BL-1249 [(5,6,7,8-tetrahydro-naphthalen-1-yl)-[2-(1H-tetrazol-5-yl)-phenyl]-amine]: A putative potassium channel opener with bladder-relaxant properties. *J Pharmacol Exp Ther* 313: 250–259.
164. Nilius, B., and A. Szallasi. 2014. Transient receptor potential channels as drug targets: From the science of basic research to the art of medicine. *Pharmacol Rev* 66: 676–814.
165. Andersson, K. E. 2016. Potential future pharmacological treatment of bladder dysfunction. *Basic Clin Pharmacol Toxicol* 119 Suppl 3: 75–85.
166. Franken, J., P. Uvin, D. De Ridder, and T. Voets. 2014. TRP channels in lower urinary tract dysfunction. *Br J Pharmacol* 171: 2537–2551.
167. Skryma, R., N. Prevarskaya, D. Gkika, and Y. Shuba. 2011. From urgency to frequency: Facts and controversies of TRPs in the lower urinary tract. *Nat Rev Urol* 8: 617–630.
168. Earley, S. 2013. TRPM4 channels in smooth muscle function. *Pflugers Arch* 465: 1223–1231.
169. Guinamard, R., M. Demion, and P. Launay. 2010. Physiological roles of the TRPM4 channel extracted from background currents. *Physiology* (*Bethesda*) 25: 155–164.
170. Nilius, B., J. Prenen, A. Janssens, G. Owsianik, C. Wang, M. X. Zhu, and T. Voets. 2005. The selectivity filter of the cation channel TRPM4. *J Biol Chem* 280: 22899–22906.
171. Nilius, B., J. Prenen, J. Tang, C. Wang, G. Owsianik, A. Janssens, T. Voets, and M. X. Zhu. 2005. Regulation of the Ca^{2+} sensitivity of the nonselective cation channel TRPM4. *J Biol Chem* 280: 6423–6433.

172. Yoo, J. C., O. V. Yarishkin, E. M. Hwang, E. Kim, D. G. Kim, N. Park, S. G. Hong, and J. Y. Park. 2010. Cloning and characterization of rat transient receptor potential-melastatin 4 (TRPM4). *Biochem Biophys Res Commun* 391: 806–811.
173. Parajuli, S. P., K. L. Hristov, M. N. Sullivan, W. Xin, A. C. Smith, S. Earley, J. Malysz, and G. V. Petkov. 2013. Control of urinary bladder smooth muscle excitability by the TRPM4 channel modulator 9-phenanthrol. *Channels (Austin)* 7: 537–540.
174. Smith, A. C., K. L. Hristov, Q. Cheng, W. Xin, S. P. Parajuli, S. Earley, J. Malysz, and G. V. Petkov. 2013. Novel role for the transient potential receptor melastatin 4 channel in guinea pig detrusor smooth muscle physiology. *Am J Physiol Cell Physiol* 304: C467–C477.
175. Smith, A. C., S. P. Parajuli, K. L. Hristov, Q. Cheng, R. P. Soder, S. A. Afeli, S. Earley, W. Xin, J. Malysz, and G. V. Petkov. 2013. TRPM4 channel: A new player in urinary bladder smooth muscle function in rats. *Am J Physiol Renal Physiol* 304: F918–F929.
176. Hamilton, K. L. 2016. New life in overactive bladder. Focus on novel regulatory mechanism in human urinary bladder: Central role of transient receptor potential melastatin 4 channels in detrusor smooth muscle function. *Am J Physiol Cell Physiol* 310: C597–C599.
177. Hristov, K. L., A. C. Smith, S. P. Parajuli, J. Malysz, E. S. Rovner, and G. V. Petkov. 2016. Novel regulatory mechanism in human urinary bladder: Central role of transient receptor potential melastatin 4 channels in detrusor smooth muscle function. *Am J Physiol Cell Physiol* 310: C600–C611.
178. Guinamard, R., T. Hof, and C. A. Del Negro. 2014. The TRPM4 channel inhibitor 9-phenanthrol. *Br J Pharmacol* 171: 1600–1613.
179. Provence, A., K. L. Hristov, E. Rovner, and G. V. Petkov. 2016. Spatial and functional interactions of the transient receptor potential melastatin-4 channels and inositol trisphosphate receptors: Novel regulatory mechanism in human detrusor smooth muscle function. *J Urol* 195: e378.
180. Provence, A., E. S. Rovner, and G. V. Petkov. 2017. Regulation of transient receptor potential melastatin 4 channel by sarcoplasmic reticulum inositol trisphosphate receptors: Role in human detrusor smooth muscle function. *Channels (Austin)* 11: 459–466.
181. Thorneloe, K. S., A. C. Sulpizio, Z. Lin, D. J. Figueroa, A. K. Clouse, G. P. McCafferty, T. P. Chendrimada et al. 2008. N-((1S)-1-{[4-((2S)-2-{[(2,4-dichlorophenyl)sulfonyl]amino}-3-hydroxypropanoyl)-1-piperazinyl]carbonyl}-3-methylbutyl)-1-benzothiophene-2-carboxamide (GSK1016790A), a novel and potent transient receptor potential vanilloid 4 channel agonist induces urinary bladder contraction and hyperactivity: Part I. *J Pharmacol Exp Ther* 326: 432–442.
182. Isogai, A., K. Lee, R. Mitsui, and H. Hashitani. 2016. Functional coupling of TRPV4 channels and BK channels in regulating spontaneous contractions of the guinea pig urinary bladder. *Pflugers Arch* 468: 1573–1585.
183. Bijos, D. A., M. J. Drake, and B. Vahabi. 2014. Anoctamin-1 in the juvenile rat urinary bladder. *PLoS One* 9: e106190.
184. Yu, W., M. L. Zeidel, and W. G. Hill. 2012. Cellular expression profile for interstitial cells of cajal in bladder—A cell often misidentified as myocyte or myofibroblast. *PLoS One* 7: e48897.
185. Li, L., C. Jiang, B. Song, J. Yan, and J. Pan. 2008. Altered expression of calcium-activated K and Cl channels in detrusor overactivity of rats with partial bladder outlet obstruction. *BJU Int* 101: 1588–1594.

15 Methods for Investigating the Regulation of Smooth Muscle Excitability by Interstitial Cells

Bernard T. Drumm and Kenton M. Sanders

CONTENTS

15.1 INTRODUCTION

In many visceral smooth muscle organs such as the gastrointestinal (GI) tract, upper and lower urinary tract (LUT) systems, lymphatic vessels, and so on, the contractility of smooth muscle cells (SMCs) can be modulated by the activity of non-contractile interstitial cells (ICs). Such integrated multi-cellular behaviors have been best characterized in the GI tract, but an abundance of studies now exist from other organs detailing the contribution of ICs to smooth muscle function. Contractions of the smooth muscle walls of the GI tract move a bolus along the system towards the anus for egestion. SMCs contract or relax within the organs of the GI tract (esophagus, stomach, small intestine, colon, and sphincters) in a coordinated manner to generate complex motor behaviors such as peristalsis, segmentation, and tone. While the contractile apparatus that underlies this motility resides within GI SMCs, the task of pacing and coordinating contractile behavior relies on a concerted effort between SMCs and specialized populations of ICs; interstitial cells of Cajal (ICC) and platelet-derived growth factor receptor alpha$^+$ (PDGFRα^+) cells. In the GI tract, SMCs and ICs comprise the **SIP** syncytium (**S**mooth muscle cell, **I**nterstitial cell of

Cajal, **P**DGFRα+ cell). The integrated activity of SIP cells tunes the excitability of GI smooth muscles and predetermines to some extent responses to enteric motor neurons and hormones.[1]

The stomach, small intestine, and colon exhibit phasic contractions to move food along the length of the GI tract for digestion. These phasic contractions result from rhythmic depolarization events termed slow waves.[2] Slow waves are generated and propagated within networks of ICC, which are electrically coupled via gap junctions to the surrounding smooth muscle layers.[1,3–8] The depolarization of the slow wave is then passively conducted to the SMCs, causing activation of voltage-dependent $Ca_v1.2$ channels and excitation-coupling.[4,9–11] Thus, while not contractile themselves, ICC act as pacemakers to generate and coordinate the underlying electrical activity required for GI smooth muscle to exhibit normal motility patterns. The GI tract is highly innervated by enteric neurons and there is evidence from many studies that ICs (ICC and PDGFRα+ ICs) act as intermediaries, receiving and transducing inputs from enteric motor neurons, and conducting responses to the electrically coupled SMCs.[1,12,13] The physiological roles of ICC and PDGFRα+ ICs have been reviewed elsewhere,[1,12–15] but in this chapter we will focus on methods to identify these cells in smooth muscle organs and strategies to probe their functions.

15.2 IDENTIFYING ICC AND PDGFRα+ POPULATIONS

The main cell population of the muscle layers of the GI tract consists of SMCs with ICs representing a small percentage of the total population (about 10%). Thus, an essential first step in investigating ICs is establishing a protocol to distinguish them from other cells and to distinguish ICC and PDGFRα+ ICs from each other. Electron microscopy (EM) has been used to characterize the morphology of ICC from the GI tract, characterizing their features such as an abundance of mitochondria, endoplasmic reticulum (ER) structures, caveolae, and an incomplete basal lamina.[16–18] ICs lack thick filaments,[17] but both ICC and PDGFRα+ ICs express an abundance of the intermediate filament vimentin and therefore can be immunolabeled with vimentin antibodies. With this in mind, investigators should be wary of only using vimentin to label for one IC type over the other as it cannot differentiate between different IC populations.

While many markers for ICC have been used previously such as rhodamine 123[19] and methylene blue,[20] these stains do not show specificity for ICC and do not work in all GI regions or species. A common method is to identify ICC based on their expression of the tyrosine kinase receptor, Kit. Kit is expressed by ICC throughout the GI tract[5–7,21] and is the most common immuno-marker that has been used for ICC. SMCs, enteric neurons and PDGFRα+ ICs show minimal expression of Kit; however, mast cells also express this receptor. In certain animal models such as rodents, mast cells are sparse in the tunica muscularis, so these cells are a minor contaminant. However, in other species mast cells may be present at a higher density, complicating the identification of ICC based only on Kit expression. In such cases, controls, such as labeling with mast cell tryptase, must be used to clearly distinguish ICC from the mast cell population. Another approach is to label ICC with antibodies to Anoctamin 1 (ANO1), a Ca^{2+}-activated-Cl^- channel, which is highly expressed in ICC throughout the GI tract.[22–25] Identifying ICC with antibodies for ANO1 may therefore be more beneficial than Kit staining and would be especially advantageous when working with human or primate muscles where Kit^+ mast cells are present throughout the musculature of the GI tract.[24]

As their name suggests, $PDGFR\alpha^+$ ICs can be identified by immuno-labeling, thanks to expression of PDGFRα and lack of Kit expression, distinguishing them from ICC.[26,27] This marker has been used in standard immunohistochemistry techniques to identify these cells throughout the rodent, primate, and human GI tract.[28–32]

Mutant mice, using the endogenous promoter for PDGFRα to drive expression of a histone 2B-eGFP fusion protein, have been an important tool for the study of $PDGFR\alpha^+$ ICs.[28] A similar approach has also been used for ICC. Mutant mice, expressing copGFP driven by the endogenous promoter for *Kit*, have provided access to these cells and easy identification by fluorescent microscopy.[21,33] Therefore, fluorescent reporter identification of these cells has allowed unequivocal identification of ICs in intact muscles or in the mixed populations of cells obtained after enzymatic dispersion of tissues.[29,31–38] This is especially advantageous, because isolated ICC and $PDGFR\alpha^+$ ICs cannot be easily distinguished from other SIP syncytium cells based on morphology alone. Therefore, identification of these cells allows investigation of their cellular physiology with techniques, such as patch clamping, to be performed with confidence. Expression of the fluorescent reporters also allows simplified use of fluorescent-activated cell sorting (FACS) to purify selected populations of cells from complex cell populations. Using these reporter strains, investigators have enriched populations of ICC and $PDGFR\alpha^+$ cells for analysis of gene expression and studied isolated cells to evaluate the function of proteins that are highly expressed. Examples include correlations between expression of ANO1,[33,34] T-type Ca^{2+} channels[35] and Ca^{2+} release channels in ICC[36–38] with membrane conductances and Ca^{2+} transients in individual cells; and expression of small-conductance potassium SK3 channels and purinergic receptors (P2Y1) in $PDGFR\alpha^+$ cells[29,31,32,39] that transduce purinergic neural inputs in the SIP syncytium.

It should be noted that there are several classes of ICC in the GI tract situated in different anatomical locations that carry out specific functions. In the small intestine and gastric antrum, myenteric ICC (ICC-MY) reside at the myenteric plexus between the longitudinal and circular smooth muscle layers.[40,41] ICC-MY generate and propagate electrical slow waves in these organs and thus are the cells of primary interest when studying GI pacemaking and rhythmicity. In the colon, one class of pacemaker ICC is located at the submucosal border (ICC-SM) of the tissue, but a second class of ICC (ICC-MY) also displays rhythmic activity that sums with the slow waves of the ICC-SM, creating complexes of electrical activity.[16,42–44] ICC-MY may also act as neuromodulators in the colon.[45] In the gastric antrum,[12,50] fundus,[49,52,53,56,57] small intestine,[51] colon and internal anal sphincter[55] classes of ICC in-between muscle bundles (ICC-IM, or ICC-DMP, Deep Muscular Plexus in the small intestine[46]), lie in close association with enteric nerve terminals and have been implicated in mediating neurotransmission to GI SMCs.

15.3 UTILIZING MUTANT ANIMALS TO STUDY ICC FUNCTION

The role of ICC as pacemakers and in mediating enteric neurotransmission in intact muscles has been investigated using a variety of mutant animal models and pharmacological approaches. Animal models lacking certain ICC populations (verified by immunolabeling and EM) such as the *W/W^v^* mouse[5–7,58] or *Sl/Sl^d^* mouse[49,52,59,60] were used to demonstrate that ICC are essential for pacemaker activity in GI muscles and in neuromodulation. In W/W^v mice, ICC-MY (but not ICC-IM) are absent in the small intestine and thus can be a useful tool for evaluating the contribution of ICC-MY to pacemaking or when one wishes to study ICC-DMP (not absent in W/W^v mutants) in isolation from other ICC populations.[5–7] The situation is reversed in the stomach: ICC-MY are present in W/W^v mutants but ICC-IM are absent. In these instances, neutralizing antibodies against Kit have been used to ablate the ICC-MY networks in the gastric antrum.[61] These animals can also be utilized in combination with electrical recordings of neural responses in intact muscles to assess the role of ICC-IM in the stomach in regards to mediating enteric neural responses to smooth muscle (reports have shown that excitatory and inhibitory neural responses evoked electrical field stimulation are reduced or altered in W/W^v mice lacking ICC-IM).[49,51,56,60,62–64] While the W/W^v mutant is well characterized in the stomach and small intestine, the role of ICC in the colon is more difficult to understand because ICC are reduced but not lost in W/W^v mice. Kit neutralizing antibodies or other techniques to lesion ICC populations might be a better means of understanding the role of ICC in the colon.

15.4 INVESTIGATING SLOW WAVES

As mentioned previously, most regions of the GI tract exhibit rhythmic or phasic contractions that are the basic movements underlying more sophisticated motor patterns, such as peristalsis and segmentation. Phasic contractions are initiated by electrical slow waves that are composed of an initial upstroke depolarization, a partial repolarization, a plateau phase that can persist for 1 to several seconds (depending upon the region of the GI tract and the species), and repolarization. Slow waves can be measured using standard intracellular microelectrode techniques. In preparation for recordings, GI tissues are opened and the mucosal layers are removed, leaving the smooth muscle layers intact. Due to the rhythmic contractile nature of the tissues, in particular the small intestine, it is advisable to perform dissections with an ice-cold physiological salt solution (e.g., Krebs) to limit tissue movement during tissue preparations. As slow waves originate and propagate in ICC-MY and conduct to SMCs, they are typically recorded from the circular smooth muscle layer in the gastric antrum and small intestine. Contractions of the muscle often make impalements impossible to obtain and therefore appropriate concentrations of dihydropyridines such as nifedipine or nicardipine (1–2 μM) are often utilized to block Ca^{2+} entry via $Ca_v1.2$ channels in SMCs, thus preventing excitation-contraction coupling. Pacemaker activity and slow waves generated by ICC are resistant to these pharmacological agents and thus movement artifacts can be removed while retaining the slow wave activity.[5,65–72] Slow waves occur at physiological temperatures and this should be maintained throughout recordings of these events (propagation of slow waves depends on activation of temperature-sensitive T-type Ca^{2+} channels in ICC).[35,73] Slow waves have been recorded using this technique in many regions of GI tracts of many species including, mouse,[5] rabbit,[67,68,72] dog,[20,42,74,76] guinea-pig,[75] and human.[70]

Recording slow waves directly from ICC (pacemaker potentials in the literature) has been accomplished in mouse and rabbit small intestinal muscles[67,68] and guinea-pig stomach.[8] These preparations require dissection to remove the longitudinal muscle from the tissue (this layer is much thinner and easier to remove than the bulkier circular smooth muscle layer), in order to expose the ICC-MY for impalement. Dual microelectrode recordings can also be utilized in this configuration to examine the relationship between the pacemaker potentials in ICC and slow waves recorded in adjacent circular SMCs.[8,67,68]

15.5 STUDYING THE ELECTROPHYSIOLOGY OF GASTROINTESTINAL INTERSTITIAL CELLS

The ionic conductances responsible for generating slow waves and neural responses in ICs can be investigated at both the tissue and single-cell level. Intracellular microelectrode recordings allow investigators to utilize pharmacological approaches to assess contributions of specific ionic conductances and intracellular pathways on slow waves in both wild type and mutant animals. For example, pharmacological agents that block ANO1 channels have shown that slow waves are critically dependent on this channel.[23,25,71,77] Some traditional Ca^{2+}-activated Cl^- channel blockers such as niflumic acid have non-specific effects, thus newer more specific ANO1 blockers should be utilized whenever possible.[77] Traditional immunohistochemical and protein analysis of whole tissues as well as molecular analysis of enriched cop-Kit$^+$ ICC populations have shown that ANO1 channels are exclusively expressed in ICC and essential for their role as pacemakers throughout the GI tract in a range of species including mouse,[22,23,25,33] monkey[23] and human.[23] This has been further supported using global knockout mouse models of ANO1 that lacked slow waves.[23,71] However, it should be noted that global knockout of ANO1 often leads to embryonic lethality and thus conditional models of ANO1 deletion are worth pursuing.[78]

ANO1 mediated spontaneous transient inward currents (STICs) can be recorded in freshly isolated single ICC using the whole cell configuration of the patch clamp technique (Cs^+ rich pipette solutions can be used to eliminate contamination from K^+ conductances).[33–38] ICC fire STICs when voltage clamped at −80 mV and exhibit spontaneous transient depolarizations (STDs) in current clamp mode. Slow wave currents can be evoked by stepping from −80 to −35 mV.[33–38]

As mentioned earlier, PDGFRα$^+$ ICs show robust expression of SK3 channels.[28,39] These channels mediate outward current, which can be recorded using whole cell patch clamp techniques. When voltage clamped at −50 mV colonic PDGFRα$^+$ cells exhibit large amplitude spontaneous transient outward currents (STOCS), which are blocked by the SK3 inhibitor apamin.[28,32] In current clamp mode, purines elicit hyperpolarization responses in PDGFRα$^+$ cells resulting from activation of Ca^{2+}-activated SK3 channels. Purinergic neurotransmission to PDGFRα$^+$ cells in intact muscles also causes hyperpolarization and inhibitory effects on contractile behavior in the colon.[28,32,79] This pathway involving purinergic effects on PDGFRα⁺ cells *in situ* has also been investigated with Ca^{2+} imaging techniques in both wild type and mutant animal stomachs lacking P2Y1 receptors.[39]

Isolation of GI ICs is typically performed by incubating tissue strips (mucosa removed) in enzymatic cocktails of various concentrations of collagenase, trypsin inhibitor, and bovine serum albumin made up in a Ca^{2+}-free physiological solution, at 37°C.[28,32–38] After enzymatic digestion (length of time needs to be calibrated for species, organ, and age of animal), tissue strips are then washed with Ca^{2+} free solution and triturated to release single cells. ICs (ICC in particular) are fragile when freshly isolated and difficult to patch clamp. This is a technical obstacle to studies of these cells. Thus, upon isolation, ICs are often incubated in culture media for 2–6 hours at 37°C before studies are initiated. After this time, the cells have stabilized and can be used for patch clamp experimentation.

Another issue is that some types of ICs have a branched morphology *in situ*, due to the presence of multiple processes (e.g., ICC-MY and PDGFRα^+ cells contain numerous processes). In murine cells these morphological features are often lost upon isolation of cells and they assume a rounded morphology that makes it difficult to distinguish one from another.[28,32–38] Thus, to perform single-cell experiments on ICs it is necessary to use cell-specific reporter strains as described previously. A further consideration when experimenting with isolated ICs is that they rarely show spontaneous activity at room temperature. Thus, to study pacemaker activity in ICC and STOCs in PDGFRα^+ cells, temperatures within the physiological range must be utilized.

As discussed previously, there is evidence supporting a role for ANO1 in the pacemaker activity of GI muscles. However, this conductance is down-regulated in cells cultured for >24 hours, and remaining activity in cultured cells appears to be due to a non-selective cation conductance.[33] Thus, while short-term culture (2–6 hours) of isolated ICC makes them hardier for patch clamp experiments, the native phenotype of the cells in culture is not stable. Remolding of cells in culture conditions renders conclusions obtained from experiments on these preparations dubious.

15.6 RECORDING Ca^{2+} SIGNALS IN GI INTERSTITIAL CELLS

The physiologically important events in GI ICs depend upon activation of Ca^{2+}-dependent conductances; that is, ANO1 in ICC and SK3 in PDGFRα^+ cells. Therefore, studies of intracellular Ca^{2+} handling mechanisms are an important topic of investigation. Loading intact muscles with cell permeable forms of Ca^{2+} indictor dyes such as Fluo-4-AM allows visualization of Ca^{2+} transients in ICs with fluorescence microscopy.[39,69–71,78,80–82] Similar to electrophysiological recording of slow waves, Ca^{2+} transients in ICs should be monitored at physiological temperatures. The protocol for dye loading often depends on the cells, organ, and species to be studied; however, incubation times of 30 mins to 1 hour are common.[39,69–71,78,80–82] Loading with Ca^{2+} indicators leads to fluorescent signals from potentially many types of cells in intact muscles. This problem can be reduced to some extent by the use of confocal microscopy; however, contamination from unwanted cellular sources can still remain problematic.

Using cell-specific reporters, such as the histone 2B-eGFP fusion protein expressed in PDGFRα$^+$ cells, allows positive identification of these cells *in situ*, even when cells other than PDGFRα$^+$ cells are loaded with Ca^{2+} indicator. The bright fluorescent properties of eGFP in the nuclei of cells allows PDGFRα$^+$ cells to be discriminated from SMCs or ICC even when utilizing a dye excited at 488 nm.[39,82] Ca^{2+} imaging can also be combined with intracellular microelectrode recordings to examine temporal relationships between Ca^{2+} signals in ICs and spontaneous or evoked electrical events in the GI tract.[69,70,82,83] In these instances, great care should be taken to ensure that the location of the recording electrode is known to the investigator and placed in a relevant area of the field of view during imaging to ensure correct interpretation of data. Similarly, muscle preparations loaded with Ca^{2+} indicators may be used to simultaneously image Ca^{2+} signals in ICs and adjacent smooth muscle layers.[84] In such experiments great care should be taken to minimze signal contamination from the different cell types. It should also be noted that photo-bleaching and signal-to-noise considerations reduce the effectiveness of confocal microscopy and may not permit fast acquisition of images at high magnifications, limiting spatial and temporal resolution of Ca^{2+} transients.

Utilizing the Cre-LoxP technique to express genetically encoded Ca^{2+} indicators, such as the GCaMPs, can circumvent many of the limitations of membrane-permeable Ca^{2+} indicators for studies of ICs *in situ*. This approach is particularly useful if the Cre-Recombinase is inducible and cell specific, as very high signal-to-noise ratios can be achieved with no background signals from other types of cells contaminating images.[83,85] Use of GCaMPs allows high magnification and faster image acquisition to enhance resolution of sub-cellular Ca^{2+} events. Pragmatically, as the indicator does not require a loading or de-esterification stage, use of genetically encoded Ca^{2+} indicators allows for quicker turnaround times when dealing with multiple tissues throughout the course of an experimental time frame.

Patch clamp techniques utilizing the whole cell configuration can be utilized together with relevant pharmacology to assess the various contributions of the intracellular Ca^{2+} machinery (ryanodine receptors [RyR], inositol-triphosphate receptors [IP_3R], endoplasmic Ca^{2+}-ATPase, plasma membrane exchangers such as Na-K-Cl cotransporter [NKCC], etc.) to either ANO1-mediated STICs and STDs in ICC or SK3-mediated STOCs in PDGFRα$^+$ ICs.[32,35–37] These same pharmacological approaches can be applied to Ca^{2+} imaging in intact muscles or intracellular microelectrode recordings of slow waves or electrically/pharmacologically evoked neural responses.[36,66,72,86] Isolating and sorting cell-specific populations of ICs using cell-specific promoters also enables the use of molecular techniques to study relevant Ca^{2+} handling genes of interest (isoforms of RyR and IP_3Rs or voltage-dependent Ca^{2+} influx channels, for example),[35,82,83] as these may be heterogeneously distributed amongst the different cell populations and be functionally significant. Knockout models of Ca^{2+} release channels can also provide insights into the Ca^{2+} handling mechanisms behind IC functions (a mouse mutant lacking the IP_3R1 isoform showed a lack of slow waves).[87] Global knockout experiments should be interpreted cautiously as many types of cells, including

SMCs, express various forms of Ca^{2+} release channels. Thus, cell-specific knockouts of Ca^{2+} release and Ca^{2+} influx channels would be advisable.

15.7 INTERSTITIAL CELLS OUTSIDE OF THE GASTROINTESTINAL TRACT

In the last two decades, there has been growing interest in studying the role of ICs in visceral smooth muscle organs outside of the GI tract. Using many of the tools first optimized in the GI tract, such as exploitation of the Kit receptor as a maker for ICC in immunolabeling experiments, the presence of ICC-like cells (ICC-LC) has been reported in smooth muscle tissues of many animal models including urethra,[88] prostate,[89] pancreas,[90] corpus cavernosum,[91] myometrium,[92] ureter,[93] and bladder.[94] ICC have been implicated as pacemakers in some of these tissues; however, the role of ICC in other tissues is currently unknown.[95]

Many of these studies have utilized immunohistochemistry techniques with antibodies against Kit to identify cells as ICC-LC. The development of transgenic animals such as the histone 2B-eGFP fusion protein driven by PDGFRα has enabled investigators to identify PDGFRα$^+$ ICs in the urinary bladder[96] and female reproductive tract.[97] The identification of ICC-LCs in the bladder has been somewhat controversial, owing to ambiguous staining of Kit$^+$ ICs in various species and preparations. In some instances, the intermediate marker Vimentin has been used to identify ICC-LC in the bladder; however, as previously mentioned, PDGFRα$^+$ ICs also stain positive for this marker and as the bladder contains an abundant PDGFRα$^+$ IC population,[96] this is not a reliable marker for ICC-LC in this organ. It has been suggested that Kit$^+$ cells in the bladder are not truly ICC-LC and Kit$^+$ immunolabeling may be contamination from Kit$^+$ mast cells. Unlike in the GI tract, mast cells are present in large numbers in the bladder even when using rodent models. Using antibodies against Kit and mast cell tryptase, a 2016 study on human, rat, mouse, and guinea pig used immunohistochemistry to demonstrate that Kit$^+$ cells in the bladder were indeed mast cells in these species and not ICC-LC, small intestine was used as a control to evaluate antibody efficacy in these studies.[98] Groups investigating ICC-LC in non-GI muscles should thus use diligent controls to assess cell identification.

There have been few studies using genetic animal models to study ICs outside of the GI tract. *WW^v* mice display decreased uterine contractions, implicating a role of ICC-LC in modulating contractility in this organ.[99] However, in the LUT, *W/W^v* show abundant IC populations and spontaneous contractile activity was unaffected; thus, this mutant may not be a suitable model to study ICC-LC function in the LUT.[100] Several groups have also tried to inhibit LUT and reproductive tract ICs with pharmacological methods, but caution must be invoked here also. Imatinib mesylate (Gleevec) has been used as an inhibitor of Kit in several studies. These experiments have shown reduced muscle contractility in the bladder,[102–104] prostate,[105] and uterus[92,99,101] using this agent. However, there is now evidence that this drug can inhibit Ca^{2+} entry in SMCs by blocking L-type Ca^{2+} channels, leading to a reduced contractile response.[106] Furthermore, Gleevec has also been reported to inhibit signaling via PDGFRα and thus may exert non-specific effects on these cells as well as smooth muscle.[106]

15.8 SINGLE-CELL EXPERIMENTATION ON NON-GASTROINTESTINAL INTERSTITIAL CELLS

ICC-LC and PDGFRα$^+$ ICs have been studied extensively at the single-cell level in non-GI tissues. In the bladder, unambiguous cell identification using the PDGFRα$^+$-eGFP mouse strain has allowed detailed investigation of the electrophysiological properties of these cells. Combining the whole-cell patch clamp configuration with qPCR studies of gene expression, (exploiting enriched PDGFRα$^+$ IC populations via FACS), it is now known that similar to PDGFRα$^+$ ICs in the colon, PDGFRα$^+$ ICs isolated from the detrusor have a functional SK3 channel conductance coupled to P2Y1 receptors.[107,108] While the physiological significance of these findings is still under scrutiny, it is hypothesized that PDGFRα$^+$ ICs may play a role in facilitating bladder filling via activation of SK3 channels.[1,108]

Using patch clamp techniques (either whole-cell or perforated patch configurations), electrophysiological recordings have been made of ICC-LC from the renal pelvis,[109] prostate,[105,110] and urethra.[88,111–114] These experiments have revealed that similar to ICC in the GI tract, ICC-LC in the prostate and urethra exhibit pacemaker-like activity through a Ca^{2+}-activated Cl^- conductance. While the Kit$^+$-copGFP mice have not yet been utilized to study ICC-LC in these tissues, they may be beneficial in performing screens of genes of interest to determine the molecular identify of the Ca^{2+} activated Cl^- channel responsible for this activity, as well as other genes of interest.

Ca^{2+} imaging studies of ICs outside of the GI tract have primarily focused on ICC-LCs on the single-cell level. While there are some studies investigating ICC-LC Ca^{2+} signaling *in situ* from the urethra[115] and renal pelvis,[116] most studies have used enzymatically-isolated cells for their experiments. Many of these studies utilized confocal microscopy to improve resolution of subcellular Ca^{2+} signals in isolated cells, utilizing Ca^{2+} dyes such as Fura 2 or Fluo3/4-AM to image Ca^{2+} dynamics ranging from Ca^{2+} spark or puff like events to propagating Ca^{2+} waves.[109,112–114,117–124] Improved temporal resolution can be achieved if confocal microscopy is combined with a spinning disc configuration to visualize temporally discrete Ca^{2+} events in these cells.[119] Ca^{2+} imaging at the single-cell level in this way allows pharmacological interventions to more accurately investigate intracellular pathways involved in physiological processes within the cell of interest. This can be difficult to do *in situ* as bath-applied drugs may be acting on multiple cell types in the tissue and thus positive or negative effects may be due to the concerted response of many cell types and not just those of interest. Single-cell Ca^{2+} imaging can also be combined with patch clamp electrophysiology to further detail the correlation of intracellular Ca^{2+} signaling to the activation of specific conductances.[112]

Many studies of isolated cells of ICC-LC derive from the rabbit and guinea-pig and they note that unlike ICC in the murine GI tract, after enzymatic isolation ICC-LC retain the cellular processes they possess *in situ* and can be kept at 4°C in Ca^{2+}-free media for up to 8 hours before experimentation.[109,112–114,117–124] For this reason, ICC-LC from rabbit and guinea-pig are often identified on the basis of a branched like morphology alone and occasionally due to a lack of contractile response to a Ca^{2+} mobilization agent such as a high dose of caffeine or high concentrations of external potassium.[119,120] This makes confident identification of isolated ICC-LC from these species difficult as they do not possess a cell-specific marker. Furthermore, single-cell molecular studies are challenging and often require the manual and laborious collection of large numbers of ICC-LC for RT-PCR work.[119] While cell-specific markers such as Kit, myosin heavy chain or mast cell tryptase can be examined after collection, this process is very time consuming and often such collections are made on the sole basis of cellular morphology. With these limitations in mind, the use of the Kit^{+}-copGFP mouse may be useful to more accurately discriminate cell identity.

15.9 SUMMARY

Non-contractile ICs play important roles in regulating the contractility of visceral smooth muscle organs, either by acting as electrical pacemakers or by acting as neuromodulators to fine-tune neurally mediated responses in electrically coupled SMCs. Investigating their function can be accomplished through a variety of techniques; however, accurate and confident identification of ICs in mixed-cell populations is essential for meaningful interpretations of data. The use of transgenic animal models with cell-specific reporters allows for this unequivocal identification and opens up a range of possibilities for exploring IC function both at the single-cell and tissue level.

REFERENCES

1. Sanders, K. M., S. M. Ward, and S. D. Koh. 2014. Interstitial cells: Regulators of smooth muscle function. *Physiol Rev.* 94: 859–907.
2. Burnstock, G., M. E. Holman, and C. L. Prosser. 1963. Electrophysiology of smooth muscle. *Physiol Rev.* 146: 496–501.
3. Langton, P., S. M. Ward, A. Carl, M. A. Norell, and K. M. Sanders. 1989. Spontaneous electrical activity of interstitial cells of Cajal isolated from canine proximal colon. *Proc Natl Acad Sci USA*. 86: 7280–7284.
4. Cousins, H. M., F. R. Edwards, H. Hickey, C. E. Hill, and G. D. Hirst. 2003. Electrical coupling between the myenteric interstitial cells of Cajal and adjacent muscle layers in the guinea-pig gastric antrum. *J Physiol.* 550 (Pt 3): 829–844.
5. Ward, S. M., A. J. Burns, S. Torihashi, and K. M. Sanders. 1994. Mutation of the proto-oncogene c-kit blocks development of interstitial cells and electrical rhythmicity in murine intestine. *J Physiol.* 480 (Pt 1): 91–97.
6. Torihashi, S., S. M. Ward, S.-I. Nishikawa, K. Nishi, S. Kobayashi, and K. M. Sanders. 1995. c-*kit*-dependent development of interstitial cells and electrical activity in the murine gastrointestinal tract. *Cell Tiss Res.* 280: 97–111.
7. Huizinga, J. D., L. Thuneberg, M. Kluppel, J. Malysz, H. B. Mikkelsen, and A. Bernstein. 1995. W/kit gene required for interstitial cells of Cajal and for intestinal pacemaker activity. *Nature* 373: 347–349.
8. Dickens, E. J., G. D. Hirst, and T. Tomita. 1999. Identification of rhythmically active cells in guinea-pig stomach. *J Physiol.* 514 (Pt 2): 515–531.
9. Somlyo, A. P. and B. Himpens. 1989. Cell calcium and its regulation in smooth muscle. *FASEB J.* 3: 2266–2276.
10. Bolton, T. B., S. A. Prestwich, A. V. Zholos, and D. V. Gordienko. 1998. Excitation-contraction coupling in gastrointestinal and other smooth muscles. *Annu Rev Physiol.* 61: 85–115.

11. Sanders, K. M. 2008. Regulation of smooth muscle excitation and contraction. *Neurogastroenterol Motil.* 20 (Suppl 1): 39–53.
12. Ward, S. M. and K. M. Sanders. 2006. Involvement of intramuscular interstitial cells of Cajal in neuroeffector transmission in the gastrointestinal tract. *J Physiol.* 576 (Pt 3): 675–682.
13. Sanders, K. M., S. J. Hwang, and S. M. Ward. 2010. Neuroeffector apparatus in gastrointestinal smooth muscle organs. *J Physiol.* 588 (Pt 23): 4621–4639.
14. Sanders, K. M., S. D. Koh, S. Ro, and S. M. Ward. 2012. Regulation of gastrointestinal motility–Insights from smooth muscle biology. *Nat Rev Gastroenterol Hepatol.* 9 (11): 633–645.
15. Sanders, K. M., Y. Kito, S. J. Hwang, and S. M. Ward. 2016. Regulation of gastrointestinal smooth muscle function by interstitial cells. *Physiology* (*Bethesda*). 31 (5): 316–326.
16. Berezin, I., J. D. Huizinga, and E. E. Daniel. 1990. Structural characterization of interstitial cells of Cajal in myenteric plexus and muscle layers of canine colon. *Can J Physiol Pharmacol.* 68 (1): 1419–1439.
17. Rumessen, J. J., H. B. Mikkelsen, and L. Thuneberg. 1992. Ultrastructure of interstitial cells of Cajal associated with deep muscular plexus of human small intestine. *Gastroenterology* 102: 56–68.
18. Komuro, T., K. Seki, and K. Horiguchi. 1999. Ultrastructural characterization of the interstitial cells of Cajal. *Arch Histol Cytol.* 62 (4): 295–316.
19. Ward, S. M., E. P. Burke, and K. M. Sanders. 1990. Use of rhodamine 123 to label and lesion interstitial cells of Cajal in canine colonic circular muscle. *Anat Embryol.* 182: 215–224.
20. Thuneberg, L., V. Johansen, J. J. Rumessen, and B. G. Andersen. 1984. Interstitial cells of Cajal (ICC): Selective uptake of methylene blue inhibits slow wave activity. *Ninth International Symposium on Gastrointestinal Motility* (Aix en Provence).
21. Ro, S., C. Park, J. Jin, H. Zheng, P. J. Blair, D. Redelman, S. M. Ward, W. Yan, and K. M. Sanders. 2010. A model to study the phenotypic changes of interstitial cells of Cajal in gastrointestinal diseases. *Gastroenterology* 138: 1068–1078.
22. Gomez-Pinilla, P. J., S. J. Gibbons, M. R. Bardsley, A. Lorincz, M. J. Pozo, P. J. Pasricha, M. Van de Rijn et al. 2009. Ano1 is a selective marker of interstitial cells of Cajal in the human and mouse gastrointestinal tract. *Am J Physiol Gastrointest Liver Physiol.* 296: G1370–G1381.
23. Hwang, S. J., P. J. Blair, F. C. Britton, K. E. O'Driscoll, G. W. Hennig, Y. R. Bayguinov, J. R. Rock, B. D. Harfe, K. M. Sanders, and S. M. Ward. 2009. Expression of anoctamin 1/TMEM16A by interstitial cells of Cajal is fundamental for slow wave activity in gastrointestinal muscles. *J Physiol.* 587: 4887–4904.
24. Blair, P. J., Y. Bayguinov, K. M. Sanders, and S. M. Ward. 2012. Interstitial cells in the primate gastrointestinal tract. *Cell Tissue Res.* 350: 199–213.
25. Cobine, C. A., E. E. Hannah, M. H. Zhu, H. E. Lyle, J. R. Rock, K. M. Sanders, S. M. Ward, and K. D. Keef. 2017. ANO1 in intramuscular interstitial cells of Cajal plays a key role in the generation of slow waves and tone in the internal anal sphincter. *J Physiol.* 595 (6): 2021–2041.
26. Iino, S. and Y. Nojyo. 2009. Immunohistochemical demonstration of c-Kit-negative fibroblast-like cells in murine gastrointestinal musculature. *Arch Histol Cytol.* 72 (2): 108–115.
27. Iino, S., K. Horiguchi, S. Horiguchi, and Y. Nojyo. 2009. C-Kit-negative fibroblast-like cells express platlet-derived growth factor alpha in the murine gastrointestinal musculature. *Histochem Cell Biol.* 131: 691–702.
28. Kurahashi, M., H. Zheng, L. Dwyer, S. M. Ward, S. D. Koh, and K. M. Sanders. 2011. A functional role for the 'fibroblast-like cells' in gastrointestinal smooth muscles. *J Physiol.* 589 (3): 697–710.

29. Kurahashi, M., Y. Nakano, G. W. Hennig, S. M. Ward, and K. M. Sanders. 2012. Platelet-derived growth factor receptor α-positive cells in the tunica muscularis of human colon. *J Cell Mol Med.* 16 (7): 1397–1404.
30. Blair, P. J., Y. Bayguinov, K. M. Sanders, and S. M. Ward. 2012. Relationship between enteric neurons and interstitial cells in the primate gastrointestinal tract. *Neurogastroenterol Motil.* 24 (9): e437–e449.
31. Kurahashi, M., Y. Nakano, L. E. Peri, J. B. Townsend, S. M. Ward, and K. M. Sanders. 2013. A novel population of sub epithelial platelet-derived growth factor receptor α-positive cells in the mouse and human colon. *Am J Physiol Gastrointest Liver Physiol.* 304 (9): G823–G834.
32. Kurahashi, M., V. Mutafova-Yambolieva, S. D. Koh, and K. M. Sanders. 2014. Platelet-derived growth factor receptor-α-positive cells and not smooth cells mediate purinergic hyperpolarization in murine colonic muscles. *Am J Physiol Cell Physiol.* 307 (6): C561–C570.
33. Zhu, M. H., T. W Kim, S. Ro, W. Yan, S. M. Ward, S. D. Koh, and K. M. Sanders. 2009. A Ca($^{2+}$)-activated Cl($^{-}$) conductance in interstitial cells of Cajal linked to slow wave currents and pacemaker activity. *J Physiol.* 587: 4905–4918.
34. Zhu, M. H., I. K. Sunh, H. Zheng, T. S. Sung, F. C. Britton, K. O'Driscoll, S. D. Koh, and K. M. Sanders. 2011. Muscarinic activation of Ca^{2+} activated Cl^- current in interstitial cells of Cajal. *J Physiol.* 589 (18): 4565–4582.
35. Zheng, H., K. S. Park, S. D. Koh, and K. M. Sanders. 2014. Expression and function of a T-type Ca^{2+} conductance in interstitial cells of Cajal of the murine small intestine. *Am. J Physiol Cell Physiol.* 306 (7): C705–C713.
36. Zhu, M. H., T. S. Sung, K. O'Driscoll, S. D. Koh, and K. M. Sanders. 2015. Intracellular Ca^{2+} release from endoplasmic reticulum regulates slow wave currents and pacemaker activity of interstitial cells of Cajal. *Am J Physiol Cell. Physiol.* 308: C608–C620.
37. Zhu, M. H., T. S. Sung, M. Kurahashi, L. E. O'Kane, K. O'Driscoll, S. D. Koh, and K. M. Sanders. 2016. Na^+-K^+-Cl^- cotransporter (NKCC) maintains the chloride gradient to sustain pacemaker activity in interstitial cells of Cajal. *Am J Physiol Gastrointest Liver Physiol.* 311 (6): G1037–G1046.
38. Sung, T. S., H. U. Kim, J. H. Kim, H. Lu, K. M. Sanders, and S. D. Koh. 2015. Protease-activated receptors modulate excitability of murine colonic smooth muscles by differential effects on interstitial cells. *J Physiol.* 593 (5): 1169–1181.
39. Baker, S. A., G. W. Hennig, A. K. Salter, M. Kurahashi, S. M. Ward, and K. M. Sanders. 2013. Distribution and Ca^{2+} signalling of fibroblast-like (PDGFRα^+) cells in the murine gastric fundus. *J Physiol.* 591: 6193–6208.
40. Komuro, T., K. Tokui, and D. S. Zhou. 1996. Identification of the interstitial cells of Cajal. *Histol Histopathol.* 11 (3): 769–786.
41. Komuro, T. 2006. Structure and organization of interstitial cells of Cajal in the gastrointestinal tract. *J Physiol.* 576 (3): 653–658.
42. Smith, T. K., J. B. Reed, and K. M. Sanders. 1987. Origin and propagation of electrical slow waves in circular muscle of canine proximal colon. *Am J Physiol Cell Physiol.* 252: C215–C224.
43. Berezin, I., J. D. Huizinga, and E. E. Daniel. 1988. Interstitial cells of Cajal in the canine colon: A special communication network at the inner border of the circular muscle. *J Comp Neurol.* 273: 42–51.
44. Sanders, K. M., R. Stevens, E. Burke, and S. W. Ward. 1990. Slow waves actively propagate at submucosal surface of circular layer in canine colon. *Am J Physiol Gastrointest Liver Physiol.* 259: G258–G263.
45. Wang, X. Y., K. M. Sanders, and S. M. Ward. 2000. Relationship between interstitial cells of Cajal and enteric motor neurons in the murine proximal colon. *Cell Tissue Res.* 302 (3): 331–342.

46. Zhou, D. S. and T. Komuro. The cellular network of interstitial cells associated with the deep muscular plexus of the guinea pig small intestine. 1992. *Anat Embryol* (*Berl*). 186 (6): 519–527.
47. Imaizumi, M. and K. Hama. 1969. An electron microscopic study on the interstitial cells of the gizzard in the love-bird (*Uroloncha domestica*). *Z Zellforsch Mik Ana*. 97: 351–357.
48. Daniel, E. E. and V. Posey-Daniel. 1984. Neuromuscular structures in opossum esophagus: Role of interstitial cells of Cajal. *Am J Physiol*. 246: G305–G315.
49. Burns, A. J., A. E. Lomax, S. Torihashi, K. M. Sanders, and S. M. Ward. 1996. Interstitial cells of Cajal mediate inhibitory neurotransmission in the stomach. *Proc Natl Acad Sci USA*. 93: 12008–12013.
50. Ward, S. M., G. J. McLaren, and K. M. Sanders. 2006. Interstitial cells of Cajal in the deep muscular plexus mediate enteric motor neurotransmission in the mouse small intestine. *J Physiol*. 573: 147–159.
51. Ward, S. M., G. Morris, L. Reese, X. Y. Wang, and K. M. Sanders. 1998. Interstitial cells of Cajal mediate enteric inhibitory neurotransmission in the lower esophageal and pyloric sphincters. *Gastroenterology* 115: 314–329.
52. Beckett, E. A., K. Horiguchi, M. Khoyi, K. M. Sanders, and S. M. Ward. 2002. Loss of enteric motor neurotransmission in the gastric fundus of Sl/Sl(d) mice. *J Physiol*. 543, 871–887.
53. Mitsui, R., and T. Komuro. 2002. Direct and indirect innervation of smooth muscle cells of rat stomach, with special reference to the interstitial cells of Cajal. *Cell Tissue Res*. 309: 219–227.
54. Klein, S., B. Seidler, A. Kettenberger, A. Sibaev, M. Rohn, R. Feil, H. D. Allescher et al. 2013. Interstitial cells of Cajal integrate excitatory and inhibitory neurotransmission with intestinal slow-wave activity. *Nature Comm*. 4: 1630.
55. Cobine, C. A., A. G. Sotherton, L. E. Peri, K. M. Sanders, S. M. Ward, and K. D. Keef. 2014. Nitrergic neuromuscular transmission in the mouse internal anal sphincter is accomplished by multiple pathways and postjunctional effector cells. *Am J Physiol Gastrointest Liver Physiol*. 307 (11): G1057–G1072.
56. Sanders, K. M., A. K. Salter, G. W. Hennig, S. D. Koh, B. A. Perrino, S. M. Ward, and S. A. Baker. 2014. Responses to enteric motor neurons in the gastric fundus of mice with reduced intramuscular interstitial cells of Cajal. *J Neurogastroenterol Motil*. 20 (2): 171–184.
57. Beckett, E. A., K. M. Sanders, and S. M. Ward. 2017. Inhibitory responses mediated by vagal nerve stimulation are diminished in stomachs of mice with reduced intramuscular interstitial cells of Cajal. *Sci Rep*. 7. 44759.
58. Maeda, H., A. Yamagata, S. Nishikawa, K. Yoshinaga, S. Kobayashi, K. Nishi, and S. I. Nishikawa. 1992. Requirement of c-kit for development of intestinal pacemaker system. *Development* 116: 369–375.
59. Mikkelsen, H. B., J. Malysz, J. D. Huizinga, and L. Thuneberg. 1998. Action potential generation, Kit receptor immunohistochemistry and morphology of steel-Dickie (*Sl/Sld*) mutant mouse small intestine. *Neurogastroenterol Motil*. 10: 11–26.
60. Ward, S. M., E. A. Beckett, X. Wang, F. Baker, M. Khoyi, and K. M. Sanders. 2000. Interstitial cells of Cajal mediate cholinergic neurotransmission from enteric motor neurons. *J Neurosci*. 20 (4): 1393–1403.
61. Ordog, T., S. M. Ward, and K. M. Sanders. 1999. Interstitial cells of cajal generate electrical slow waves in the murine stomach. *J Physiol*. 518 (Pt 1): 257–269.
62. Ward, S. M., A. J. Burns, S. Torihashi, S. C. Harney, and K. M. Sanders. 1995. Impaired development of interstitial cells and intestinal electrical rhythmicity in steel mutants. *Am J Physiol Cell Physiol*. 269: C1577–C1585.
63. Suzuki, H., S. M. Ward, Y. R. Bayguinov, F. R. Edwards, and G. D. Hirst. 2003. Involvement of intramuscular interstitial cells in nitrergic inhibition in the mouse gastric antrum. *J Physiol*. 546: 751–763.

64. Bhetwal, B. P., K. M. Sanders, C. An, D. M. Trappanese, R. S. Moreland, and B. A. Perrino. 2013. Ca^{2+} sensitization pathways accessed by cholinergic neurotransmission in the murine gastric fundus. *J Physiol.* 591 (12): 2971–2986.
65. Malysz, J., D. Richardson, L. Farraway, M. O. Christen, and J. D. Huizinga. 1995. Generation of slow wave type action potentials in the mouse small intestine involves a non-L-type calcium channel. *Can J Physiol Pharmacol.* 73 (10): 1502–1511.
66. Malysz, J., G. Donnelly, and J. D. Huizinga. 2001. Regulation of slow wave frequency by IP_3 sensitive calcium release in the murine small intestine. *Am J Physiol Gastrointest. Liver Physiol.* 280: G439–G448.
67. Kito, Y. and H. Suzuki. 2003. Properties of pacemaker potentials recorded from myenteric interstitial cells of Cajal distributed in the mouse small intestine. *J Physiol.* 15 (553): 803–818.
68. Kito, Y., S. M. Ward, and K. M. Sanders. 2005. Pacemaker potentials generated by interstitial cells of Cajal in the murine small intestine. *Am J Physiol Cell Physiol.* 288 (3): C710–C720.
69. Park, K. J., G. W. Hennig, H. T. Lee, N. J. Spencer, S. M. Ward, T. K. Smith, and K. M. Sanders. 2006. Spatial and temporal mapping of pacemaker activity in interstitial cells of Cajal in mouse ileum in situ. *Am J Physiol Cell Physiol.* 290: C1411–1427.
70. Lee, H. T., G. W. Hennig, N. W. Fleming, K. D. Keef, N. J. Spencer, S. M. Ward, K. M. Sanders, and T. K. Smith. 2007. The mechanism and spread of pacemaker activity through myenteric interstitial cells of Cajal in human small intestine. *Gastroenterology* 132 (5): 1852–1865.
71. Singh, R. D., S. J. Gibbons, S. A. Saravanaperumal, P. Du, G. W. Hennig, S. T. Eisenman, A. Mazzoneet al. 2014. ANO1, a Ca^{2+} activated Cl^- channel, coordinates contractility in mouse intestine by Ca^{2+} transient coordination between interstitial cells of Cajal. *J Physiol.* 592 (18): 4051–4068.
72. Kito, Y., R. Mitsui, S. M. Ward, and K. M. Sanders. 2015. Characterization of slow waves generated by myenteric interstitial cells of Cajal of the rabbit small intestine. *Am J Physiol Gastrointest Liver Physiol.* 308 (5): G378–38.
73. Gibbons, S. J., P. R. Strege, S. Lei, J. L. Roeder, A. Mazzone, Y. Ou, A. Rich, and G. Farrugia. 2009. The alpha 1H Ca^{2+} channel subunit is expressed in mouse jejunal interstitial cells of Cajal and myocytes. *J Cell Mol Med.* 13: 4422–4431.
74. Connor, J. A., C. L. Prosser, and W. A. Weems. 1974. A study of pace-maker activity in intestinal smooth muscle. *J Physiol.* 240 (3): 671–701.
75. Ohba, M., Y. Sakamoto, and T. Tomita. 1975. The slow wave in the circular muscle of the guinea-pig stomach. *J Physiol.* 253 (2): 505–516.
76. el-Sharkawy, T. Y., K. G. Morgan, and J. H. Szurszewski. 1978. Intracellular electrical activity of canine and human gastric smooth muscle. *J Physiol.* 279: 291–307.
77. Hwang, S. J., N. Basma, K. M. Sanders, and S. M. Ward. 2016. Effects of new-generation inhibitors of the calcium-activated chloride channel anoctamin 1 on slow waves in the gastrointestinal tract. *Br J Pharmacol.* 173 (8): 1339–1349.
78. Malysz, J., S. J. Gibbons, S. A. Saravanaperumal, P. Du, S. T. Eisenman, C. Cao, U. Oh et al. 2016. Conditional genetic deletion of Ano1 in interstitial cells of Cajal impairs Ca2+ transients and slow waves in adult mouse small intestine. *Am J Physiol Gastrointest Liver Physiol.* 312 (3): G228–G245.
79. Durnin, L., S. J. Hwang, M. Kurahashi, B. T. Drumm, S. M. Ward, K. C. Sasse, K. M. Sanders, and V. N. Mutafova-Yambolieva. 2014. Uridine adenosine tetraphosphate is a novel neurogenic P2Y1 receptor activator in the gut. *Proc Natl Acad Sci USA.* 111 (44): 15821–15826.

80. Hennig, G. W., N. J. Spencer, S. Jokela-Willis, P. O. Bayguinov, H. T. Lee, L. A. Ritchie, S. M. Ward, T. K. Smithand, and K. M. Sanders. 2010. ICC-MY coordinate smooth muscle electrical and mechanical activity in the murine small intestine. *Neurogastroenterol Motil.* 22 (5): e138–e151.
81. Lowie, B. J., X. Y. Wang, E. J. White, and J. D. Huizinga. 2011. On the origin of rhythmic calcium transients in the ICC-MP of the mouse small intestine. *Am J Physiol Gastrointest Liver Physiol.* 301: G835–G845.
82. Baker, S. A., G. W. Hennig, S. M. Ward, and K. M. Sanders. 2015. Temporal sequence of activation of cells involved in purinergic neurotransmission in the colon. *J Physiol.* 593: 1945–1963.
83. Drumm, B. T., G. W. Hennig, M. J. Battersby, E. Cunningham, T. Sung, S. M. Ward, K. M. Sanders, and S. A. Baker. 2017. Clustering of Ca^{2+} transients in interstitial cells of Cajal define slow wave duration. *J Gen Physiol.* 149 (7): 703–725.
84. Yamazawa, T. and M. Iino. 2002. Simultaneous imaging of Ca^{2+} signals in interstitial cells of Cajal and longitudinal smooth muscle cells during rhytmic activity in mouse ileum. *J Physiol.* 538 (3): 823–835.
85. Baker, S. A., B. T. Drumm, D. Sauer, G. W. Hennig, S. M. Ward, and K. M. Sanders. 2016. Spontaneous Ca^{2+} transients in interstitial cells of Cajal located within the deep muscular plexus of the murine small intestine. *J Physiol.* 594.12: 3317–3338.
86. Bayguinov, O., S. M. Ward, J. L. Kenyon, and K. M. Sanders. 2007. Voltage-gated Ca^{2+} currents are necessary for slow-wave propagation in the canine gastric antrum. *Am J Physiol Cell Physiol.* 293 (5): C1645–C1659.
87. Suzuki, H., H. Takano, Y. Yamamoto, T. Komuro, M. Saito, K. Kato, and K. Mikoshiba. 2000. Properties of gastric smooth muscles obtained from mice which lack inositol triphosphate receptor. *J Physiol.* 525 (1): 105–111.
88. Sergeant, G. P., M. A. Hollywood, K. D. McCloskey, K. D. Thornbury, and N. G. McHale. 2000. Specialised pacemaking cells in the rabbit urethra. *J Physiol.* 526 (2): 359–366.
89. Exintaris, B., M. F. Klemm, and R. J. Lang. 2002. Spontaneous slow wave and contractile activity of the guinea pig prostate. *J Urol.* 168: 315–322.
90. Popescu, L. M., S. M. Ciontea, D. Cretoiu, M. E. Hinescu, E. Radu, N. Ionescu, M. Ceausuet al. 2005. Novel type of interstitial cells (Cajal-like) in human fallopian tube. *J Cell Mol Med.* 9 (2): 479–523.
91. Hashitani, H. and H. Suzuki. 2004. Identification of interstitial cells of Cajal in corporal tissues of the guinea-pig penis. *Br J Pharmacol.* 141: 199–204.
92. Popescu, L. M., C. Vidulescu, A. Curici, L. Caravia, A. A. Simionescu, S. M. Ciontea, and S. Simion. 2006. Imatinib inhibits spontaneous rhythmic contractions of human uterus and intestine. *Eur J Pharmacol.* 546 (1–3): 177–181.
93. Pezzone, M. A., S. C. Watkins, S. M. Alber, W. E. King, W. C. de Groat, M. B. Chancellor, and M. O. Fraser. 2003. Identification of c-kit positive cells in the mouse ureter: The interstitial cells of Cajal of the urinary tract. *Am J Physiol Cell Physiol.* 284: C925–C929.
94. McCloskey, K. D and A. M. Gurney. 2002. Kit positive cells in the guinea pig bladder. *J Urol.* 168: 832–836.
95. Hashitani, H. 2006. Interaction between interstitial cells and smooth muscles in the lower urinary tract and penis. *J Physiol.* 576 (3): 707–714.
96. Koh, B. H., R. Roy, M. A. Hollywood, K. D. Thornbury, N. G. McHale, G. P. Sergeant, W. J. Hatton, S. M. Ward, K. M. Sanders, and S. D. Koh. 2012. Platelet-derived growth factor receptor-α cells in mouse urinary bladder: A new class of interstitial cells. *J Cell Mol Med.* 16 (4): 691–700.

97. Peri, L. E., B. H. Koh, G. K. Ward, Y. Bayguinov, S. J. Hwang, T. W. Gould, C. J. Mullan, K. M. Sanders, and S. M. Ward. 2015. A novel class of interstitial cells in the mouse and monkey female reproductive tracts. *Biol Reprod.* 92 (4): 102, 1–17.
98. Gevaert, T., D. Ridder, E. Vanstreels, D. Daelemans, W. Everaerts, F. V. Aa, I. Pintelon et al. 2016. The stem cell growth factor receptor KIT is not expressed on interstitial cells in bladder. *J Cell Mol Med.* 21 (6): 1206–1216. doi:10.1111/jcmm.13054.
99. Allix, S., E. Reyes-Gomez, G. Aubin-Houzelstein, D. Noël, L. Tiret, J. J. Panthier, and F. Bernex. 2008. Uterine contractions depend on KIT-positive interstitial cells in the mouse: Genetic and pharmacological evidence. *Biol Reprod.* 79 (3): 510–517.
100. McCloskey, K. D., U. A. Anderson, R. A. Davidson, Y. R. Bayguinov, K. M. Sanders, and S. M. Ward. 2009. Comparison of mechanical and electrical activity and interstitial cells of Cajal in urinary bladders from wild-type and *W/W^v^* mice. *Br J Pharmacol.* 156 (2): 273–283.
101. Hutchings, G., J. Deprest, B. Nilius, T. Roskams, and D. De Ridder. 2006. The effect of imatinib mesylate on the contractility of isolated rabbit myometrial strips. *Gynecol Obstet Invest.* 62 (2): 79–83.
102. Biers, S. M., J. M. Reynard, T. Doore, and A. F. Brading. 2006. The functional effects of a c-kit tyrosine inhibitor on guinea-pig and human detrusor. *BJU Int.* 97: 612–616.
103. Kubota, Y., S. M. Biers, K. Kohri, and A. F. Brading. 2006. Effects of imatinib mesylate (Glivec) as a c-kit tyrosine kinase inhibitor in the guinea-pig urinary bladder. *Neurourol Urodyn.* 25 (3): 205–210.
104. Gevaert, T., G. Hutchings, W. Everaerts, H. Prenen, T. Roskams, B. Nilius, and D. De Ridder. 2013. Administration of imatinib mesylate in rats impairs the neonatal development of intramuscular interstitial cells of Cajal in bladder and results in altered contractile properties. *Neurourol Urodyn.* 33: 461–468.
105. Lam, M., A. Dey, R. J. Lang, and B. Exintaris. 2013. Effects of imatinib mesylate on the spontaneous activity generated by the guinea-pig prostate. *BJU Int.* 112: E398–E405.
106. Hashitani, H., M. Hayase, and H. Suzuki. 2008. Effects of imatinib mesylate on spontaneous electrical and mechanical activity in smooth muscle of the guinea-pig stomach. *Br J Pharmacol.* 154 (2): 451–4459.
107. Lee, H., B. H. Koh, L. E. Peri, K. M. Sanders, and S. D. Koh. 2013. Functional expression of SK channels in murine detrusor PDGFRα^+ cells. *J Physiol.* 591 (2): 503–513.
108. Lee, H., B. H. Koh, L. E. Peri, K. M. Sanders, and S. D. Koh. 2014. Purinergic inhibitory regulation of murine detrusor muscles mediated by PDGFRα^+ interstitial cells. *J Physiol.* 592 (6): 1283–1293.
109. Lang, R. J., H. Hashitani, M. A. Tonta, H. C. Parkington, and H. Suzuki. 2007. Spontaneous electrical and Ca^{2+} signals in typical and atypical smooth muscle cells and interstitial cell of Cajal-like cells of mouse renal pelvis. *J Physiol.* 583 (3): 1049–1068.
110. Shigemasa, Y., M. Lam, R. Mitsui, and H. Hashitani. 2014. Voltage-dependency of slow wave frequency in the guinea pig prostate. *J Urol.* 192 (4): 1286–1292.
111. Sergeant, G. P., M. A. Hollywood, K. D. McCloskey, N. G. McHale, and K. D. Thornbury. 2001. Role of IP3 in modulation of spontaneous activity in pacemaker cells of rabbit urethra. *Am J Physiol Cell Physiol.* 280: C1349–C1356.
112. Sergeant, G. P., L. Johnston, N. G. McHale, K. D. Thornbury, and M. A. Hollywood. 2006. Activation of cGMP/PKG pathway inhibits electrical activity in rabbit urethral interstitial cells of Cajal by reducing the spread of Ca^{2+} waves. *J Physiol.* 574 (1): 167–181.
113. Sergeant, G. P., J. E. Bradley, K. D. Thornbury, N. G. McHale, and M. A. Hollywood. 2008. Role of mitochondria in modulation of spontaneous Ca^{2+} waves in freshly dispersed interstitial cells of Cajal from the rabbit urethra. *J. Physiol.* 586 (19): 4631–4642.

114. Bradley, J. E., S. Kadima, B. Drumm, M. A. Hollywood, K. D. Thornbury, N. G. McHale, and G. P. Sergeant. 2010. Novel excitatory effects of adenosine triphosphate on contractile and pacemaker activity in rabbit urethral smooth muscle. *J Urol.* 183: 801–811.
115. Hashitani, H. and H. Suzuki. 2007. Properties of spontaneous Ca^{2+} transients recorded from interstitial cells of Cajal-like cells of the rabbit urethra in situ. *J Physiol.* 583 (2): 505–519.
116. Hashitani, H., M. J. Nguyen, H. Noda, R. Mitsui, R. Higashi, K. Ohta, K. I. Nakamura, and R. J. Lang. 2017. Interstitial cell modulation of pyeloureteric peristalsis in the mouse renal pelvis examined using FIBSEM tomography and calcium indicators. *Pflugers Arch.* 469 (5–6): 797–813. doi:10.1007/s00424-016-1930-6.
117. Bradley, J. E., M. A. Hollywood, L. Johnston, R. J. Large, T. Matsuda, A. Baba, N. G. McHale, K. D. Thornbury, and G. P. Sergeant. 2006. Contribution of reverse Na^{+}-Ca^{2+} exchange to spontaneous activity in interstitial cells of Cajal in the rabbit urethra. *J Physiol.* 574 (3): 651–661.
118. Drumm, B. T., S. D. Koh, K. E. Andersson, and S. M. Ward. 2014. Calcium signalling in Cajal-like interstitial cells of the lower urinary tract. *Nat Rev Urol.* 11 (10): 555–564.
119. Drumm, B. T., R. J. Large, M. A. Hollywood, K. D. Thornbury, S. A. Baker, B. J. Harvey, N. G. McHale, and G. P. Sergeant. 2015. The role of Ca^{2+} influx in spontaneous Ca^{2+} wave propagation in interstitial cells of Cajal from the rabbit urethra. *J Physiol.* 593 (1): 3333–3350.
120. Drumm, B. T., G. P. Sergeant, M. A. Hollywood, K. D. Thornbury, T. Matsuda, A. Baba, B. J. Harvey, and N. G. McHale. 2014. The effect of high $[K^{+}]_{o}$ on spontaneous Ca^{2+} waves in freshly isolated interstitial cells of Cajal from the rabbit urethra. *Physiol Rep.* 2 (1): e00203. doi:10.1002/phy2.203.
121. Drumm, B. T., G. P. Sergeant, M. A. Hollywood, K. D. Thornbury, N. G. McHale, and B. J. Harvey. 2014. The role of cAMP dependent protein kinase in modulating spontaneous intracellular Ca^{2+} waves in interstitial cells of Cajal from the rabbit urethra. *Cell Calcium.* 56 (3): 181–187.
122. Lam, H., Y. Shigemasa, B. Exintaris, R. J. Lang, and H. Hasitani. 2011. Spontaneous Ca^{2+} signalling of interstitial cells in the guinea pig prostate. *J Urol.* 186 (6): 2478–2486.
123. Johnston, L., G. P. Sergeant, M. A. Hollywood, K. D. Thornbury, and N. G. McHale. 2005. Calcium oscillations in interstitial cells of the rabbit urethra. *J Physiol.* 565 (2): 449–61.
124. Hashitani, H., R. J. Lang, and H. Suzuki. 2010. Role of perinuclear mitochondria in the spatiotemporal dynamics of spontaneous Ca^{2+} waves in interstitial cells of Cajal-like cells of the rabbit urethra. *Br J Pharmacol.* 161: 680–694.

16 Electrical Pacemaking in Lymphatic Vessels

Scott D. Zawieja, Jorge A. Castorena-Gonzalez, Kim H. T. To, Peichun Gui, Timothy Domeier, and Michael J. Davis

CONTENTS

16.1 INTRODUCTION

The lymphatic system is an essential component of overall fluid homeostasis. The majority of lymph transport depends on active, propulsive contractions produced by networks of collecting lymphatic vessels. These contractions originate from the spontaneous generation of action potentials (APs) that spread rapidly and in a coordinated manner through the lymphatic smooth muscle (SM) layer. Contractions are normally entrained over the vessel segment between two valves, called a lymphangion, but also sometimes over regions spanning multiple lymphangions. Although the electrical pacemaker that triggers these APs is not completely understood, the underlying ionic mechanisms are potential targets for therapy directed at restoring a dysfunctional lymphatic system. This chapter examines the evidence for the cell type responsible for lymphatic pacemaking, making frequent comparisons to pacemaking cells in the gastrointestinal (GI) tract and, to a lesser extent, cells in the cardiac conduction system. The current hypothesis for lymphatic pacemaking is then discussed and the evidence supporting and contradicting that hypothesis is reviewed. Finally, an alternative mechanism is considered that may be working in pressurized vessels under physiologically relevant conditions.

16.2 THE IMPORTANCE OF UNDERSTANDING MECHANISMS UNDERLYING LYMPHATIC PACEMAKING

A primary function of the lymphatic system is fluid homeostasis, whereby initial lymphatics in the interstitium capture excess fluid and protein filtered from blood vessels, while collecting lymphatic vessels pump it uphill and return it to the great veins [1]. Lymph transport in humans amounts to >3 L of fluid and 120 g of protein per day [1,2], which if not returned to the vascular system, remains in the interstitium and results eventually in edema [3–5]. Edema due in part or in whole from lymphatic dysfunction is termed lymphedema.

Dysfunction of the lymphatic system affects over 10 million people annually in the USA [6]. Primary lymphedemas include congenital defects such as Milroy's disease (*VEGFR3/FLT4* mutations) and lymphedema distichiasis (*FOXC2* mutations) [7,8], but these occur at a relatively low frequency of less than 1:10,000 live births. The majority of lymphedema cases develop secondary to other disease processes or the therapies they entail. Worldwide, lymphedema is observed in over 120 million patients with filariasis, in which the mosquito-borne parasitic nematode *Wuchereria bancrofti* infects the host lymphatic tissues. In developed countries, secondary lymphedema is associated with peripheral artery disease, surgery-related trauma, morbid obesity, and cancer therapies such as surgery or radiation treatment; in these cases, lymphedema results from the failure of the lymphatic system to cope with an abnormally high imposed pressure load [9]. There are no cures for either primary or secondary lymphedema as decongestive/massage therapies do not address the underlying cause(s).

In contrast to the blood vascular system, lymph transport in many parts of the body normally occurs *against* a hydrostatic pressure gradient and this transport is largely an active process accomplished by the spontaneous contractions of collecting lymphatic vessels that propel lymph centrally. The hydrostatic pressure in initial lymphatics of the rat is stable near 0 cmH_2O, approximating interstitial hydrostatic pressure [10]. As these vessels converge to form collectors, they are invested with SM cells and exhibit spontaneous, robust, coordinated contractions that generate intralymphatic pressure fluctuations with each contraction cycle [11]. Mean pressure rises to 10–15 cmH_2O in pre-nodal afferent collectors before falling toward central venous pressure levels (2–3 cmH_2O) in the thoracic duct [4,12–15]. Although the cited pressure measurements were recorded in supine quadrupeds (rats), they establish the concept that lymph is actively transported against an adverse pressure gradient even in the absence of a superimposed hydrostatic column. Other measurements in the legs of upright bipeds (humans) reveal that intralymphatic pressures normally reach much higher values [16] in order to propel lymph against standing hydrostatic columns as high as 100 cmH_2O. The presence of one-way valves every few mm is critical to interrupting the standing lymph column and preventing backward movement of lymph. Lymph transport thus relies critically on the intrinsic, phasic contractions of lymphatic muscle cells, along with patent, unidirectional valves, which together comprise a series of *lymph pumps*. Dysfunction of the lymphatic pump system can lead to lymphedema.

Surprisingly little is known about how, and why, lymphatic vessels become dysfunctional in lymphedema, but the most relevant clinical information comes from the work of Olszewski and colleagues in Poland, who cannulated lymphatics in the legs and arms of normal [16] and lymphedema patients [17] in both supine and upright positions. Diastolic pressure in lymphatic collectors did not rise substantially in normal limbs upon standing because the lymphatic valve system was patent, but in lymphedematous limbs diastolic pressure rose from ~0 to 60 mmHg [16–18], suggestive of possible valve dysfunction. Subsequent measurements of lymph flow under similar conditions [19] confirmed that lymph transport depended critically on the phasic activity of the lymphatic muscle in the walls of the collecting vessels and to a lesser degree on passive compression [20]. Clinical observations in lymphedematous limbs revealed that (1) spontaneous contractions ceased or were attenuated [17], (2) lymphatic diastolic pressure was elevated, (3) vessel diameters were enlarged, and (4) many valves apparently were incompetent [17,19]. Collectively, these findings point to an underlying problem in chronic lymphedema with both the lymphatic pacemaker and the intraluminal valves. It should be obvious that the cellular mechanisms underlying pacemaker and valve dysfunction must be understood before effective strategies can be devised to reverse lymphatic pump dysfunction.

16.3 THE CELL TYPE RESPONSIBLE FOR ELECTRICAL PACEMAKING IN LYMPHATIC VESSELS

A first step toward understanding the mechanisms underlying lymphatic pacemaking is to identify the cell type responsible for initiating spontaneous electrical activity. Current concepts about the ionic mechanisms underlying the pacemaker are based largely on studies of rhythmicity in the GI tract [21], where it is now well established that interstitial cells of Cajal (ICCs) serve as pacemakers and coordinate pacemaking activity.

16.3.1 Parallels to Pacemaking in the Gastrointestinal Tract

Several types of ICC networks are distributed within and around the circular and longitudinal SM cell layers of the stomach and intestine [22] and their function is in turn modulated by other interstitial cells that mediate responses to neural input [23]. ICCs have a resting membrane potential of ~−60 mV and exhibit periodic, relatively long-lasting depolarizations to ~−30 mV (slow waves; Figure 16.1) [22]. It is not clear to what extent ICCs in the GI tract share a common ionic mechanism for pacemaking. Slow waves in myenteric ICCs (ICC-MY) of the gastric antrum appear to be driven largely by IP_3-mediated activation of the calcium-activated Cl^- channel *ANO1* (also called TMEM16A), whereas in ICCs of other regions or layers of the GI tract [24] this channel may act in concert with the activation of a cation channel, possibly TRPC4 or TRPM7 [22,25,26], to generate slow waves. The underlying ion currents are easier to dissect with patch clamp studies of isolated, cultured ICCs, but some of those studies are complicated by changes in ion channel expression, even after only short-term culture of the cells.

Activation of the Cl^- and/or cation channels in ICC-MY cells leads to the generation of slow waves that are 2–10 s in duration in the gastric antrum, but 1–2 s in duration in the small intestine [28]. These depolarizing waves persist in the presence of dihydropyridines (DHP) [21,22], distinguishing them from other pacemaker cells (Figure 16.1), such as those in the central nervous system. Once generated, slow waves are conducted electrotonically through gap junctions to other ICCs and to SM cells in the circular and longitudinal layers, triggering APs and Ca^{2+} influx in the latter cells to drive contraction [29]. As shown in Figure 16.2, the onset of the pacemaker potential (PP) in ICC-MY cells precedes the onset of the slow wave in visceral SM cells, which also display a secondary, regenerative component. In the absence of ICC-generated PPs, spontaneous AP firing can be initiated by visceral SM cells, but the electrical activity and resulting contractions are not sufficiently coordinated to generate peristaltic waves.

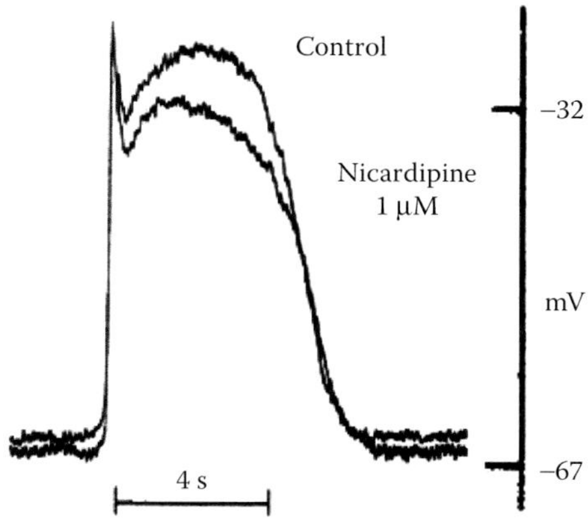

FIGURE 16.1 Slow waves in guinea pig gastric antrum with and without nicardipine present; traces are superimposed. (From Ozaki, H. et al., *Am. J. Physiol. Cell Physiol.*, 260, C917–C925, 1991.)

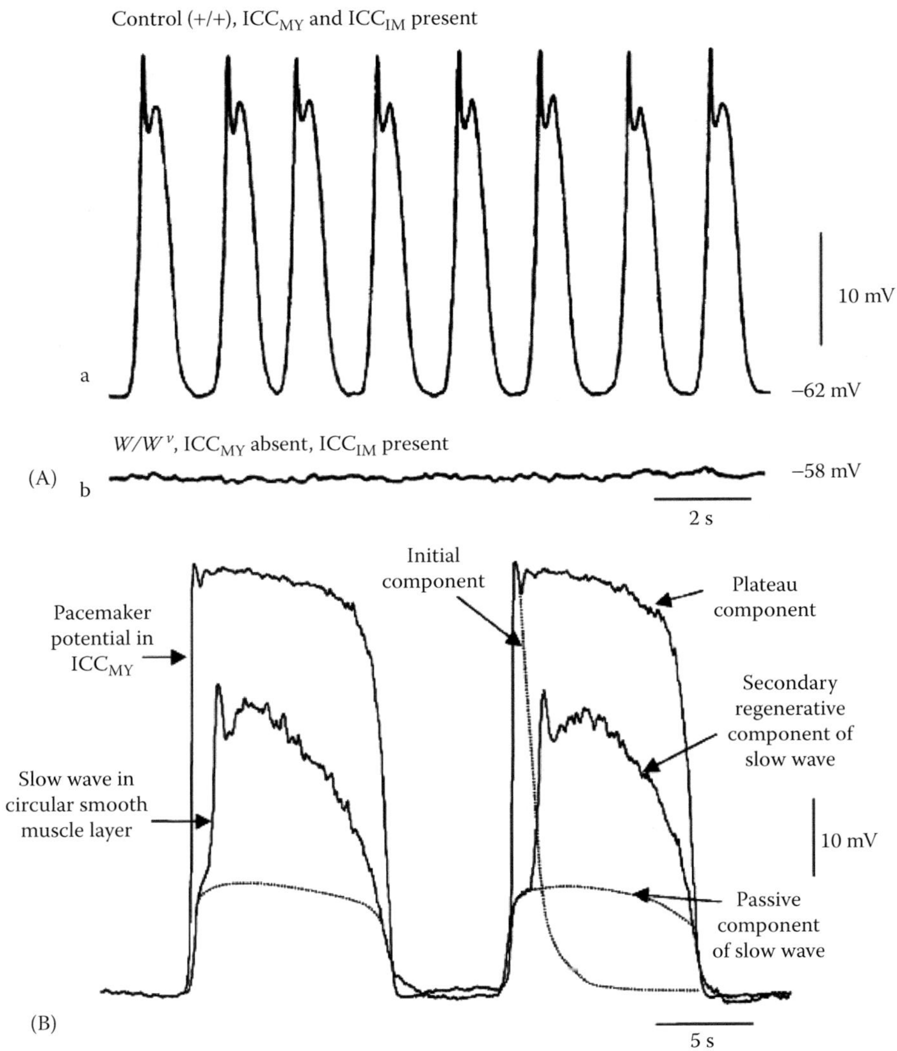

FIGURE 16.2 (A) Comparison of slow waves in ICC-MY from control and *W/W*^{v} mice. (B) Simultaneous recordings of the PP in an ICC-MY and the associated Vm change in a nearby SM cell. (From Hirst, G.D.S. and Ward, S.M., *J. Physiol.*, 550, 337–346, 2003.)

16.3.2 c-Kit(+) Cells and ICC

The prototypical marker for ICCs in the GI tract is the receptor tyrosine kinase c-Kit [30]. In mouse and primate GI tracts, ICCs (and almost exclusively ICCs) are c-Kit(+), as confirmed by staining with c-Kit antibodies and by nuclear GFP expression in Kit$^{+/copGFP}$ mice (Figure 16.3). *W/W*^{v} mice, in which ICC-MY cells in the small intestine are absent or dysfunctional due to a c-Kit mutation [30,31], fail to generate slow waves (Figure 16.2Ab). Chronic treatment of normal mice with a

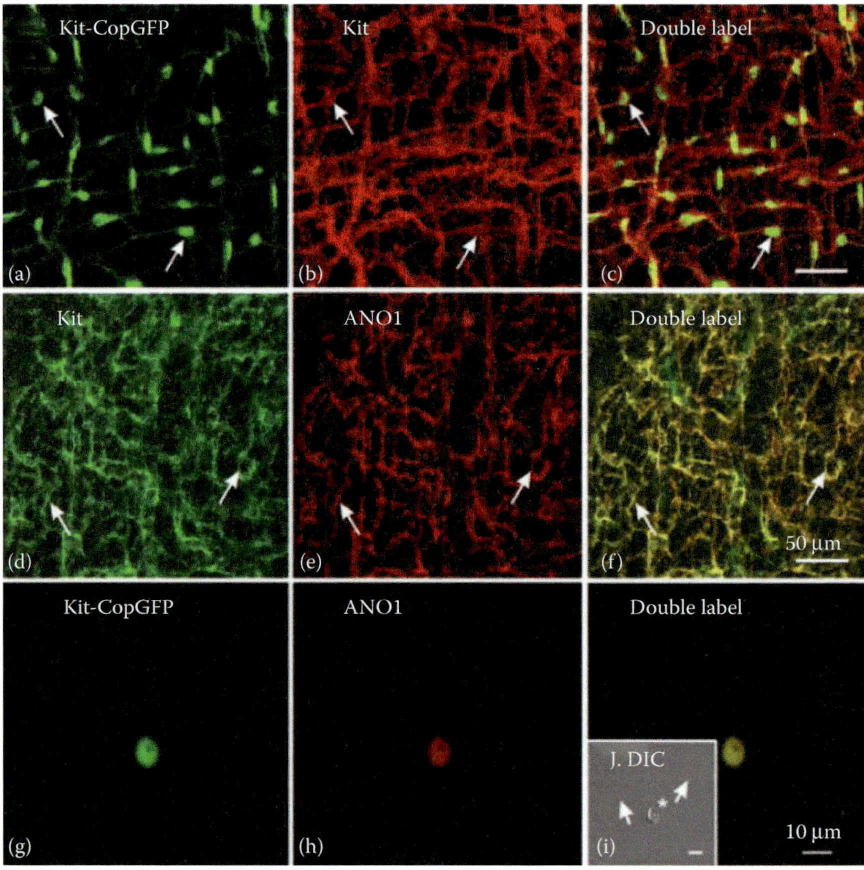

FIGURE 16.3 (a) ICC-MY cells in the mouse small intestine are c-Kit(+) as revealed by GFP expression in $Kit^{+/copGFP}$ mice. (b) Fixed tissue shows ICC staining with c-Kit antibody and (c) overlap of the two signals. (d–f) c-Kit and ANO1 labeling of ICCs in the small intestine. (g–i) and from isolated single cells after enzymatic dispersion. (From Zhu, M.H. et al., *J. Physiol.*, 587, 4905–4918, 2009.)

c-Kit blocking antibody also leads to the abolition of slow waves [32,33]. c-Kit and TMEM16A colocalize substantially in ICC-MY cells [34], consistent with expression of TMEM16A in those cells (Figure 16.3). Mice with global knockout of *Ano1*, the gene that encodes the calcium activated chloride channel (CaCC) TMEM16A, fail to generate slow waves and spontaneous contractions in both the gastric antrum and small intestine [35] (Figure 16.4), strongly supporting the idea that TMEM16A is a critical component of the pacemaker in c-Kit(+) ICCs. Most $Ano1^{-/-}$ mice die shortly after birth but a small percentage survive for 2–3 weeks, allowing limited electrophysiological studies of the GI tissues from those animals.

Recent evidence indicates that the rhythmic, electrical excitation of SM contractions in other hollow organs such as the bladder [36], urethra [37], ureter [38], gallbladder [39], and oviduct [40] are also mediated and/or coordinated by interstitial

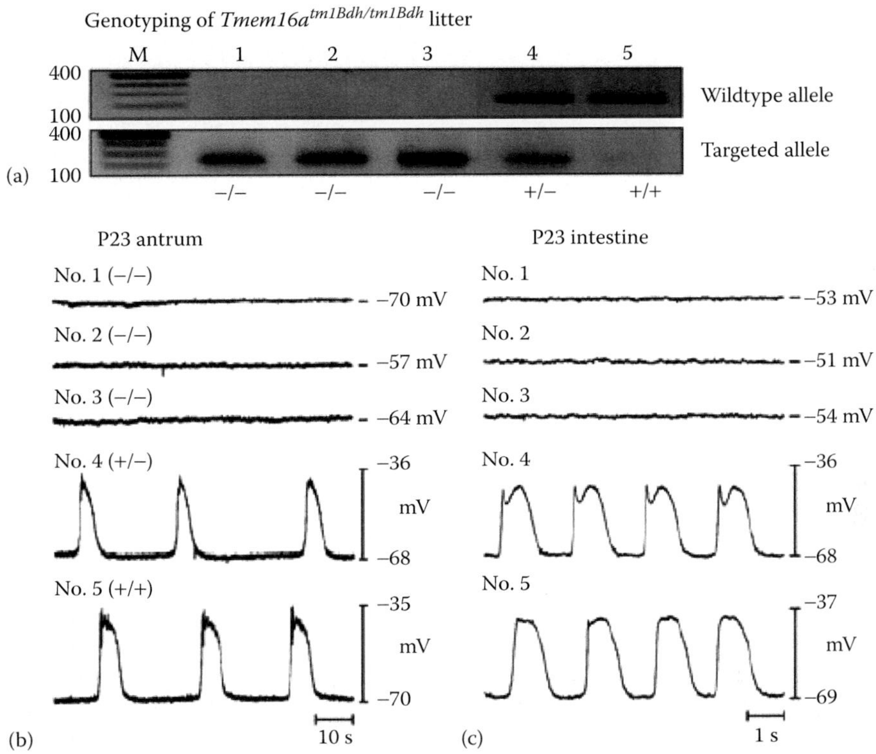

FIGURE 16.4 (a) Genotyping of TMEM16A (ANO1) global KO mice and WT and heterozygous littermates, showing absence of TMEM16A message from the former. (b and c) Slow waves in antrum and jejunal muscle are absent in TMEM16A KO mice but not WT and heterozygous littermates. (From Hwang, S.J. et al., *J. Physiol.*, 587, 4887–4904, 2009.)

cells. These cells are often referred to as *ICC like cells* because some, but not all, are c-Kit(+) [41]. For example, the pacemaking cells in bladder SM may be PDGFRα(+) but not c-Kit(+), and in that tissue TRP and/or TREK channels rather than TMEM16A may be critical in generating or modulating pacemaking current [42,43]. In renal pelvis, which contains a layer of SM cells that develop spontaneous contractions, pacemaking appears to be initiated not by c-Kit(+) cells, but by a subpopulation of SM cells with atypical morphology [44,45] that may be PDGFRα(+) and may receive input from other cell types [46]). Similar mechanisms may be operating in ICC-like cells in other urogenital organs [41]. In veins and terminal arteries supplying the submucosal plexus of the GI tract, pericytes have been implicated in pacemaking [47,48]. The longitudinally-oriented pericytes along terminal arterioles/capillaries generate Ca^{2+} transients that are sometimes synchronized with, but precede, Ca^{2+} transients in circumferentially-oriented SM cells along parent arterioles. The close correspondence between the global Ca^{2+} signals in the two cell types, with a short time delay, presumably reflects the conduction of an underlying electrical signal.

16.3.3 ICC-Like Cells in Lymphatic Vessels

Due to an emerging appreciation of the roles for interstitial cells as pacemakers in the aforementioned tissues, some groups have hypothesized that ICC-like cells in the lymphatic wall control lymphatic pacemaking. A teleological need for ICCs could be envisioned in the GI tract to allow coordination of pacemaker activity over large distances and between the two multi-cellular SM layers. In contrast, they may not be required in a lymphatic vessel with only a single layer of SM cells (as in the rat and mouse). In rat and mouse vessels electrical activity needs to be entrained only over a distance of ~1 mm, the length of a single lymphangion. However, teleological arguments can be misleading. McCloskey, McHale et al. [49] reported that sheep mesenteric lymphatics contained a layer of electron-dense cells immediately outside of the endothelium that shared several anatomical characteristics with ICCs from the GI tract. The cells stained positive for c-Kit and the intermediate filament vimentin (Figure 16.5). Ultrastructurally, the cells in this layer contained both thin and intermediate filaments, but lacked thick filaments, in contrast to the multiple layers of SM cells that contained thin and thick filaments [49,50]. Enzymatic dispersion produced a subpopulation of highly-branched, vimentin(+) cells distinct from the typical spindle-shaped SM cells, consistent with ICC-like cells identified in other tissues [39,51]. Interestingly, in one of the four published patch clamp papers using dispersed lymphatic SM cells, McHale's group found that only a subpopulation of cells (1%) exhibited spontaneous APs [52]; whether that low percentage reflected their identity as ICC-like cells or simply cells that were more resistant to enzymatic damage was not established. Boedtkjer et al. [50] identified cells with similar morphological characteristics in human thoracic duct, that is, cells positive for c-Kit and the hematopoietic progenitor cell antigen CD34, and which were stained by methylene blue (another putative ICC marker) (Figure 16.6). Although the methylene blue (+) cells were stellate-shaped with fine processes extending between SM cells, the authors were unable to find evidence that they formed gap junctions with the SM cells, which presumably would be required for electrical coupling, but instead

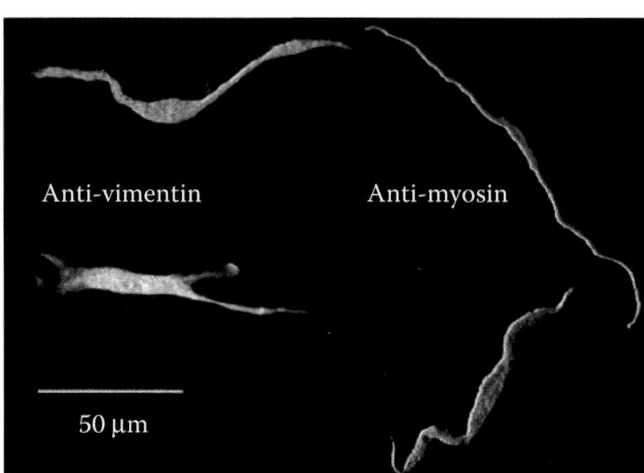

FIGURE 16.5 Vimentin(+) cells isolated by enzymatic dispersion from sheep mesenteric lymphatics; these cells are more branched than myosin (+) SM cells from the same vessels. (From McCloskey, K.D. et al., *Cell Tissue Res.*, 310, 77–84, 2002.)

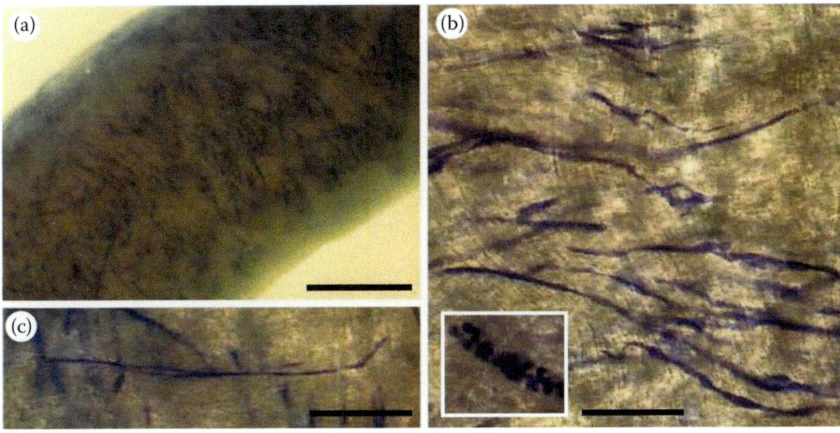

FIGURE 16.6 Elongated, methylene blue (+) cells in the wall of human thoracic duct. (a)–(c) Low doses of wortmannin blunt contractions without substantially altering the electrical properties of lymphatic SM. (From Briggs Boedtkjer, D. et al., *Cells Tissues Organs*, 197, 145–158, 2013.)

made connections to SM cells with peg-and-socket junctions [50]. The function of that cell population has not yet been established.

In summary, although ICC-like cells have been proposed to drive lymphatic pacemaking, the supporting evidence is based exclusively on morphology and staining for known ICC markers, for which the fidelity may be tissue-specific. Many types of immune cells as well as their progenitor cells, some of which are also c-Kit(+) [53,54], may reside on and/or within the lymphatic wall; those cells may play roles in immune responses or be necessary for reparative responses to injury. Indeed, comparisons of the functional characteristics of ICCs (ICC-MY) with those of the lymphatic pacemaker suggest that the two are quite different, as summarized in Table 16.1. For example, ICCs in the GI tract have stable resting membrane potentials around −60 mV with pacemaking potentials that are insensitive to DHPs, whereas lymphatic SM cells exhibit a repeating diastolic depolarization (DD, Section 16.5) between −45 and −40 mV, triggering APs that are completely abolished by DHPs. However, differences between lymphatic vessels from various regions may exist, with large vessels—those containing multiple SM cell layers—perhaps needing an ICC-like cell network to coordinate electrical activity within a lymphangion. Definitive identification of ICC-like cells in lymphatic vessels and tests of their critical role in pacemaking will likely require the development and use of genetic methods (reporters, knock-outs), for example, in the mouse, analogous to the GI field where such models have revolutionized thinking about the mechanisms of pacemaking.

In the absence of functional evidence for ICC-like cells acting as pacemakers in the lymphatic wall, our assumption is that the SM cells themselves are the pacemakers. Imtiaz et al. [55,56] argue that the fundamental concepts with respect to the pacemaking mechanism in lymphatic vessels are independent of the type of pacemaker cell, whether it is an ICC or a lymphatic SM cell. However, the eventual development of therapies to target a dysfunctional lymphatic pacemaker selectively will require identification of the pacemaking cell type in order to avoid non-specific effects on other cells/tissues.

TABLE 16.1
Comparison of Pacemaking Characteristics in the GI Tract and Lymphatic Smooth Muscle

Characteristics of Pacemaking Events/Cells	GI Tract	References	Lymphatic SM	References
Duration of pacemaking event	2–10 s	[22]	~1–1.5 s	[57], Figures 16.9 and 16.12
Resting Vm of pacemaker cell	−60 mV	[58]	−40 to −45 mV	[57], Figures 16.9, 16.11, and 16.12
Stability of resting Vm	Stable		DD	Figures 16.9 and 16.16
Effect of DHPs on PP	Insensitive	[21]	Blocked at <100 nM	[57], Figures 16.11 and 16.12
Contractility of pacemaker cell	Non-contractile	[22]	Contractile	Figures 16.9 and 16.12
Expression of TMEM16A by pacemaker cell	In ICCs only	[35]	In SMCs, mast cells	[59]
Consequence of ANO1/TMEM16A deletion	No slow wave	[35]	Contractions slower but persist	[59]
Cells expressing GFP in $Kit^{+/copGFP}$ mice	ICCs only	[34]	Adventitial mast cells	[59]
Consequence of c-Kit deletion/mutation	No slow wave	[60]	Spontaneous APs persist	*

* Unpublished observations made in our lab.

16.4 THE IONIC BASIS FOR ELECTRICAL PACEMAKING IN LYMPHATIC VESSELS

16.4.1 The Proposed Ionic Mechanism for Pacemaking in Lymphatic Smooth Muscle

The proposed ionic mechanism for pacemaking in lymphatic SM was derived primarily from data recorded from guinea pig and sheep mesenteric lymphatic vessels [55,56] and shares several characteristics with the ionic mechanisms in GI pacemaking cells. According to the current scheme for lymphatic SM shown in Figure 16.7, agonists (and perhaps also stretch/pressure) activate PLC, producing IP_3, which acts on IP_3Rs to release Ca^{2+} from the sarcoplasmic reticulum. That localized Ca^{2+} increase (a Ca^{2+} puff) and/or its propagated form within the cell (a Ca^{2+} wave), activates Cl^- channels to elicit depolarization. The discrete depolarization event is 1–2 mV in amplitude and is termed a spontaneous transient depolarization (STD). STDs are proposed to summate to produce larger (5–20 mV) PPs that bring the membrane potential from a resting value of −50 mV (guinea pig), −45 mV (rat, human), or −35 (mouse) to the threshold for activation of voltage-gated cation channels. Voltage-gated Na^+ channels (Na_V1) and T-type

Agonists/stretch
↓
PLC
↓
IP_3
↓
Localized Ca^{2+} puff
↓
ICl_{Ca} activation
↓
STD
↓
Summation to PP
↓
Vm crosses threshold
↓
Ca_V3/Na_V1-initiated upstroke?
↓
$Ca_V1.2$-mediated AP
↓
Ca^{2+} entry and global Ca^{2+} flash
↓
Contraction

FIGURE 16.7 Existing model of the sequence of events proposed to control lymphatic pacemaking. Ca_V3 = T-type VGCCs, $Ca_V1.2$ = L-type VGCCs, Na_V1 = VGSCs. For details see text. (Based on Hollywood, M.A. et al., *J. Physiol.*, 503, 13–21, 1997; Hwang, S.J. et al., *Br. J. Pharmacol.*, 173, 1339–1349, 2016; Imtiaz, M.S., *Biophysical J.*, 92, 3843–3861, 2007; Imtiaz, M.S. et al., *Febs. J.*, 277, 278–285, 2010; Kamishima, T. and Quayle, J.M., *Biochem. Soc. Trans.*, 31, 943–946, 2003; von der Weid, P.Y. et al., *Am. J. Physiol. Heart Circ. Physiol.*, 295, H1989-2000, 2008.)

voltage-gated Ca^{2+} channels (Ca_V3), if expressed, should be activated first, enhancing and accelerating depolarization, and triggering the opening of L-type Ca^{2+} channels ($Ca_V1.2$). It is the synchronized activation of $Ca_V1.2$ channels that generates an AP that peaks around 0 mV, lasts for 1–1.5 s, produces a global Ca^{2+} rise (Ca^{2+} flash [61]) in the lymphatic SM cell, and triggers a robust, rapid contraction. The AP is conducted to neighboring LSM cells via gap junctions to produce the nearly-synchronous contraction of a lymphangion. This scheme has been incorporated into a numerical model that elegantly reproduces much of the behavior observed with respect to initiation of APs in lymphatic muscle [55,56]. The evidence for each of these steps (the Ca^{2+} puff-CaCC-STD-AP hypothesis) will now be examined in detail, although not in sequential order. We begin with the elementary electrical event in lymphatic SM: the STD.

16.4.2 STDs Result from ICl_{Ca} Activation

The STD is thought to reflect the activation of a Cl^- channel to produce an inward, depolarizing Cl^- current (i.e. efflux Cl^- movement down the prevailing Cl^- electrochemical gradient). In ICCs and also many SMs, net Cl^- current is inward due to a high intracellular Cl^- concentration accomplished largely by the activity of the

Na^+-K^+-$2Cl^-$ co-transporter NKCC1 [65]. The sodium gradient established by the rapid removal of Na^+ ions through the sodium ATPase pump drives the NKCC1 activity and the accumulation of Cl^-, which results in a more positive Cl^- equilibrium potential. Pharmaceutical inhibition or genetic deletion of NKCC1, or reduction of intracellular Cl^- by dialysis with low Cl^- solutions, reduces the pacemaking frequency and the amplitude of unitary potentials in ICCs [66,67]. Candidate Cl^- channels include TMEM16 family members and cGMP-activated Cl^- channels. As mentioned previously, TMEM16A is the key Cl^- channel underlying pacemaking in ICC-MY [34,35]. Mouse and rat lymphatic SM cells express TMEM16A [59] as does arterial SM [68,69], and the latter also express a functional bestrophin/cGMP-activated Cl^- channel [68,70]. STDs could also be explained by the activation of cation channels, allowing inward movement of Na^+ and/or Ca^{2+}. Candidate cation channels include TRPC3 and TRPC6, both of which are Na^+ and Ca^{2+} permeable [71], and TRPM4, which is Na^+ permeable and activated by intracellular Ca^{2+} [72]. Alternatively, multiple TRPs activated in tandem, or combinations of different TRP isoforms assembled as heterotetramers, potentially could mediate the STD, similar to the depolarization mechanism proposed to underlie the myogenic response in arterioles [72,73].

A major problem with distinguishing between the roles of Cl^- and cation channels in lymphatic pacemaking is the lack of selective inhibitors for either class of channel. Niflumic acid (NFA) has been widely used to block Cl^- channels in this [57] and other contexts [34], but the concentrations often used (≥50 μM) also block cation channels as well as voltage-gated calcium channel (VGCCs) [74,75]. Thus, the observation that niflumic acid (NFA) prevents AP generation by lymphatic SM (Figure 16.8)—a key piece of evidence supporting a role for ICl_{Ca} activation—could be explained by inhibition of TRP channels or $Ca_V1.2$ VGCCs, rather than blockade of ICl_{Ca}-mediated STDs. Likewise, DIDS, NPPB and 9-AC, and other commonly-used Cl^- channel blockers that have been shown to block STDs [57], are effective only at relatively high concentrations (300 μM, 50 μM, 1 mM, respectively) [57]—concentrations also known to inhibit cation channels and/or VGCCs [52,57,74,75]. An example of this off-target effect is illustrated by Vm recordings from a pressurized rat mesenteric lymphatic vessel (Figure 16.9a). The application of NFA (50 μM, the same dose used in Figure 16.8) inhibits spontaneous AP firing and contractions and

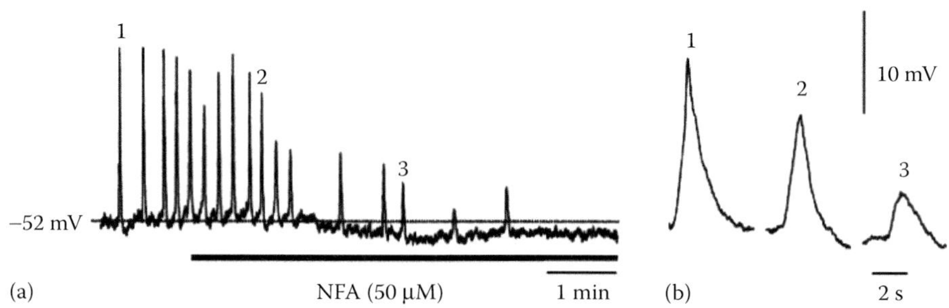

FIGURE 16.8 (a) Inhibition of APs in lymphatic SM by NFA and (b) expanded traces show three types of events, corresponding to 1 = AP, 2 = PP, 3 = STD. (From von der Weid, P.Y. et al., *Am. J. Physiol. Heart Circ. Physiol.*, 295, H1989–H2000, 2008.)

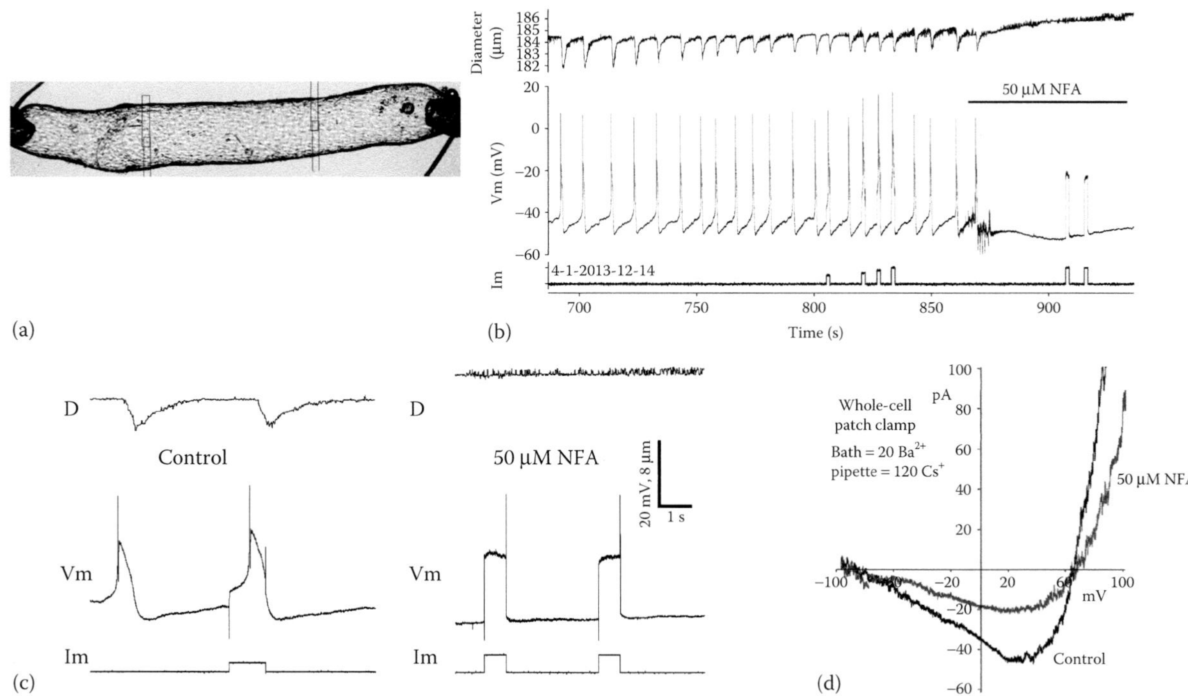

FIGURE 16.9 (a) A pressurized rat mesenteric lymphatic vessel, with two diameter tracking windows, just before impalement of a SM cell with a sharp microelectrode. (b) The application of NFA (same dose as in Figure 16.8) stops spontaneous APs and contractions and transiently hyperpolarizes; however, depolarizing current pulses show that APs cannot be evoked in the presence of NFA, suggesting that VGCCs are inhibited. Capacitance artifacts associated with the current pulses have been removed for clarity. (c) Expanded time scale showing: a spontaneous AP and an evoked AP (by a depolarizing pulse) prior to NFA (D = diameter), but failure of even larger depolarizing pulses to evoke APs when NFA is present. (d) NFA inhibits L-type current in isolated LSM cells.

leads to dilation (Figure 16.9b). The dilation itself is suggestive of VGCC inhibition and is also supported by the inability to electrically evoke APs: depolarizing currents injected into the cell before NFA is applied trigger premature APs (Figure 16.9c) but fail to do so in the presence of NFA (Figure 16.9d). If NFA inhibited only Cl^- channels, APs should still be evoked by sufficiently strong depolarizing stimuli. Furthermore, Figure 16.9e shows that 50 μM NFA inhibits >50% of VGCC current in an isolated, patch-clamped rat mesenteric lymphatic cell. We find that other reported inhibitors of CACCs block VGCCs in LSM at equal or lower concentrations than are effective for blocking CACCs: DIDS and 9-AC block VGCCs at 100 μM (not shown) whereas they do not block STDs until the concentrations exceed 100 and 250 μM, respectively [52,57]. Even more recently developed inhibitors of CACCs still suffer from critical selectivity issues. The purportedly specific TMEM16A inhibitor, TMEM16A-A01, has been shown to block VGCCs and may confound the effects attributed to TMEM16A on blood vascular tone [76]. TMEM16A-A01 along with other commonly used TMEM16A inhibitors, CACC-A01 and MONNA [63], all demonstrate vasoactive properties in vascular preparations with $[Cl^-]$ reduced to minimize the contribution of Cl^- channels [76]. A more definitive way to delineate the individual roles of these channels in LSM would be genetic deletion of the specific Cl^- or cation channel(s) in the mouse, but to date no published studies have used this approach to assess their potential roles in lymphatic contractile function.

As mentioned in the previous section, the precedent for involvement of a Cl^- channel in lymphatic pacemaking originates in the GI literature. Pacemaker potentials underlying slow waves in the gastric antrum and the small intestine consist of two components: a fast, primary one and a slower, plateau component [22,28]. The primary component depends on Ca^{2+} influx and is thought to involve activation of T-type VGCCs [28], which is a mechanism still under debate. The plateau component of the slow wave is composed of elementary currents, called unitary potentials, that first were apparent from noise analysis of sharp electrode recordings of Vm [77]. Evidence that unitary potentials were generated by Cl^- efflux included the following: (1) an inhibition of the plateau component by Cl^- channel blockers (albeit non-selective blockers); (2) an increase in the plateau level when external Cl^- was reduced (until prolonged reduction depleted internal Cl^-) [28]; and (3) subsequent demonstration of Cl^--selective current in ICCs freshly isolated from the mouse small intestine [78], which was later identified as TMEM16A (Figure 16.2b).

van Helden found evidence for a Cl^- channel with similar characteristics underlying pacemaking in guinea pig mesenteric veins [79] and mesenteric lymphatics [80]. As shown in Figure 16.10a and b, STDs were often observed at the feet of APs, suggesting that their summation might trigger the rapid upstroke phase of the AP. STDs were inhibited by classic Cl^- channel blockers and both STDs and lymphatic contractions were suppressed after prolonged exposure to low $[Cl^-]$ solutions (>5 min) [80]). A voltage clamp study in sheep mesenteric lymphatic SM subsequently found evidence for a Cl^- current in enzymatically dispersed sheep lymphatic SM cells [52]): a whole-cell Cl^- current was activated by both Ca^{2+} influx and release and inhibited by the Cl^- channel blocker 9-AC, with a reversal potential that shifted appropriately with changes in the Cl^- equilibrium potential. In current clamp mode, dispersed lymphatic SM cells exhibited STDs that were a few mV in amplitude and some

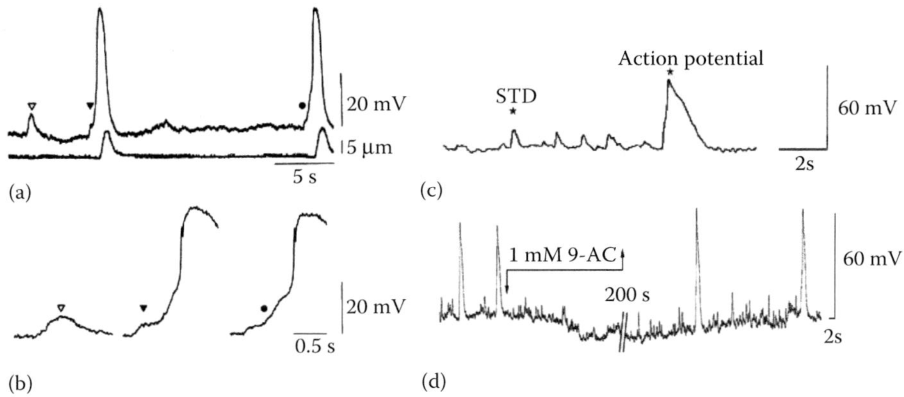

FIGURE 16.10 (a and b) STDs (arrowheads) sometimes can be observed at the foot of APs (circles) in guinea pig lymphatic SM. (From van Helden, D.F., *J. Physiol.*, 471, 465–479, 1993.) (c) STDs and spontaneous APs in isolated sheep lymphatic SM held under current clamp. (d) APs are reversibly inhibited by the Cl^- channel blocker 9-AC; recording is a combination of 6 stacked traces. (From Toland, H.M. et al., *Am. J. Physiol. Heart Circ. Physiol.*, 279, C1327–C1335, 2000.)

cells showed spontaneous APs (Figure 16.10c). The spontaneous APs were reversibly inhibited by 9-AC (Figure 16.10d), but whether 1 mM 9-AC blocks VGCCs was not determined. The molecular identity of the Cl^- channel remains unknown but will likely be needed in order to selectively target that channel.

16.4.3 Summation of STDs to PPs

The single-channel amplitude of TMEM16A (a homo-dimer) is 8pS, consistent with that predicted from noise analysis of unitary potentials [77], and is too low for a single opening to generate a 1–2 mV STD. Thus, the amplitude of an STD likely reflects the activation of multiple Cl^- channels, perhaps in a cluster [81,82], rather than a single channel. Regardless of the number of channels activated, 1–2 mV STDs are thought to summate into PPs (5–15 mV) if STDs occur near-simultaneously in several (perhaps adjacent) SM cells. To the extent that it occurs, synchronization of STDs between cells may depend on Ca^{2+} influx through L-type VGCCs [56]. When summation of STDs is sufficient to bring the membrane potential to the threshold for recruitment of VGCCs, an AP fires [55]. Recordings of APs, PPS, and STDs in strips of guinea pig lymphatic muscle are shown in (Figure 16.11).

One line of supporting evidence for STD summation comes from simultaneous Vm recordings and confocal Ca^{2+} imaging in guinea pig lymphatic muscle, as shown in Figure 16.12. Here, the top trace is the Vm recording and each of the other traces is the Fluo-4 signal from an ROI in a different cell (see map of sites on the left). Each Ca^{2+} event is associated with an STD of variable amplitude. Ca^{2+} events that are not synchronized between cells (left half of traces) are associated with STDs <2 mV in

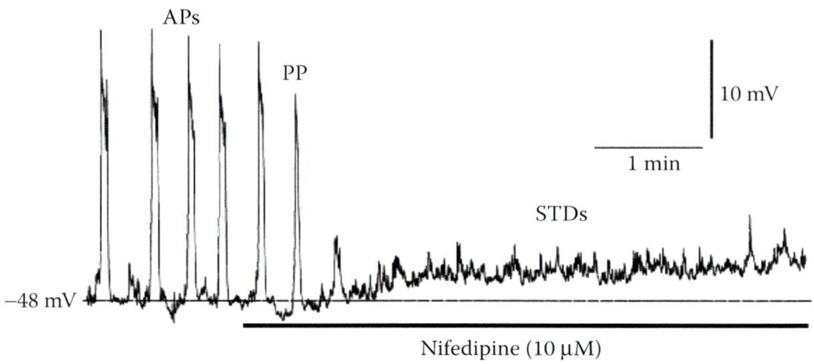

FIGURE 16.11 Vm traces, showing that addition of nifedipine leads to inhibition of APs, apparent unmasking of PPs with eventually only STDs remaining. (Reproduced from von der Weid, P.Y. et al., *Am. J. Physiol. Heart Circ. Physiol.*, 295, H1989–H2000, 2008.)

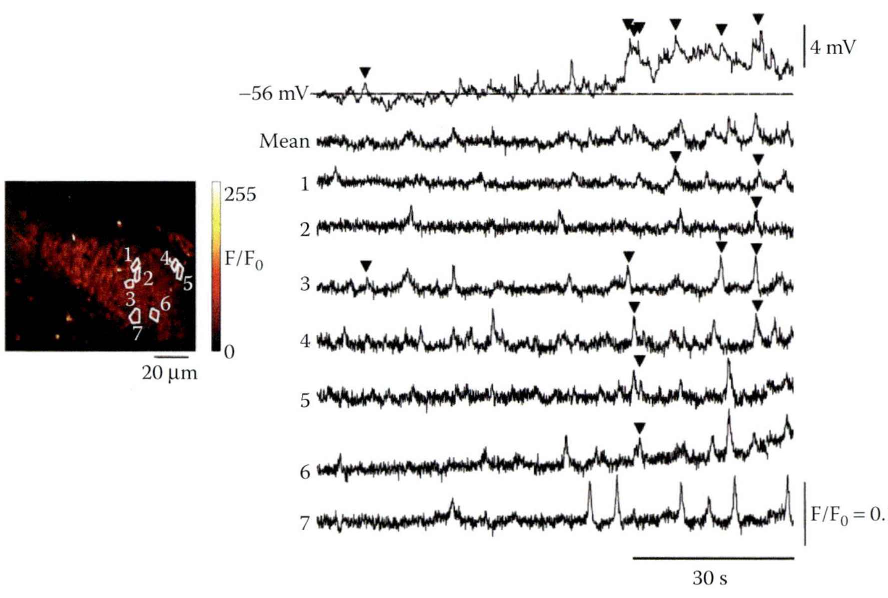

FIGURE 16.12 STD summation correlates with entrained Ca^{2+} transients in multiple lymphatic SM cells. (From von der Weid, P.Y. et al., *Am. J. Physiol. Heart Circ. Physiol.*, 295, H1989–H2000, 2008.) See text for details.

amplitude, in contrast to STDs that are 3–4 mV in amplitude when Ca^{2+} events in several cells coincide (aligned arrowheads, right half of traces). (Note: The Vm and Ca^{2+} measurements were not made in the same cell. The only simultaneous recordings of Vm and confocal Ca^{2+} changes [only confocal imaging provides sufficiently detailed information about subcellular Ca^{2+} changes] are studies in which Vm was recorded on the upper surface and Ca^{2+} on the lower surface of a sheet of lymphatic muscle [55–57]).

These observations strongly suggest that STDs can summate at least to some degree, but can they summate to produce the types of events (PPs) shown in Figure 16.11? It is curious that PPs can seemingly be demonstrated only immediately after nifedipine or NFA application (Figures 16.8, 16.11, and 16.4 in [56]), after which they disappear. The discrimination of APs from PPs on the basis of their amplitudes and presence/absence of a spike component is often compounded by low-resolution recordings [57,79,83] made with only 100 Hz sampling and 50 Hz filtering (APs in Figures 16.8 and 16.10a–b). An alternative explanation is that the so-called *pacemaker potential* is a transitional event recorded during the gradual addition of VGCC inhibitor to the bath (Figures 16.11 and 16.13a) or during recovery from washout of the same, and actually reflects a blunted AP in which a substantial fraction of VGCCs are blocked. To achieve higher resolution, the recordings in Figure 16.13 were made in pressurized rat mesenteric lymphatic SM cells at 10 KHz with 1 KHz filtering. Note that normal APs (in the absence of inhibitors) always have a clear spike component and plateau phase (compare the first AP in Figure 16.13c to those in Figures 16.8 and 16.11), but the gradual addition of a VGCC blocker (mibefradil, nifedipine or NNC 55-0396) progressively converts these to 20 mV oscillations (note also the gradual decrease in upstroke velocity in Figure 16.13c). A so-called pacemaker potential that is actually a blunted AP would explain why: (1) PPs that are sufficiently large to cross the threshold often do not trigger APs (Figure 16.11); (2) PPs are seldom (or never) recorded in the absence of VGCC inhibitors (Figure 16.11); and (3) PPs eventually disappear when fully effective concentrations of DHPs are reached (Figures 16.11 and 16.13) [56]. In spite of the higher-resolution recording shown in Figure 16.13b, no STDs were observed during the oscillations, suggesting

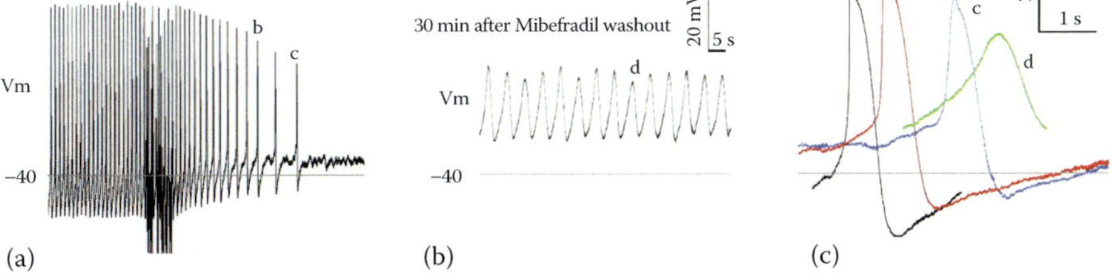

FIGURE 16.13 (a) Mibefradil (MIB) inhibits APs and contractions, dilates, and induces Vm oscillations resembling PPs in pressurized rat mesenteric lymphatic SM. During washout (b), regular oscillations ~15 mV in amplitude are observed. (c) Comparison of single events labeled in (a) and (b) reveal that all events are APs, with progressive blunting of the spike component and lowering of upstroke velocity. Although mibefradil is reputed to be a T-type VGCC inhibitor, we find that it inhibits L-type VGCCs even at low concentrations and that the same effects are observed with the L-type VGCC inhibitor nifedipine and another reputed T-type calcium channel inhibitor, NNC-55-0396. D = inner diameter.

that the oscillations are VGCC-mediated events rather than the summation of STDs. This would also explain why the intracellular Ca^{2+} events associated with apparent PPs are synchronized between cells (Figure 16.2 in [56]), as they always are in a normal AP.

16.4.4 Ca^{2+} Release Events Underlie STDs

In lymphatic SM, Ca^{2+} release from intracellular stores is hypothesized to be the trigger that activates Cl^- channels. Supporting evidence includes the observations that STDs are: (1) not blocked in Ca^{2+}-free solution (until the stores completely run down) but are blocked by (2) BAPTA to chelate Ca^{2+} and thereby deplete Ca^{2+} stores, or (3) cyclopiazonic acid (CPA) or thapsigargin (TG) treatment to inhibit Ca^{2+} reuptake by SERCA and thereby deplete Ca^{2+} stores [79,83]. Store release of Ca^{2+} has now emerged as a common mechanism in the initiation of pacemaker activity in a number of different types of SM [55].

16.4.5 IP_3R-Mediated Ca^{2+} Release

In general, Ca^{2+} release events are considered to be either IP_3R-mediated Ca^{2+} puffs or ryanodine receptor (RYR)-mediated Ca^{2+} sparks [84]. Both events can result in sensitization of the respective sarcoplasmic reticulum receptors to further enhance Ca^{2+} release and result in a Ca^{2+} wave that can travel throughout the cell [85]. Many of these concepts are based on the original experiments and models developed by Berridge and colleagues [86,87].

The triggering event for store release of Ca^{2+} and STD production in lymphatic SM is thought to be exclusively via IP_3 generation (Figure 16.7), that is, Ca^{2+} puffs. Supporting evidence includes the following observations. First, the slower kinetics of the Ca^{2+} signaling events are characteristic of Ca^{2+} puffs (as opposed to Ca^{2+} sparks). Second, 2-APB and xestospongin-C, purported IP_3R blockers, reduce STD amplitude. Third, thimerosal, which increases the sensitivity of IP_3R for IP_3 and Bt3 [42,74,88] IP_3/AM, a membrane-permeant analog of IP_3, increases STD amplitude [57]. Although Ca^{2+} puffs and waves are localized events that do not require VGCC-mediated Ca^{2+} influx, their synchronization between cells seems to require some degree of VGCC activation [55,56]. In strips of guinea pig lymphatic SM loaded with Fluo-4, only a few spontaneous intracellular Ca^{2+} events are recorded until the IP_3-generating agonist ET-1 is applied (Figure 16.14a–b); in the presence of ET-1 the events increase in frequency and begin to synchronize between some cells (* in Figure 16.14c) until eventually they are masked by Ca^{2+} flashes occurring at rhythmic intervals in all the cells within the field of view (Figure 16.14b and d).

The existing evidence suggests little or no role for Ca^{2+} sparks in lymphatic contractions, in contrast to their roles in vascular SM and cardiac myocytes [64,89]. In arterial SM, Ca^{2+} sparks are tightly coupled to the regulation of large-conductance, Ca^{2+}-activated K^+ channels whose activation produces vasodilation

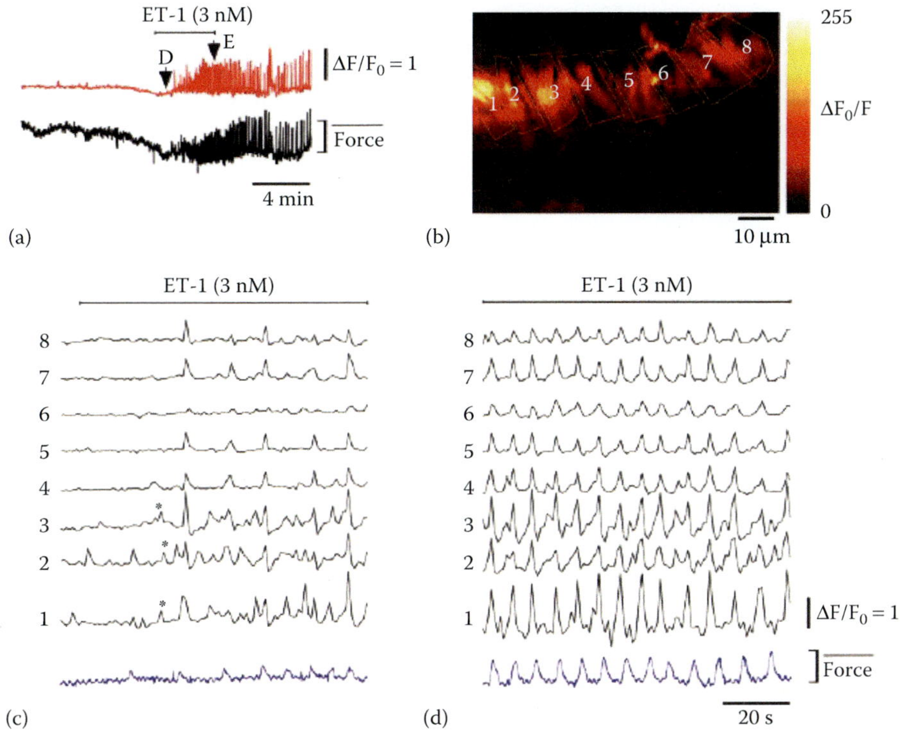

FIGURE 16.14 Synchronous Ca^{2+} events can occur in the absence of APs (panel (c)*) but they do not cause constriction (note lack of corresponding deflection in bottom force trace). Wire myograph-mounted guinea pig mesenteric lymphatic. (From Imtiaz, M.S. et al., *Biophys. J.*, 92, 3843–3861, 2007.)

[90]. In cardiac myocytes, Ca^{2+} sparks can trigger APs under some conditions [64,89]. Zhao et al. found no significant effects of ryanodine (1 or 20 μM) on the rhythmic contractions induced by ET-1 in guinea pig lymphatics, although they did not directly measure Ca^{2+} transients or STDs [91]. Likewise, von der Weid et al. [57] confirmed a lack of effect of ryanodine (30 μM) on STDs using a similar preparation and also found that tetracaine, which blocks RYRs, was ineffective at inhibiting STDs. The lack of action of these agents was in contrast to those of 2-aminoethoxydiphenyl borate (2-APB) and xestospongin-C [57] (commonly used to inhibit IP_3R), which reduced or blocked STD activity in guinea pig lymphatics. However, the off-target effects of 2-APB are widely documented [92,93] and accumulating evidence suggests that xestospongin-C may act on Ca^{2+} stores rather than IP_3Rs [94,95]. Again, genetic deletion or siRNA knock-down approaches could provide more definitive answers, however, neither approach has been used to date to address this issue. For example, there have been no tests of lymphatic contractile function or pacemaking in IP_3R1 KO mice in which slow waves in the circular muscle of gastric antrum are absent [96].

16.4.6 PLC-Mediated IP_3 Production

Inositol 1,4,5 trisphosphate (IP_3) and diacylglycerol (DAG) are generated as products of phospholipase C (PLC) activation [97] (Figure 16.7). Agonists such as ET-1 and the thromboxane analog U46619, which are known to stimulate PLC and enhance synthesis of IP_3, increase the frequency and number of STDs in guinea pig lymphatic SM; further, these effects persist long after removal of the agonist, which is consistent with activation of an enzyme. Zhao and van Helden found that the PLC inhibitor U73122 (5 μM) blocked the increases in both the number and frequency of lymphatic contractions stimulated by ET-1 [91]. An important and related question is whether STDs per se are blocked by PLC inhibition, which would be predicted from the scheme shown in Figure 16.7; however, to our knowledge this idea has not been tested in any of the published studies focusing on Ca^{2+} events and STDs [55,57].

16.4.7 Agonist versus Stretch-Activation of PLC

One implication of the hypothesis stated in Figure 16.7 is that increased stretch, similar to the effects of agonists that activate PLC, generates IP_3, leading to an increase in STD activity and AP firing. Indeed, PLC is an enzyme linked to mechanotransduction pathways [98] and pressurization of small arteries is associated with increased IP_3 generation by arterial SM [99–101] (because circumferential stretch of the muscle layer produces longitudinal stretch of individual SM cells). Although similar biochemical data are lacking for lymphatic SM, one piece of evidence supporting a role for stretch-activation of PLC in lymphatic SM is the observation that U73122 (5 μM) reduces the spontaneous contractions of guinea pig mesenteric lymphatics by 75% [91]). This finding suggests that basal PLC activation (by whatever combination of stretch and/or endogenous agonists) results in IP_3 levels that produce some baseline degree of spontaneous contractile activity.

The paucity of additional supporting evidence for a stretch-induced increase in STD activity may be a consequence of the use of unpressurized/unstretched preparations of lymphatic SM for electrophysiology studies. In this context it is important to note that the existing model for lymphatic pacemaking (Figure 16.7) is derived almost exclusively from electrophysiological data obtained from such preparations (mostly strips of guinea pig lymphatic SM), which do not allow for testing the effects of increasing stretch. To date, there is only a single published study regarding the effect of stretch per se on STDs and AP generation. In that study, von der Weid et al. measured the Vm of lymphatic SM in rat mesenteric lymphatic vessels mounted in a wire myograph [102]. Vm was more depolarized than that of unstretched guinea pig lymphatic SM, in part because of the removal, or partial removal, of the tonic hyperpolarizing influence of the endothelium (nitric oxide and/or prostanoids), resulting from the passage of the wires through the lumen of the vessel. Increasing stretch from the minimal value at which the vessel developed spontaneous force transients resulted in an increase in STD frequency, but not amplitude, and a very slight membrane hyperpolarization. The hyperpolarization is surprising not only because it is at odds with the scheme in Figure 16.7 but also because arteriolar SM depolarizes in response to stretch [103,104]. Another surprising finding was that NFA did not alter AP frequency in wire-myograph mounted vessels [102], as it does in unstretched guinea-pig lymphatic vessels [57], pointing to possible differences in the underlying ionic mechanisms leading to AP generation in the two preparations. The differential effects of inhibitors on stretched (wire-mounted) and unstretched (strips) of lymphatic SM [105] suggest the influence of a stretch-activated current whose activity can dramatically influence the interplay of other ion channels, which is a topic addressed in Section 16.5. In this context a limitation of the wire myograph method is that vessel segments must be very short to facilitate maintenance of isopotentiality and, as a result, the axial length, which also an important determinant of basal pacemaking rate, cannot be controlled and in fact may even decrease as wire separation (and preload) increases.

16.4.8 Activation of VGCCs

According to the scheme in Figure 16.7, after summation of STDs to PPs, the membrane will be sufficiently depolarized to trigger the activation of voltage-gated cation channels. The voltage-gated channel with the most negative activation threshold is presumably a Na_V1 channel, as this family of channels is known to be important in other pacemaking cells, for example, neurons and cardiomyocytes [106]. Hollywood et al. first identified a rapidly-activated and inactivated, TTX-sensitive Na^+ ($Na_V1.5$) current in sheep lymphatic muscle cells studied using patch-clamp methods [61]. In isometric rings of sheep lymphatic muscle, TTX lowered the upstroke velocity of the AP, abolished spontaneous contractions in some vessels, and substantially reduced the frequency of spontaneous contractions in others [107]. We have observed similar effects of TTX on the frequency of contractions in pressurized lymphatics from both rat and mouse (unpublished observations). Telinius and colleagues identified message for $Na_V1.3$ in a majority of samples of human thoracic duct and mesenteric lymphatics (by PCR), and Na_V1 isoforms in some samples, with $Na_V1.7$ being the next most frequently expressed isoform [108]. Spontaneous contractions in 6 of 10 preparations were inhibited by TTX, which blocks $Na_V1.1$ 4 and $Na_V1.7$ 8, but not by ranolazine, which blocks $Na_V1.4,5,7,8$. Further, contraction frequency was remarkably increased by the Na_v activator veratridine [108]. Collectively, these findings suggest that multiple Na_V1 channels are expressed in lymphatic muscle and that $Na_V1.3$ activation facilitates AP generation; however, Na_V1 channels may not absolutely be required for AP firing because VGCCs can take over to varying degrees when they are blocked.

Like Na_V channels, Ca_V3 channels have the potential to play a facilitating role in AP initiation because their activation threshold is also more negative than that of L-type VGCCs. The best electrophysiological evidence for Ca_V3 current in lymphatic SM was presented in an abstract describing patch clamp recordings of a fast-inactivating, inward Ca^{2+} current in sheep mesenteric SM cells [109]; unfortunately, that work was never published in full. Evidence for a functional role of Ca_V3 channels in pacemaking was provided by Beckett et al. [107], who found that 100 μM Ni^{2+}, a commonly-used Ca_V3 inhibitor, reduced the pacemaking frequency of sheep mesenteric lymphatic SM. However, the effects of only a single dose of Ni^{2+} were reported—a dose that should have effectively blocked $Ca_V3.2$ channels but also may have partially blocked $Ca_V3.1$ channels and Ca_V1 channels (see Table 2 in [110]). In a subsequent study, Lee et al. [105] identified message for both $Ca_V3.1$ and $Ca_V3.2$ in guinea-pig mesenteric lymphatics and found that 100 μM Ni^{2+} or 100 nM mibefradil (another reputedly selective Ca_V3 blocker) inhibited pressure-induced changes in contraction frequency, but not contraction amplitude. Both inhibitors reduced AP firing frequency and produced a slight hyperpolarization of SM in (unpressurized) vessel strips; interestingly, no hyperpolarizations were recorded in wire-mounted vessels in response to Ni^{2+} or mibefradil [105]. The collective conclusion from these two studies is that Ca_V3 channels play a role in setting pacemaking frequency while Ca^{2+} influx through $Ca_V1.2$ channels determines contraction amplitude. A problem with this interpretation is, again, the non-specificity of the inhibitors: both Ni^{2+} and mibefradil, at the single concentrations used, are likely to have effects on $Ca_V1.2$ channels [110], which are well known to be expressed in lymphatic SM and play a critical role in AP generation (Figure 16.11). Thus, an alternative explanation is that partial inhibition of $Ca_V1.2$ by Ni^{2+} or nifedipine leads to changes in pacemaking frequency, whereas more complete inhibition leads to changes in contraction amplitude. To move forward, it will be important to assess the full concentration-response relationships on both frequency and amplitude for all of the inhibitors used to block these channels and to implement selective knock-down or knock out strategies.

16.4.9 Ca^{2+} Flashes and Conduction of APs

When the membrane potential of lymphatic SM, through whatever combination of ionic mechanisms, reaches the threshold for activation of $Ca_V1.2$ VGCCs, an AP is triggered. During the upstroke of the AP, a large, global increase in intracellular Ca^{2+} occurs, often termed a Ca^{2+} flash, followed by activation of the contractile machinery and shortening of the contractile apparatus. Full contractions are ones in which the amplitude is a substantial fraction of maximal diameter and is conducted among, and coordinated between, adjacent SM cells. In studies of rat mesenteric and mouse popliteal lymphatic SM, we find that Ca^{2+} puffs are not associated with even the slightest measureable local contraction, that Ca^{2+} waves produce small-amplitude (1–2 μm), localized vasomotion (Figure 16.14d), and that only Ca^{2+} flashes produce full, synchronized contractions that are conducted at a high rate (~10 mm/s, unpublished data). In pressurized vessels, Ca^{2+} flashes are conducted over the entire length of the (normal) vessel at the same rate as the contraction wave [111]. This is not a Ca^{2+} *wave* per se, as occurs intracellularly, but reflects the rapid conduction of a depolarization (AP) wave, which then triggers Ca^{2+} influx through $Ca_V1.2$ channels in each cell through which it passes. Lee et al. [105] reported that guinea pig mesenteric lymphatics express message for both $Ca_V1.2$ and $Ca_V1.3$, but not other isoforms of Ca_V1 that are more typical of neurons. We have confirmed this finding in rat and mouse lymphatics [112] but cannot be certain if the $Ca_V1.3$ message comes from non-SM cell types, since our PCR analyses (like those of Lee et al.) were performed on whole vessels containing SMs, ECs, dendritic cells, mast cells, macrophages, ICCs, and so on. $Ca_V1.2$ seems to be the most critical VGCC isoform, at least in mouse lymphatic SM, because SM-specific knock out in mice (leaving $Ca_V1.3$ and Ca_V3 isoforms untouched) completely abolishes large-amplitude contractions (unpublished observations).

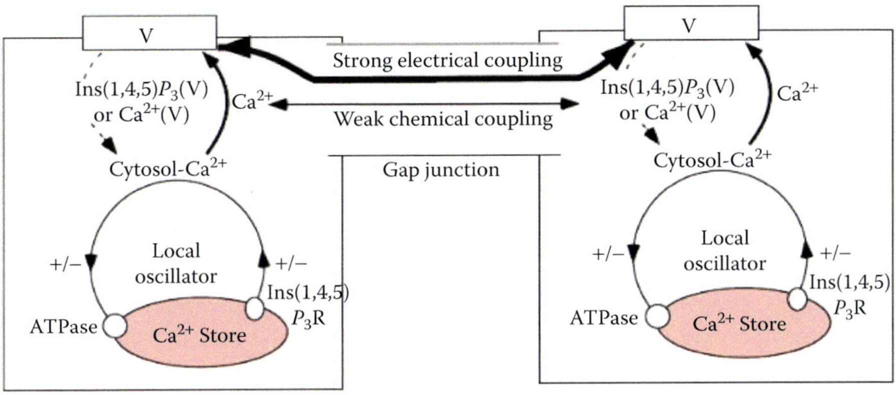

FIGURE 16.15 Diagram showing a local oscillator involving IP_3-IP_3R-Ca^{2+} release in two cells that are coupled electrically via gap junctions. (From Imtiaz, M.S. et al., *FEBS J.*, 277, 278–285, 2010.)

16.4.10 Coupled Oscillators

Imtiaz and colleagues propose that rhythmic pacemaking in lymphatic vessels results from the interactions of two coupled oscillators [55] that produce entrainment of activity in adjacent cells. The first oscillator is composed of the interaction of IP_3 with IP_3R and the subsequent sensitization of IP_3R by Ca^{2+} that is released internally and Ca^{2+} that enters the cell through VGCC activation (Figure 16.15). This is the sequence of events underlying an intracellular Ca^{2+} wave [113]. The second oscillator is represented by the electrical coupling of adjacent cells through gap junctions, so that when an AP fires in one cell, the depolarization spreads and triggers AP firing in the next cell. Coupling could also be mediated by diffusion of Ca^{2+} or IP_3 through gap junctions between cells, but that is a much slower process (<0.1 mm/s) with a very limited range (a few microns), in contrast to electrical coupling, for which the length constant is known to be a few millimeters. A critical role for gap junctions in maintaining coordinated intra- and inter-lymphangion contractions is well documented in pressurized lymphatic vessel studies using non-selective gap junction inhibitors [114–116], but the identities of the various connexins mediating coupling between lymphatic SM cells are unknown at this time.

16.5 AN ALTERNATIVE DEPOLARIZATION MECHANISM

16.5.1 Vm Depolarization Patterns Observed in Pressurized Lymphatic Vessels

Our laboratory has recently devised methods for recording Vm in the SM layer of pressurized rat, mouse, and human lymphatics. Thus, the SM cells retain their natural geometry and their length depends on the intraluminal pressure level and, to a lesser extent, on the degree of axial stretch. After confirming that robust, spontaneous contractions occur in each preparation, wortmannin is used to impair contractions (1–2 μM, a dose that blocks MLCK) and permit stable Vm recordings using sharp, intracellular microelectrodes. Wortmannin does not alter the resting Vm or the characteristics of the AP (Figure 16.16b). We analyzed multiple aspects of the AP: peak amplitude, width, plateau amplitude duration, and magnitude of the after-hyperpolarization, as shown in Figure 16.16c–f. Here, the effects of wortmannin were determined using wire-mounted vessels (Figure 16.16a) because impalements of cells under near-isometric conditions could be made and maintained before/after addition of wortmannin. In subsequent pressure myograph experiments the wortmannin dose is titrated to preserve small contractions (2–10 μm, Figure 16.17) and some degree of basal tone, allowing evaluation of the relative timing of electrical and mechanical events and/or determination of differential effects that agonists/inhibitors may have on the electrical and mechanical activity. Vm recordings in pressurized lymphatic vessels have a different appearance from those in unstretched vessel strips or wire-myograph-mounted vessels (Figure 16.17). At physiological pressures (3–5 cmH_2O for rat mesenteric lymphatics [4]), Vm undergoes a biphasic DD during diastole, with an initial ramp-like linear slope that is followed by an exponential phase just prior to reaching the threshold for firing an AP, somewhat similar to the pacemaking potential observed in cardiac SA nodal cells (SANCs) [117], and similar to the more subtle depolarization that precedes slow wave firing in some GI cells [118]. In pressurized lymphatic vessels this ramp-like DD is sufficiently exaggerated that it is difficult to determine a *resting Vm* (Figure 16.17a). STDs, if they are observed at all, are infrequent. Their rare appearance at the foot of the AP upstroke appears to be a random occurrence. It is possible but highly unlikely that STDs would be summating seamlessly to create such a smooth DD. The relatively constant rate of DD that evokes a highly regular pattern of AP firing gives the impression that a time-dependent, depolarizing current, or combination of currents, is acting to bring Vm to a threshold to trigger APs rather than APs being triggered by the summation of stochastically-generated STDs. Only when pressure is lowered to very low levels (<0.5 cmH_2O) such that the firing rate is quite low, do STDs begin to appear, or are unmasked.

A central role ascribed to STDs in initiating APs (Figure 16.7) becomes even more perplexing in pressurized vessels when the pattern of Vm changes over time is examined as a function of pressure. We have been able to record Vm continuously in the presence of wortmannin while changing pressure over the range of ~0.5 to 5 cmH_2O—the range over which the majority of the change in contraction frequency occurs in the rat. These are difficult experiments because the electrode always dislodges at some point due to expansion/collapse of the vessel, but in a few cases, pressure can be changed over a reasonable range before that occurs. As shown in Figure 16.18, the rate of DD (i.e., slope of change in Vm in diastole) in a pressurized lymphatic vessel decreases substantially as pressure is lowered from 4.5 to 0.2 cmH_2O. Curiously, at the two lowest pressures shown, the slope changes even during diastole. Note the absence of STDs from any of these recordings. Further, examination of similar recordings, some containing STDs, suggests that the rate of DD clearly changes with pressure/stretch, but neither the frequency, amplitude nor degree of summation of STDs seems to correlate with pressure (which is different than what was observed in wire-myograph-mounted rat mesenteric lymphatics [102]). These observations suggest that STDs may be at most only a minor contributor to stretch/pressure-induced depolarizing current as opposed to playing a major role in unpressurized vessels.

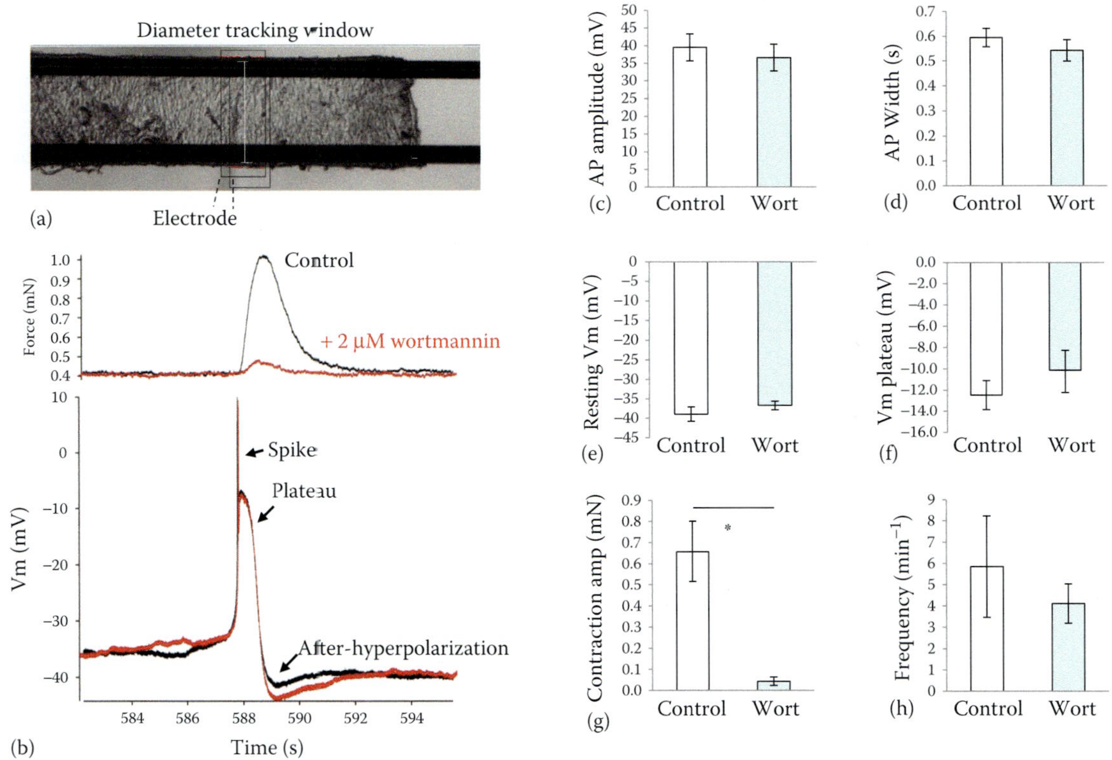

FIGURE 16.16 (a) Wire-myograph mounted rat mesenteric lymphatic showing diameter tracking window and shadow of intracellular Vm recording electrode. (b) AP recorded in a SM cell and the corresponding force transient of the vessel before and after addition of 2 μM wortmannin (traces are superimposed). (c–f) Analysis of resting Vm and AP characteristics before/after 2 μM wortmannin. (g and h) Wortmannin blunts spontaneous contraction amplitude with only a slight, non-significant change in frequency.

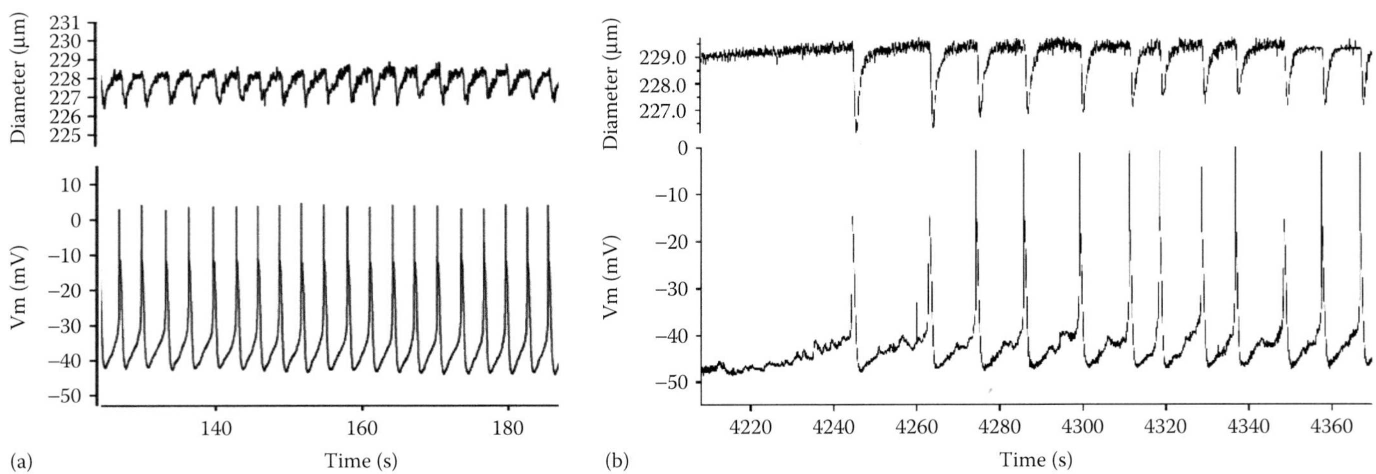

FIGURE 16.17 Examples of Vm recordings in the SM layers of two pressurized rat mesenteric lymphatic vessels. (a) At 3 cmH_2O luminal pressure no STDs are apparent; Frequency = 21/min. (b) At 1 cmH_2O the basal frequency is lower (Frequency = 3–7/min) and some STDs can be observed. The STDs are superimposed on a gradual DD whose slope is increased at the higher pressure.

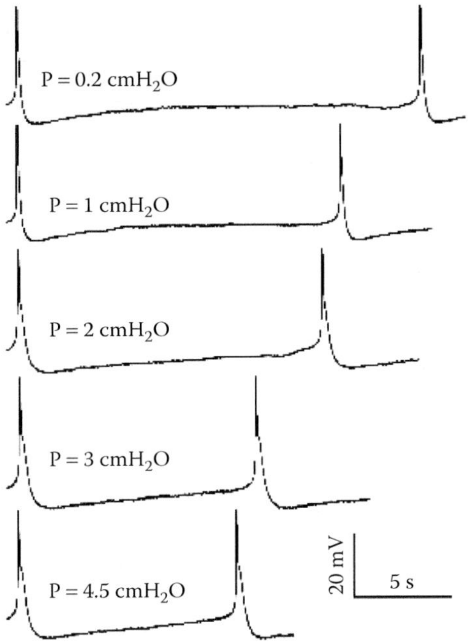

FIGURE 16.18 Vm recordings in the SM layer of a rat's mesenterics lymphatic vessel at 5 different luminal pressures. Note the change in the length and shape of the DD phases at the higher pressures. The resting Vm changed less than 3 mV between these recordings.

16.5.2 Effects of Store Depletion on Vm and [Ca^{2+}]

Another observation that is problematic for the Ca^{2+} puff-CACC-STD-AP hypothesis is the effect of Ca^{2+} store depletion. Thapsigargin (TG) is a SERCA inhibitor commonly used to test the effect of depleting Ca^{2+} stores [119]; typically, TG first empties the stores, then prevents refilling [119]. In strips of guinea pig lymphatic muscle loaded with Fluo-4, TG silences intracellular Ca^{2+} events and abolishes ET-1-induced vasomotion [91], as predicted by the scheme in Figure 16.7. In separate experiments with Vm measurements on similar preparations, TG inhibited both the amplitude and frequency of STDs [57,83]. The effects on STDs of CPA, another SERCA inhibitor, were similar [83].

In confocal imaging experiments on pressurized rat mesenteric or mouse popliteal lymphatic vessels loaded with Fluo-4 (Figure 16.19), we confirm the existence of at least 3 types of spontaneous intracellular Ca^{2+} events: (1) localized events with spatial and temporal characteristics of Ca^{2+} puffs (*), (2) events that propagate (waves) within cells, and (3) large, global Ca^{2+} *flashes* that travel rapidly between cells (~10 mm/s) over the entire length of the vessel (MDJ unpublished observations). The flashes follow APs and precede contractions, as observed in guinea pig lymphatics [56]. These recordings are made using wortmannin (2 μM) to inhibit contractions, rather than using nifedipine to silence APs. TG (1 μM) inhibits Ca^{2+}

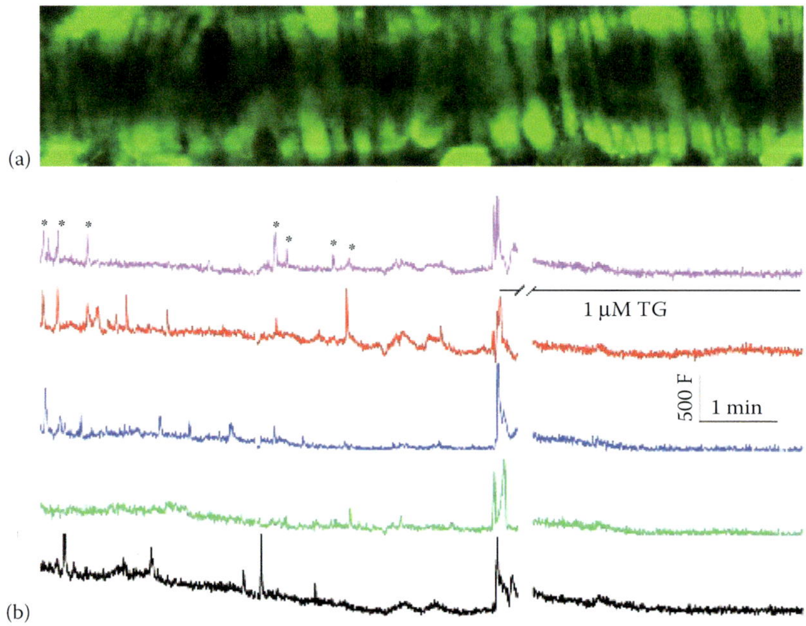

FIGURE 16.19 (a) Confocal image of the SM layer of a mouse popliteal lymphatic vessel loaded with Fluo-4 AM. (b) Spontaneous Ca^{2+} events (*) in 5 adjacent cells before (left) and after treatment with TG (right). Vessel was treated with 2 μM wortmannin to prevent movement.

puffs and Ca^{2+} waves, similar to its effect on strips of guinea pig lymphatic SM. In separate experiments measuring Vm in the SM layer of pressurized rat mesenteric lymphatics, the same doses of TG (1, 5 or 10 μM) have virtually no effect on resting Vm, AP firing rate, or AP amplitude (Figure 16.20a); TG certainly does not block spontaneous APs or contractions in the rat. There is one striking effect of TG on the AP (Figure 16.20b): it shortens the AP plateau (note also the reduced contraction amplitude). This effect may be due to off-target inhibition of VGCCs, which TG has been reported to block [120]. In the mouse, CPA causes depolarization and high frequency contractions but eventually leads to a flaccid paralysis of the lymphatic vessels as opposed to a rigor paralysis brought on by ryanodine, despite similar effects of CPA and ryanodine on the electrical behavior of lymphatic muscle (SDZ unpublished observations).

Collectively, these observations point to an alternative underlying mechanism whereby STDs occur secondarily, or are unrelated, to a different pacemaker current, at least at physiological pressures. They suggest that CACC activation/STD production is only one of several depolarizing influences. While we cannot exclude a role for CACCs in setting the resting membrane potential at very low pressures when the muscle cells are unstretched, and possibly modulating the pacemaker at other pressures, the earlier observations argue against their central role in pacemaking in pressurized lymphatic vessels.

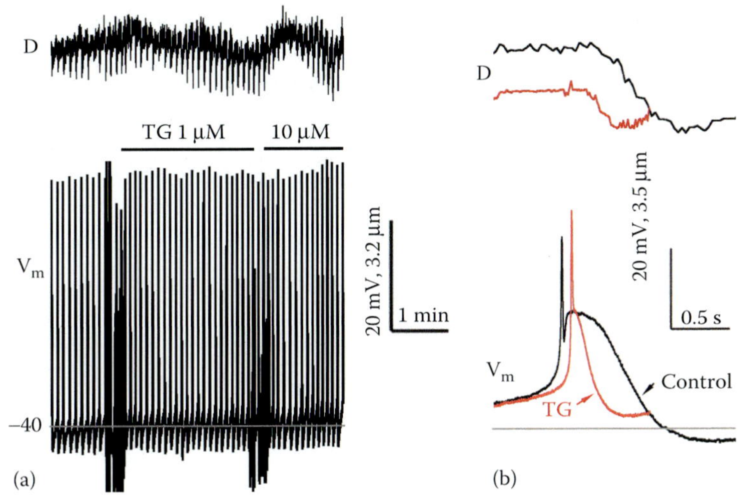

FIGURE 16.20 In a pressurized rat mesenteric lymphatic vessel, TG (1 μM) has ~no effect on (a) the frequency of spontaneous APs recorded in the SM layer but (b) dramatically shortens the plateau phase of the AP. D = inner diameter. Vessel was treated with 2 μM wortmannin to prevent movement.

16.5.3 Other Depolarizing Currents

What other mechanisms might account for spontaneous pacemaking? As mentioned in Section 16.5.1, lymphatic SM cells in pressurized vessels exhibit cardiac-like DD leading to AP firing (Figure 16.6), suggesting a pacemaker mechanism that could be similar to that in the cardiac conduction system. In cardiac cells, it is widely believed that pacemaking results from the interaction of a number of different ion channels [121,122]. There are both major and minor contributors whose ionic components appear to vary between species and among the specific cell types in the conduction system [122]. In sino-atrial node cells (SANCs) the major inward currents are contributed by $Ca_V1.2$ and Ca_V3 channels and the so-called *funny current*, I_f (HCN channels) [88]; these are opposed by two K_V currents: a rapid repolarizing current, I_{Kr}, contributed by ERG (K_V11) channels, and a slower component, I_{Ks}, contributed by KCNQ1 (K_V7) channels [121–124]. Our preliminary studies suggest that all of these currents are present and important for lymphatic SM pacemaking. Other currents in cardiac pacemaker cells include a Na^+-Ca^{2+} exchange current (NCX1), an inward rectifier K^+ current (Kir2.x), TTX-sensitive ($Na_V1.5$) and TTX-insensitive Na^+ currents [122]. Another important element of the cardiac pacemaker is intracellular Ca^{2+}. Some studies show that elementary Ca^{2+} release events such as sparks can significantly influence the balance of inward/outward current during DD and thus can potentially trigger APs [121,124]. Ca^{2+} levels regulate some channels directly ($Ca_V1.2$, Ca_V3, K_{Ca}) and others indirectly through modulating cAMP production (HCN, K_V). Studies in rabbit SANCs suggest that the late exponential DD is due to the activity of NCX brought on by local subsarcolemmal ryanodine receptor–mediated Ca^{2+} release [121–124]. Although Ca^{2+} sparks are not classically observed in lymphatic SM (ryanodine is reported to be without effect [9,57]), Ca^{2+} puffs [55,56] could influence the lymphatic pacemaker in a similar way.

The roles of the ion channels mentioned earlier in lymphatic pacemaking need to be explored further. In some cases, inhibitors for these channels are relatively specific (TTX, nifedipine) but in others (NCX, HCN) issues with off-target effects on other channels will complicate interpretation. Mouse models with tissue-specific deletion of the channels will likely make important contributions to our understanding of lymphatic pacemaking, as have similar models in the heart.

16.6 CONCLUSION

In conclusion, although much progress has been made toward understanding the ionic basis of pacemaking in lymphatic vessels, there is still much that needs to be understood, with the hope of being able to rescue dysfunctional lymphatic pacemakers in collecting vessels of lymphedema patients by specific targeting of the appropriate ion channels.

ACKNOWLEDGMENTS

This work was supported by NIH grants HL-120867, HL-125608, and HL-122578 to MJD.

REFERENCES

1. Wiig, H., and M. A. Swartz. 2012. Interstitial fluid and lymph formation and transport: Physiological regulation and roles in inflammation and cancer. *Physiol Rev* 92: 1005–1060.
2. Schmid-Schönbein, G. W. 1990. Microlymphatics and lymph flow. *Physiol Rev* 70: 987–1028.
3. Aukland, K., and R. K. Reed. 1993. Interstitial-lymphatic mechanisms in the control of extracellular fluid volume. *Physiol Rev* 73: 1–78.
4. Benoit, J. N., D. C. Zawieja, A. H. Goodman, and H. J. Granger. 1989. Characterization of intact mesenteric lymphatic pump and its responsiveness to acute edemagenic stress. *Am J Physiol Heart Circu Physiol* 257: H2059–H2069.
5. Chen, H. I., H. J. Granger, and A. E. Taylor. 1976. Interaction of capillary, interstitial, and lymphatic forces in the canine hindpaw. *Circ Res* 39: 245–254.
6. Rockson, S. G., and K. K. Rivera. 2008. Estimating the population burden of lymphedema. *Ann N Y Acad Sci* 1131: 147–154.
7. Brice, G., A. H. Child, A. Evans, R. Bell, S. Mansour, K. Burnand, M. Sarfarazi, S. Jeffery, and P. Mortimer. 2005. Milroy disease and the VEGFR-3 mutation phenotype. *J Med Genet* 42: 98–102.
8. Mellor, R., N. Tate, A. Stanton, C. Hubert, T. Makinen, A. Smith, K. Burnard, S. Jeffery, J. R. Levick, and P. S. Mortimer. 2011. Mutations in *FOXC2* in humans (Lymphoedema Distichiasis Syndrome) cause lymphatic dysfunction on dependency. *J Vasc Res* 48: 397–407.
9. Rockson, S. G. 2008. Secondary lymphedema: Is it a primary disease? *Lymphat Res Biol* 6: 63–64.

10. Hargens, A. R., and B. W. Zweifach. 1976. Transport between blood and peripheral lymph in intestine. *Microvasc Res* 11: 89–101.
11. von der Weid, P. Y., and D. C. Zawieja. 2004. Lymphatic smooth muscle: The motor unit of lymph drainage. *Int J Biochem Cell Biol* 36: 1147–1153.
12. Clough, G., and L. H. Smaje. 1978. Simultaneous measurement of pressure in the interstitium and the terminal lymphatics of the cat mesentery. *J Physiol* 283: 457–468.
13. Hargens, A. R., and B. W. Zweifach. 1977. Contractile stimuli in collecting lymph vessels. *Am J Physiol Heart Circ Physiol* 233: H57–H65.
14. Unthank, J. L., and R. D. Hogan. 1988. Modulation of the spontaneous contractions of the initial lymphatics of the bat's wing by arterial and venous occlusion. *Blood Vessels* 25: 115–121.
15. Zweifach, B. W., and H. H. Lipowsky. 1984. Pressure-flow relations in blood and lymph microcirculation. In: *Handbook of Physiology Section 2: The Cardiovascular System,* Volume IV, Chapter 7, p. 297, E. M. Renkin and C. C. Michel (Eds.). Bethesda, MD: American Physiological Society, pp. 251–307.
16. Olszewski, W. L., and A. Engeset. 1980. Intrinsic contractility of prenodal lymph vessels and lymph flow in human leg. *Am J Physiol Heart Circ Physiol* 239: H775–H783.
17. Olszewski, W. L. 2002. Contractility patterns of normal and pathologically changed human lymphatics. *Ann N Y Acad Sci* 979: 52–63.
18. Olszewski, W. L., and A. Engeset. 1979. Intrinsic contractility of leg lymphatics in man: Preliminary communication. *Lymphology* 12: 81–84.
19. Stanton, A. W., S. Modi, R. H. Mellor, J. R. Levick, and P. S. Mortimer. 2009. Recent advances in breast cancer-related lymphedema of the arm: Lymphatic pump failure and predisposing factors. *Lymphat Res Biol* 7: 29–45.
20. Engeset, A., W. Olszewski, P. M. Jaeger, J. Sokolowski, and L. Theodorsen. 1977. Twenty-four-hour variation in flow and composition of leg lymph in normal men. *Acta Physiol Scandinavica* 99: 140–148.
21. Van Helden, D. F., M. S. Imtiaz, K. Nurgaliyeva, P. Y. von der Weid, and P. J. Dosen. 2000. Role of calcium stores and membrane voltage in the generation of slow wave action potentials in guinea-pig gastric pyloris. *J Physiol* 524: 245–265.
22. Hirst, G. D. S., and S. M. Ward. 2003. Interstitial cells: Involvement in rhythmicity and neural control of gut smooth muscle. *J Physiol* 550: 337–346.
23. Hirst, G. D., E. J. Dickens, and F. R. Edwards. 2002. Pacemaker shift in the gastric antrum of guinea-pigs produced by excitatory vagal stimulation involves intramuscular interstitial cells. *J Physiol* 541: 917–928.
24. Suzuki, H., Y. Yamamoto, and G. D. S. Hirst. 1999. Properties of spontaneous activity of gastric smooth muscle. *Korean J Physiol Pharmacol* 3: 119–125.
25. Kim, B. J., I. So, and K. W. Kim. 2006. The relationship of TRP channels to the pacemaker activity of interstitial cells of Cajal in the gastrointestinal tract. *J Smooth Muscle Res* 42: 1–7.
26. Walker, R. L., S. D. Koh, G. P. Sergeant, K. M. Sanders, and B. Horowitz. 2002. TRPC4 currents have properties similar to the pacemaker current in interstitial cells of Cajal. *Am J Physiol Cell Physiol* 283: C1637–C1645.
27. Ozaki, H., R. J. Stevens, D. P. Blondfield, N. G. Publicover, and K. M. Sanders. 1991. Simultaneous measurement of membrane potential, cytosolic Ca^{2+}, and tension in intact smooth muscles. *Am J Physiol Cell Physiol* 260: C917–C925.
28. Hirst, G. D., and F. R. Edwards. 2004. Role of interstitial cells of Cajal in the control of gastric motility. *J Pharmacol Sci* 96: 1–10.
29. Sanders, K. M., and S. M. Ward. 2008. Interstitial cells of Cajal: A new perspective on smooth muscle function. *J Physiol* 576: 721–726.
30. Iino, S., S. Horiguchi, and K. Horiguchi. 2011. Interstitial cells of Cajal in the gastrointestinal musculature of W(jic) c-kit mutant mice. *J Smooth Muscle Res* 47: 111–121.

31. Iino, S., K. Horiguchi, and Y. Nojyo. 2009. W(sh)/W(sh) c-Kit mutant mice possess interstitial cells of Cajal in the deep muscular plexus layer of the small intestine. *Neurosci Lett* 459: 123–126.
32. Torihashi, S., S. M. Ward, S. Nishikawa, K. Nishi, S. K. obayashi, and K. M. Sanders. 1995. c-kit-dependent development of interstitial cells and electrical activity in the murine gastrointestinal tract. *Cell Tissue Res* 280: 97–111.
33. Ward, S. M., S. C. Harney, J. R. Bayguinov, G. J. McLaren, and K. M. Sanders. 1997. Development of electrical rhythmicity in the murine gastrointestinal tract is specifically encoded in the tunica muscularis. *J Physiol* 505 (Pt 1): 241–258.
34. Zhu, M. H., T. W. Kim, S. Ro, W. Yan, S. M. Ward, S. D. Koh, and K. M. Sanders. 2009. A Ca^{2+}-activated Cl^- conductance in interstitial cells of Cajal linked to slow wave currents and pacemaker activity. *J Physiol* 587: 4905–4918.
35. Hwang, S. J., P. J. A. Blair, F. C. Britton, K. E. O'Driscoll, G. Hennig, Y. R. Bayguinov, J. R. Rock, B. D. Harfe, K. M. Sanders, and S. M. Ward. 2009. Expression of anoctamin 1/TMEM16A by interstitial cells of Cajal is fundamental for slow wave activity in gastrointestinal muscles. *J Physiol* 587: 4887–4904.
36. Koh, B. H., R. Roy, M. A. Hollywood, K. D. Thornbury, N. G. McHale, G. P. Sergeant, W. J. Hatton, S. M. Ward, K. M. Sanders, and S. D. Koh. 2012. Platelet-derived growth factor receptor-alpha cells in mouse urinary bladder: A new class of interstitial cells. *J Cell Mol Med* 16: 691–700.
37. Drumm, B. T., R. J. Large, M. A. Hollywood, K. D. Thornbury, S. A. Baker, B. J. Harvey, N. G. McHale, and G. P. Sergeant. 2015. The role of Ca(2+) influx in spontaneous Ca(2+) wave propagation in interstitial cells of Cajal from the rabbit urethra. *J Physiol* 593: 3333–3350.
38. Drumm, B. T., S. D. Koh, K. E. Andersson, and S. M. Ward. 2014. Calcium signalling in Cajal-like interstitial cells of the lower urinary tract. *Nat Rev Urol* 11: 555–564.
39. Lavoie, B., O. B. Balemba, M. T. Nelson, S. M. Ward, and G. M. Mawe. 2007. Morphological and physiological evidence for interstitial cell of Cajal-like cells in the guinea pig gallbladder. *J Physiol* 579: 487–501.
40. Dixon, R. E., G. W. Hennig, S. A. Baker, F. C. Britton, B. D. Harfe, J. R. Rock, K. M. Sanders, and S. M. Ward. 2012. Electrical slow waves in the mouse oviduct are dependent upon a calcium activated chloride conductance encoded by Tmem16a. *Biol Reprod* 86: 1–7.
41. Hashitani, H., and R. J. Lang. 2010. Functions of ICC-like cells in the urinary tract and male genital organs. *J Cell Mol Med* 14: 1199–1211.
42. Baker, S. A., W. J. Hatton, J. Han, G. W. Hennig, F. C. Britton, and S. D. Koh. 2010. Role of TREK-1 potassium channel in bladder overactivity after partial bladder outlet obstruction in mouse. *J Urol* 183: 793–800.
43. Baker, S. A., G. W. Hennig, J. Han, F. C. Britton, T. K. Smith, and S. D. Koh. 2008. Methionine and its derivatives increase bladder excitability by inhibiting stretch-dependent K(+) channels. *Br J Pharmacol* 153: 1259–1271.
44. Iqbal, J., M. A. Tonta, R. Mitsui, Q. Li, M. Kett, J. Li, H. C. Parkington, H. Hashitani, and R. J. Lang. 2012. Potassium and ANO1/ TMEM16A chloride channel profiles distinguish atypical and typical smooth muscle cells from interstitial cells in the mouse renal pelvis. *Br J Pharmacol* 165: 2389–2408.
45. Lang, R. J., H. Hashitani, M. A. Tonta, H. C. Parkington, and H. Suzuki. 2007. Spontaneous electrical and Ca^{2+} signals in typical and atypical smooth muscle cells and interstitial cell of Cajal-like cells of mouse renal pelvis. *J Physiol* 583: 1049–1068.
46. Hashitani, H., M. J. Nguyen, H. Noda, R. Mitsui, R. Higashi, K. Ohta, K. I. Nakamura, and R. J. Lang. 2017. Interstitial cell modulation of pyeloureteric peristalsis in the mouse renal pelvis examined using FIBSEM tomography and calcium indicators. *Pflugers Arch* 469: 797–813.

47. Hashitani, H., R. Mitsui, S. Masaki, and D. F. Van Helden. 2015. Pacemaker role of pericytes in generating synchronized spontaneous Ca^{2+} transients in the myenteric microvasculature of the guinea-pig gastric antrum. *Cell Calcium* 58: 442–456.
48. Hashitani, H., D. F. Van Helden, and H. Suzuki. 1996. Properties of spontaneous depolarizations in circular smooth muscle cells of rabbit urethra. *Br J Pharmacol* 118: 1627–1632.
49. McCloskey, K. D., M. A. Hollywood, K. D. Thornbury, S. M. Ward, and N. G. McHale. 2002. Kit-like immunopositive cells in sheep mesenteric lymphatic vessels. *Cell Tissue Res* 310: 77–84.
50. Briggs Boedtkjer, D., J. Rumessen, U. Baandrup, M. Skov Mikkelsen, N. Telinius, H. Pilegaard, C. Aalkjaer, and V. Hjortdal. 2013. Identification of interstitial Cajal-like cells in the human thoracic duct. *Cells Tissues Organs* 197: 145–158.
51. Dixon, R. E., S. J. Hwang, G. W. Hennig, K. H. Ramsey, J. H. Schripsema, K. M. Sanders, and S. M. Ward. 2009. Chlamydia infection causes loss of pacemaker cells and inhibits oocyte transport in the mouse oviduct. *Biol Reprod* 80: 665–673.
52. Toland, H. M., K. D. McCloskey, K. D. Thornbury, N. G. McHale, and M. A. Hollywood. 2000. Ca^{2+}-activated Cl^- current in sheep lymphatic smooth muscle. *Am J Physiol Heart Circ Physiol* 279: C1327–C1335.
53. Ray, P., N. Krishnamoorthy, T. B. Oriss, and A. Ray. 2010. Signaling of c-kit in dendritic cells influences adaptive immunity. *Ann N Y Acad Sci* 1183: 104–122.
54. Ray, P., N. Krishnamoorthy, and A. Ray. 2008. Emerging functions of c-kit and its ligand stem cell factor in dendritic cells: Regulators of T cell differentiation. *Cell Cycle* 7: 2826–2832.
55. Imtiaz, M. S., P. Y. von der Weid, and D. F. van Helden. 2010. Synchronization of Ca^{2+} oscillations: A coupled oscillator-based mechanism in smooth muscle. *FEBS J* 277: 278–285.
56. Imtiaz, M. S., J. Zhao, K. Hosaka, P. Y. von der Weid, M. Crowe, and D. F. van Helden. 2007. Pacemaking through Ca^{2+} stores interacting as coupled oscillators via membrane depolarization. *Biophys J* 92: 3843–3861.
57. von der Weid, P. Y., M. Rahman, M. S. Imtiaz, and D. F. van Helden. 2008. Spontaneous transient depolarizations in lymphatic vessels of the guinea pig mesentery: Pharmacology and implication for spontaneous contractility. *Am J Physiol Heart Circ Physiol* 295: H1989–H2000.
58. Ordog, T., S. M. Ward, and K. M. Sanders. 1999. Interstitial cells of cajal generate electrical slow waves in the murine stomach. *J Physiol* 518 (Pt 1): 257–269.
59. Gui, P., S. D. Zawieja, M. Li, S. Sulley, J. H. Jaggar, J. R. Rock, and M. J. Davis. 2016. The Ca^{2+}-activated Cl^- channel TMEM16A(ANO1) modulates, but is not required for, pacemaking in mouse lymphatic vessels. *FASEB J* 30: 726.723.
60. Sanders, K. M., and S. M. Ward. 2007. Kit mutants and gastrointestinal physiology. *J Physiol* 578: 33–42.
61. Heppner, T. J., M. E. Werner, B. Nausch, C. Vial, R. J. Evans, and M. T. Nelson. 2009. Nerve-evoked purinergic signalling suppresses action potentials, Ca^{2+} flashes and contractility evoked by muscarinic receptor activation in mouse urinary bladder smooth muscle. *J Physiol* 587: 5275–5288.
62. Hollywood, M. A., K. D. Cotton, K. D. Thornbury, and N. G. McHale. 1997. Tetrodotoxin-sensitive sodium current in sheep lymphatic smooth muscle. *J Physiol* 503: 13–21.
63. Hwang, S. J., N. Basma, K. M. Sanders, and S. M. Ward. 2016. Effects of new-generation inhibitors of the calcium-activated chloride channel anoctamin 1 on slow waves in the gastrointestinal tract. *Br J Pharmacol* 173: 1339–1349.
64. Kamishima, T., and J. M. Quayle. 2003. Ca^{2+}-induced Ca^{2+} release in cardiac and smooth muscle cells. *Biochem Soc Trans* 31: 943–946.

65. Kito, Y., R. Mitsui, S. M. Ward, and K. M. Sanders. 2015. Characterization of slow waves generated by myenteric interstitial cells of Cajal of the rabbit small intestine. *Am J Physiol Gastrointest Liver Physiol* 308: G378– G388.
66. Wouters, M., A. De Laet, L. V. Donck, E. Delpire, P. P. van Bogaert, J. P. Timmermans, A. de Kerchove d'Exaerde, K. Smans, and J. M. Vanderwinden. 2006. Subtractive hybridization unravels a role for the ion cotransporter NKCC1 in the murine intestinal pacemaker. *Am J Physiol Gastrointest Liver Physiol* 290: G1219–G1227.
67. Zhu, M. H., T. S. Sung, M. Kurahashi, L. E. O'Kane, K. O'Driscoll, S. D. Koh, and K. M. Sanders. 2016. Na^+-K^+-Cl^- cotransporter (NKCC) maintains the chloride gradient to sustain pacemaker activity in interstitial cells of Cajal. *Am J Physiol Gastrointest Liver Physiol* 311: G1037–G1046.
68. Dam, V. S., D. M. Boedtkjer, C. Aalkjaer, and V. Matchkov. 2014. The bestrophin- and TMEM16A-associated Ca(2+)- activated Cl(-) channels in vascular smooth muscles. *Channels (Austin)* 8: 361–369.
69. Davis, A. J., A. S. Forrest, T. A. Jepps, M. L. Valencik, M. Wiwchar, C. A. Singer, W. R. Sones, I. A. Greenwood, and N. Leblanc. 2010. Expression profile and protein translation of TMEM16A in murine smooth muscle. *Am J Physiol Cell Physiol* 299: C948–959.
70. Dam, V. S., D. M. Boedtkjer, J. Nyvad, C. Aalkjaer, and V. Matchkov. 2014. TMEM16A knockdown abrogates two different Ca(2+)-activated Cl (-) currents and contractility of smooth muscle in rat mesenteric small arteries. *Pflugers Arch* 466: 1391–1409.
71. Welsh, D. G., A. D. Morielli, M. T. Nelson, and J. E. Brayden. 2002. Transient receptor potential channels regulate myogenic tone of resistance arteries. *Circ Res* 90: 248–250.
72. Earley, S., B. J. Waldron, and J. E. Brayden. 2004. Critical role for transient receptor potential channel TRPM4 in myogenic constriction of cerebral arteries. *Circ Res* 95: 922–929.
73. Earley, S. 2013. TRPM4 channels in smooth muscle function. *Pflugers Arch* 465: 1223–1231.
74. Balderas, E., R. Ateaga-Tlecuitl, M. Rivera, J. C. Gomora, and A. Darszon. 2011. Niflumic acid blocks native and recombinant T-type channels. *J Cell Physiol* 227: 2542–2555.
75. Walsh, K. B., and C. Wang. 1996. Effect of chloride channel blockers on the cardiac CFTR chloride and L-type calcium currents. *Cardiovasc Res* 32: 391–399.
76. Matchkov, V. V., D. M. Boedtkjer, and C. Aalkjaer. 2015. The role of Ca(2+) activated Cl(-) channels in blood pressure control. *Curr Opin Pharmacol* 21: 127–137.
77. Edwards, F. R., G. D. Hirst, and H. Suzuki. 1999. Unitary nature of regenerative potentials recorded from circular smooth muscle of guinea-pig antrum. *J Physiol* 519 (Pt 1): 235–250.
78. Tokutomi, N., H. Maeda, Y. Tokutomi, D. Sato, M. Sugita, S. Nishikawa, S. Nishikawa, J. Nakao, T. Imamura, and K. Nishi. 1995. Rhythmic Cl^- current and physiological roles of the intestinal c-kit-positive cells. *Pflugers Arch* 431: 169–177.
79. van Helden, D. F. 1991. Spontaneous and noradrenaline-induced transient depolarizations in the smooth muscle of guinea-pig mesenteric vein. *J Physiol* 437: 511–541.
80. van Helden, D. F. 1993. Pacemaker potentials in lymphatic smooth muscle of the guinea-pig mesentery. *J Physiol* 471: 465–479.
81. Dixon, R. E., C. M. Moreno, C. Yuan, X. Opitz-Araya, M. D. Binder, M. F. Navedo, and L. F. Santana. 2015. Graded Ca(2)(+)/calmodulin-dependent coupling of voltage-gated $Ca_V1.2$ channels. *Elife* 4.
82. Navedo, M. F., E. P. Cheng, C. Yuan, S. Votaw, J. D. Molkentin, J. D. Scott, and L. F. Santana. 2010. Increased coupled gating of L-type Ca^{2+} channels during hypertension and Timothy syndrome. *Circ Res* 106: 748–756.

83. Ferrusi, I., J. Zhao, D. F. van Helden, and M. Crowe. 2004. Cyclopiazonic acid decreases spontaneous transient depolarizations in guinea pig mesenteric lymphatic vessels in endothelium-dependent and -independent manners. *Am J Physiol Heart Circ Physiol* 286: H2287–H2295.
84. Berridge, M. J. 1997. Elementary and global aspects of calcium signalling. *J Exp Biol* 200: 315–319.
85. McCarron, J. G., J. W. Craig, K. N. Bradley, and T. C. Muir. 2002. Agonist-induced phasic and tonic responses in smooth muscle are mediated by InsP_3. *J Cell Sci* 115: 2207–2218.
86. Berridge, M. J. 1991. Cytoplasmic calcium oscillations: A two pool model. *Cell Calcium* 12: 63–72.
87. Dupont, G., M. J. Berridge, and A. Goldbeter. 1991. Signal-induced Ca^{2+} oscillations: Properties of a model based on Ca^{2+}-induced Ca^{2+} release. *Cell Calcium* 12: 73–85.
88. Accili, E. A., C. Proenza, M. Baruscotti, and D. DiFrancesco. 2002. From funny current to HCN channels: 20 years of excitation. *News Physiol Sci* 17: 32–37.
89. Guatimosim, S., K. Dilly, L. F. Santana, M. Saleet Jafri, E. A. Sobie, and W. J. Lederer. 2002. Local Ca(2+) signaling and EC coupling in heart: Ca(2+) sparks and the regulation of the [Ca(2+)](i) transient. *J Mol Cell Cardiol* 34: 941–950.
90. Brayden, J. E., and M. T. Nelson. 1992. Regulation of arterial tone by activation of calcium-dependent potassium channels. *Science* 256: 532–535.
91. Zhao, J., and D. F. van Helden. 2003. ET-1-associated vasomotion and vasospasm in lymphatic vessels of the guinea-pig mesentery. *British J Pharmacol* 140: 1399–1413.
92. Bootman, M. D., T. J. Collins, L. Mackenzie, H. L. Roderick, M. J. Berridge, and C. M. Peppiatt. 2002. 2-aminoethoxydiphenyl borate (2-APB) is a reliable blocker of store-operated Ca^{2+} entry but an inconsistent inhibitor of InsP3-induced Ca^{2+} release. *FASEB J* 16: 1145–1150.
93. Prakriya, M., and R. S. Lewis. 2001. Potentiation and inhibition of Ca^{2+} release-activated Ca^{2+} channels by 2-aminoethyldiphenyl borate (2-APB) occurs independently of IP_3 receptors. *J Physiol* 536: 3–19.
94. Ozaki, H., M. Hori, Y. S. Kim, S. C. Kwon, D. S. Ahn, H. Nakazawa, M. Kobayashi, and H. Karaki. 2002. Inhibitory mechanism of xestospongin-C on contraction and ion channels in the intestinal smooth muscle. *Br J Pharmacol* 137: 1207–1212.
95. Saleem, H., S. C. Tovey, T. F. Molinski, and C. W. Taylor. 2014. Interactions of antagonists with subtypes of inositol 1,4,5-trisphosphate (IP3) receptor. *Br J Pharmacol* 171: 3298–3312.
96. Suzuki, H., H. Takano, Y. Yamamoto, T. Komuro, M. Saito, K. Kato, and K. Mikoshiba. 2000. Properties of gastric smooth muscles obtained from mice which lack inositol trisphosphate receptor. *J Physiol* 525 (Pt 1): 105–111.
97. Berridge, M. J. 1987. Inositol trisphosphate and diacylglycerol: Two interacting second messengers. *Annu Rev Biochem* 56: 159–193.
98. Liu, C., and C. Montell. 2015. Forcing open TRP channels: Mechanical gating as a unifying activation mechanism. *Biochem Biophys Res Commun* 460: 22–25.
99. Narayanan, J., M. Imig, R. J. Roman, and D. R. Harder. 1994. Pressurization of isolated renal arteries increases inositol trisphosphate and diacylglycerol. *Am J Physiol Heart Circ Physiol* 266: H1840–H1845.
100. Laher, I., P. Vorkapic, A. L. Dowd, and J. A. Bevan. 1990. Protein kinase C potentiates stretch-induced cerebral artery tone by increasing intracellular sensitivity to Ca^{2+}. *Biochem Biophys Res Commun* 165: 312–318.
101. Osol, G., I. Laher, and M. Kelley. 1993. Myogenic tone is coupled to phospholipase C and G protein activation in small cerebral arteries. *Am J Physiol Heart Circ Physiol* 265: H415–H420.

102. von der Weid, P. Y., S. Y. Lee, M. S. Imtiaz, D. C. Zawieja, and M. J. Davis. 2014. Electrophysiological properties of rat mesenteric lymphatic vessels and their regulation by stretch. *Lymphat Res Biol* 12: 66–75.
103. Harder, D. R. 1984. Pressure-dependent membrane depolarization in cat middle cerebral artery. *Circ Res* 55: 197–202.
104. Knot, H. J., and M. T. Nelson. 1998. Regulation of arterial diameter and wall [Ca^{2+}] in cerebral arteries of rat by membrane potential and intravascular pressure. *J Physiol* 508 (Pt 1): 199–209.
105. Lee, S., S. Roizes, and P. Y. von der Weid. 2014. Distinct roles of L- and T-type voltage-dependent Ca^{2+} channels in regulation of lymphatic vessel contractile activity. *J Physiol* 592: 5409–5427.
106. Liu, M., K. C. Yang, and S. C. Dudley, Jr. Cardiac sodium channel mutations: Why so many phenotypes? *Nat Rev Cardiol* 11: 607–615, 2014.
107. Beckett, E. A., M. A. Hollywood, K. D. Thornbury, and N. G. McHale. 2007. Spontaneous electrical activity in sheep mesenteric lymphatics. *Lymphat Res Biol* 5: 29–43.
108. Telinius, N., J. Majgaard, S. Kim, N. Katballe, E. Pahle, J. Nielsen, V. Hjortdal, C. Aalkjaer, and D. B. Boedtkjer. 2015. Voltage-gated sodium channels contribute to action potentials and spontaneous contractility in isolated human lymphatic vessels. *J Physiol* 593: 3109–3122.
109. Hollywood, M. A., K. D. Cotton, K. D. Thornbury, and N. G. McHale. 1997. Isolated sheep mesenteric lymphatic smooth muscle cells possess both T- and L-type calcium currents. *J Physiol* 501: P109–P110.
110. Jensen, L. J., and N. H. Holstein-Rathlou. 2009. Is there a role for T-type Ca^{2+} channels in regulation of vasomotor tone in mesenteric arterioles? *Can J Physiol Pharmacol* 87: 8–20.
111. Castorena-Gonzalez, J. A., S. D. Zawieja, M. Li, S. Srinivasan, A. M. Simon, G. Hennig, R. de la Torre, L. A. Martinez-Lemus, and M. J. Davis. 2017. Uncovering mechanisms of connexin-related lymphedema. *Circulation Research* (in revision).
112. To, K. H. T., P. Gui, M. Li, S. D. Zawieja, J. A. Castorena-Gonzalez, and M. J. Davis. 2017. T-type, but not L-type, voltage-gated calcium channels are dispensable for spontaneous lymphatic contractions. *Sci Rep* (still under review).
113. McCarron, J. G., S. Chalmers, D. MacMillan, and M. L. Olson. 2010. Agonist-evoked Ca(2+) wave progression requires Ca(2+) and IP(3). *J Cell Physiol* 224: 334–344.
114. Crowe, M. J., P. Y. von der Weid, J. A. Brock, and D. F. Van Helden. 1997. Co-ordination of contractile activity in guinea-pig mesenteric lymphatics. *J Physiol* 500 (Pt 1): 235–244.
115. McHale, N. G., and M. K. Meharg. 1992. Co-ordination of pumping in isolated bovine lymphatic vessels. *J Physiol* 450: 503–512.
116. Zawieja, D. C., K. L. Davis, R. Schuster, W. M. Hinds, and H. J. Granger. 1993. Distribution, propagation, and coordination of contractile activity in lymphatics. *Am J Physiol Heart Circ Physiol* 264: H1283–H1291.
117. Bogdanov, K. Y., V. A. Maltsev, T. M. Vinogradova, A. E. Lyashkov, H. A. Spurgeon, M. D. Stern, and E. G. Lakatta. 2006. Membrane potential fluctuations resulting from submembrane Ca^{2+} releases in rabbit sinoatrial nodal cells impart an exponential phase to the late diastolic depolarization that controls their chronotropic state. *Circ Res* 99: 979–987.
118. Kito, Y., S. M. Ward, and K. M. Sanders. 2005. Pacemaker potentials generated by interstitial cells of Cajal in the murine intestine. *Am J Physiol Cell Physiol* 288: C710–C720.

119. Thastrup, O., A. P. Dawson, O. Scharff, B. Foder, P. J. Cullen, B. K. Drobak, P. J. Bjerrum, S. B. Christensen, and M. R. Hanley. 1994. Thapsigargin, a novel molecular probe for studying intracellular calcium release and storage. *Agents Actions* 43: 187–193.
120. Rossier, M. F., C. P. Python, M. M. Burnay, W. Schlegel, M. B. Vallotton, and A. M. Capponi. 1993. Thapsigargin inhibits voltage-activated calcium channels in adrenal glomerulosa cells. *Biochem J* 296: 309–312.
121. Maltsev, V. A., and E. G. Lakatta. 2008. Dynamic interactions of an intracellular Ca^{2+} clock and membrane ion channels clock underlie robust initiation and regulation of cardiac pacemaker function. *Cardiovasc Res* 77: 274–284.
122. Satoh, H. 2003. Sino-atrial nodal cells of mammalian hearts: Ionic currents and gene expression of pacemaker ionic channels. *J Smooth Muscle Res* 39: 175–193.
123. Charpentier, F., J. Merot, G. Loussouarn, and I. Baro. 2010. Delayed rectifier currents and cardiac repolarization. *J Mol Cell Cardiol* 48: 37–44.
124. Vinogradova, T. M., and E. G. Lakatta. 2009. Regulation of basal and reserve cardiac pacemaking function by interactions of cAMP-mediated PKA-dependent Ca^{2+} cycling with surface membrane channels. *J Mol Cell Cardiol* 47: 456–474.

17 NADPH Oxidase, Redox Signaling, and Vascular Smooth Muscle Function

Daniel S. de Jesus, Eugenia Cifuentes-Pagano, Sanghamitra Sahoo, and Patrick J. Pagano

CONTENTS

17.1 INTRODUCTION

Aside from the elemental role of large vessels to conduct blood flow toward organ systems, two of the most fundamental characteristics of the vascular system is the adaptive reactivity and distensibility of arterioles in the microcirculation. The ability of veins and arteries to adapt to fluctuations in flow and pressure is essential in the control of blood supply and thus oxygen and nutrient delivery to distinct organ systems. The vascular wall responds dynamically to local nutritional and metabolic demands as well as paracrine and endocrine influences depending on the rate of environmental changes or degree of effectivity and potency of constrictor and dilator agents. The intima, media, and adventitia all combine to elicit a coordinated response to these factors. Normal elasticity of blood vessels also plays a crucial role in dampening the wide and potentially deleterious pressure pulses from cardiac contraction and ensuring a steady flow of blood throughout the circulatory system. As the central processor of vascular tone modulation, the medial smooth muscle cell (SMC) is regulated by numerous endogenous factors, such as hormones, reactive oxygen species (ROS), and mechanical stress. Changes in the normal functioning of SMCs contribute to the development of myriad pathologies ranging from hypertension to tumor growth.

In the following sections, we will survey how a family of proteins classically defined by its role in the immune response, the NADPH oxidase (Nox), is being appreciated as an increasingly important regulator of normal vascular function as well as pathological dysfunction.

17.2 VASCULAR SMOOTH MUSCLE CELL STRUCTURE AND FUNCTION

Vascular smooth muscle cells (VSMCs) are highly specialized cells that play a critical role in maintaining structural and functional integrity of blood vessels, *via* contraction and regulation of vessel tone, blood flow and pressure. SMCs in adult mammals display low rates of proliferation, markedly reduced synthetic activity, and express a unique set of contractile proteins, ion channels, and specialized signaling molecules required for the cells' contractile function. Unlike what is broadly appreciated for skeletal and cardiac myocytes, SMCs do not terminally differentiate and can undergo profound and reversible transitioning between a quiescent, contractile phenotype to a proliferative synthetic phenotype in response to various physiological and pathological stimuli. These include but are not limited to growth factors and inhibitors, extracellular matrix components, mechanical factors such as stretch and shear stress, cell-cell interactions, and various inflammatory mediators [1]. This phenotypic plasticity of SMCs is essential for vascular development during morphogenesis of the blood vessels as well as a latent potential switch triggered by injury for the purposes of repair. Unrestrained plasticity, however, including very high rates of SMC proliferation, migration and proteolytic activity can lead to uncontrolled remodeling, abnormal vessel architecture and integrity and susceptibility to extravascular insults. VSMC phenotypic switching is known to play a key role in the etiology of many cardiovascular diseases including atherosclerosis, hypertension [2,3], angioplasty-induced restenosis, vascular aneurysms, chronic kidney disease (CKD) and vascular aging.

SMC differentiation and phenotype switching are characterized by shifts in expression of genes encoding smooth muscle (SM)-specific contractile proteins and extracellular matrix proteins. Fully differentiated SMCs express a range of contractile proteins, that is, SM α-actin, SM22α, SM myosin light chain kinase (SM-MLCK), SM myosin heavy chain, SM-calponin, smoothelin and leiomodin among others [1,4], which are also often relied upon as markers of SMC contractile phenotype. Many of these proteins are involved in SMC contraction either as a structural component of the contractile apparatus or as a regulator of contraction. A decrease in the expression levels of these proteins associated with the contractile phenotype is, therefore, taken as representative of the *synthetic phenotype*. In addition, studies have shown that there are phenotypic distinctions among arterial and venous VSMCs [5,6]. Wong et al. [5] showed that in response to 10% fetal calf serum, venous SMC was more dedifferentiated, proliferative, and synthetic than arterial SMC, which could implicate a role for venous SMC in atherosclerosis occurring in vein grafts. The modulators influencing these arterial and venous SMC differences are still poorly defined. For the purposes of this chapter, we will focus our attention on arterial SMCs.

Expression of SMC contractile genes depends on a cis-acting DNA sequence known as a CArG box ($CC(A/T)_6GG$), which serves as the binding site for serum response factor (SRF), a MADS (MCM1, Agamous, Deficiens, SRF) box transcription factor [7]. Seminal studies from various laboratories have shown that myocardin is a potent coactivator of SRF that is sufficient and necessary for transcription of SM contractile genes and is, therefore, a likely target for the signaling pathways that modulates SMC phenotype. Mechanistically speaking, in promoting SMC differentiation SRF homodimers interact with a segment of the myocardin protein including the basic and Q-rich domain to form a ternary complex on DNA. SRF, on its own, is a very weak activator of transcription. Myocardin confers to the ternary complex its strong transcriptional activation domain. Another well-defined domain on myocardin, SAP, reportedly confers specificity of target gene expression [8–11]. Other transcription factors known to regulate vascular SM-specific genes include myocyte enhancer factor (MEF 2B and 2C), myocardin-related factor-A (MRTFA, MLK1, MAL), GATA family members, GATA 4/5/6, Kruppel-like factor (KLF)-family of zinc finger proteins such as Sp1/3, KLF4, KLF5, and homeodomain protein Nkx3.1, Nkx3.2, and Hoxb7 [12]. A subset of homeodomain proteins (MHox, Nkx3.1, Nkx3.2, Barx1b, and Barx2b) and Gata-6 are expressed in certain SMC subtypes and have been illustrated to interact with SRF to regulate SMC differentiation marker gene expression [13–16].

Since SMCs with distinct phenotypes express varying combinations of contractile proteins [17–19] rather than a completely different sets of proteins, it would appear rational to examine changes in an array of contractile proteins before deeming the cell population contractile or synthetic. In addition, these data should preferably be complemented with assessments of SMC morphology, proliferation, and migration. Although previous studies imply that a loss of differentiation markers was required for proliferation of VSMCs, recent studies show that these processes are not mutually exclusive and that many factors other than VSMC's proliferation status influence differentiation [1].

Although similar in their primary function of contraction, SMCs show considerable phenotypic diversity and plasticity depending on their longitudinal location along the arterial tree (large conduit vs. small resistance vessels), and their tissue or organ microenvironment (heart, brain, kidney, etc.), developmental stage or pathophysiological status. Indeed, a major factor in the variation in SMC populations can be explained by the variable embryonic origin of SMCs [20–23]. This aspect of SMC diversity has extensively been reviewed by many groups [1,7,22,24]; however, the relative contribution of each of the aforementioned modulators to these diverse phenotypic states remain unclear [25–27].

17.2.1 Regulation of Smooth Muscle Cell Function

A growing complexity of signaling pathways, including phosphoinositide-3-kinase (PI3K)/Akt, extracellular signal-related kinase (ERK) and p38 mitogen-activated protein kinase (MAPK) [28–32], have been posited in the control of SMC phenotype. Numerous stimuli including cytokines or growth factors (TGF-β, PDGF-BB, fibroblast growth factor (FGF-2), insulin growth factor (IGF)-I and II) [29,33–37], extracellular matrix (MMPs), vasoactive molecules (endothelin-1, angiotensin II, histamine, norepinephrine), microRNAs, chromosomal structure modifiers, mechanical forces, vasodilators, such as nitric oxide (NO) and ROS among others are known to modulate SMC differentiation. Overall, most of these factors have been extensively studied and are reviewed in detail elsewhere [1,38,39]. Nonetheless, we will discuss a few of the factors here, in the context of ROS-dependent signaling.

Vascular SMC contraction is mediated in part by calcium influx through L-type voltage-gated Ca^{2+} channels (VGCC)-induced activation of the Rho/ROCK signaling cascade [40]. Wamhoff and colleagues [41] coupled VGCC-induced SMC contraction to the regulation of SMC differentiation marker genes and demonstrated that activation of L-type VGCC stimulates CArG-dependent increased expression of SMC differentiation marker genes SM α-actin, SM 22α and SM MHC. Consistent with previous studies, they also showed that regulation of these genes by VGCC is dependent on Rho/ROCK signaling and myocardin, resulting in increased binding of SRF to CArG cis-regulatory elements in the SM-α actin/SM-MHC promoters [41].

Chromosome structure modifiers, such as histone deacetylases (HDACs), regulate gene expression *via* the control of chromatin structure and function. Tang et al. [42] showed that class I HDAC inhibition acting *via* suppressed Notch signaling attenuates expression of differentiation markers including SM MHC, SM α-actin, and calponin. Notably, in response to *oxidized* phospholipids, HDAC2 and HDAC5 recruitment and hypoacetylation of histone 4 at the SM-actin promoter can act to promote de-differentiation [43]. On the other hand, HDAC7 has been linked to increased recruitment of SRF-myocardin complex to SM22α promoter and activation of SMC marker gene expression, which eventually leads to induction of stem cell differentiation towards an SMC lineage [44]. While the effect of oxidized lipids is intriguing and in line with the general theory of

ROS-mediated differentiation, the effects of HDACs to positively and negatively control SMC contractile gene expression is expected to be far more complex.

MicroRNAs (miRNAs) are endogenous, single-stranded, short, non-coding 19-22-nucleotide RNAs, which are highly conserved and regulate gene expression either by mRNA degradation or by inhibiting mRNA translation. Since miRNAs are expressed in a developmental stage- and/or tissue-specific manner, they are presumed to play a key role in cell differentiation, proliferation, and apoptosis [45]. Several reports demonstrate that miR-143 and miR-145 are the most abundant miRNAs in differentiated SMCs and are essential for maintaining a contractile phenotype [46,47]. They are acknowledged to be positively regulated by SRF and myocardin and to promote SMC differentiation by downregulating factors such as KLF4 and Elk-1 [47]. Similarly, PDGF, another important repressor of SMC differentiation, can induce the expression of miR-221, which in turn suppresses c-Kit expression and evokes the downregulation of myocardin and downstream SMC marker genes, promoting SMC regression from a contractile to a synthetic phenotype [48]. More recently, our laboratory demonstrated that hypoxia-induced upregulation of miR-214 can pleiotropically downregulate MEF2C and leiomodin 1 (LMOD1), a SMC-specific contractile protein, resulting in multi-tiered control of SMC dedifferentiation and proliferation in pulmonary arterial hypertension (PAH) [3]. By extension, it could be postulated that a reported ROS-mediated activation of MEF2C [49] would serve as a counterregulatory mechanism to neutralize this effect. Other microRNAs, including miR-663 [50], miR-23b [51], and miR-133 [52] have also been implicated in vascular SM phenotype and remodeling. Accordingly, miRNAs are proving to be vital players modulating SMC differentiation and are discussed *extensively* in Chapter 14.

Well established as important signaling molecules in the cardiovascular system, ROS are increasingly being recognized as key upstream modulators of SMC differentiation [53,54]. Among the several cellular sources of ROS, including mitochondria, lipoxygenases, and cyclooxygenases, Nox make up a predominant source in a majority of cell types including VSMCs [55]. In addition to mediating SMC proliferation, migration, apoptosis, and the secretion of inflammatory cytokines [56], Nox-derived ROS have for some time been proposed to transmodulate SMC phenotype [35,57]. Studies from our laboratory and others have revealed that mechanical forces stimulate Nox [58–60]. Indeed, cyclic stretch induces MEF2B that, in turn, promotes Nox1 gene expression and ROS in transitioning VSMCs to a synthetic phenotype [2]. Al Ghouleh et al. [61] also identified a novel facilitative role for scaffolding protein EBP50 (aka NHERF1) as a positive modulator of the Nox1 oxidase complex. On AngII stimulation of VSMCs, EBP50 binds via its PDZ domains to the Nox organizer subunit $p47^{phox}$ at a unique PDZ-binding C-terminal domain (4 aa), and, in turn, enhances Nox1-derived ROS and attendant SMC hypertrophy and resistance artery vasoconstriction. EBP50 is also known for its influential roles concerning G-protein coupled receptors and actin cytoskeleton reorganization [62]. In fact, EBP50 could be an intermediary between growth factors like AngII and the cytoskeletal proteins. Similarly, Janiszewski et al. [63] showed that the thiol oxidoreductase protein, protein disulfide isomerase (PDI), is also a modulator of the Nox1 oxidase complex in VSMC. PDI is a chaperone from the endoplasmic reticulum that is involved in protein trafficking, cell adhesion, and NO internalization [64–66]. Upon AngII stimulation of VSMC, PDI translocates to the plasma membrane, which can associate with Nox1 oxidase and regulate Nox1-derived $O_2^{\bullet-}$ and promote vascular dysfunction [63,67]. Indeed, diverse signaling molecules and cross-talk between pathways are expected to influence SMC differentiation and function [68–70]. As acceptance of the role of ROS as a common intermediary in these pathways grows, we aim to present an overview of the current state of knowledge on (1) Nox isozyme distribution, structure, and expression in arterial SMC; (2) Nox function in SMC phenotypic switching and vascular remodeling; and (3) Nox inhibitors and widely employed methods to assay Nox-derived ROS.

17.3 THE NADPH OXIDASE FAMILY AND VASCULAR SMOOTH MUSCLE CELL FUNCTION

Noxs comprise an enzyme class whose main function described to date is reportedly the production of ROS [71,72]. The first Nox to be identified and defined by its catalytic core, cytochrome b_{558} subunit gp91*phox* (now Nox2) is highly expressed in phagocytic cells. It is widely held that Nox2-derived ROS are instrumental in host defense and killing of bacterial and fungal pathogens [73]. This was later broadened by multiple groups to include the ability of the Nox to acidify phagosomes [74]. Defects in the Nox2 function result in a severe immunodeficiency, and individuals suffering from chronic granulomatous disease (CGD), a rare genetic disorder caused by mutation in Nox2 subunits, are highly susceptible to life-threatening infection by bacteria and fungi [75,76]. Phylogenetic analyses shed light on highly conserved core Nox motifs in evolution from its earliest orthologues in fungus to those in higher primates [77]. In mammalian systems, Noxs are ubiquitous and implicated in physiological functions ranging from self-renewal to cell death. Noxs in the vasculature (i.e., endothelial cells, fibroblasts and VSMCs) subserve key roles in cytoskeletal rearrangement, cell migration, proliferation, hypertrophy, and apoptosis, as well as endothelium-dependent vasodilation and constriction to name a few of an ever-growing list of cellular functions [78–81]. In vascular cells, ROS are generated by mitochondrial electron transport, heme oxygenase, cyclooxygenase, lipoxygenase, NO synthase, xanthine oxidase, cytochrome P-450, and Nox [82–87]. Of these, Nox is commonly termed the major source of cellular ROS and a *professional* producer of superoxide anion and hydrogen peroxide (H_2O_2). That is, its only ascribed catalytic function to date is the generation of these ROS as agents of physiological redox signaling, antimicrobial action, and oxidative stress. Indeed, with respect to the latter, at high levels of production Nox has been extensively implicated in a variety of vascular diseases [72,88].

17.3.1 The NADPH Oxidase Family: Structure and Activation

The Nox isozymes belong to a family of transmembrane proteins responsible for catalyzing the reduction of molecular oxygen (O_2) to generate superoxide ($O_2^{\bullet-}$) and/or H_2O_2 expending NADPH as an electron donor [72,73,89]. The oxidase family is composed of seven isozymes, Nox1 through 5, as well as DUOXs 1 and 2. The isozymes take the name of their main heme-containing catalytic subunit. All share common characteristics which include a catalytic core in the transmembrane portion of the protein and NADPH and FAD binding sites on their C-terminal tail [89]. For

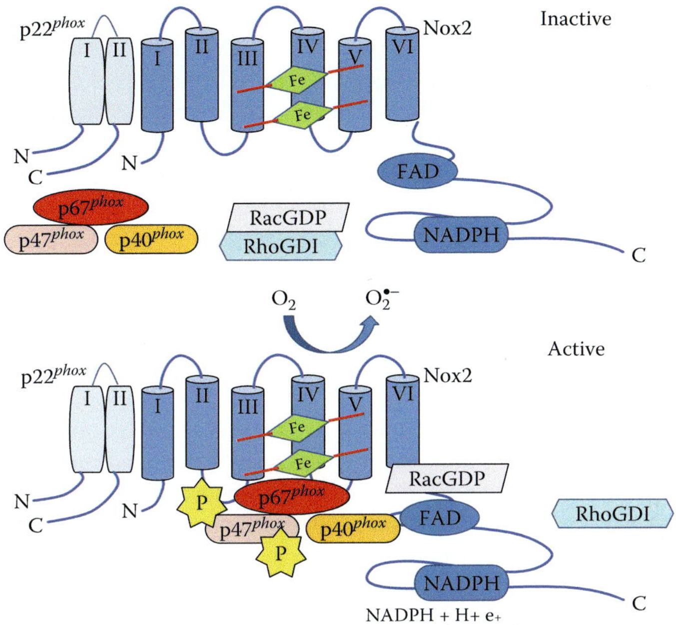

FIGURE 17.1 Structure and activation of phagocyte Nox2*. Nox2 oxidase consists of membrane-bound cytochrome b_{558} (Nox2 and p22*phox*) and cytosolic subunits p47*phox*, p67*phox*, p40*phox*, and small GTPase Rac. Upon stimulation, cytosolic subunits translocate to the membrane and interact with Nox2 and p22*phox*, activating the replete Nox2 enzyme complex and generating $O_2^{\bullet-}$. *The complex, as with all Nox isozymes, is defined by its core subunit.

the purposes and scope of this chapter, we will delve more deeply into Nox1, Nox2, Nox4, and Nox5, which are well described to be expressed in vascular cells [79]. These isoforms differ in their tissue distribution, subcellular localization, regulatory mechanisms, levels, nature of ROS produced, and specific cytosolic components (Figure 17.1 illustrates Nox2 isozyme composition).

17.3.1.1 NADPH Oxidase 2

The prototypical Nox is the phagocyte Nox2 (aka gp91*phox*, wherein *phox* refers to phagocytic oxidase). It was first described in phagocytes [90–92] and an NADPH-driven oxidase system was subsequently reported in the human fibroblasts and the vasculature [79,93–96]. Following these seminal discoveries in non-phagocytic cells, the Nox field grew rapidly to include findings of a similar $O_2^{\bullet-}$ -generating activity in myriad cell types ranging from cancer cells [97–99] to those of the central nervous system [100–102]. Given that Nox2 was the first isozyme to be identified, was well characterized, and was elementally related to other Nox isozymes described after it, it befittingly assumed its place as the prototype of the Nox family. Nox2 is a heteromultimeric enzyme consisting of two membrane-bound components, Nox2 (aka gp91*phox*) and p22*phox*, that together form a heterodimeric flavoprotein (cytochrome b_{558}). Nox2 is composed prototypically of six predicted transmembrane helices based on hydropathy plots [91] containing two iron-heme groups at its core that make up the electron transport machinery of the enzyme complex and binding sites for NADPH and flavine adenine dinucleotide (FAD) on its cytosolic C-terminus dehydrogenase domain (DHR). This composite structure funnels the flow of electrons from NADPH to FAD (*via* specific binding motifs on the DHR) on to iron-heme groups of the cytochrome, and lastly to O_2 to generate $O_2^{\bullet-}$ (Figure 17.1 illustrates Nox2 structure and activation). Nox2's enzymatic activity is tightly regulated by complex interactions among its various membrane-bound as well as cytosolic subunits [89,91]. For a fully assembled and active Nox2 oxidase, the cytochrome b_{558} associates with cytosolic components p47*phox* (organizer subunit), p67*phox* (activator subunit), p40*phox*, and the small Rho family GTP-binding protein Rac1 or Rac2 [103]. Phosphorylation of p47*phox* at key serines (ser303, ser304, ser359, and ser 370) [104,105] is affected by a variety of kinases, that is, protein kinase C (PKC) isoforms (α, β, δ, and ζ), MAPK, p21-activated kinases (PAK), and AKT/PKB [106,107]. Phosphorylation of p47*phox*, in turn, disinhibits a well-characterized auto-inhibitory region (AIR) on latent p47*phox*, which under normal conditions prevents its SH3 domain from association with a crucial proline-rich region on p22*phox*. p47*phox* phosphorylation is a vital part of Nox assembly and activation, since mutation of ser359 and ser370 abolish p47*phox* association with p67*phox* and its translocation to the plasma membrane [104,105]. Interaction of p47*phox* with p67*phox*, p40*phox*, Rac and p22*phox* subunits results in formation of the newly formed complex to the plasma membrane and its binding to cytochrome [89,103,108]. This step is also essential for proper orientation of the activation domain on p67*phox* with Nox2's C-terminal DHR allowing for electron transfer from NADPH to FAD, the rate-limiting function in the generation of $O_2^{\bullet-}$. Exchange of GDP for GTP on the small protein Rac, with subsequent activation and translocation to the membrane, occurs simultaneously and is coordinated with the p40*phox*/p47*phox*/p67*phox* complex [103]. These mechanisms allow for a tightly regulated enzyme and controlled production of ROS upon induction by specific stimuli.

17.3.1.2 NADPH Oxidase 1

Following the discovery of Nox2 in the vasculature, its homolog Nox1 (previously Mox1) was identified, displaying 56% sequence identity to Nox2 [109,110]. An alternatively spliced shorter form of Nox1 lacking exon 11 (Nox1v) is expressed in colon cells [111–113]; however, this spliced variant is reportedly translated into a protein incapable of producing $O_2^{\bullet-}$ [112]. Similar to Nox2, Nox1 is predicted to contain six transmembrane α-helical domains, motifs for the binding of NADPH and FAD moieties, and two iron-heme clusters, which coordinate generation of $O_2^{\bullet-}$. Like Nox2, Nox1 is directly associated with and stabilized by p22*phox* [114] and forms an active complex with organizing NOXO1 subunit (homolog of p47*phox*), activating NOXA1 subunit (homolog of p67*phox*), and Rac1 [115] (Figure 17.2 illustrates Nox 1, 4, and 5 protein structure). Unlike the p47*phox* subunit, the lack of an AIR in NOXO1 allows for its constitutive interaction with p22*phox* [89]. Indeed, the co-expression of Nox1-p22*phox* heterodimer with NOXO1 and NOXA1 components lead to spontaneous $O_2^{\bullet-}$ production, which is potentiated by phorbol ester (PMA) stimulation [116]. Also, evidence suggests that Nox1-derived $O_2^{\bullet-}$ can also be mediated by p47*phox* and p67*phox* [117,118], but not as efficiently as by NOXO1 and NOXA1 [119].

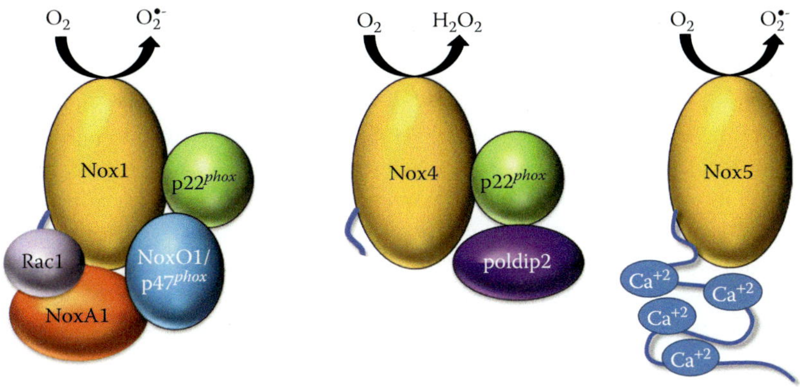

FIGURE 17.2 Schematic diagram of Noxs in VSMCs. In addition to Nox2 (primarily ascribed to microvessels), Nox1, Nox4, Nox5, and cytosolic subunits NoxO1 and NoxA1, which are homologs of p47*phox* and p67*phox*, respectively, are all putatively expressed in diverse permutations in SMCs. Of note, Nox4 and Nox5 do not require known cytosolic components to induce oxidase activation.

17.3.1.3 NADPH Oxidase 4

Nox4, originally dubbed *Renox* based on its relatively high expression in the kidney, shares 39% homology with Nox2 [120]. Like Nox1 and Nox2, Nox4 possesses NADPH binding sites, FAD domains, and two iron-heme groups [89] (Figure 17.2). Intriguingly, HEK293 cells expressing Nox4 produce large amounts of H_2O_2, but arguably no detectable $O_2^{\bullet-}$ [91]. As such, abundant Nox4-derived H_2O_2 may be consequential of a greater enrichment of this isoform in intracellular compartments colocalized with high levels of superoxide dismutases (SODs) that dismute $O_2^{\bullet-}$ into H_2O_2. Interestingly, unlike that in Nox1 and Nox2 the third extracytosolic loop (E-loop) of Nox4 is 28 amino acids longer containing a highly conserved histidine predicted to donate protons that *ramp up* the spontaneous dismutation of $O_2^{\bullet-}$ into H_2O_2 [121]. Importantly, heterodimerization with p22*phox* is well known to be necessary and sufficient for Nox4 oxidase's stability and ROS-generating activity [91,108]. In fact, Nox4 activity seems to be transcriptionally modulated as there is a strong correlation between Nox4 expression and ROS production [122,123]. In keeping with this less complex configuration, Nox4 activity does not appear to be dependent on the currently identified regulatory subunits as p47*phox* and/or NOXO1, p67*phox* and/or NOXA1, and Rac. These findings suggest that Nox4:p22*phox* complexation alone functions sufficiently, but more recent evidence elegantly showed that it can be regulated by polymerase-δ-interacting protein (Poldip2) [124]. Whether the latter is through a direct interaction or *via* cytoskeletal changes is not entirely clear. That being said, the verdict is still not out on other potential regulatory elements in Nox4.

17.3.1.4 NADPH Oxidase 5

Nox5 is a 565-amino acid protein with 27% identity to Nox2. Like Nox1 and Nox2, Noxs 4 and 5 possess the requisite DHR and iron-heme motifs which are responsible for electron transfer [89] (Figure 17.2). However, unlike other Noxs, Nox5 contains a long intracellular N-terminal domain with four calcium-binding EF-hand motifs that render this oxidase highly sensitive to calcium [125]. Moreover, Nox5 does not require p22*phox* or other known regulatory subunits for activation [126,127]. A rise in intracellular calcium levels activates Nox5 by binding to the N-terminal EF-hand motifs, which, in turn, allosterically alters the conformation of this domain enabling an intramolecular interaction between N-terminal and C-terminal domains [128]. Phosphorylation of T494 and S498 of Nox5 augments the sensitivity of the enzyme for calcium leading to an increase in $O_2^{\bullet-}$ generation [129]. This is consistent with Nox5 being activated in either a one- or two-step process. Interestingly, distinct products of the Nox5 gene: Nox5α, Nox5β, Nox5γ, Nox5δ, and Nox5ε (a.k.a. Nox5s) differ in the primary sequences of their calcium-binding regions and tissue distribution [130].

17.3.2 NADPH Oxidase Family and Vascular Smooth Muscle Cell Function

The production of ROS by vascular cells can play pivotal signaling roles essential for the maintenance of vascular homeostasis, regulating salutary processes such as vascular remodeling, migration, and self-renewal (Figure 17.3). In response to pathological stimuli, an unabated rise in ROS levels promotes VSMC dysfunction *via* activation of anti-inflammatory pathways, upregulation of adhesion molecules, and abnormal cell proliferation. In this way, Nox isozymes are important players in the physiology and pathophysiology of blood vessels. Vascular cells express Nox1, Nox2, Nox4, Nox5, and p22*phox* as well as cytosolic components p47*phox*, p67*phox*, NOXA1, NOXO1, and Rac. Importantly, expression profiles (i.e., unique combinations of membrane-integrated and cytosolic subunits) are expected to vary widely depending on the vascular bed. Evidence supports that Nox2 is expressed more commonly in VSMC from resistance vessels and not conduit arteries [55], whereas p67*phox* expression is still somewhat elusive. The vascular expression of Nox subunits is species-specific. In other words, whereas Nox1 and Nox4 are expressed in SMC in rodents and higher mammals and primates, Nox5 is missing from rats and mice and expressed in higher mammals including primates [56]. Moreover, Nox proteins are differentially distributed within the VSMC, with Nox1 being expressed in plasma membrane, caveolae, and endosomes [131,132], Nox2 primarily in the plasma membrane [71,108], and Nox4 in the nucleus, endoplasmic reticulum, mitochondrion, and focal adhesions [133]. Nox5 has been traced to the plasma membrane and endoplasmatic reticulum [130].

The application of Nox inhibitors is key to elucidating our understanding of the contribution of Nox to various physiological and pathophysiological functions in vascular cells. In this way, Ang II-induced activation of Nox-derived ROS has been well characterized to set in motion a number of factors that promote hypertension [134] (Figure 17.3). AngII stimulates Nox1, Nox2, Nox4, and p22*phox* expression and attendant ROS production, which seems to be pivotal for vascular dysfunction and hypertension [135]. Early in the characterization of the vascular Nox, it was widely accepted that Nox-derived $O_2^{\bullet-}$ elicited rises in blood pressure and impairments in vasomotor tone by direct interference with NO bioactivity.

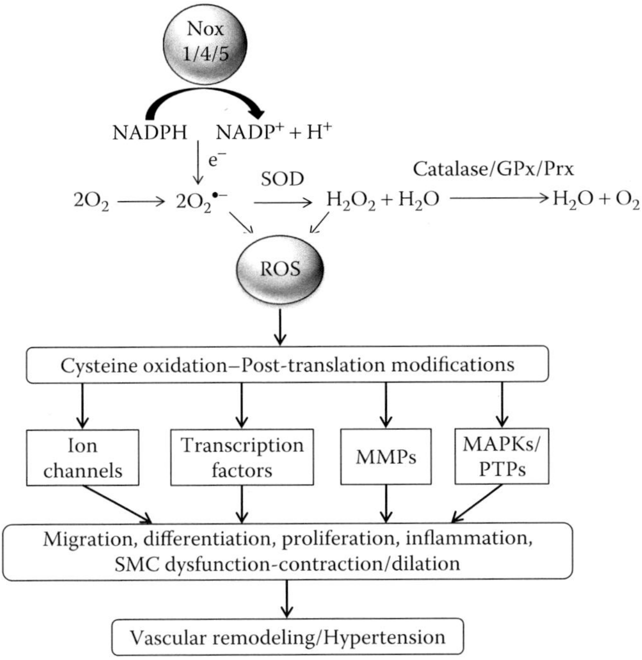

FIGURE 17.3 Redox-sensitive pathways activated by NADPH oxidase.

Over the years, it has become increasingly clear that the Nox-derived ROS play myriad roles in vascular abnormalities contributing to hypertension via ROS- and reactive nitrogen species (RNS)-specific activation of a host of factors ranging from MAPKs to a wide array of transcription factors (see earlier in the chapter for more detail). Interestingly, studies showed that localized overexpression of Nox1 in VSMC promotes an increase in systemic blood pressure and hypertrophy in response to Ang II [136], whereas deletion of Nox1 prevents Ang II-induced hypertrophy and averts increases in blood pressure of mice [137]. Also, p22^{phox} overexpression in VSMC elevates Nox1 in the vessel wall (perhaps via stabilization), as well as enhancing the Ang II-mediated VSMC hypertrophy and blood pressure response [138,139]. On the other hand, systemic suppression of p22^{phox} prevents this phenotype [140].

As should be evident from previous sections, Noxs play a major role in mediating contractility, proliferation, and migration as well as participating in the development of various vascular disorders like atherosclerosis, hypertension, and ischemia-reperfusion injury. Genetically modified animal models are powerful tools for understanding the signaling/cellular mechanism involved in disease pathogenesis. However, isoform-specific inhibitors (1) have the added advantage of providing temporal and acute silencing of a pathway without physiological compensation; (2) selectively interfere with a specific and delimited function of the Nox; and (3) form the foundation for the development of therapeutics. Recently, Nox1 was validated as a therapeutic target for ischemic retinopathy

[141] and retinas from Nox1 KO mice, but not from Nox2 and Nox4 KOs, are protected from neovascularization, capillary vaso-obliteration, vascular leakage, and vascular adherence of leukocytes, due to a decrease in ROS-induced VEGF expression [141]. Furthermore, consistent with this, retinal vascular cells isolated from a rodent model exposed to hyperoxia treated with GSK137831, a Nox1/Nox4 inhibitor, revealed a reduction in Nox1-induced ROS production and the progression of the vasculopathy [141]. A few other studies have illustrated that GSK137831 prevents Ang II-induced cardiac fibroblast proliferation and migration *via* inhibition of Nox4-derived ROS [142], and can attenuate hypoxia-induced right ventricular hypertrophy, vascular cell proliferation, and pulmonary hypertension development [143].

With the advent of highly specific Nox inhibitors such as Nox2ds-*tat* and NoxA1ds, their use is proving to be an excellent strategy to intervene in the progression of animal model of disease such as PAH. For instance, PAH is a rapidly degenerating disease characterized by elevated pulmonary vascular resistance (PVR) and a chronic rise in pulmonary arterial pressure (PAP) [144–146]. Moreover, in animal models, chronic hypoxia recapitulates rises in PVR and hallmark vascular changes occurring in human PAH [147]. Furthermore, VSMC and endothelial cell dysfunction appear to play an important role in vascular remodeling characterized by cell proliferation, migration, and vascular muscularization [148,149]. Nox1-induced ROS was reported to play pivotal role in PAH progression [150], and recently, Ranayhossaini et al. [151] showed that NoxA1ds is able to reduce hypoxia-induced endothelial cell proliferation via Nox1-derived $O_2^{\bullet-}$ production, which is a major characteristic of PAH development [151].

Nox2ds-*tat* inhibits $O_2^{\bullet-}$ production in endothelial cells in response to various stimuli, that is, hypoxia [152], atrial natriuretic peptide [153], nutrient deprivation [154], calcineurin inhibitors [155], and shear stress [156]. Likewise, Nox2ds-*tat* also inhibits Ang II-induced ROS production in human artery resistance SMC [55]. Moreover, infusion of Nox2ds-*tat* delays elevation in blood pressure and prevents Ang-II-induced hypertrophy, suggesting a bimodal effect of AngII on VSMC hypertrophy; i.e. one as a response to ROS-mediated rises in pressure and another an intracellular direct response via ROS-sensitive signaling [157,158]. The situation appears distinct in spontaneously hypertensive rats [159,160], wherein treatment with Nox2ds-*tat* exhibited a more profound lowering of blood pressure by way of baroreflex action *via* Nox2-induced ROS [159]. Furthermore, like Nox1, several studies have reported that Nox2 participates in vascular disorders, i.e. endothelial cell dysfunction, atherosclerosis, and restenosis after arterial injury in experimental models employing Nox2 knockout models and by the inhibition of Nox2 with Nox2ds-*tat* [161–165].

17.3.2.1 NADPH Oxidase 1 and Vascular Smooth Muscle Cell Function

Nox1 is characterized by basal protein expression and activity capable of generating relatively low amounts of $O_2^{\bullet-}$ in VSMC [166]. However, in response to growth factors (i.e., PDGF), vasoactive agents (i.e., angiotensin II), and mechanical forces (i.e., cyclic stretch), Nox1 produces high levels of $O_2^{\bullet-}$, which can participate in the development of vascular disease [2,167–169].

Platelet-derived growth factor (PDGF) is well-known to play an important role in the VSMC, promoting a phenotypic switching from contractile to a synthetic state [170]. Briefly, the binding of PDGF to its receptor induces receptor tyrosine kinase activation and autophosphorylation as well as of several target proteins including PKC. This activation, in turn, enhances Nox1-induced ROS production culminating in VSMC migration [171,172]. Moreover, PDGF-induced Nox1 activity stimulates adhesion and extracellular matrix production, which associated with cell proliferation contributes to neointima formation [173], a key hallmark of atherosclerosis. Reportedly, PDGF modulates subcellular redistribution of the endoplasmic reticulum PDI and interaction with RhoGDI, which enhances stress fiber organization, focal adhesion number and VSMC migration [174]. Along these lines, evidence supports that PDGF stimulates Nox1-dependent ROS-mediated cofilin, which regulates actin-filament turnover [175,176]. In addition, PDGF mediates activation of slingshot phosphatase via oxidation and release of 14-3-3. This, in turn, permits slingshot phosphatase and activation of cofilin [167,171], required for VSMC migration [171]. In this context, Nox1 can also propagate its signal through the c-Src-mediated activation of the epidermal growth factor receptor (EGFR), ERK 1/2, and matrix metalloproteinase 9 (MM9), all acting in concert to promote VSMC migration (Figure 17.3) [177]. Furthermore, studies have linked Nox1-mediated VSMC function through other growth stimulants such as serum and basic FGF (bFGF). That is, serum and bFGF-induced Nox1 activity promote VSMC proliferation and migration [168,178].

Another important regulator of Nox1 expression and activity in the VSMC is the pro-constrictor hormone angiotensin II (Ang II). Ang II exerts its action via Ang II type 1 (AT_1R), a G-protein coupled receptor (Gq), which relays its effect on Noxs via PKC and phospholipase C (PLC) [179], phospholipase A_2 (PLA_2) and phospholipase D (PLD) [180,181]. Through the AT1R, Ang II is a potent regulator of Nox1, Nox2, and Nox5 expression and activity [180], as well as p22*phox* and the cytosolic subunits p47*phox* and p67*phox* in VSMC [56,180,182]. However, the precise biological role of different Nox enzymes activated by Ang II remains unclear. Unique differences in tissue distribution, cellular localization, and subcellular compartmentalization are likely to play key roles in Nox specific action induced by Ang II [180,183]. In VSMC, AngII promotes hypertrophy, hyperplasia, and migration, which can all combine to reduce blood flow and participate in the development of hypertension [184–186]. Indeed, Ang II-induced hypertension is enhanced in transgenic mice overexpressing Nox1 in VSMC [136]. Consistent with this effect, the depletion of Nox1 prevented the development of hypertension in response to AngII [137,187]. Indeed, Ang II is a well-known constrictor of VSMC. Ang II-induced H_2O_2 promotes an increase in the intracellular calcium (Ca^{2+}) concentration by means of the opening of voltage-operated calcium channels on the plasma membrane and release of calcium from intracellular stores (Figure 17.3). The subsequent rise in cytosolic calcium stimulates calmodulin-dependent myosin light chain kinase. Parallel to this effect, myosin-light chain may be phosphorylated secondarily to oxidation-induced MAPK activation. In aggregate, a combined effect of these pathways elevates myosin light chain kinase phosphorylation and VSMC contraction [188,189]. Accordingly, these and other mechanical forces may promote VSMC proliferation, migration, and extracellular matrix remodeling [190,191] and contribute to the development of a wide array of vascular diseases.

Rises in blood pressure and blood flow are well-known mechanical forces capable of altering the phenotype of the VSMC from differentiated to a synthetic phenotype [190,192]. The synthetic phenotype is characterized by an increase in VSMC migration, abnormal extracellular matrix production, and loss of contractility, which can induce the development of vascular injury and diseases such as hypertension and arteriosclerosis [193]. Recently, Rodriguez et al. [2] showed that cyclic stretch in vitro, under conditions mimicking high blood pressure in humans, can upregulate Nox1 expression and activity in a time-dependent manner, *via* myocyte enhance factor 2B (MEF2B). Nox1-induced ROS production was subsequently shown to modulate several proteins related to the switch of VSMC to a synthetic phenotype, as well as to promote VSMC migration [2].

17.3.2.2 NADPH Oxidase 4 and Vascular Smooth Muscle Cell Function

Nox4 is reportedly the isoform with the highest profile of expression in the vascular wall [79,194]. In VSMC, like Nox1, Nox4-induced ROS is implicated in a variety of physiological effects, such as vascular tone development [195], differentiation [35], as well as pathophysiological processes including proliferation and migration [56,196]. In VSMC, Nox4 is upregulated by several stimuli such as IGF-1 [197], hypoglycemia [198], and hypoxia [199]. Although Nox4 reportedly participates in SMC differentiation *in vitro*, scant information exists for Nox4 and differentiation *in vivo*.

One of the most widely accepted roles of Nox4 in the VSMC function is its participation in cell differentiation. The differentiated phenotype of VSMC in the vascular wall is crucial to maintaining contraction and the patency of a healthy blood vessel [1]. Furthermore, VSMCs are susceptible to a phenotypic switching in response to vascular injury, which is enabled by ROS [56,133]. Nox4-induced ROS play an essential role in the maintenance of SM-MHC, SM α-actin, H-caldesmon, and calponin, as well as the preservation of the differentiated phenotype of VSMC [35]. Interestingly, Clempus et al. [35] showed that expression of Nox4 is reduced by almost 50% in late passage cells, which inversely correlated with expression of contractile proteins, such as SM α-actin, calponin, and SM-MHC. They also suggested that Nox4-derived ROS, that is, H_2O_2, *via* regulating SRF was essential for the maintenance of the differentiated phenotype of VSMC. In differentiated VSMC in early cell passages Nox4 co-localizes with SM α-actin-based stress fibers and with focal adhesions in the late cell passage [35]. Therefore, dependence on not only the source but the localization of ROS in cellular compartments underscores the importance of compartmentalized signaling. On the other hand, it seems plausible to postulate that Nox4-induced ROS in focal adhesions could activate a different pathway leading to inhibition of SRF or others transcription factor that represses VSMC gene expression. In brief, additional studies are essential to understand the role of compartmentalized ROS signaling in regulating VSMC gene expression.

It is well recognized that Nox4 plays a role in VSMC migration [200]. Recently, Poldip2 was shown to associate with Nox4 and increase its activity and to be involved in cytoskeletal remodeling and migration [124,167]. In addition, Datla et al. [196] showed that the interaction between Poldip2 and Nox4 regulates focal adhesion turnover, traction force generation, and force polarization, which, in turn, promote

VSMC migration [124,196]. Recently, Matsushima et al. [201] showed that in cardiomyocytes Nox4 activity could also be regulated post-transcriptionally through phosphorylation. It was demonstrated that Fyn, a non-receptor tyrosine kinase, directly interacts with the C-terminal domain of Nox4 and phosphorylates tyrosine 566 and negatively regulates Nox 4-derived ROS production and apoptosis. Venema et al. [202] illustrated that upon AngII stimulation of VSMC, Fyn is critical to the activation of the Janus-activated kinase/STAT pathway. Fyn present in VSMC could also play a key role mediating Nox4 activity and further regulation of VSMC migration and differentiation. Moreover, IGF-1 appears to be a potent mitogen and anti-apoptotic factor for VSMC [197,203] and modulates Nox4-induced ROS to promote VSMC migration [197]. Another IGF binding protein IGFBP-3 associates with TGF-β in persistent hypoxia leading to an increase in TGF-β signaling, which promotes cell proliferation [204]. As mentioned, hypoxia is a recognized stimulus for Nox4 expression and activity in VSMC [199]. Briefly, hypoxia induces an increase in TGF-β expression, which can regulate Nox4-induced ROS and PI3K/Akt signaling leading to VSMC proliferation and remodeling [204]. Not only does hypoxia-inducible factor-1α (HIF-1α) play a major role in the transcriptional regulation of Nox4 [199], Nox4-induced ROS reportedly promotes stability of HIF-2α [205]. Thus, it appears that Nox4 can participate upstream or downstream of HIF signaling pathways promoting VSMC proliferation. Accordingly, Nox4 has been shown to be sensitive to pulmonary oxygen levels [206], and Nox4 protein levels were increased in pulmonary aortic SMC isolated from mice exposed to chronic hypoxia [205]. In fact, findings suggest that patients with idiopathic PAH present high levels of Nox4 in the vessel media [207]; thus, functional abrogation of Nox4 may be expected to reduce pulmonary arterial SMC proliferation and may provide a novel therapeutic target for hyperproliferative vascular disease. Keeping this in mind, it is equally important to understand that the role of Nox4 in vascular function is likely to depend on various factors including inducers (i.e., growth factors and oxygen tension), cellular localization, as well as on the physiologic state of the cell [208,209].

17.3.2.3 NADPH Oxidase 5 and Vascular Smooth Muscle Cell Function

Nox5 is expressed in human VSMC and is induced by PDGF, c-abl, Ca^{2+}-calmodulin-dependent protein kinases, MAPKs, thrombin, cAMP-response-element-binding protein (CREB), and PKC [128,210–212], and can be found in a variety of human vessels including coronary arteries, umbilical veins, pulmonary arteries [128,213], aortic SMC [210], and microvascular endothelial cells [128]. Like Nox1 and Nox4, Nox5 plays a major role in VSMC proliferation. PDGF-induced VSMC proliferation is heavily dependent on JAK/STAT activation [210]. Briefly, the interaction between the ligand and JAK receptor leads to a phosphorylation in its tyrosine kinase residues, which also phosphorylates and associates with STAT proteins—latent transcription factors localized in the cytoplasm [214]. Once phosphorylated, JAK/STAT enter into the nucleus to promote gene expression and VSMC proliferation [210]. However, the precise mechanism by which Nox5 might modify PDGF-induced proliferation is unknown; but given the sensitivity of protein tyrosine phosphatases (PTPs) to ROS, it seems that Nox5-derived ROS could inactivate PTPs and hence increase JAK activity [210]. Furthermore, Nox5-induced oxidant production may promote cysteine oxidative post-translation modification (Oxi-PTM) including S-glutathionylation, N-nitrosylation, and sulfhydration, leading to a change in VSMC function [215]. Of these PTMs, sulfhydration has been shown to decrease PTP1B [216]. Moreover, an increase in Nox5 in patients with coronary artery disease compared with healthy subjects [213], and temporal and spatial changes in its expression suggest that Nox5-derived ROS promotes atherosclerosis, perhaps in a stage-dependent manner. These findings imply that Nox5-derived ROS is necessary for the development of atherosclerosis [213]. Consistent with this notion, Gole et al. [217] demonstrated that Nox5-derived ROS increases VSMC migration by way of intermediate-conductance Ca^{2+}-activated K^+ channel (KCNN4) activation [217]. This field of study awaits the generation of SMC-specific overexpression and knockdown of individual Nox isoforms to more clearly delineate the physiological role of each isozyme in SMC function.

17.4 METHODS FOR REACTIVE OXYGEN SPECIES AND REACTIVE NITROGEN SPECIES DETECTION

Given the importance of ROS in physiology and pathophysiology of vascular cells, it is necessary to identify the specific reactive species being produced, their localization, quantity, and temporal generation in response to a stimulus [218]. There are a variety of methods available to detect ROS, each with advantages and pitfalls (Table 17.1), and despite progress with the development of new probes, the detection and quantification of these species remain a challenge. In this section, we will survey the current and most-commonly employed methods to monitor ROS production in vascular cells.

17.4.1 HYDROETHIDINE AND MITO-HYDROETHIDINE

Hydroethidine (HE; also known as dihydroethidium—DHE) and its mitochondria-targeted analog (Mito-HE or MitoSOX) are redox-sensitive probes used to measure $O_2^{\bullet-}$ production at the whole cell- and mitochondria-level, respectively. HE and Mito-HE share the same basic chemical structure, and both are cell-permeant and rapidly react with $O_2^{\bullet-}$. Once inside the cell, HE is found predominantly in the cytosol, while Mito-HE accumulates primarily in mitochondria. The reaction of HE and Mito-HE with $O_2^{\bullet-}$ yields the formation of a specific metabolite, 2-hydroxyethidium (2-OH-E^+) and 2-hydroxy-Mito-ethidium (2-OH-Mito-E^+), respectively. Despite the specificity of these metabolites as indicators of $O_2^{\bullet-}$, HE and Mito-HE undergo oxidation by peroxynitrite ($ONOO^-$), H_2O_2, and hydroxyl radical (OH) yielding nonspecific byproducts, that is, ethidium (E^+), ethidium dimer (E^+-E^+), Mito-E^+, and Mito-E^+-Mito-E^+ dimer. Nonetheless, E^+, E^+-E^+ dimer, and Mito-E^+, Mito-E^+-Mito-E^+ dimer are non-fluorescent products.

Fluorescence microscopy is a convenient and thus often exploited methodology to detect products derived from HE and Mito-He oxidation in intact cells or tissue. An excitable red fluorescence emitted as a consequence of their oxidation is used to quantify $O_2^{\bullet-}$. However, considering that levels of E^+ and Mito-E^+ detected in biological samples are ten-fold higher than 2-OH-E^+ and 2-OH-Mito-E^+, it is

TABLE 17.1
ROS Probes

Probe	Application	Advantages	Disadvantages
Hydroethidine (HE)	Detects $O_2^{\bullet-}$	• High specificity for $O_2^{\bullet}-$ • Cell membrane permeant	• Can be oxidized by other ROS and RNS species • Fluorescence-based techniques should be employed with caution and validated with HPLC
Mito-HE (Mito-Sox)	Detects mitochondrial $O_2^{\bullet-}$	• High specificity for $O_2^{\bullet}-$ • Accumulates in mitochondria	• Can be oxidized by other ROS and RNS • Fluorescence-based techniques should be employed with caution and validated with HPLC
Hydropropidine	Quantifies $O_2^{\bullet-}$ production in cell/tissue lysates, subcellular organelles or cell-free assays	• High specificity for $O_2^{\bullet}-$ • Cell membrane impermeant	• Can be oxidized by other ROS and RNS
Cytochrome C	Quantifies $O_2^{\bullet-}$ production in cell/tissue lysates, subcellular organelles or cell-free assays	• High specificity for $O_2^{\bullet}-$	• Relatively low sensitivity • Membrane impermeable
Amplex Red	Detects H_2O_2 production	• High specificity and sensitivity for H_2O_2	• Light sensitive: photo-oxidation can artificially enhance fluorescence
DCFH-DA	Detects oxidation	• Cell-permeant • May be used as a marker for oxidative stress in cells and tissues	• Light sensitive • Low selectivity: interacts with a variety of ROS and RNS, particularly $ONOO^-$
CBA	Detects intracellular H_2O_2 and $ONOO^-$	• Cell-permeant • Can be used as a marker for H_2O_2 and $ONOO^-$ production	• Relatively low sensitivity to H_2O_2
HyPer	Detects H_2O_2	• High specificity for H_2O_2 • Fast reaction rate • Reversibility to oxidation and reduction • May be targeted to specific cellular compartments • Measures real-time H_2O_2 production	• Highly sensitive to pH: pH needs to be monitored

often difficult to discern the latter oxidized metabolites by fluorescence microscopy or even flow cytometry of HE-stained cells [219]. Due to eclipsing and often inextricable fluorescence spectra obtained from 2-OH-E^+/2-OH-Mito-E^+ and E^+/Mito-E^+, data obtained by fluorescence-based imaging techniques should be set aside in preference of other more qualitative and quantitative techniques. With that shortfall in mind, new protocols are being developed to minimize the fluorescence emission by E^+. It has been proposed that a change in the excitation of HE from 480 to 405 nm will reduce the emission intensity of E^+ and allow for a more precise identification of 2-OH-E^+ [220]. On the other hand, the use of high-performance liquid chromatography (HPLC) has become a far more attractive alternative aimed at differentially eluting and quantifying 2-OH-E^+ and 2-OH-Mito-E^+ from their precursors in cell or tissue lysates [219,221]. In short, HE and Mito-HE are reliable fluorescent probes due to their specificity for $O_2^{\bullet-}$. Nevertheless, the use of HPLC separation and detection of metabolites, rather than by fluorescence microscopy and flow cytometry, is recommended.

17.4.2 Hydropropidine

Recently, Michalski et al. synthesized a new redox-sensitive membrane-impermeant fluorogenic probe, hydropropidine (HPr^+). HPr^+ is a positively-charged, water-soluble analog of HE, used to detect $O_2^{\bullet-}$ [222]. A two-electron reduction product of propidium iodide (Pr^{2+}-$2I^-$), HPr^+ is a cell-impermeant and thus an extracellular fluorescent probe. The reaction between HPr^+ and $O_2^{\bullet-}$ forms a specific product, 2-hydroxipropidium (2-OH-Pr^+) in cell-free and cell-based systems, with a reaction stoichiometry of 1:2 (HPr+: $O_2^{\bullet-}$) [222]. Unlike HE, the product formed by reaction between HPr^+ and $O_2^{\bullet-}$ is predominantly 2-OH-Pr^+, which has proven specificity for $O_2^{\bullet-}$. However, HPr^+ oxidation by one-electron donor agents such as $ONOO^-$ and HO, gives rise to a heterodimer (HPr^+-Pr^{2+}) and/or homodimer (Pr^{2+}-Pr^{2+}).

17.4.3 Cytochrome C Reduction

The cytochrome c reduction assay is considered a *gold standard* methodology to quantify $O_2^{\bullet-}$ production. The technique is based on the reduction of the ferricytochrome c to ferrocytochrome c by a donor electron derived from $O_2^{\bullet-}$. Reduced ferrocytochrome c exhibits increased spectrophotometric absorbance at 550 nm, whereas no change in its absorbance at 540 and 560 nm serve as isosbestic points. However, ferricytochrome c may be subject to some interference by experimental chemical reagents or be directly reduced by electrons directly donated by enzymes. Due to ferricytochrome c's susceptibility to reduction, the cytochrome c assay must be performed in the presence and absence of SOD so that the difference between signal without SOD and with SOD can be used to tabulate $O_2^{\bullet-}$-mediated reduction, *per se*; $O_2^{\bullet-}$ is quantified using cytochrome's c extinction coefficient [223].

The cytochrome c assay is very specific for the detection of $O_2^{\bullet-}$ but lacks sensitivity to detect low levels of $O_2^{\bullet-}$. Thus, it is very useful for the detection of Nox-derived ROS from phagocytes due to the high levels of $O_2^{\bullet-}$ produced by these cells. However, the use of this technique for $O_2^{\bullet-}$ production by nonphagocytic cells such as SMC and endothelial cells (recognized to produce smaller amounts of $O_2^{\bullet-}$) can be challenging. Also, this assay can be used only to detect extracellular $O_2^{\bullet-}$ or that derived from broken cell preparations. For instance, it cannot be employed to measure cytoplasmic or mitochondrial $O_2^{\bullet-}$ in intact cells.

17.4.4 Electron Paramagnetic Resonance

Electron paramagnetic resonance (EPR) is also considered a gold standard methodology for the detection of an unpaired electron of free radicals in biological systems. The combination of EPR and an appropriated spin trap is a powerful and reliable technique to measure $O_2^{\bullet-}$, •OH, and NO. Spin traps can be used to both identify and quantify free radicals. The most widely used spin trap to detect $O_2^{\bullet-}$ or •OH is DMPO (5,5-dimethyl-1-1-pyrroline-N-oxide). The product formed between the reaction with $O_2^{\bullet-}$ is DMPO/•OOH, which has a short half-life and is rapidly decomposed to form a DMPO/•OH adduct. Thus, polyethylene glycol-conjugated SOD (PEG-SOD), a mimetic of SOD, and DMSO (dimethyl sulfoxide), a •OH scavenger, must be used to certify that DMPO/•OH adduct formation is dependent on the formation of $O_2^{\bullet-}$ and not from metal-catalyzed •OH formation.

Other compounds used to detect free radicals are the cyclic hydroxylamines. Cyclic hydroxylamines-based probes, i.e. 1-hydroxy-3-carboxy-2,2,5,-tetramethyl-pyrrolidine hydrochloride (CPH) and its analog 1-hydroxy-3-methoxycarbonyl-2,2,5,5-tetramethyl pyrrolidine (CMH) are not spin traps, which means that they cannot "trap" radicals for later detection. However, they are oxidized by free radicals to form a stable nitroxide radical with a half-life of several hours, which can readily be detected by EPR. CPH and CMH can provide a quantitative measurement of $O_2^{\bullet-}$, and their product, 3-carboxy-proxyl radical (derived from the reaction between $O_2^{\bullet-}$ and CPH) has a half-life of ~300 minutes in smooth muscle cells to approximately 90 minutes in the presence of ascorbate [223,224]. CMH is a plasma membrane-permeant derivative of CPH, and has been described to measure total cellular and tissue $O_2^{\bullet-}$, and CPH, being mitochondria -impermeant, is well-known to measure $O_2^{\bullet-}$ in the same way with the exception of the mitochondrion. The limitation of these compounds is their susceptibility to being oxidized by several ROS such as peroxynitrite. However, to avoid non-specific detection, the use of scavengers of $O_2^{\bullet-}$, •OH, and ONOO- is recommended, which is necessary to specifically identify the oxidant generating the signal.

17.4.5 Amplex Red

The Amplex Red (N-acetyl-3,7-dihydroxyphenoxazine) assay is a fluorescent-enhanced method used to quantify H_2O_2 [225]. Amplex Red is a colorless and nonfluorescent compound used as a probe for detecting extracellular H_2O_2, and because H_2O_2 is freely diffusible through lipid membranes, the Amplex Red assay can be used as an indicator of both intra- and extracellular H_2O_2. This technique is based on the reaction of Amplex Red with H_2O_2 in a1:1 stoichiometric ratio catalyzed by horseradish peroxidase (HRP) to form a highly fluorescent compound resorufin [225], which can be detected colorimetrically at $\lambda ex = 530$ nm and $\lambda em = 560$ nm or by fluorescence at $\lambda ex = 530$ nm and λem = 590 nm. Amplex Red has proved to be a reliable technique to measure H_2O_2 due to its high specificity and sensitivity to low H_2O_2 levels. PEG-catalase (catalase-polyethylene glycol) is frequently used to determine fluorescence background and to confirm selectivity of the assay for H_2O_2. Despite its specificity for H_2O_2 detection, Amplex Red is sensitive to light. Photo-oxidation of Amplex Red can yield H_2O_2 and consequent artifactual augmentation in fluorescence [226]. Precautions with ambient room light are necessary when using Amplex Red assay. Moreover, $O_2^{\bullet-}$ can react with HRP leading to a decrease in its activity, thus interfering with the H_2O_2 measurement. This interference can be obviated by the inclusion of PEG-SOD in the preparation (superoxide dismutase-polyethylene glycol) [227]. Indeed, Amplex Red is very specific and sensitive to freely diffusing H_2O_2 and can be used as well in both cell and tissue lysates.

17.4.6 2′,7′-Dichlorofluorescein Diacetate

2′,7′-dichlorofluorescein diacetate (DCFH-DA) is a fluorescent probe widely used for detecting intracellular H_2O_2 production. It is a cell-permeant probe which, after diffusion into the cell, is deacetylated by intracellular esterases to a cell impermeant and non-fluorescent compound, DCFH, which is retained in the cell. Once accumulated in the cytosol, DCFH will undergo two-electron oxidation to form a fluorescent compound 2′,7′-dichlorofluorescein (DCF), which is measured by a λex = 495 nm and λem = 529 nm using fluorescent-based techniques (e.g., flow cytometry, microplate readers, and confocal microscopy) [228]. Despite attempts to quantify H_2O_2 with this probe, DCFH can be oxidized by a variety of ROS and RNS and thus may be considered an indirect and non-specific indicator of oxidative stress in cells and tissues [229,230]. A number of studies have demonstrated that DCFH does not directly react with H_2O_2 to form the fluorescent product, DCF. Instead, the oxidation of DCFH is mediated by several one-electron donors, such as $OH^\bullet$, compounds I and II that are formed from peroxidase activity or heme groups on interaction with H_2O_2 [231,232], hypochlorous acid, and $ONOO^-$. The oxidation of DCFH or reduction of DCF forms an intermediate radical $DCF^{\bullet-}$ [233], which rapidly reacts with oxygen (O_2) to form $O_2^{\bullet-}$ [231,233] and artificially amplify fluorescence. Moreover, DCFH can undergo a nonspecific oxidation by cellular mitochondrial cytochrome c leading to oxidized DCF. For all of these reasons, the DCFH-DA assay cannot be reliably used to measure intracellular H_2O_2 or other ROS or RNS [229,234].

17.4.7 Coumarin-7 Boronic Acid

Coumarin-7 boronic acid (CBA) is a boronate-based fluorogenic probe recently developed for detecting intracellular H_2O_2 and $ONOO^-$ [235]. After diffusion into the cell, the reaction between either H_2O_2 and $ONOO^-$ with CBA yields a fluorescent product 7-hydroxycoumarin (COH) [235]. Stoichiometric and kinetic analyses have demonstrated that CBA reacts a million times faster with $ONOO^-$ than H_2O_2 [236]. For $ONOO^-$ detection, CBA must be used at low micromolar concentration to monitor real-time $ONOO^-$ formation, whereas for H_2O_2, CBA should be used at millimolar concentrations to provide reliable results [236]. Despite a low rate of reactivity, H_2O_2 can efficiently oxidize CBA. In addition, H_2O_2 and $ONOO^-$ scavengers are necessary as experimental baseline controls.

17.4.8 HyPer

HyPer is a genetically encoded biosensor of intracellular H_2O_2 production, consisting of circularly permuted YFP (cpYFP) inserted in the regulatory domain of Escherichia coli OxyR (OxyR-RD), a natural H_2O_2 sensor/transcription factor [237]. The OxyRD contains numerous cysteine residues, two of which are crucial for sensing H_2O_2: Cys199, and Cys208 [238]. The oxidation of Cys199 leads to the formation of sulfenic acid, which is repelled by the hydrophobic region due to its negative charge, inducing the approximation of Cys199 to Cys208, an intermediary disulfide bond [239,240] and an activating conformational change in the OxyRD domain [237]. HyPer is a fluorescent indicator specific for H_2O_2 [237,240], with two excitation peaks at $\lambda = 420$ nm (protonated) and 500 nm (deprotonated) and one emission peak at $\lambda = 516$ nm. Thus, HyPer is a ratiometric sensor that upon oxidation by H_2O_2 results in a characteristic decrease in the 420 nm excitation peak and an increase in the 500 nm excitation peak. HyPer displays high sensitivity to H_2O_2 in the nanomolar range [237]. A major shortcoming of HyPer is its high sensitivity to pH. The 500/420 nm excitation ratio is significantly shifted by pH oscillation, for example, 0.2 pH units, which is sufficient to mediate changes in the excitation ratio [241]. Therefore, it is necessary to monitor pH changes during the experiment to avoid artifactual measurements and misinterpretation of the results. Parenthetically, one way to monitor intracellular pH changes is the use of a pH-indicator, called SypHer. SypHer is a mutated form of HyPer engineered by mutation of one of the two H_2O_2-sensing cysteine residues of the OxyRD domain [242,243]. HyPer may be used to measure total cytosolic H_2O_2. On addition of a subcellular localization tag to the HyPer sequence, it is possible to target a specific compartment in the cell, for example, cytosol, mitochondrial intermembrane space, peroxisome, and nucleus [244] and thus determine delimited ROS generation. Thus, HyPer proves to be an important tool to measure real-time intracellular H_2O_2 production in living cells.

17.5 NADPH OXIDASE INHIBITORS AS VALUABLE TOOLS TO STUDY VASCULAR PATHOLOGY

Dissection of the independent roles of each vascular Nox in a given signaling pathway or in the development of a specific pathology requires the application of isoform-specific Nox inhibitors. The enzymatic complexity of Nox and the high homology among members of the family have presented both intrinsic opportunities and challenges. Historically, many of the inhibitors purported to be Nox inhibitors have displayed shortcomings in their affinity for not only one or more Noxs but also other oxidases. Other limitations include direct ROS scavenging effects, poor pharmacokinetics, and/or the need for enzymatic processing for their activation [245]. To date, only a handful of drugs shows selectivity towards a specific isoform and of those only a few can be considered truly isoform specific. A detailed discussion of the many described Nox inhibitors, and their strengths and pitfalls, can be found elsewhere [246–250] and is summarized in Table 17.2. We will center our attention

TABLE 17.2
NADPH Oxidase Inhibitors: Mechanism of Action, Selectivity, and Off-Target Effects

Inhibitor	Targets and Potency	Mechanism of Action	Selectivity and Off-Target Effects
Nox2ds-tat	Nox2 $IC_{50} = 0.74\ \mu M$ [250]	Binds to p47^{phox} and prevents p47^{phox} translocation to plasma membrane and interaction with Nox2 [250]	Highly selective for Nox2 with no inhibitory effects on Nox1, Nox4 or xanthine oxidase activity [250]
NoxA1ds	Nox1 $IC_{50} = 19$ nM [151]	Blocks Nox1-NoxA1 binding and isozyme activation [151]	Highly selective for Nox1 with no inhibitory effects on Nox2, Nox4, Nox5 or xanthine oxidase activity [151]
DPI	Broad spectrum Nox inhibitor [247]	Binds FAD and prevents electron flow through the flavocytochrome conduit [247]	Irreversibly and non-selectively inhibits a wide array of flavin-dependent enzymes, for example, NO synthase and xanthine oxidase [247]
Apocynin	Broad spectrum Nox inhibitor [247]	Prevents p47^{phox} translocation [251]	Direct scavenging activity of non-radical oxidant species such as $HOCl^-$ and H_2O_2. Inhibits other Nox isozymes that do not require p47^{phox} [252,253]
VAS2870	Broad spectrum Nox inhibitor $IC_{50} = 10.6\ \mu M$ [254]	None characterized	Pan-Nox inhibitor. No antioxidant properties. No inhibition of xanthine oxidase. Off-target effects on cellular thiol redox state, that is, thiol alkylation [255]
VAS3947	Broad spectrum Nox inhibitor $IC_{50} = 1–12\ \mu M$ in different cells lines [245,256]	None characterized	No antioxidant or ROS scavenging effects. No inhibition of xanthine oxidase or eNOS activity [245,256]
ML171	Nox1 $IC_{50} = 0.25\ \mu M$ [257]	None characterized	Considerably selective for Nox1 with a potency 20-fold higher than that for other Noxs. No inhibition of mitochondrial ROS generation. No effects on xanthine oxidase activity [257]

(Continued)

TABLE 17.2 (*Continued*)
NADPH Oxidase Inhibitors: Mechanism of Action, Selectivity, and Off-target Effects

Inhibitor	Targets and Potency	Mechanism of Action	Selectivity and Off-Target Effects
GKT136901	Nox1/Nox4 Ki = 165 nM [258,259]	None characterized	10-fold lower potency for Nox2. No inhibition of other ROS-generating and redox-sensitive enzymes [258,259]. Potential interference with peroxidase-dependent assays
Tetra-hydroquinolines (11g, 11h)	Nox2 11g: $IC_{50} = 20\ \mu M$ 11h: $IC_{50} = 20\ \mu M$ [260]	Modeled to interfere with $p47^{phox}$ and $p22^{phox}$ interaction	Highly selective for Nox2. No inhibitory effects on Nox1, Nox4, Nox5 or xanthine oxidase activity. No free radical scavenging properties [260]

primarily on the more selective and recently identified inhibitors including both peptidic and small molecule in the context of their application to prevent or treat cardiovascular disease.

17.5.1 Nox2ds-tat

Nox2ds-tat, a.k.a. gp91ds-tat, was the first isoform-specific inhibitor of the Nox family and is a peptidic inhibitor designed to target the Nox2 canonical isozyme system [251]. Indeed, our laboratory showed that it specifically inhibits canonical Nox2 but not hybrid Nox2 or Nox1, Nox4, Nox5 or xanthine oxidase [251]. As such, Nox2ds-tat has garnered broad attention as a Nox2-selective inhibitor [55,152,153,246–248]. Nox2ds tat mimics a region of the intracellular loop B from the membrane-integrated Nox2 catalytic subunit and contains a 9 amino acid *tat* sequence, which confers to it the ability to permeate the cell membrane. Its mechanism of action is the blockade of the interaction of the cytosolic subunit $p47^{phox}$ with the B-loop of Nox2, thus interfering with subunit assembly and enzyme activation. Nox2ds-tat effectively inhibits Nox2-dependent superoxide production *in vivo* and *in vitro* and has been extensively used to assess the contribution of Nox2 in a variety of models including cancer, cardiovascular and neurodegenerative diseases [157,265–267].

17.5.2 NoxA1ds

Another isoform-specific peptidic inhibitor was designed by our laboratory to mimic a putative activation domain on NoxA1. Its mechanism of action is *via* binding to the DHR of Nox1, and as a consequence to interfere with a critical NoxA1-Nox1 interaction for the generation of ROS [151]. NoxA1ds displays high efficacy to inhibit

Nox1-dependent ROS production with virtually no effect on Nox2, Nox4, Nox5 or xanthine oxidase activity. Furthermore, this peptide can translocate across the plasma membrane and associate with the Nox1 subunit [151]. In all, these properties render NoxA1ds a desirable tool to study Nox1's role in VSMCs.

17.5.3 Diphenylene Iodonium and Apocynin

Both compounds are traditionally characterized and widely used as broad-spectrum Nox inhibitors [19,252,268,269]. Additionally, a large body of literature has proved that both agents elicit other off-target effects [270–272]. Therefore, findings from studies using diphenylene iodonium (DPI) or apocynin must be more carefully examined for re-interpretation. In particular and apart from its inhibitory actions on Nox, DPI non-selectively inhibits a wide variety of flavin-dependent enzymes including NO synthase and xanthine oxidase [248]. Apocynin (4-hydroxy-3-methoxyacetophenone, common names: apocynin, acetovanillone) was also reported to directly scavenge non-radical oxidant species, including H_2O_2 and HOCl [255]. Apocynin has been used as an effective inhibitor of the Nox in phagocytic and non-phagocytic cells [273,274]. However, the mechanism of inhibition is not entirely known. Nevertheless, apocynin is described as impairing $p47^{phox}$ translocation to the membrane blocking Nox activation in several cell types such as leukocytes, monocytes, and endothelial cells [253,275–277]. An important feature of apocynin's mechanism of action is the peroxidase-mediated oxidation to form di-apocynin from apocynin, which has been shown to be more potent than the monomer alone [276]. Nonetheless, the reported non-specific actions diminish the value of DPI and apocynin considerably as research tools or for *in vivo* and translational application. Ideally, in order to dissect the roles of Noxs in redox signaling pathways, one or more isoform-selective inhibitors or gene silencing techniques ought to be employed.

17.5.4 Triazolo Pyrimidine Derivatives: VAS2870 and VAS3947

Two small molecules introduced by Vasopharm GmbH (VAS2870 and VAS3947) have been described as effective Nox inhibitors with no intrinsic antioxidant activity [257]. Importantly, their Nox inhibitory specificity has not been directly assessed for comparison across Nox isoforms, and they are, therefore, referred to as *pan-inhibitors* [278]. Across native and Nox-overexpressing cells, they have been proven to collectively inhibit Nox1, Nox2, Nox4, and Nox5 [279]. Despite not being isoform-specific, VAS2870 and VAS3947 may be considered a useful tool to understand the broader role of Nox in a given pathology as long as possible off-target effects, including effects on cellular thiol redox status are taken into account [259].

17.5.5 ML171

The first small molecule to our knowledge to be characterized as an isoform-specific Nox inhibitor is ML171. It was reported that ML171 is 20-fold more effective at inhibiting Nox1 isoform versus any of the other major isoforms [261]. However, more recently ML171, was reported to interfere with peroxidase-dependent ROS-measuring assays, bringing into question the validity of this 2-acetylphenothiazine as a true Nox1-selective agent [280]. Yet, upon modification of the phenothiazine back bone by an N-substitution with a $CH3S^-$ group, Seredenina et al. [280] reported a loss of interference with the assay and yielded analogs active against a range of Noxs.

17.5.6 GKT-136901 and GKT137831

Another set of small molecules that can be utilized as tools for the study of Nox-related effects on VSMC are two pyrazolopyridine derivatives, GKT-136901 and GKT137831 [262,281]. They are dual Nox1/4 inhibitors and, to our knowledge, represent the first proven orally-active Nox inhibitors. Both of these compounds have been used to evaluate the contribution of Nox1 or Nox4 in a particular disease process [246]. GKT136901 reportedly scavenges peroxynitrite [282]. Thus, when wanting to dissect the role of one or the other Nox, the use of genetic approaches or Nox1-specific inhibitors is necessary. Perhaps most importantly, the potential for these compounds to target the well-established and ubiquitous pro-differentiating role of Nox4 remains a significant concern.

17.5.7 Tetrahydroquinolines

Two isoform-specific inhibitors of Nox2 have recently been introduced, 11g and 11h, that belong to a group of bridge tetrahydroquinolines. These molecules display no inhibitory action on Noxs other than Nox2 when tested in cell-based assays [264]. They do not interfere with ROS assays and do not have an effect on xanthine oxidase or display ROS scavenging properties. Although to date there is no definitive empirical information on their mechanism of action, *in silico* analysis suggests they exert their effect by interfering with the assembly of the active enzyme by blocking $p47^{phox}$ SH3 supergroove with a hallmark binding motif on $p22^{phox}$ (unpublished data).

17.6 CONCLUSION

Considerable evidence gathered over the past few decades supports the notion that upregulation of Noxs in the vascular wall is a key factor for the development of vascular disease. Nox family Noxs and their regulatory proteins localize in specific compartments within the cell including the plasma membrane, caveolae/lipid rafts, focal adhesions, and the nucleus. These and other subcellular locations are required for ROS production and purported activation of specific redox signaling pathways that mediate physiological and pathophysiological responses. Noxs in VSMC are increasingly being recognized as pivotal in contraction and regulation of vascular tone, blood flow, and pressure. As well, in response to pathological stimuli, a rise in ROS production leads to VSMC dysfunction via induction of proliferation, migration and proteolytic activity, which can give rise to an increase in vascular remodeling, abnormal vessel architecture, and integrity and combine to promote cardiovascular diseases.

Considering the lack of success in clinical trials of antioxidant vitamins in the treatment of cardiovascular disease in part due to the poor study design but also to deficient bioactivity of these compounds and the potential for excessive interference in physiological redox signaling, attention has turned to more nuanced measures to tamp down the production of ROS. The selective targeting of Nox isoforms, or their subunits, provide a way to understand the role of Noxs in vascular cells, as well as offer a new therapeutic approach to several pathological vascular conditions. Major strides are being made towards the development of new drugs to inhibit Nox isoforms and limit oxidative stress in VSMC, but there is much work yet to be done to generate more potent and selective inhibitors and therapeutics.

REFERENCES

1. Owens, G. K., M. S. Kumar, and B. R. Wamhoff. 2004. Molecular regulation of vascular smooth muscle cell differentiation in development and disease. *Physiol Rev* 84 (3): 767–801. doi:10.1152/physrev.00041.2003.
2. Rodriguez, A. I., G. Csanyi, D. J. Ranayhossaini, D. M. Feck, K. J. Blose, L. Assatourian, D. A. Vorp, and P. J. Pagano. 2015. MEF2B-Nox1 signaling is critical for stretch-induced phenotypic modulation of vascular smooth muscle cells. *Arterioscler Thromb Vasc Biol* 35 (2): 430–438. doi:10.1161/ATVBAHA.114.304936.
3. Sahoo, S., D. N. Meijles, I. Al Ghouleh, M. Tandon, E. Cifuentes-Pagano, J. Sembrat, M. Rojas, E. Goncharova, and P. J. Pagano. 2016. MEF2C-MYOCD and leiomodin1 suppression by miRNA-214 promotes smooth muscle cell phenotype switching in pulmonary arterial hypertension. *PLoS One* 11 (5): e0153780. doi:10.1371/journal.pone.0153780.
4. Nanda, V., and J. M. Miano. 2012. Leiomodin 1, a new serum response factor-dependent target gene expressed preferentially in differentiated smooth muscle cells. *J Biol Chem* 287 (4): 2459–2467. doi:10.1074/jbc.M111.302224.
5. Wong, A. P., N. Nili, and B. H. Strauss. 2005. In vitro differences between venous and arterial-derived smooth muscle cells: Potential modulatory role of decorin. *Cardiovasc Res* 65 (3): 702–710. doi:10.1016/j.cardiores.2004.10.012.
6. Yang, Z., B. S. Oemar, T. Carrel, B. Kipfer, F. Julmy, and T. F. Lüscher. 1998. Different proliferative properties of smooth muscle cells of human arterial and venous bypass vessels: Role of PDGF receptors, mitogen-activated protein kinase, and cyclin-dependent kinase inhibitors. *Circulation* 97 (2): 181–187.
7. Miano, J. M. 2003. Serum response factor: Toggling between disparate programs of gene expression. *J Mol Cell Cardiol* 35 (6): 577–593.
8. Chen, J., C. M. Kitchen, J. W. Streb, and J. M. Miano. 2002. Myocardin: A component of a molecular switch for smooth muscle differentiation. *J Mol Cell Cardiol* 34 (10): 1345–1356.
9. Posern, G., and R. Treisman. 2006. Actin' together: Serum response factor, its cofactors and the link to signal transduction. *Trends Cell Biol* 16 (11): 588–596. doi:10.1016/j.tcb.2006.09.008.
10. Wang, D. Z., and E. N. Olson. 2004. Control of smooth muscle development by the myocardin family of transcriptional coactivators. *Curr Opin Genet Dev* 14 (5): 558–566. doi:10.1016/j.gde.2004.08.003.
11. Wang, Z., D. Z. Wang, D. Hockemeyer, J. McAnally, A. Nordheim, and E. N. Olson. 2004. Myocardin and ternary complex factors compete for SRF to control smooth muscle gene expression. *Nature* 428 (6979): 185–189. doi:10.1038/nature02382.

12. Kumar, M. S., and G. K. Owens. 2003. Combinatorial control of smooth muscle-specific gene expression. *Arterioscler Thromb Vasc Biol* 23 (5): 737–747. doi:10.1161/01.ATV.0000065197.07635.BA.
13. Carson, J. A., R. A. Fillmore, R. J. Schwartz, and W. E. Zimmer. 2000. The smooth muscle gamma-actin gene promoter is a molecular target for the mouse bagpipe homologue, mNkx3-1, and serum response factor. *J Biol Chem* 275 (50): 39061–39072. doi:10.1074/jbc.M006532200.
14. Herring, B. P., A. M. Kriegel, and A. M. Hoggatt. 2001. Identification of Barx2b, a serum response factor-associated homeodomain protein. *J Biol Chem* 276 (17): 14482–14489. doi:10.1074/jbc.M011585200.
15. Nakamura, M., W. Nishida, S. Mori, K. Hiwada, K. Hayashi, and K. Sobue. 2001. Transcriptional activation of beta-tropomyosin mediated by serum response factor and a novel Barx homologue, Barx1b, in smooth muscle cells. *J Biol Chem* 276 (21): 18313–18320.
16. Nishida, W., M. Nakamura, S. Mori, M. Takahashi, Y. Ohkawa, S. Tadokoro, K. Yoshida, K. Hiwada, K. Hayashi, and K. Sobue. 2002. A triad of serum response factor and the GATA and NK families governs the transcription of smooth and cardiac muscle genes. *J Biol Chem* 277 (9): 7308–7317. doi:10.1074/jbc.M111824200.
17. Hsieh, C. M., M. Yoshizumi, W. O. Endege, C. J. Kho, M. K. Jain, S. Kashiki, R. de los Santos, W. S. Lee, M. A. Perrella, and M. E. Lee. 1996. APEG-1, a novel gene preferentially expressed in aortic smooth muscle cells, is down-regulated by vascular injury. *J Biol Chem* 271 (29): 17354–17359.
18. Sobue, K., K. Hayashi, and W. Nishida. 1999. Expressional regulation of smooth muscle cell-specific genes in association with phenotypic modulation. *Mol Cell Biochem* 190 (1–2): 105–118.
19. Xue, B., T. G. Beltz, Y. Yu, F. Guo, C. E. Gomez-Sanchez, M. Hay, and A. K. Johnson. 2011. Central interactions of aldosterone and angiotensin II in aldosterone- and angiotensin II-induced hypertension. *Am J Physiol Heart Circ Physiol* 300 (2): H555–H564. doi:10.1152/ajpheart.00847.2010.
20. Gittenberger-de Groot, A. C., M. C. DeRuiter, M. Bergwerff, and R. E. Poelmann. 1999. Smooth muscle cell origin and its relation to heterogeneity in development and disease. *Arterioscler Thromb Vasc Biol* 19 (7): 1589–1594.
21. Hungerford, J. E., and C. D. Little. 1999. Developmental biology of the vascular smooth muscle cell: Building a multilayered vessel wall. *J Vasc Res* 36 (1): 2–27. doi:10.1159/000025622.
22. Majesky, M. W. 2007. Developmental basis of vascular smooth muscle diversity. *Arterioscler Thromb Vasc Biol* 27 (6): 1248–1258. doi:10.1161/ATVBAHA.107.141069.
23. Sinha, S., D. Iyer, and A. Granata. 2014. Embryonic origins of human vascular smooth muscle cells: Implications for in vitro modeling and clinical application. *Cell Mol Life Sci* 71 (12): 2271–2288. doi:10.1007/s00018-013-1554-3.
24. Yoshida, T., and G. K. Owens. 2005. Molecular determinants of vascular smooth muscle cell diversity. *Circ Res* 96 (3): 280–291. doi:10.1161/01.RES.0000155951.62152.2e.
25. Crisan, M., S. Yap, L. Casteilla, C. W. Chen, M. Corselli, T. S. Park, G. Andriolo et al. 2008. A perivascular origin for mesenchymal stem cells in multiple human organs. *Cell Stem Cell* 3 (3): 301–313. doi:10.1016/j.stem.2008.07.003.
26. Nguyen, A. T., D. Gomez, R. D. Bell, J. H. Campbell, A. W. Clowes, G. Gabbiani, C. M. Giachelli et al. 2013. Smooth muscle cell plasticity: Fact or fiction? *Circ Res* 112 (1): 17–22. doi:10.1161/CIRCRESAHA.112.281048.
27. Zhang, L., and Q. Xu. 2017. Vascular progenitors and smooth muscle cells chicken and egg? *Circ Res* 120 (2): 246–248. doi:10.1161/CIRCRESAHA.116.310341.

28. Duan, C., J. R. Bauchat, and T. Hsieh. 2000. Phosphatidylinositol 3-kinase is required for insulin-like growth factor-I-induced vascular smooth muscle cell proliferation and migration. *Circ Res* 86 (1): 15–23.
29. Hayashi, K., H. Saga, Y. Chimori, K. Kimura, Y. Yamanaka, and K. Sobue. 1998. Differentiated phenotype of smooth muscle cells depends on signaling pathways through insulin-like growth factors and phosphatidylinositol 3-kinase. *J Biol Chem* 273 (44): 28860–28867.
30. Hayashi, K., M. Takahashi, K. Kimura, W. Nishida, H. Saga, and K. Sobue. 1999. Changes in the balance of phosphoinositide 3-kinase/protein kinase B (Akt) and the mitogen-activated protein kinases (ERK/p38MAPK) determine a phenotype of visceral and vascular smooth muscle cells. *J Cell Biol* 145 (4): 727–740.
31. Tsai, I. C., Z. C. Pan, H. P. Cheng, C. H. Liu, B. T. Lin, and M. J. Jiang. 2016. Reactive oxygen species derived from NADPH oxidase 1 and mitochondria mediate angiotensin II-induced smooth muscle cell senescence. *J Mol Cell Cardiol* 98: 18–27. doi:10.1016/j.yjmcc.2016.07.001.
32. Wang, C. C., I. Gurevich, and B. Draznin. 2003. Insulin affects vascular smooth muscle cell phenotype and migration via distinct signaling pathways. *Diabetes* 52 (10): 2562–2569.
33. Adam, P. J., C. P. Regan, M. B. Hautmann, and G. K. Owens. 2000. Positive- and negative-acting Kruppel-like transcription factors bind a transforming growth factor beta control element required for expression of the smooth muscle cell differentiation marker SM22alpha in vivo. *J Biol Chem* 275 (48): 37798–37806. doi:10.1074/jbc.M006323200.
34. Chen, P. Y., L. Qin, G. Li, G. Tellides, and M. Simons. 2016. Smooth muscle FGF/TGFβ cross talk regulates atherosclerosis progression. *EMBO Mol Med* 8 (7): 712–728. doi:10.15252/emmm.201506181.
35. Clempus, R. E., D. Sorescu, A. E. Dikalova, L. Pounkova, P. Jo, G. P. Sorescu, H. H. Schmidt, B. Lassegue, and K. K. Griendling. 2007. Nox4 is required for maintenance of the differentiated vascular smooth muscle cell phenotype. *Arterioscler Thromb Vasc Biol* 27 (1): 42–48. doi:10.1161/01.ATV.0000251500.94478.18.
36. Hao, H., P. Ropraz, V. Verin, E. Camenzind, A. Geinoz, M. S. Pepper, G. Gabbiani, and M. L. Bochaton-Piallat. 2002. Heterogeneity of smooth muscle cell populations cultured from pig coronary artery. *Arterioscler Thromb Vasc Biol* 22 (7): 1093–1099.
37. Orlandi, A., P. Ropraz, and G. Gabbiani. 1994. Proliferative activity and alpha-smooth muscle actin expression in cultured rat aortic smooth muscle cells are differently modulated by transforming growth factor-beta 1 and heparin. *Exp Cell Res* 214 (2): 528–536. doi:10.1006/excr.1994.1290.
38. Alexander, M. R., and G. K. Owens. 2012. Epigenetic control of smooth muscle cell differentiation and phenotypic switching in vascular development and disease. *Annu Rev Physiol* 74: 13–40. doi:10.1146/annurev-physiol-012110-142315.
39. Mack, C. P. 2011. Signaling mechanisms that regulate smooth muscle cell differentiation. *Arterioscler Thromb Vasc Biol* 31 (7): 1495–1505. doi:10.1161/ATVBAHA.110.221135.
40. Sakurada, S., N. Takuwa, N. Sugimoto, Y. Wang, M. Seto, Y. Sasaki, and Y. Takuwa. 2003. Ca^{2+}-dependent activation of Rho and Rho kinase in membrane depolarization-induced and receptor stimulation-induced vascular smooth muscle contraction. *Circ Res* 93 (6): 548–556. doi:10.1161/01.RES.0000090998.08629.60.
41. Wamhoff, B. R., D. K. Bowles, O. G. McDonald, S. Sinha, A. P. Somlyo, A. V. Somlyo, and G. K. Owens. 2004. L-type voltage-gated Ca^{2+} channels modulate expression of smooth muscle differentiation marker genes via a rho kinase/myocardin/SRF-dependent mechanism. *Circ Res* 95 (4): 406–414. doi:10.1161/01.RES.0000138582.36921.9e.

42. Tang, Y., J. M. Boucher, and L. Liaw. 2012. Histone deacetylase activity selectively regulates notch-mediated smooth muscle differentiation in human vascular cells. *J Am Heart Assoc* 1 (3): e000901. doi:10.1161/JAHA.112.000901.
43. Yoshida, T., Q. Gan, and G. K. Owens. 2008. Kruppel-like factor 4, Elk-1, and histone deacetylases cooperatively suppress smooth muscle cell differentiation markers in response to oxidized phospholipids. *Am J Physiol Cell Physiol* 295 (5): C1175–C1182. doi:10.1152/ajpcell.00288.2008.
44. Margariti, A., Q. Xiao, A. Zampetaki, Z. Zhang, H. Li, D. Martin, Y. Hu, L. Zeng, and Q. Xu. 2009. Splicing of HDAC7 modulates the SRF-myocardin complex during stem-cell differentiation towards smooth muscle cells. *J Cell Sci* 122 (Pt 4): 460–470. doi:10.1242/jcs.034850.
45. Bartel, D. P., and C. Z. Chen. 2004. Micromanagers of gene expression: The potentially widespread influence of metazoan microRNAs. *Nat Rev Genet* 5 (5): 396–400. doi:10.1038/nrg1328.
46. Boettger, T., N. Beetz, S. Kostin, J. Schneider, M. Krüger, L. Hein, and T. Braun. 2009. Acquisition of the contractile phenotype by murine arterial smooth muscle cells depends on the Mir143/145 gene cluster. *J Clin Invest* 119 (9): 2634–2647. doi:10.1172/JCI38864.
47. Cordes, K. R., N. T. Sheehy, M. P. White, E. C. Berry, S. U. Morton, A. N. Muth, T. H. Lee, J. M. Miano, K. N. Ivey, and D. Srivastava. 2009. miR-145 and miR-143 regulate smooth muscle cell fate and plasticity. *Nature* 460 (7256): 705–710. doi:10.1038/nature08195.
48. Davis, B. N., A. C. Hilyard, P. H. Nguyen, G. Lagna, and A. Hata. 2009. Induction of microRNA-221 by platelet-derived growth factor signaling is critical for modulation of vascular smooth muscle phenotype. *J Biol Chem* 284 (6): 3728–3738. doi:10.1074/jbc.M808788200.
49. Li, J., M. Stouffs, L. Serrander, B. Banfi, E. Bettiol, Y. Charnay, K. Steger, K. H. Krause, and M. E. Jaconi. 2006. The NADPH oxidase NOX4 drives cardiac differentiation: Role in regulating cardiac transcription factors and MAP kinase activation. *Mol Biol Cell* 17 (9): 3978–3988. doi:10.1091/mbc.E05-06-0532.
50. Li, P., N. Zhu, B. Yi, N. Wang, M. Chen, X. You, X. Zhao, C. C. Solomides, Y. Qin, and J. Sun. 2013. MicroRNA-663 regulates human vascular smooth muscle cell phenotypic switch and vascular neointimal formation. *Circ Res* 113 (10): 1117–1127. doi:10.1161/CIRCRESAHA.113.301306.
51. Iaconetti, C., S. De Rosa, A. Polimeni, S. Sorrentino, C. Gareri, A. Carino, J. Sabatino, M. Colangelo, A. Curcio, and C. Indolfi. 2015. Down-regulation of miR-23b induces phenotypic switching of vascular smooth muscle cells in vitro and in vivo. *Cardiovasc Res* 107 (4): 522–533. doi:10.1093/cvr/cvv141.
52. Torella, D., C. Iaconetti, D. Catalucci, G. M. Ellison, A. Leone, C. D. Waring, A. Bochicchio et al. 2011. MicroRNA-133 controls vascular smooth muscle cell phenotypic switch in vitro and vascular remodeling in vivo. *Circ Res* 109 (8): 880–893. doi:10.1161/CIRCRESAHA.111.240150.
53. Lee, M., A. San Martin, A. Valdivia, A. Martin-Garrido, and K. K. Griendling. 2016. Redox-sensitive regulation of myocardin-related transcription factor (MRTF-A) phosphorylation via palladin in vascular smooth muscle cell differentiation marker gene expression. *PLoS One* 11 (4): e0153199. doi:10.1371/journal.pone.0153199.
54. Su, B., S. Mitra, H. Gregg, S. Flavahan, M. A. Chotani, K. R. Clark, P. J. Goldschmidt-Clermont, and N. A. Flavahan. 2001. Redox regulation of vascular smooth muscle cell differentiation. *Circ Res* 89 (1): 39–46.
55. Touyz, R. M., X. Chen, F. Tabet, G. Yao, G. He, M. T. Quinn, P. J. Pagano, and E. L. Schiffrin. 2002. Expression of a functionally active gp91phox-containing neutrophil-type NAD(P)H oxidase in smooth muscle cells from human resistance arteries: Regulation by angiotensin II. *Circ Res* 90 (11): 1205–1213.

56. Clempus, R. E., and K. K. Griendling. 2006. Reactive oxygen species signaling in vascular smooth muscle cells. *Cardiovasc Res* 71 (2): 216–225. doi:10.1016/j.cardiores.2006.02.033.
57. Xiao, Q., Z. Luo, A. E. Pepe, A. Margariti, L. Zeng, and Q. Xu. 2009. Embryonic stem cell differentiation into smooth muscle cells is mediated by Nox4-produced H2O2. *Am J Physiol Cell Physiol* 296 (4): C711–C723. doi:10.1152/ajpcell.00442.2008.
58. Grote, K., I. Flach, M. Luchtefeld, E. Akin, S. M. Holland, H. Drexler, and B. Schieffer. 2003. Mechanical stretch enhances mRNA expression and proenzyme release of matrix metalloproteinase-2 (MMP-2) via NAD(P)H oxidase-derived reactive oxygen species. *Circ Res* 92 (11): e80–e86. doi:10.1161/01.RES.0000077044.60138.7C.
59. Mata-Greenwood, E., A. Grobe, S. Kumar, Y. Noskina, and S. M. Black. 2005. Cyclic stretch increases VEGF expression in pulmonary arterial smooth muscle cells via TGF-beta1 and reactive oxygen species: A requirement for NAD(P)H oxidase. *Am J Physiol Lung Cell Mol Physiol* 289 (2): L288–L289. doi:10.1152/ajplung.00417.2004.
60. Rodríguez, A. I., G. Csányi, D. J. Ranayhossaini, D. M. Feck, K. J. Blose, L. Assatourian, D. A. Vorp, and P. J. Pagano. 2015. MEF2B-Nox1 signaling is critical for stretch-induced phenotypic modulation of vascular smooth muscle cells. *Arterioscler Thromb Vasc Biol* 35 (2): 430–438. doi:10.1161/ATVBAHA.114.304936.
61. Al Ghouleh, I., D. N. Meijles, S. Mutchler, Q. Zhang, S. Sahoo, A. Gorelova, J. Henrich Amaral et al. 2016. Binding of EBP50 to Nox organizing subunit p47phox is pivotal to cellular reactive species generation and altered vascular phenotype. *Proc Natl Acad Sci USA* 113 (36): E5308–E5317. doi:10.1073/pnas.1514161113.
62. Sun, C., J. Zheng, S. Cheng, D. Feng, and J. He. 2013. EBP50 phosphorylation by Cdc2/Cyclin B kinase affects actin cytoskeleton reorganization and regulates functions of human breast cancer cell line MDA-MB-231. *Mol Cells* 36 (1): 47–54. doi:10.1007/s10059-013-0014-0.
63. Janiszewski, M., L. R. Lopes, A. O. Carmo, M. A. Pedro, R. P. Brandes, C. X. Santos, and F. R. Laurindo. 2005. Regulation of NAD(P)H oxidase by associated protein disulfide isomerase in vascular smooth muscle cells. *J Biol Chem* 280 (49): 40813–40819. doi:10.1074/jbc.M509255200.
64. Pariser, H. P., J. Zhang, and R. E. Hausman. 2000. The cell adhesion molecule retina cognin is a cell surface protein disulfide isomerase that uses disulfide exchange activity to modulate cell adhesion. *Exp Cell Res* 258 (1): 42–52. doi:10.1006/excr.2000.4931.
65. Terada, K., P. Manchikalapudi, R. Noiva, H. O. Jauregui, R. J. Stockert, and M. L. Schilsky. 1995. Secretion, surface localization, turnover, and steady state expression of protein disulfide isomerase in rat hepatocytes. *J Biol Chem* 270 (35): 20410–20416.
66. Zai, A., M. A. Rudd, A. W. Scribner, and J. Loscalzo. 1999. Cell-surface protein disulfide isomerase catalyzes transnitrosation and regulates intracellular transfer of nitric oxide. *J Clin Invest* 103 (3): 393–399. doi:10.1172/JCI4890.
67. Laurindo, F. R., D. C. Fernandes, A. M. Amanso, L. R. Lopes, and C. X. Santos. 2008. Novel role of protein disulfide isomerase in the regulation of NADPH oxidase activity: Pathophysiological implications in vascular diseases. *Antioxid Redox Signal* 10 (6): 1101–1113. doi:10.1089/ars.2007.2011.
68. Alexander, M. R., M. Murgai, C. W. Moehle, and G. K. Owens. 2012. Interleukin-1β modulates smooth muscle cell phenotype to a distinct inflammatory state relative to PDGF-DD via NF-κB-dependent mechanisms. *Physiol Genomics* 44 (7): 417–429. doi:10.1152/physiolgenomics.00160.2011.
69. Morrow, D., S. Guha, C. Sweeney, Y. Birney, T. Walshe, C. O'Brien, D. Walls, E. M. Redmond, and P. A. Cahill. 2008. Notch and vascular smooth muscle cell phenotype. *Circ Res* 103 (12): 1370–1382. doi:10.1161/CIRCRESAHA.108.187534.
70. Trebak, M., R. Ginnan, H. A. Singer, and D. Jourd'heuil. 2010. Interplay between calcium and reactive oxygen/nitrogen species: An essential paradigm for vascular smooth muscle signaling. *Antioxid Redox Signal* 12 (5): 657–674. doi:10.1089/ars.2009.2842.
71. Babior, B. M. 1999. NADPH oxidase: An update. *Blood* 93 (5): 1464–1476.

72. Babior, B. M. 2004. NADPH oxidase. *Curr Opin Immunol* 16 (1): 42–47.
73. Segal, A. W. 2005. How neutrophils kill microbes. *Annu Rev Immunol* 23: 197–223. doi:10.1146/annurev.immunol.23.021704.115653.
74. El Chemaly, A., and N. Demaurex. 2012. Do Hv1 proton channels regulate the ionic and redox homeostasis of phagosomes? *Mol Cell Endocrinol* 353 (1–2): 82–87. doi:10.1016/j.mce.2011.10.005.
75. Dinauer, M. C., and S. H. Orkin. 1992. Chronic granulomatous disease. *Annu Rev Med* 43: 117–124. doi:10.1146/annurev.me.43.020192.001001.
76. Roos, D., R. van Bruggen, and C. Meischl. 2003. Oxidative killing of microbes by neutrophils. *Microbes Infect* 5 (14): 1307–1315.
77. Kawahara, T., and J. D. Lambeth. 2007. Molecular evolution of Phox-related regulatory subunits for NADPH oxidase enzymes. *BMC Evol Biol* 7: 178. doi:10.1186/1471-2148-7-178.
78. Basuroy, S., S. Bhattacharya, C. W. Leffler, and H. Parfenova. 2009. Nox4 NADPH oxidase mediates oxidative stress and apoptosis caused by TNF-alpha in cerebral vascular endothelial cells. *Am J Physiol Cell Physiol* 296 (3): C422–C432. doi:10.1152/ajpcell.00381.2008.
79. Drummond, G. R., S. Selemidis, K. K. Griendling, and C. G. Sobey. 2011. Combating oxidative stress in vascular disease: NADPH oxidases as therapeutic targets. *Nat Rev Drug Discov* 10 (6): 453–471. doi:10.1038/nrd3403.
80. Lassegue, B., and R. E. Clempus. 2003. Vascular NAD(P)H oxidases: Specific features, expression, and regulation. *Am J Physiol Regul Integr Comp Physiol* 285 (2): R277–R297. doi:10.1152/ajpregu.00758.2002.
81. Pendyala, S., P. V. Usatyuk, I. A. Gorshkova, J. G. Garcia, and V. Natarajan. 2009. Regulation of NADPH oxidase in vascular endothelium: The role of phospholipases, protein kinases, and cytoskeletal proteins. *Antioxid Redox Signal* 11 (4): 841–860. doi:10.1089/ARS.2008.2231.
82. Kou, B., J. Ni, M. Vatish, and D. R. Singer. 2008. Xanthine oxidase interaction with vascular endothelial growth factor in human endothelial cell angiogenesis. *Microcirculation* 15 (3): 251–267. doi:10.1080/10739680701651495.
83. Li, Q., M. Mao, Y. Qiu, G. Liu, T. Sheng, X. Yu, S. Wang, and D. Zhu. 2016. Key role of ROS in the process of 15-lipoxygenase/15-hydroxyeicosatetraenoiccid-induced pulmonary vascular remodeling in hypoxia pulmonary hypertension. *PLoS One* 11 (2): e0149164. doi:10.1371/journal.pone.0149164.
84. Tang, X., Y. X. Luo, H. Z. Chen, and D. P. Liu. 2014. Mitochondria, endothelial cell function, and vascular diseases. *Front Physiol* 5: 175. doi:10.3389/fphys.2014.00175.
85. Toniolo, A., C. Buccellati, C. Pinna, R. M. Gaion, A. Sala, and C. Bolego. 2013. Cyclooxygenase-1 and prostacyclin production by endothelial cells in the presence of mild oxidative stress. *PLoS One* 8 (2): e56683. doi:10.1371/journal.pone.0056683.
86. Widlansky, M. E., and D. D. Gutterman. 2011. Regulation of endothelial function by mitochondrial reactive oxygen species. *Antioxid Redox Signal* 15 (6): 1517–1530. doi:10.1089/ars.2010.3642.
87. Yaghini, F. A., C. Y. Song, E. N. Lavrentyev, H. U. Ghafoor, X. R. Fang, A. M. Estes, W. B. Campbell, and K. U. Malik. 2010. Angiotensin II-induced vascular smooth muscle cell migration and growth are mediated by cytochrome P450 1B1-dependent superoxide generation. *Hypertension* 55 (6): 1461–1467. doi:10.1161/HYPERTENSIONAHA.110.150029.
88. Konior, A., A. Schramm, M. Czesnikiewicz-Guzik, and T. J. Guzik. 2014. NADPH oxidases in vascular pathology. *Antioxid Redox Signal* 20 (17): 2794–2814. doi:10.1089/ars.2013.5607.
89. Selemidis, S., C. G. Sobey, K. Wingler, H. H. Schmidt, and G. R. Drummond. 2008. NADPH oxidases in the vasculature: Molecular features, roles in disease and pharmacological inhibition. *Pharmacol Ther* 120 (3): 254–291. doi:10.1016/j.pharmthera.2008.08.005.

90. Babior, B. M., R. S. Kipnes, and J. T. Curnutte. 1973. Biological defense mechanisms. The production by leukocytes of superoxide, a potential bactericidal agent. *J Clin Invest* 52 (3): 741–744. doi:10.1172/JCI107236.
91. Bedard, K., and K. H. Krause. 2007. The NOX family of ROS-generating NADPH oxidases: Physiology and pathophysiology. *Physiol Rev* 87 (1): 245–313. doi:10.1152/physrev.00044.2005.
92. Rossi, F., and M. Zatti. 1964. Biochemical aspects of phagocytosis in polymorphonuclear leucocytes. NADH and NADPH oxidation by the granules of resting and phagocytizing cells. *Experientia* 20 (1): 21–23.
93. Griendling, K. K., C. A. Minieri, J. D. Ollerenshaw, and R. W. Alexander. 1994. Angiotensin II stimulates NADH and NADPH oxidase activity in cultured vascular smooth muscle cells. *Circ Res* 74 (6): 1141–1148.
94. Meier, B., A. R. Cross, J. T. Hancock, F. J. Kaup, and O. T. Jones. 1991. Identification of a superoxide-generating NADPH oxidase system in human fibroblasts. *Biochem J* 275 (Pt 1): 241–245.
95. Pagano, P. J., J. K. Clark, M. E. Cifuentes-Pagano, S. M. Clark, G. M. Callis, and M. T. Quinn. 1997. Localization of a constitutively active, phagocyte-like NADPH oxidase in rabbit aortic adventitia: Enhancement by angiotensin II. *Proc Natl Acad Sci USA* 94 (26): 14483–14488.
96. Pagano, P. J., Y. Ito, K. Tornheim, P. M. Gallop, A. I. Tauber, and R. A. Cohen. 1995. An NADPH oxidase superoxide-generating system in the rabbit aorta. *Am J Physiol* 268 (6 Pt 2): H2274–H2280.
97. Dho, S. H., J. Y. Kim, K. P. Lee, E. S. Kwon, J. C. Lim, C. J. Kim, D. Jeong, and K. S. Kwon. 2017. STAT5A-mediated NOX5-L expression promotes the proliferation and metastasis of breast cancer cells. *Exp Cell Res* 351 (1): 51–58. doi:10.1016/j.yexcr.2016.12.020.
98. Han, M., T. Zhang, L. Yang, Z. Wang, J. Ruan, and X. Chang. 2016. Association between NADPH oxidase (NOX) and lung cancer: A systematic review and meta-analysis. *J Thorac Dis* 8 (7): 1704–1711. doi:10.21037/jtd.2016.06.31.
99. Roy, K., Y. Wu, J. L. Meitzler, A. Juhasz, H. Liu, G. Jiang, J. Lu, S. Antony, and J. H. Doroshow. 2015. NADPH oxidases and cancer. *Clin Sci (Lond)* 128 (12): 863–875. doi:10.1042/CS20140542.
100. Choi, S. R., S. G. Kwon, H. S. Choi, H. J. Han, A. J. Beitz, and J. H. Lee. 2016. Neuronal NOS activates spinal NADPH oxidase 2 contributing to central sigma-1 receptor-induced pain hypersensitivity in mice. *Biol Pharm Bull* 39 (12): 1922–1931. doi:10.1248/bpb.b16-00326.
101. Infanger, D. W., R. V. Sharma, and R. L. Davisson. 2006. NADPH oxidases of the brain: Distribution, regulation, and function. *Antioxid Redox Signal* 8 (9–10): 1583–1596. doi:10.1089/ars.2006.8.1583.
102. Nayernia, Z., V. Jaquet, and K. H. Krause. 2014. New insights on NOX enzymes in the central nervous system. *Antioxid Redox Signal* 20 (17): 2815–2837. doi:10.1089/ars.2013.5703.
103. Nauseef, W. M. 2004. Assembly of the phagocyte NADPH oxidase. *Histochem Cell Biol* 122 (4): 277–291. doi:10.1007/s00418-004-0679-8.
104. Inanami, O., J. L. Johnson, J. K. McAdara, J. E. Benna, L. R. Faust, P. E. Newburger, and B. M. Babior. 1998. Activation of the leukocyte NADPH oxidase by phorbol ester requires the phosphorylation of p47PHOX on serine 303 or 304. *J Biol Chem* 273 (16): 9539–9543.
105. Johnson, J. L., J. W. Park, J. E. Benna, L. P. Faust, O. Inanami, and B. M. Babior. 1998. Activation of p47(PHOX), a cytosolic subunit of the leukocyte NADPH oxidase. Phosphorylation of ser-359 or ser-370 precedes phosphorylation at other sites and is required for activity. *J Biol Chem* 273 (52): 35147–35152.

106. Groemping, Y., K. Lapouge, S. J. Smerdon, and K. Rittinger. 2003. Molecular basis of phosphorylation-induced activation of the NADPH oxidase. *Cell* 113 (3): 343–355.
107. Leusen, J. H., A. J. Verhoeven, and D. Roos. 1996. Interactions between the components of the human NADPH oxidase: Intrigues in the phox family. *J Lab Clin Med* 128 (5): 461–476.
108. Lambeth, J. D. 2004. NOX enzymes and the biology of reactive oxygen. *Nat Rev Immunol* 4 (3): 181–189. doi:10.1038/nri1312.
109. Geiszt, M. 2006. NADPH oxidases: New kids on the block. *Cardiovasc Res* 71 (2): 289–299. doi:10.1016/j.cardiores.2006.05.004.
110. Suh, Y. A., R. S. Arnold, B. Lassegue, J. Shi, X. Xu, D. Sorescu, A. B. Chung, K. K. Griendling, and J. D. Lambeth. 1999. Cell transformation by the superoxide-generating oxidase Mox1. *Nature* 401 (6748): 79–82. doi:10.1038/43459.
111. Bánfi, B., A. Maturana, S. Jaconi, S. Arnaudeau, T. Laforge, B. Sinha, E. Ligeti, N. Demaurex, and K. H. Krause. 2000. A mammalian H+ channel generated through alternative splicing of the NADPH oxidase homolog NOH-1. *Science* 287 (5450): 138–142.
112. Geiszt, M., K. Lekstrom, and T. L. Leto. 2004. Analysis of mRNA transcripts from the NAD(P)H oxidase 1 (Nox1) gene. Evidence against production of the NADPH oxidase homolog-1 short (NOH-1S) transcript variant. *J Biol Chem* 279 (49): 51661–51668. doi:10.1074/jbc.M409325200.
113. Harper, R. W., C. Xu, K. Soucek, H. Setiadi, and J. P. Eiserich. 2005. A reappraisal of the genomic organization of human Nox1 and its splice variants. *Arch Biochem Biophys* 435 (2): 323–330. doi:10.1016/j.abb.2004.12.021.
114. Ambasta, R. K., P. Kumar, K. K. Griendling, H. H. Schmidt, R. Busse, and R. P. Brandes. 2004. Direct interaction of the novel Nox proteins with p22phox is required for the formation of a functionally active NADPH oxidase. *J Biol Chem* 279 (44): 45935–45941. doi:10.1074/jbc.M406486200.
115. Cheng, G., and J. D. Lambeth. 2005. Alternative mRNA splice forms of NOXO1: Differential tissue expression and regulation of Nox1 and Nox3. *Gene* 356: 118–126. doi:10.1016/j.gene.2005.03.008.
116. Takeya, R., N. Ueno, K. Kami, M. Taura, M. Kohjima, T. Izaki, H. Nunoi, and H. Sumimoto. 2003. Novel human homologues of p47phox and p67phox participate in activation of superoxide-producing NADPH oxidases. *J Biol Chem* 278 (27): 25234–25246. doi:10.1074/jbc.M212856200.
117. Ago, T., F. Kuribayashi, H. Hiroaki, R. Takeya, T. Ito, D. Kohda, and H. Sumimoto. 2003. Phosphorylation of p47phox directs phox homology domain from SH3 domain toward phosphoinositides, leading to phagocyte NADPH oxidase activation. *Proc Natl Acad Sci USA* 100 (8): 4474–4479. doi:10.1073/pnas.0735712100.
118. Geiszt, M., K. Lekstrom, J. Witta, and T. L. Leto. 2003. Proteins homologous to p47phox and p67phox support superoxide production by NAD(P)H oxidase 1 in colon epithelial cells. *J Biol Chem* 278 (22): 20006–20012. doi:10.1074/jbc.M301289200.
119. Bánfi, B., R. A. Clark, K. Steger, and K. H. Krause. 2003. Two novel proteins activate superoxide generation by the NADPH oxidase NOX1. *J Biol Chem* 278 (6): 3510–3513. doi:10.1074/jbc.C200613200.
120. Shiose, A., J. Kuroda, K. Tsuruya, M. Hirai, H. Hirakata, S. Naito, M. Hattori, Y. Sakaki, and H. Sumimoto. 2001. A novel superoxide-producing NAD(P)H oxidase in kidney. *J Biol Chem* 276 (2): 1417–1423. doi:10.1074/jbc.M007597200.
121. Takac, I., K. Schröder, L. Zhang, B. Lardy, N. Anilkumar, J. D. Lambeth, A. M. Shah, F. Morel, and R. P. Brandes. 2011. The E-loop is involved in hydrogen peroxide formation by the NADPH oxidase Nox4. *J Biol Chem* 286 (15): 13304–13313. doi:10.1074/jbc.M110.192138.

122. Katsuyama, M., H. Hirai, K. Iwata, M. Ibi, K. Matsuno, M. Matsumoto, and C. Yabe-Nishimura. 2011. Sp3 transcription factor is crucial for transcriptional activation of the human NOX4 gene. *FEBS J* 278 (6): 964–972. doi:10.1111/j.1742-4658.2011.08018.x.
123. Serrander, L., L. Cartier, K. Bedard, B. Banfi, B. Lardy, O. Plastre, A. Sienkiewicz, L. Fórró, W. Schlegel, and K. H. Krause. 2007a. NOX4 activity is determined by mRNA levels and reveals a unique pattern of ROS generation. *Biochem J* 406 (1): 105–114. doi:10.1042/BJ20061903.
124. Lyle, A. N., N. N. Deshpande, Y. Taniyama, B. Seidel-Rogol, L. Pounkova, P. Du, C. Papaharalambus, B. Lassègue, and K. K. Griendling. 2009. Poldip2, a novel regulator of Nox4 and cytoskeletal integrity in vascular smooth muscle cells. *Circ Res* 105 (3): 249–259. doi:10.1161/CIRCRESAHA.109.193722.
125. Bánfi, B., G. Molnár, A. Maturana, K. Steger, B. Hegedûs, N. Demaurex, and K. H. Krause. 2001. A Ca(2+)-activated NADPH oxidase in testis, spleen, and lymph nodes. *J Biol Chem* 276 (40): 37594–37601. doi:10.1074/jbc.M103034200.
126. Fulton, D. J. 2009. Nox5 and the regulation of cellular function. *Antioxid Redox Signal* 11 (10): 2443–2452. doi:10.1089/ars.2009.2587.
127. Kawahara, T., D. Ritsick, G. Cheng, and J. D. Lambeth. 2005. Point mutations in the proline-rich region of p22phox are dominant inhibitors of Nox1- and Nox2-dependent reactive oxygen generation. *J Biol Chem* 280 (36): 31859–31869. doi:10.1074/jbc.M501882200.
128. Bánfi, B., F. Tirone, I. Durussel, J. Knisz, P. Moskwa, G. Z. Molnár, K. H. Krause, and J. A. Cox. 2004. Mechanism of Ca^{2+} activation of the NADPH oxidase 5 (NOX5). *J Biol Chem* 279 (18): 18583–18591. doi:10.1074/jbc.M310268200.
129. Jagnandan, D., J. E. Church, B. Banfi, D. J. Stuehr, M. B. Marrero, and D. J. Fulton. 2007. Novel mechanism of activation of NADPH oxidase 5. Calcium sensitization via phosphorylation. *J Biol Chem* 282 (9): 6494–6507. doi:10.1074/jbc.M608966200.
130. BelAiba, R. S., T. Djordjevic, A. Petry, K. Diemer, S. Bonello, B. Banfi, J. Hess, A. Pogrebniak, C. Bickel, and A. Görlach. 2007. NOX5 variants are functionally active in endothelial cells. *Free Radic Biol Med* 42 (4): 446–459. doi:10.1016/j.freeradbiomed.2006.10.054.
131. Hanna, I. R., L. L. Hilenski, A. Dikalova, Y. Taniyama, S. Dikalov, A. Lyle, M. T. Quinn, B. Lassègue, and K. K. Griendling. 2004. Functional association of nox1 with p22phox in vascular smooth muscle cells. *Free Radic Biol Med* 37 (10): 1542–1549. doi:10.1016/j.freeradbiomed.2004.08.011.
132. Hilenski, L. L., R. E. Clempus, M. T. Quinn, J. D. Lambeth, and K. K. Griendling. 2004. Distinct subcellular localizations of Nox1 and Nox4 in vascular smooth muscle cells. *Arterioscler Thromb Vasc Biol* 24 (4): 677–683. doi:10.1161/01.ATV.0000112024.13727.2c.
133. Deliri, H., and C. A. McNamara. 2007. Nox 4 regulation of vascular smooth muscle cell differentiation marker gene expression. *Arterioscler Thromb Vasc Biol* 27 (1): 12–14. doi:10.1161/01.ATV.0000254154.43871.50.
134. Ding, L., A. Chapman, R. Boyd, and H. D. Wang. 2007. ERK activation contributes to regulation of spontaneous contractile tone via superoxide anion in isolated rat aorta of angiotensin II-induced hypertension. *Am J Physiol Heart Circ Physiol* 292 (6): H2997–H3005. doi:10.1152/ajpheart.00388.2006.
135. Rajagopalan, S., S. Kurz, T. Münzel, M. Tarpey, B. A. Freeman, K. K. Griendling, and D. G. Harrison. 1996. Angiotensin II-mediated hypertension in the rat increases vascular superoxide production via membrane NADH/NADPH oxidase activation. Contribution to alterations of vasomotor tone. *J Clin Invest* 97 (8): 1916–1923. doi:10.1172/JCI118623.
136. Dikalova, A., R. Clempus, B. Lassègue, G. Cheng, J. McCoy, S. Dikalov, A. San Martin et al. 2005. Nox1 overexpression potentiates angiotensin II-induced hypertension and vascular smooth muscle hypertrophy in transgenic mice. *Circulation* 112 (17): 2668–2676. doi:10.1161/CIRCULATIONAHA.105.538934.

137. Gavazzi, G., B. Banfi, C. Deffert, L. Fiette, M. Schappi, F. Herrmann, and K. H. Krause. 2006. Decreased blood pressure in NOX1-deficient mice. *FEBS Lett* 580 (2): 497–504. doi:10.1016/j.febslet.2005.12.049.
138. Laude, K., H. Cai, B. Fink, N. Hoch, D. S. Weber, L. McCann, G. Kojda et al. 2005. Hemodynamic and biochemical adaptations to vascular smooth muscle overexpression of p22phox in mice. *Am J Physiol Heart Circ Physiol* 288 (1): H7–H12. doi:10.1152/ajpheart.00637.2004.
139. Weber, D. S., P. Rocic, A. M. Mellis, K. Laude, A. N. Lyle, D. G. Harrison, and K. K. Griendling. 2005. Angiotensin II-induced hypertrophy is potentiated in mice overexpressing p22phox in vascular smooth muscle. *Am J Physiol Heart Circ Physiol* 288 (1): H37–H42. doi:10.1152/ajpheart.00638.2004.
140. Modlinger, P., T. Chabrashvili, P. S. Gill, M. Mendonca, D. G. Harrison, K. K. Griendling, M. Li et al. 2006. RNA silencing in vivo reveals role of p22phox in rat angiotensin slow pressor response. *Hypertension* 47 (2): 238–244. doi:10.1161/01.HYP.0000200023.02195.73.
141. Wilkinson-Berka, J. L., D. Deliyanti, I. Rana, A. G. Miller, A. Agrotis, R. Armani, C. Szyndralewiez et al. 2014. NADPH oxidase, NOX1, mediates vascular injury in ischemic retinopathy. *Antioxid Redox Signal* 20 (17): 2726–2740. doi:10.1089/ars.2013.5357.
142. Somanna, N. K., A. J. Valente, M. Krenz, W. P. Fay, P. Delafontaine, and B. Chandrasekar. 2016. The Nox1/4 dual inhibitor GKT137831 or Nox4 knockdown inhibits angiotensin-II-induced adult mouse cardiac fibroblast proliferation and migration. AT1 physically associates with Nox4. *J Cell Physiol* 231 (5): 1130–1141. doi:10.1002/jcp.25210.
143. Green, D. E., T. C. Murphy, B. Y. Kang, J. M. Kleinhenz, C. Szyndralewiez, P. Page, R. L. Sutliff, and C. M. Hart. 2012. The Nox4 inhibitor GKT137831 attenuates hypoxia-induced pulmonary vascular cell proliferation. *Am J Respir Cell Mol Biol* 47 (5): 718–726. doi:10.1165/rcmb.2011-0418OC.
144. McLaughlin, V. V., S. J. Shah, R. Souza, and M. Humbert. 2015. Management of pulmonary arterial hypertension. *J Am Coll Cardiol* 65 (18): 1976–1997. doi:10.1016/j.jacc.2015.03.540.
145. Paulin, R., and E. D. Michelakis. 2014. The metabolic theory of pulmonary arterial hypertension. *Circ Res* 115 (1): 148–164. doi:10.1161/CIRCRESAHA.115.301130.
146. Vonk-Noordegraaf, A., F. Haddad, K. M. Chin, P. R. Forfia, S. M. Kawut, J. Lumens, R. Naeije et al. 2013. Right heart adaptation to pulmonary arterial hypertension: Physiology and pathobiology. *J Am Coll Cardiol* 62 (25 Suppl): D22–D33. doi:10.1016/j.jacc.2013.10.027.
147. Stenmark, K. R., K. A. Fagan, and M. G. Frid. 2006. Hypoxia-induced pulmonary vascular remodeling: Cellular and molecular mechanisms. *Circ Res* 99 (7): 675–691. doi:10.1161/01.RES.0000243584.45145.3f.
148. Budhiraja, R., R. M. Tuder, and P. M. Hassoun. 2004. Endothelial dysfunction in pulmonary hypertension. *Circulation* 109 (2): 159–165. doi:10.1161/01.CIR.0000102381.57477.50.
149. Sakao, S., K. Tatsumi, and N. F. Voelkel. 2009. Endothelial cells and pulmonary arterial hypertension: Apoptosis, proliferation, interaction and transdifferentiation. *Respir Res* 10: 95. doi:10.1186/1465-9921-10-95.
150. Veit, F., O. Pak, B. Egemnazarov, M. Roth, D. Kosanovic, M. Seimetz, N. Sommer et al. 2013. Function of NADPH oxidase 1 in pulmonary arterial smooth muscle cells after monocrotaline-induced pulmonary vascular remodeling. *Antioxid Redox Signal* 19 (18): 2213–2231. doi:10.1089/ars.2012.4904.
151. Ranayhossaini, D. J., A. I. Rodriguez, S. Sahoo, B. B. Chen, R. K. Mallampalli, E. E. Kelley, G. Csanyi, M. T. Gladwin, G. Romero, and P. J. Pagano. 2013. Selective recapitulation of conserved and nonconserved regions of putative NOXA1 protein activation domain confers isoform-specific inhibition of Nox1 oxidase and attenuation of endothelial cell migration. *J Biol Chem* 288 (51): 36437–36450. doi:10.1074/jbc.M113.521344.

152. Al-Shabrawey, M., M. Bartoli, A. B. El-Remessy, D. H. Platt, S. Matragoon, M. A. Behzadian, R. W. Caldwell, and R. B. Caldwell. 2005. Inhibition of NAD(P)H oxidase activity blocks vascular endothelial growth factor overexpression and neovascularization during ischemic retinopathy. *Am J Pathol* 167 (2): 599–607. doi:10.1016/S0002-9440(10)63001-5.
153. Fürst, R., C. Brueckl, W. M. Kuebler, S. Zahler, F. Krötz, A. Görlach, A. M. Vollmar, and A. K. Kiemer. 2005. Atrial natriuretic peptide induces mitogen-activated protein kinase phosphatase-1 in human endothelial cells via Rac1 and NAD(P)H oxidase/Nox2-activation. *Circ Res* 96 (1): 43–53. doi:10.1161/01.RES.0000151983.01148.06.
154. Lopes, N. H., S. S. Vasudevan, D. Gregg, B. Selvakumar, P. J. Pagano, H. Kovacic, and P. J. Goldschmidt-Clermont. 2002. Rac-dependent monocyte chemoattractant protein-1 production is induced by nutrient deprivation. *Circ Res* 91 (9): 798–805.
155. Krötz, F., M. Keller, S. Derflinger, H. Schmid, T. Gloe, F. Bassermann, J. Duyster et al. 2007. Mycophenolate acid inhibits endothelial NAD(P)H oxidase activity and superoxide formation by a Rac1-dependent mechanism. *Hypertension* 49 (1): 201–208. doi:10.1161/01.HYP.0000251162.14782.d4.
156. Duerrschmidt, N., C. Stielow, G. Muller, P. J. Pagano, and H. Morawietz. 2006. NO-mediated regulation of NAD(P)H oxidase by laminar shear stress in human endothelial cells. *J Physiol* 576 (Pt 2): 557–567. doi:10.1113/jphysiol.2006.111070.
157. Liu, J., F. Yang, X. P. Yang, M. Jankowski, and P. J. Pagano. 2003. NAD(P)H oxidase mediates angiotensin II-induced vascular macrophage infiltration and medial hypertrophy. *Arterioscler Thromb Vasc Biol* 23 (5): 776–782. doi:10.1161/01.ATV.0000066684.37829.16.
158. Liu, J., A. Ormsby, N. Oja-Tebbe, and P. J. Pagano. 2004. Gene transfer of NAD(P)H oxidase inhibitor to the vascular adventitia attenuates medial smooth muscle hypertrophy. *Circ Res* 95 (6): 587–594. doi:10.1161/01.RES.0000142317.88591.e6.
159. Sellers, K. W., C. Sun, C. Diez-Freire, H. Waki, C. Morisseau, J. R. Falck, B. D. Hammock, J. F. Paton, and M. K. Raizada. 2005. Novel mechanism of brain soluble epoxide hydrolase-mediated blood pressure regulation in the spontaneously hypertensive rat. *FASEB J* 19 (6): 626–628. doi:10.1096/fj.04-3128fje.
160. Zhou, X., H. G. Bohlen, S. J. Miller, and J. L. Unthank. 2008. NAD(P)H oxidase-derived peroxide mediates elevated basal and impaired flow-induced NO production in SHR mesenteric arteries in vivo. *Am J Physiol Heart Circ Physiol* 295 (3): H1008–H1016. doi:10.1152/ajpheart.00114.2008.
161. Barry-Lane, P. A., C. Patterson, M. van der Merwe, Z. Hu, S. M. Holland, E. T. Yeh, and M. S. Runge. 2001. p47phox is required for atherosclerotic lesion progression in ApoE(-/-) mice. *J Clin Invest* 108 (10): 1513–1522. doi:10.1172/JCI11927.
162. Dourron, H. M., G. M. Jacobson, J. L. Park, J. Liu, D. J. Reddy, M. L. Scheel, and P. J. Pagano. 2005. Perivascular gene transfer of NADPH oxidase inhibitor suppresses angioplasty-induced neointimal proliferation of rat carotid artery. *Am J Physiol Heart Circ Physiol* 288 (2): H946–H953. doi:10.1152/ajpheart.00413.2004.
163. Judkins, C. P., H. Diep, B. R. Broughton, A. E. Mast, E. U. Hooker, A. A. Miller, S. Selemidis, G. J. Dusting, C. G. Sobey, and G. R. Drummond. 2010. Direct evidence of a role for Nox2 in superoxide production, reduced nitric oxide bioavailability, and early atherosclerotic plaque formation in ApoE-/- mice. *Am J Physiol Heart Circ Physiol* 298 (1): H24–H32. doi:10.1152/ajpheart.00799.2009.
164. Vendrov, A. E., Z. S. Hakim, N. R. Madamanchi, M. Rojas, C. Madamanchi, and M. S. Runge. 2007. Atherosclerosis is attenuated by limiting superoxide generation in both macrophages and vessel wall cells. *Arterioscler Thromb Vasc Biol* 27 (12): 2714–2721. doi:10.1161/ATVBAHA.107.152629.

165. Weaver, M., J. Liu, D. Pimentel, D. J. Reddy, P. Harding, E. L. Peterson, and P. J. Pagano. 2006. Adventitial delivery of dominant-negative p67phox attenuates neointimal hyperplasia of the rat carotid artery. *Am J Physiol Heart Circ Physiol* 290 (5): H1933–H1941. doi:10.1152/ajpheart.00690.2005.
166. Kawahara, T., M. T. Quinn, and J. D. Lambeth. 2007. Molecular evolution of the reactive oxygen-generating NADPH oxidase (Nox/Duox) family of enzymes. *BMC Evol Biol* 7: 109. doi:10.1186/1471-2148-7-109.
167. Brown, D. I., and K. K. Griendling. 2009. Nox proteins in signal transduction. *Free Radic Biol Med* 47 (9): 1239–1253. doi:10.1016/j.freeradbiomed.2009.07.023.
168. Lassègue, B., and K. K. Griendling. 2010. NADPH oxidases: Functions and pathologies in the vasculature. *Arterioscler Thromb Vasc Biol* 30 (4): 653–661. doi:10.1161/ATVBAHA.108.181610.
169. San Martin, A., R. Foncea, F. R. Laurindo, R. Ebensperger, K. K. Griendling, and F. Leighton. 2007. Nox1-based NADPH oxidase-derived superoxide is required for VSMC activation by advanced glycation end-products. *Free Radic Biol Med* 42 (11): 1671–1679. doi:10.1016/j.freeradbiomed.2007.02.002.
170. Zhao, Y., S. K. Biswas, P. H. McNulty, M. Kozak, J. Y. Jun, and L. Segar. 2011. PDGF-induced vascular smooth muscle cell proliferation is associated with dysregulation of insulin receptor substrates. *Am J Physiol Cell Physiol* 300 (6): C1375–C1385. doi:10.1152/ajpcell.00670.2008.
171. Maheswaranathan, M., H. K. Gole, I. Fernandez, B. Lassègue, K. K. Griendling, and A. San Martín. 2011. Platelet-derived growth factor (PDGF) regulates Slingshot phosphatase activity via Nox1-dependent auto-dephosphorylation of serine 834 in vascular smooth muscle cells. *J Biol Chem* 286 (41): 35430–35437. doi:10.1074/jbc.M111.268284.
172. Streeter, J., B. M. Schickling, S. Jiang, B. Stanic, W. H. Thiel, L. Gakhar, J. C. Houtman, and F. J. Miller. 2014. Phosphorylation of Nox1 regulates association with NoxA1 activation domain. *Circ Res* 115 (11): 911–918. doi:10.1161/CIRCRESAHA.115.304267.
173. Lee, M. Y., A. San Martin, P. K. Mehta, A. E. Dikalova, A. M. Garrido, S. R. Datla, E. Lyons et al. 2009. Mechanisms of vascular smooth muscle NADPH oxidase 1 (Nox1) contribution to injury-induced neointimal formation. *Arterioscler Thromb Vasc Biol* 29 (4): 480–487. doi:10.1161/ATVBAHA.108.181925.
174. Pescatore, L. A., D. Bonatto, F. L. Forti, A. Sadok, H. Kovacic, and F. R. Laurindo. 2012. Protein disulfide isomerase is required for platelet-derived growth factor-induced vascular smooth muscle cell migration, Nox1 NADPH oxidase expression, and RhoGTPase activation. *J Biol Chem* 287 (35): 29290–29300. doi:10.1074/jbc.M112.394551.
175. Chen, H., B. W. Bernstein, and J. R. Bamburg. 2000. Regulating actin-filament dynamics in vivo. *Trends Biochem Sci* 25 (1): 19–23.
176. San Martín, A., M. Y. Lee, H. C. Williams, K. Mizuno, B. Lassègue, and K. K. Griendling. 2008. Dual regulation of cofilin activity by LIM kinase and Slingshot-1L phosphatase controls platelet-derived growth factor-induced migration of human aortic smooth muscle cells. *Circ Res* 102 (4): 432–438. doi:10.1161/CIRCRESAHA.107.158923.
177. Jagadeesha, D. K., M. Takapoo, B. Banfi, R. C. Bhalla, and F. J. Miller. 2012. Nox1 transactivation of epidermal growth factor receptor promotes N-cadherin shedding and smooth muscle cell migration. *Cardiovasc Res* 93 (3): 406–413. doi:10.1093/cvr/cvr308.
178. Schröder, K., I. Helmcke, K. Palfi, K. H. Krause, R. Busse, and R. P. Brandes. 2007. Nox1 mediates basic fibroblast growth factor-induced migration of vascular smooth muscle cells. *Arterioscler Thromb Vasc Biol* 27 (8): 1736–1743. doi:10.1161/ATVBAHA.107.142117.

179. Zafari, A. M., M. Ushio-Fukai, C. A. Minieri, M. Akers, B. Lassègue, and K. K. Griendling. 1999. Arachidonic acid metabolites mediate angiotensin II-induced NADH/NADPH oxidase activity and hypertrophy in vascular smooth muscle cells. *Antioxid Redox Signal* 1 (2): 167–179. doi:10.1089/ars.1999.1.2-167.
180. Nguyen Dinh Cat, A., A. C. Montezano, D. Burger, and R. M. Touyz. 2013. Angiotensin II, NADPH oxidase, and redox signaling in the vasculature. *Antioxid Redox Signal* 19 (10): 1110–1120. doi:10.1089/ars.2012.4641.
181. Touyz, R. M., and E. L. Schiffrin. 1999. Ang II-stimulated superoxide production is mediated via phospholipase D in human vascular smooth muscle cells. *Hypertension* 34 (4 Pt 2): 976–982.
182. Virdis, A., M. F. Neves, F. Amiri, R. M. Touyz, and E. L. Schiffrin. 2004. Role of NAD(P)H oxidase on vascular alterations in angiotensin II-infused mice. *J Hypertens* 22 (3): 535–542.
183. Brandes, R. P., and K. Schröder. 2008. Differential vascular functions of Nox family NADPH oxidases. *Curr Opin Lipidol* 19 (5): 513–518. doi:10.1097/MOL.0b013e32830c91e3.
184. de Leeuw, P. W. 1999. How do angiotensin II receptor antagonists affect blood pressure? *Am J Cardiol* 84 (2A): 5K–6K.
185. Hughes, A. D. 1998. Molecular and cellular mechanisms of action of angiotensin II (AT1) receptors in vascular smooth muscle. *J Hum Hypertens* 12 (5): 275–281.
186. Muthalif, M. M., N. A. Karzoun, L. Gaber, Z. Khandekar, I. F. Benter, A. E. Saeed, J. H. Parmentier, A. Estes, and K. U. Malik. 2000. Angiotensin II-induced hypertension: Contribution of Ras GTPase/Mitogen-activated protein kinase and cytochrome P450 metabolites. *Hypertension* 36 (4): 604–609.
187. Matsuno, K., H. Yamada, K. Iwata, D. Jin, M. Katsuyama, M. Matsuki, S. Takai, K. Yamanishi, M. Miyazaki, H. Matsubara, and C. Yabe-Nishimura. 2005. Nox1 is involved in angiotensin II-mediated hypertension: A study in Nox1-deficient mice. *Circulation* 112 (17): 2677–2685. doi:10.1161/CIRCULATIONAHA.105.573709.
188. Ardanaz, N., and P. J. Pagano. 2006. Hydrogen peroxide as a paracrine vascular mediator: Regulation and signaling leading to dysfunction. *Exp Biol Med (Maywood)* 231 (3): 237–251.
189. Somlyo, A. P., and A. V. Somlyo. 2000. Signal transduction by G-proteins, rho-kinase and protein phosphatase to smooth muscle and non-muscle myosin II. *J Physiol* 522 Pt 2: 177–185.
190. Anwar, M. A., J. Shalhoub, C. S. Lim, M. S. Gohel, and A. H. Davies. 2012. The effect of pressure-induced mechanical stretch on vascular wall differential gene expression. *J Vasc Res* 49 (6): 463–478. doi:10.1159/000339151.
191. Lemarié, C. A., P. L. Tharaux, and S. Lehoux. 2010. Extracellular matrix alterations in hypertensive vascular remodeling. *J Mol Cell Cardiol* 48 (3): 433–439. doi:10.1016/j.yjmcc.2009.09.018.
192. Lehoux, S., Y. Castier, and A. Tedgui. 2006. Molecular mechanisms of the vascular responses to haemodynamic forces. *J Intern Med* 259 (4): 381–392. doi:10.1111/j.1365-2796.2006.01624.x.
193. Gomez, D., and G. K. Owens. 2012. Smooth muscle cell phenotypic switching in atherosclerosis. *Cardiovasc Res* 95 (2): 156–64. doi:10.1093/cvr/cvs115.
194. Lassègue, B., A. San Martín, and K. K. Griendling. 2012. Biochemistry, physiology, and pathophysiology of NADPH oxidases in the cardiovascular system. *Circ Res* 110 (10): 1364–1390. doi:10.1161/CIRCRESAHA.111.243972.
195. Lassègue, B., and R. E. Clempus. 2003. Vascular NAD(P)H oxidases: Specific features, expression, and regulation. *Am J Physiol Regul Integr Comp Physiol* 285 (2): R277–R297. doi:10.1152/ajpregu.00758.2002.

196. Datla, S. R., D. J. McGrail, S. Vukelic, L. P. Huff, A. N. Lyle, L. Pounkova, M. Lee et al. 2014. Poldip2 controls vascular smooth muscle cell migration by regulating focal adhesion turnover and force polarization. *Am J Physiol Heart Circ Physiol* 307 (7): H945–H957. doi:10.1152/ajpheart.00918.2013.
197. Meng, D., D. D. Lv, and J. Fang. 2008. Insulin-like growth factor-I induces reactive oxygen species production and cell migration through Nox4 and Rac1 in vascular smooth muscle cells. *Cardiovasc Res* 80 (2): 299–308. doi:10.1093/cvr/cvn173.
198. Xi, G., X. Shen, C. Wai, C. K. Vilas, and D. R. Clemmons. 2015. Hyperglycemia stimulates p62/PKCζ interaction, which mediates NF-κB activation, increased Nox4 expression, and inflammatory cytokine activation in vascular smooth muscle. *FASEB J* 29 (12): 4772–4782. doi:10.1096/fj.15-275453.
199. Diebold, I., A. Petry, J. Hess, and A. Görlach. 2010b. The NADPH oxidase subunit NOX4 is a new target gene of the hypoxia-inducible factor-1. *Mol Biol Cell* 21 (12): 2087–2096. doi:10.1091/mbc.E09-12-1003.
200. Aguado, A., T. Fischer, C. Rodríguez, A. Manea, J. Martínez-González, R. M. Touyz, R. Hernanz et al. 2016. Hu antigen R is required for NOX-1 but not NOX-4 regulation by inflammatory stimuli in vascular smooth muscle cells. *J Hypertens* 34 (2): 253–265. doi:10.1097/HJH.0000000000000801.
201. Matsushima, S., J. Kuroda, P. Zhai, T. Liu, S. Ikeda, N. Nagarajan, S. Oka, T. Yokota, S. Kinugawa, C. P. Hsu, H. Li, H. Tsutsui, and J. Sadoshima. 2016. Tyrosine kinase FYN negatively regulates NOX4 in cardiac remodeling. *J Clin Invest* 126 (9): 3403–3416. doi:10.1172/JCI85624.
202. Venema, R. C., V. J. Venema, D. C. Eaton, and M. B. Marrero. 1998. Angiotensin II-induced tyrosine phosphorylation of signal transducers and activators of transcription 1 is regulated by Janus-activated kinase 2 and Fyn kinases and mitogen-activated protein kinase phosphatase 1. *J Biol Chem* 273 (46): 30795–30800.
203. Delafontaine, P., Y. H. Song, and Y. Li. 2004. Expression, regulation, and function of IGF-1, IGF-1R, and IGF-1 binding proteins in blood vessels. *Arterioscler Thromb Vasc Biol* 24 (3): 435–444. doi:10.1161/01.ATV.0000105902.89459.09.
204. Ismail, S., A. Sturrock, P. Wu, B. Cahill, K. Norman, T. Huecksteadt, K. Sanders, T. Kennedy, and J. Hoidal. 2009. NOX4 mediates hypoxia-induced proliferation of human pulmonary artery smooth muscle cells: The role of autocrine production of transforming growth factor-{beta}1 and insulin-like growth factor binding protein-3. *Am J Physiol Lung Cell Mol Physiol* 296 (3): L489–L499. doi:10.1152/ajplung.90488.2008.
205. Diebold, I., D. Flügel, S. Becht, R. S. Belaiba, S. Bonello, J. Hess, T. Kietzmann, and A. Görlach. 2010a. The hypoxia-inducible factor-2alpha is stabilized by oxidative stress involving NOX4. *Antioxid Redox Signal* 13 (4): 425–436. doi:10.1089/ars.2009.3014.
206. Djordjevic, T., R. S. BelAiba, S. Bonello, J. Pfeilschifter, J. Hess, and A. Görlach. 2005. Human urotensin II is a novel activator of NADPH oxidase in human pulmonary artery smooth muscle cells. *Arterioscler Thromb Vasc Biol* 25 (3): 519–525. doi:10.1161/01.ATV.0000154279.98244.eb.
207. Mittal, M., M. Roth, P. König, S. Hofmann, E. Dony, P. Goyal, A. C. Selbitz et al. 2007. Hypoxia-dependent regulation of nonphagocytic NADPH oxidase subunit NOX4 in the pulmonary vasculature. *Circ Res* 101 (3): 258–267. doi:10.1161/CIRCRESAHA.107.148015.
208. Kuroda, J., T. Ago, S. Matsushima, P. Zhai, M. D. Schneider, and J. Sadoshima. 2010. NADPH oxidase 4 (Nox4) is a major source of oxidative stress in the failing heart. *Proc Natl Acad Sci USA* 107 (35): 15565–15570. doi:10.1073/pnas.1002178107.
209. Schröder, K., M. Zhang, S. Benkhoff, A. Mieth, R. Pliquett, J. Kosowski, C. Kruse et al. 2012. Nox4 is a protective reactive oxygen species generating vascular NADPH oxidase. *Circ Res* 110 (9): 1217–1225. doi:10.1161/CIRCRESAHA.112.267054.

210. Jay, D. B., C. A. Papaharalambus, B. Seidel-Rogol, A. E. Dikalova, B. Lassègue, and K. K. Griendling. 2008. Nox5 mediates PDGF-induced proliferation in human aortic smooth muscle cells. *Free Radic Biol Med* 45 (3): 329–335. doi:10.1016/j.freeradbiomed.2008.04.024.
211. Montezano, A. C., D. Burger, G. S. Ceravolo, H. Yusuf, M. Montero, and R. M. Touyz. 2011. Novel Nox homologues in the vasculature: Focusing on Nox4 and Nox5. *Clin Sci (Lond)* 120 (4): 131–141. doi:10.1042/CS20100384.
212. Serrander, L., V. Jaquet, K. Bedard, O. Plastre, O. Hartley, S. Arnaudeau, N. Demaurex, W. Schlegel, and K. H. Krause. 2007b. NOX5 is expressed at the plasma membrane and generates superoxide in response to protein kinase C activation. *Biochimie* 89 (9): 1159–1167. doi:10.1016/j.biochi.2007.05.004.
213. Guzik, T. J., W. Chen, M. C. Gongora, B. Guzik, H. E. Lob, D. Mangalat, N. Hoch et al. 2008. Calcium-dependent NOX5 nicotinamide adenine dinucleotide phosphate oxidase contributes to vascular oxidative stress in human coronary artery disease. *J Am Coll Cardiol* 52 (22): 1803–1809. doi:10.1016/j.jacc.2008.07.063.
214. Rawlings, J. S., K. M. Rosler, and D. A. Harrison. 2004. The JAK/STAT signaling pathway. *J Cell Sci* 117 (Pt 8): 1281–1283. doi:10.1242/jcs.00963.
215. Montezano, A. C., S. Tsiropoulou, M. Dulak-Lis, A. Harvey, L. de L. Camargo, and R. M. Touyz. 2015. Redox signaling, Nox5 and vascular remodeling in hypertension. *Curr Opin Nephrol Hypertens* 24 (5): 425–433. doi:10.1097/MNH.0000000000000153.
216. Dalle-Donne, I., D. Giustarini, R. Colombo, A. Milzani, and R. Rossi. 2005. S-glutathionylation in human platelets by a thiol-disulfide exchange-independent mechanism. *Free Radic Biol Med* 38 (11): 1501–1510. doi:10.1016/j.freeradbiomed.2005.02.019.
217. Gole, H. K., D. L. Tharp, and D. K. Bowles. 2014. Upregulation of intermediate-conductance Ca^{2+}-activated K^{+} channels (KCNN4) in porcine coronary smooth muscle requires NADPH oxidase 5 (NOX5). *PLoS One* 9 (8): e105337. doi:10.1371/journal.pone.0105337.
218. Winterbourn, C. C. 2014. The challenges of using fluorescent probes to detect and quantify specific reactive oxygen species in living cells. *Biochim Biophys Acta* 1840 (2): 730–738. doi:10.1016/j.bbagen.2013.05.004.
219. Zielonka, J., J. Vasquez-Vivar, and B. Kalyanaraman. 2008. Detection of 2-hydroxyethidium in cellular systems: A unique marker product of superoxide and hydroethidine. *Nat Protoc* 3 (1): 8–21. doi:10.1038/nprot.2007.473.
220. Nazarewicz, R. R., A. Bikineyeva, and S. I. Dikalov. 2013. Rapid and specific measurements of superoxide using fluorescence spectroscopy. *J Biomol Screen* 18 (4): 498–503. doi:10.1177/1087057112468765.
221. Kalyanaraman, B., B. P. Dranka, M. Hardy, R. Michalski, and J. Zielonka. 2014. HPLC-based monitoring of products formed from hydroethidine-based fluorogenic probes—the ultimate approach for intra- and extracellular superoxide detection. *Biochim Biophys Acta* 1840 (2): 739–744. doi:10.1016/j.bbagen.2013.05.008.
222. Michalski, R., J. Zielonka, M. Hardy, J. Joseph, and B. Kalyanaraman. 2013. Hydropropidine: A novel, cell-impermeant fluorogenic probe for detecting extracellular superoxide. *Free Radic Biol Med* 54: 135–147. doi:10.1016/j.freeradbiomed.2012.09.018.
223. Dikalov, S., K. K. Griendling, and D. G. Harrison. 2007. Measurement of reactive oxygen species in cardiovascular studies. *Hypertension* 49 (4): 717–727. doi:10.1161/01.HYP.0000258594.87211.6b.
224. Dikalov, S., M. Skatchkov M, B. Fink B, and E. Bassenge E. 1997. Quantification of superoxide radicals and peroxynitrite in vascular cells using oxidation of sterically hindered hydroxylamines and electron spin resonance. *Nitric Oxide* 1 (5): 423–431
225. Zhou, M., Z. Diwu, N. Panchuk-Voloshina, and R. P. Haugland. 1997. A stable nonfluorescent derivative of resorufin for the fluorometric determination of trace hydrogen peroxide: Applications in detecting the activity of phagocyte NADPH oxidase and other oxidases. *Anal Biochem* 253 (2): 162–168. doi:10.1006/abio.1997.2391.

226. Summers, F. A., B. Zhao, D. Ganini, and R. P. Mason. 2013. Photooxidation of Amplex Red to resorufin: Implications of exposing the Amplex Red assay to light. *Methods Enzymol* 526: 1–17. doi:10.1016/B978-0-12-405883-5.00001-6.
227. Ibrahim, W. H., H. M. Habib, H. Kamal, D. K. St Clair, and C. K. Chow. 2013. Mitochondrial superoxide mediates labile iron level: Evidence from Mn-SOD-transgenic mice and heterozygous knockout mice and isolated rat liver mitochondria. *Free Radic Biol Med* 65: 143–149. doi:10.1016/j.freeradbiomed.2013.06.026.
228. Kalyanaraman, B., V. Darley-Usmar, K. J. Davies, P. A. Dennery, H. J. Forman, M. B. Grisham, G. E. Mann, K. Moore, L. J. Roberts, and H. Ischiropoulos. 2012. Measuring reactive oxygen and nitrogen species with fluorescent probes: Challenges and limitations. *Free Radic Biol Med* 52 (1): 1–6. doi:10.1016/j.freeradbiomed.2011.09.030.
229. LeBel, C. P., H. Ischiropoulos, and S. C. Bondy. 1992. Evaluation of the probe 2′,7′-dichlorofluorescin as an indicator of reactive oxygen species formation and oxidative stress. *Chem Res Toxicol* 5 (2): 227–231.
230. Mody, N., F. Parhami, T. A. Sarafian, and L. L. Demer. 2001. Oxidative stress modulates osteoblastic differentiation of vascular and bone cells. *Free Radic Biol Med* 31 (4): 509–519.
231. Rota, C., C. F. Chignell, and R. P. Mason. 1999a. Evidence for free radical formation during the oxidation of 2′-7′-dichlorofluorescin to the fluorescent dye 2′-7′-dichlorofluorescein by horseradish peroxidase: Possible implications for oxidative stress measurements. *Free Radic Biol Med* 27 (7–8): 873–881.
232. Rota, C., Y. C. Fann, and R. P. Mason. 1999b. Phenoxyl free radical formation during the oxidation of the fluorescent dye 2′,7′-dichlorofluorescein by horseradish peroxidase. Possible consequences for oxidative stress measurements. *J Biol Chem* 274 (40): 28161–28168.
233. Wrona, M., and P. Wardman. 2006. Properties of the radical intermediate obtained on oxidation of 2′,7′-dichlorodihydrofluorescein, a probe for oxidative stress. *Free Radic Biol Med* 41 (4): 657–667. doi:10.1016/j.freeradbiomed.2006.05.006.
234. Karlsson, M., T. Kurz, U. T. Brunk, S. E. Nilsson, and C. I. Frennesson. 2010. What does the commonly used DCF test for oxidative stress really show? *Biochem J* 428 (2): 183–190. doi:10.1042/BJ20100208.
235. Zielonka, J., A. Sikora, M. Hardy, J. Joseph, B. P. Dranka, and B. Kalyanaraman. 2012. Boronate probes as diagnostic tools for real time monitoring of peroxynitrite and hydroperoxides. *Chem Res Toxicol* 25 (9): 1793–1799. doi:10.1021/tx300164j.
236. Zielonka, J., A. Sikora, J. Joseph, and B. Kalyanaraman. 2010. Peroxynitrite is the major species formed from different flux ratios of co-generated nitric oxide and superoxide: Direct reaction with boronate-based fluorescent probe. *J Biol Chem* 285 (19): 14210–14216. doi:10.1074/jbc.M110.110080.
237. Belousov, V. V., A. F. Fradkov, K. A. Lukyanov, D. B. Staroverov, K. S. Shakhbazov, A. V. Terskikh, and S. Lukyanov. 2006. Genetically encoded fluorescent indicator for intracellular hydrogen peroxide. *Nat Methods* 3 (4): 281–286. doi:10.1038/nmeth866.
238. Zheng, M., F. Aslund, and G. Storz. 1998. Activation of the OxyR transcription factor by reversible disulfide bond formation. *Science* 279 (5357): 1718–1721.
239. Lee, C., S. M. Lee, P. Mukhopadhyay, S. J. Kim, S. C. Lee, W. S. Ahn, M. H. Yu, G. Storz, and S. E. Ryu. 2004. Redox regulation of OxyR requires specific disulfide bond formation involving a rapid kinetic reaction path. *Nat Struct Mol Biol* 11 (12): 1179–1185. doi:10.1038/nsmb856.
240. Lukyanov, K. A., and V. V. Belousov. 2014. Genetically encoded fluorescent redox sensors. *Biochim Biophys Acta* 1840 (2): 745–756. doi:10.1016/j.bbagen.2013.05.030.
241. Meyer, A. J., and T. P. Dick. 2010. Fluorescent protein-based redox probes. *Antioxid Redox Signal* 13 (5): 621–650. doi:10.1089/ars.2009.2948.
242. Kaludercic, N., S. Deshwal, and F. Di Lisa. 2014. Reactive oxygen species and redox compartmentalization. *Front Physiol* 5: 285. doi:10.3389/fphys.2014.00285.

243. Poburko, D., J. Santo-Domingo, and N. Demaurex. 2011. Dynamic regulation of the mitochondrial proton gradient during cytosolic calcium elevations. *J Biol Chem* 286 (13): 11672–11684. doi:10.1074/jbc.M110.159962.
244. Malinouski, M., Y. Zhou, V. V. Belousov, D. L. Hatfield, and V. N. Gladyshev. 2011. Hydrogen peroxide probes directed to different cellular compartments. *PLoS One* 6 (1): e14564. doi:10.1371/journal.pone.0014564.
245. Aldieri, E., C. Riganti, M. Polimeni, E. Gazzano, C. Lussiana, I. Campia, and D. Ghigo. 2008. Classical inhibitors of NOX NAD(P)H oxidases are not specific. *Curr Drug Metab* 9 (8): 686–696.
246. Altenhofer, S., K. A. Radermacher, P. W. Kleikers, K. Wingler, and H. H. Schmidt. 2015. Evolution of NADPH oxidase inhibitors: Selectivity and mechanisms for target engagement. *Antioxid Redox Signal* 23 (5): 406–427. doi:10.1089/ars.2013.5814.
247. Cifuentes-Pagano, E., G. Csanyi, and P. J. Pagano. 2012. NADPH oxidase inhibitors: A decade of discovery from Nox2ds to HTS. *Cell Mol Life Sci* 69 (14): 2315–2325. doi:10.1007/s00018-012-1009-2.
248. Cifuentes-Pagano, E., D. N. Meijles, and P. J. Pagano. 2014. The quest for selective Nox inhibitors and therapeutics: Challenges, triumphs and pitfalls. *Antioxid Redox Signal* 20 (17): 2741–2754. doi:10.1089/ars.2013.5620.
249. Cifuentes-Pagano, M. E., D. N. Meijles, and P. J. Pagano. 2015. Nox inhibitors & therapies: Rational design of peptidic and small molecule inhibitors. *Curr Pharm Des* 21 (41): 6023–6035.
250. Jaquet, V., L. Scapozza, R. Clark, K. H. Krause, and J. D. Lambeth. 2009. Small molecule NOX inhibitors: ROS-generating NADPH oxidases as therapeutic targets. *Antioxid Redox Signal* 11 (10): 2535–2552. doi:10.1089/ARS.2009.2585.
251. Rey, F. E., M. E. Cifuentes, A. Kiarash, M. T. Quinn, and P. J. Pagano. 2001. Novel competitive inhibitor of NAD(P)H oxidase assembly attenuates vascular O(2)(-) and systolic blood pressure in mice. *Circ Res* 89 (5): 408–414.
252. Csányi, G., E. Cifuentes-Pagano, I. Al Ghouleh, D. J. Ranayhossaini, L. Egaña, L. R. Lopes, H. M. Jackson, E. E. Kelley, and P. J. Pagano. 2011. Nox2 B-loop peptide, Nox2ds, specifically inhibits the NADPH oxidase Nox2. *Free Radic Biol Med* 51 (6): 1116–1125. doi:10.1016/j.freeradbiomed.2011.04.025.
253. Stolk, J., T. J. Hiltermann, J. H. Dijkman, and A. J. Verhoeven. 1994. Characteristics of the inhibition of NADPH oxidase activation in neutrophils by apocynin, a methoxy-substituted catechol. *Am J Respir Cell Mol Biol* 11 (1): 95–102. doi:10.1165/ajrcmb.11.1.8018341.
254. Heumüller, S., S. Wind, E. Barbosa-Sicard, H. H. Schmidt, R. Busse, K. Schröder, and R. P. Brandes. 2008. Apocynin is not an inhibitor of vascular NADPH oxidases but an antioxidant. *Hypertension* 51 (2): 211–217. doi:10.1161/HYPERTENSIONAHA.107.100214.
255. Petrônio, M. S., M. L. Zeraik, L. M. Fonseca, and V. F. Ximenes. 2013. Apocynin: Chemical and biophysical properties of a NADPH oxidase inhibitor. *Molecules* 18 (3): 2821–2839. doi:10.3390/molecules18032821.
256. El-Naga, R. N. 2015. Apocynin protects against ethanol-induced gastric ulcer in rats by attenuating the upregulation of NADPH oxidases 1 and 4. *Chem Biol Interact.* 242: 317–326.
257. ten Freyhaus, H., M. Huntgeburth, K. Wingler, J. Schnitker, A. T. Bäumer, M. Vantler, M. M. Bekhite, M. Wartenberg, H. Sauer, and S. Rosenkranz. 2006. Novel Nox inhibitor VAS2870 attenuates PDGF-dependent smooth muscle cell chemotaxis, but not proliferation. *Cardiovasc Res* 71 (2): 331–341. doi:10.1016/j.cardiores.2006.01.022.

258. Stielow, C., R. A. Catar, G. Muller, et al. 2006. Novel Nox inhibitor of oxLDL-induced reactive oxygen species formation in human endothelial cells. *Biochem Biophys Res Commun.* 344 (1): 200–205.
259. Sun, Q. A., D. T. Hess, B. Wang, M. Miyagi, and J. S. Stamler. 2012. Off-target thiol alkylation by the NADPH oxidase inhibitor 3-benzyl-7-(2-benzoxazolyl)thio-1,2,3-triazolo[4,5-d]pyrimidine (VAS2870). *Free Radic Biol Med* 52 (9): 1897–1902. doi:10.1016/j.freeradbiomed.2012.02.046.
260. Wind, S., K. Beuerlein, T. Eucker, H. Müller, P. Scheurer, M. E. Armitage, H. Ho, H. H. Schmidt, and K. Wingler. 2010. Comparative pharmacology of chemically distinct NADPH oxidase inhibitors. *Br J Pharmacol* 161 (4): 885–898. doi:10.1111/j.1476-5381.2010.00920.x.
261. Gianni, D., N. Taulet, H. Zhang, C. DerMardirossian, J. Kister, L. Martinez, W. R. Roush, S. J. Brown, G. M. Bokoch, and H. Rosen. 2010. A novel and specific NADPH oxidase-1 (Nox1) small-molecule inhibitor blocks the formation of functional invadopodia in human colon cancer cells. *ACS Chem Biol* 5 (10): 981–993. doi:10.1021/cb100219n.
262. Laleu, B., F. Gaggini, M. Orchard, L. Fioraso-Cartier, L. Cagnon, S. Houngninou-Molango, A. Gradia et al. 2010. First in class, potent, and orally bioavailable NADPH oxidase isoform 4 (Nox4) inhibitors for the treatment of idiopathic pulmonary fibrosis. *J Med Chem* 53 (21): 7715–7730. doi:10.1021/jm100773e.
263. Sedeek, M., G. Callera, A. Montezano, A. Gutsol, F. Heitz, C. Szyndralewiez, P. Page et al. 2010. Critical role of Nox4-based NADPH oxidase in glucose-induced oxidative stress in the kidney: Implications in type 2 diabetic nephropathy. *Am J Physiol Renal Physiol* 299 (6): F1348–F1358. doi:10.1152/ajprenal.00028.2010.
264. Cifuentes-Pagano, E., J. Saha, G. Csányi, I. A. Ghouleh, S. Sahoo, A. Rodríguez, P. Wipf, P. J. Pagano, and E. M. Skoda. 2013. Bridged tetrahydroisoquinolines as selective NADPH oxidase 2 (Nox2) inhibitors. *MedChemComm* 4 (7): 1085–1092. doi:10.1039/C3MD00061C.
265. Jacobson, G. M., H. M. Dourron, J. Liu, O. A. Carretero, D. J. Reddy, T. Andrzejewski, and P. J. Pagano. 2003. Novel NAD(P)H oxidase inhibitor suppresses angioplasty-induced superoxide and neointimal hyperplasia of rat carotid artery. *Circ Res* 92 (6): 637–643. doi:10.1161/01.RES.0000063423.94645.8A.
266. Jung, O., J. G. Schreiber, H. Geiger, T. Pedrazzini, R. Busse, and R. P. Brandes. 2004. gp91phox-containing NADPH oxidase mediates endothelial dysfunction in renovascular hypertension. *Circulation* 109 (14): 1795–1801. doi:10.1161/01.CIR.0000124223.00113.A4.
267. Shelat, P. B., M. Chalimoniuk, J. H. Wang, J. B. Strosznajder, J. C. Lee, A. Y. Sun, A. Simonyi, and G. Y. Sun. 2008. Amyloid beta peptide and NMDA induce ROS from NADPH oxidase and AA release from cytosolic phospholipase A2 in cortical neurons. *J Neurochem* 106 (1): 45–55. doi:10.1111/j.1471–4159.2008.05347.x.
268. Miesel, R., M. Kurpisz, and H. Kroger. 1996. Suppression of inflammatory arthritis by simultaneous inhibition of nitric oxide synthase and NADPH oxidase. *Free Radic Biol Med* 20 (1): 75–81.
269. Paravicini, T. M., S. Chrissobolis, G. R. Drummond, and C. G. Sobey. 2004. Increased NADPH-oxidase activity and Nox4 expression during chronic hypertension is associated with enhanced cerebral vasodilatation to NADPH in vivo. *Stroke* 35 (2): 584–589. doi:10.1161/01.STR.0000112974.37028.58.
270. O'Donnell, B. V., D. G. Tew, O. T. Jones, and P. J. England. 1993. Studies on the inhibitory mechanism of iodonium compounds with special reference to neutrophil NADPH oxidase. *Biochem J* 290 (Pt 1): 41–49.

271. Stuehr, D. J., O. A. Fasehun, N. S. Kwon, S. S. Gross, J. A. Gonzalez, R. Levi, and C. F. Nathan. 1991. Inhibition of macrophage and endothelial cell nitric oxide synthase by diphenyleneiodonium and its analogs. *FASEB J* 5 (1): 98–103.
272. Williams, H. C., and K. K. Griendling. 2007. NADPH oxidase inhibitors: New antihypertensive agents? *J Cardiovasc Pharmacol* 50 (1): 9–16. doi:10.1097/FJC.0b013e318063e820.
273. Lafeber, F. P., C. J. Beukelman, E. van den Worm, J. L. van Roy, M. E. Vianen, J. A. van Roon, H. van Dijk, and J. W. Bijlsma. 1999. Apocynin, a plant-derived, cartilage-saving drug, might be useful in the treatment of rheumatoid arthritis. *Rheumatology (Oxford)* 38 (11): 1088–1093.
274. Zhang, Y., M. M. Chan, M. C. Andrews, T. A. Mori, K. D. Croft, K. U. McKenzie, C. G. Schyvens, and J. A. Whitworth. 2005. Apocynin but not allopurinol prevents and reverses adrenocorticotropic hormone-induced hypertension in the rat. *Am J Hypertens* 18 (7): 910–916. doi:10.1016/j.amjhyper.2005.02.017.
275. Barbieri, S. S., V. Cavalca, S. Eligini, M. Brambilla, A. Caiani, E. Tremoli, and S. Colli. 2004. Apocynin prevents cyclooxygenase 2 expression in human monocytes through NADPH oxidase and glutathione redox-dependent mechanisms. *Free Radic Biol Med* 37 (2): 156–165. doi:10.1016/j.freeradbiomed.2004.04.020.
276. Johnson, D. K., K. J. Schillinger, D. M. Kwait, C. V. Hughes, E. J. McNamara, F. Ishmael, R. W. O'Donnell et al. 2002. Inhibition of NADPH oxidase activation in endothelial cells by ortho-methoxy-substituted catechols. *Endothelium* 9 (3): 191–203.
277. Peters, E. A., J. T. Hiltermann, and J. Stolk. 2001. Effect of apocynin on ozone-induced airway hyperresponsiveness to methacholine in asthmatics. *Free Radic Biol Med* 31 (11): 1442–1447.
278. Wingler, K., S. A. Altenhoefer, P. W. Kleikers, K. A. Radermacher, C. Kleinschnitz, and H. H. Schmidt. 2012. VAS2870 is a pan-NADPH oxidase inhibitor. *Cell Mol Life Sci* 69 (18): 3159–3160. doi:10.1007/s00018-012-1107-1.
279. Altenhöfer, S., P. W. M. Kleikers, K. A. Radermacher, P. Scheurer, J. J. R. Hermans, P. Schiffers, H. Ho, K. Wingler, and H. H. H. W. Schmidt. 2012. The NOX toolbox: Validating the role of NADPH oxidases in physiology and disease. *Cell Mol Life Sci* 69 (14): 2327–2343. doi:10.1007/s00018-012-1010-9.
280. Seredenina, T., G. Chiriano, A. Filippova, Z. Nayernia, Z. Mahiout, L. Fioraso-Cartier, O. Plastre, L. Scapozza, K. H. Krause, and V. Jaquet. 2015. A subset of N-substituted phenothiazines inhibits NADPH oxidases. *Free Radic Biol Med* 86: 239–249. doi:10.1016/j.freeradbiomed.2015.05.023.
281. Gaggini, F., B. Laleu, M. Orchard, L. Fioraso-Cartier, L. Cagnon, S. Houngninou-Molango, A. Gradia et al. 2011. Design, synthesis and biological activity of original pyrazolo-pyrido-diazepine, -pyrazine and -oxazine dione derivatives as novel dual Nox4/Nox1 inhibitors. *Bioorg Med Chem* 19 (23): 6989–6999. doi:10.1016/j.bmc.2011.10.016.
282. Schildknecht, S., A. Weber, H. R. Gerding, R. Pape, M. Robotta, M. Drescher, A. Marquardt, A. Daiber, B. Ferger, and M. Leist. 2014. The NOX1/4 inhibitor GKT136901 as selective and direct scavenger of peroxynitrite. *Curr Med Chem* 21 (3): 365–376.

Index

Note: Page numbers followed by f and t refer to figures and tables respectively.

G

H

I

K

L

M

N

O

T

U